THIRD EDITION

COMMERCIAL REFRIGERATION

FOR AIR CONDITIONING TECHNICIANS

DICK WIRZ

Australia • Brazil • Mexico • Singapore • United Kingdom • United States

Commercial Refrigeration for Air Conditioning Technicians, Third Edition
Author(s) Dick Wirz

SVP, GM Skills & Global Product Management: Jonathan Lau

Product Team Manager: Katie McGuire

Senior Director, Development: Marah Bellegarde

Senior Product Development Manager: Larry Main

Associate Content Developer: Jenn Alverson

Product Assistant: Jason Koumourdas

Vice President, Marketing Services: Jennifer Ann Baker

Marketing Manager: Scott Chrysler

Senior Production Director: Wendy Troeger

Production Director: Andrew Crouth

Senior Content Project Manager: Glenn Castle

Senior Art Director: Jack Pendleton

Cover image(s): Adisa/Shutterstock.com

For product information and technology assistance, contact us at
Cengage Learning Customer & Sales Support, 1-800-354-9706
For permission to use material from this text or product,
submit all requests online at **www.cengage.com/permissions.**
Further permissions questions can be e-mailed to
permissionrequest@cengage.com

Library of Congress Control Number: 2016955108

ISBN: 978-1-305-50643-5

Cengage Learning
20 Channel Center Street
Boston, MA 02210
USA

Cengage Learning is a leading provider of customized learning solutions with office locations around the globe, including Singapore, the United Kingdom, Australia, Mexico, Brazil, and Japan. Locate your local office at: **www.cengage.com/global**

Cengage Learning products are represented in Canada by Nelson Education, Ltd.

To learn more about Cengage Learning, visit **www.cengage.com**

Purchase any of our products at your local college store or at our preferred online store **www.cengagebrain.com**

Notice to the Reader

Printed in the United States of America
Print Number: 01   Print Year: 2016

CONTENTS

PREFACE

COMMERCIAL REFRIGERATION FOR TECHNICIANS

by Dick Wirz

This book is written for refrigeration technicians, air-conditioning (A/C) technicians, and advanced A/C students. Commercial refrigeration, as a specific subject, is taught in only about 10 percent of the schools that have HVACR programs. As a result, there are very few A/C students or technicians exposed to this related technology. The industry certainly needs technicians trained in A/C. However, if A/C technicians also understand commercial refrigeration, they will be more valuable to their companies and their customers.

The subject matter and format of this book are the result of a 60-hour course I developed for the Washington, D.C., chapter of the National Association of Power Engineers (NAPE). Doug Smarte, executive director of the NAPE Education Foundation, recognized the need to offer training in commercial refrigeration to building engineers and property managers.

ORGANIZATION

The first six of the fourteen chapters combine both a review of refrigeration theory and an introduction to how those principles are applied specifically to commercial refrigeration equipment. Wherever possible, refrigeration concepts are compared with those of A/C. This makes it easier for HVACR students and technicians to relate what they already know to the field of commercial refrigeration.

Chapter 7 applies the information in the first six chapters to the troubleshooting of nine common refrigeration system problems. It also includes a diagnostic chart for the reader to use in class, as well as on the job. Chapters 4 and 8, on compressors and motor controls, also have their share of troubleshooting instructions. The commercial refrigeration industry has been at the forefront of developing retrofitting practices for replacing chlorofluorocarbon (CFC) and hydrochlorofluorocarbon (HCFC) refrigerants. Therefore, Chapter 9 covers retrofitting, along with the other related practices of recovery, evacuation, and charging in greater detail and with more practical applications than found in other HVACR textbooks.

Chapter 10 is an introduction to the fascinating world of supermarket refrigeration. By popular demand it has been greatly expanded in the third edition. Chapters 11 and 12 deal with walk-ins, reach-ins, and ice machines. Chapter 13 is a relatively short but important look at the role of the refrigeration technician in food preservation and health issues.

Because many HVACR technicians eventually start their own companies, Chapter 14 deals with the business side of this industry. Even if most students have no immediate intention of becoming self-employed, this chapter provides a glimpse of what their employers deal with on a daily basis. By gaining these insights, technicians can become a more supportive and vital part of their organizations.

FEATURES

Throughout the book, there are tips called Technician's Rules of Thumb (TROT). These practical bits of information are a collection of practices that experienced technicians use to help them service equipment better and faster. There is a complete list of these rules of thumb included in the appendix. The appendix also contains pressure/temperature (P/T) charts for the refrigerants currently used in most refrigeration applications.

The glossary has definitions and explanations of technical words and phrases used in the book. The use of boldface type and italics is an indication that the word may not be familiar to many of the readers and that there is an explanation of its meaning in the glossary.

SUPPLEMENTS

Most HVACR technicians are very visual learners. Therefore, this book includes many pictures and drawings to illustrate what has been written in the text. If this book is used as part of a course on refrigeration, there is an Instructor Companion site, which is free to qualified teachers. The online Instructor Companion site has everything you need for your course in one place. This collection of book-specific lecture and class tools is available online via www.cengage.com/login. Access and download PowerPoint® presentations, an Instructor Guide with answers to the end of chapter review questions, and test banks. Contact Cengage Learning or your local sales representative to obtain an instructor account. The PowerPoint® presentations are designed to be used by the instructor in the classroom. These visual teaching aids help explain refrigeration concepts and increase student understanding. Students can also access the PowerPoint® presentations for free via the Student Companion Site located at www.cengagebrain.com.

WHAT IS NEW IN THE THIRD EDITION

Of course, there were some corrections that needed to be made. Thanks so much to all of you who took the time to email your comments and suggestions to me at teacherwirz@cox.net. I hope that you will continue to do so. Special thanks goes to Jon Hamel who uncovered more mistakes than I had any right to make. He also asked very insightful questions. Those questions resulted in adding explanations and making clarifications in several sections. Thanks Jon.

Chapter 10, "Supermarket Refrigeration," has received the most new material. Requests for more on this subject than was in earlier editions have resulted in extensive research on the latest and greatest in the most progressive area of commercial refrigeration. I appreciate the input and pictures from so many manufacturers on this subject. Special thanks to Holly Villareal for her technical help and text revisions, which helped me tremendously in updating and expanding Chapter 10. Her knowledge of rack systems, as well as their controls and controllers, is exceptional. Thanks as well go to Jess Lukin, who provided many supermarket service scenarios with pictures. The real-life service scenarios help to explain many of the difficult concepts, not only in supermarket refrigeration, but in refrigeration and electrical service in general. Other instructive service scenarios where provided by Gary Purdue, Bill McDonald, Mike Haynes, and even a few really old ones from me.

The following is a brief description of each chapter's new or improved subject matter and pictures:

Chapter 1: *Refrigeration*—No, R134a is not an azeotrope. Most of you caught this. Bad slip on my part, and of course it had to be at the beginning of the book. There is a brief description of the July 2015 EPA refrigerant guidelines and their impact on commercial refrigeration.

Chapter 2: *Evaporators*—Replaced two pictures of evaporators with better ones.

Chapter 3: *Condensers*—Made one correction when the term *HPR valve* was used when describing a water regulating valve.

Chapter 4: *Compressors*—Added two sections about compressor blow-by and the basics of reading model and serial numbers. Plus, 16 more pictures and 5 more review questions have been added.

Chapter 5: *Metering Devices*—Expanded on pressure-limiting expansion valves and balanced port valves. Also added a service tip and picture on cap tubes.

Chapter 6: *Controls and Accessories*—Removed section on one-time pump-down. It was confusing, and the process is very seldom used in the field. Added a second diagram of a standard pump-down for more clarity. Also, added three more diagrams to show the steps of oil safety control and how it works with a current sensing relay. Added a picture of a receiver with a queen valve as well as the common king valve. Added four service scenarios of real-life service calls, one with six pictures.

Chapter 7: *Refrigeration System Troubleshooting*—Made some corrections and replaced two figures of the diagnostic chart. Added two service scenarios just before the summary.

Chapter 8: *Compressor Motor Controls*—Added two service scenarios just before summary.

Chapter 9: *Retrofitting, Recovery, Evacuation, and Charging*—Added a picture of the third step of triple evacuation. On evacuation slides a micron gauge was installed on the receiver to show the evacuation level in the system rather than just at the vacuum pump. This was recommended by Joe Moravek as being the proper location according to most manufacturers.

Chapter 10: *Supermarket Refrigeration*—Greatly expanded sections concerning rack system oil return, controllers, evaporator pressure/temperature regulation, as well as transcritical and subcritical CO_2 systems. Even if you are not currently working on supermarket equipment, there is plenty of information added to this chapter that applies to smaller refrigeration systems. Even the new section on CO_2 is relevant because Coca-Cola is using transcritical CO_2 systems in all their vending machines. This chapter material has nearly 50 more pictures and 19 more review questions. There are 8 service scenarios with 73 more pictures.

Chapter 11: *Walk-in Refrigerators and Freezers*—A service scenario has been added.

In summary, the third edition has approximately 200 additional pages with over 150 more pictures and diagrams.

ABOUT THE AUTHOR

My career in HVACR started with a summer job in 1963. For the first eight years, I installed ductwork and serviced residential A/C and heating equipment. Over the next 30 years, I enjoyed the world of commercial refrigeration. I am a licensed Master HVACR Technician and Master Electrician in several states. I am certified in all categories offered by Industry Competency Exams (ICE) and North American Training Excellence (NATE). For more than 25 years, I was president and co-owner of a successful commercial refrigeration company. I now enjoy teaching and publishing training materials. In this way, I am able to repay some of the many benefits I have received from a very rewarding career in the HVACR industry.

I graduated from Virginia Tech with a degree in business management and a minor in mechanical engineering. Twenty years later, I returned to school to earn my master's in business administration in order to qualify as a community college professor upon retiring from my refrigeration company. To help me become a more effective instructor, I spent two years in a postgraduate program at George Mason University, where I earned a certificate in community college teaching. In 2014, I retired from full-time teaching after fifteen years at Northern Virginia Community College, Woodbridge Campus. My wife and I continue to produce teaching aids under the corporate name of *Refrigeration Training Services* (www.hvacteaching.com) for HVACR programs in the form of instructor CDs similar to those available for this book.

ACKNOWLEDGMENTS

I would like to thank my wife, Irene, for her tremendous help in this project. She provided all the graphics used in the book and in the instructor and student CDs. Without her editing, graphics, software expertise, and support, this book never would have become a reality.

I would also like to thank the many people who have contacted me since the first and second editions of the book were published. Your appreciation has enforced my belief that the tremendous amount of work that went into this book has been beneficial to thousands of students, teachers, and technicians. Your comments helped shape the changes and updates in the third edition.

For this third edition a special thanks goes out to Holly Villareal for her technical expertise and writing, and to Jess Lukin for so many service scenarios and pictures. Also to Gary Purdue, Bill McDonald, and Mike Haynes for their service scenarios. These real-life service situations have made the material in the book come alive and become more relevant. I hope some of you reading this will contribute for the next edition.

Teaching and training has been a rewarding, yet humbling, experience for me. Although I thought I knew HVACR well, I soon realized that there was much more I needed to master in order to teach the subjects. Someone once told me, "You can never know everything about anything, but you have to keep trying." Now I realize how true those words are. I have become a lifelong student, and I encourage the same thirst for knowledge in those I teach. I believe you must agree because you have taken the time to pick up this book and read at least this much of it. Thank you.

FEEDBACK

There was considerable editorial effort to eliminate mistakes from this book. However, there may be something we missed. If you find some apparent inaccuracy or have a question concerning the content, please contact the publisher, or use my personal email address below. A special thanks goes out to Jon Hamel. He found some errors, but also asked insightful questions. As a result, those sections are more accurate and clear in this edition.

Thank you for using this book. I am sure what you learn will be of great benefit to your success in this industry. I also hope this text will become an important part of your technical library.

Dick Wirz
teacherwirz@cox.net

REFRIGERATION

1

CHAPTER OVERVIEW

This chapter begins by explaining what this book is about and for whom it is written. This is followed by a thorough review of the refrigeration cycle. Next, air conditioning is compared with commercial refrigeration; both their similarities and their differences are explained. The newer refrigerants used in commercial refrigeration are also covered. Finally, the four basic components of a refrigeration system are discussed.

OBJECTIVES

After completing this chapter, you should be able to

- Describe temperature ranges of refrigeration
- Describe the refrigeration cycle
- Relate refrigeration to air conditioning
- Describe the relationship between a refrigerant's pressure and temperature
- Describe the newer refrigerants used in commercial refrigeration systems
- Describe the relationship among the four basic components of a refrigeration system

INTRODUCTION

Most technicians tend to specialize in a single type of air conditioning (AC) application, such as residential, light commercial, or heavy commercial systems. However, very often, opportunities arise outside a technician's primary area of expertise, so it is smart to be knowledgeable in more than one specialty. For instance, a company that provides good service on a restaurant's AC may be asked to service its commercial refrigeration equipment. Likewise, a building engineer who competently handles the large chillers of a commercial building may have his or her responsibility expanded to include maintaining refrigeration and ice machines in the building's cafeteria.

The primary objective of this book is to help AC technicians understand commercial refrigeration. Someone once said, "Luck is preparation meeting opportunity." The more knowledge areas technicians master, the better they can take advantage of any opportunities that arise.

Therefore, *Commercial Refrigeration for Air Conditioning Technicians* is written for both experienced AC technicians and students who have a firm basis in AC theory. This first chapter is intended to be a review of basic refrigeration as well as an introduction to the similarities and differences between AC and commercial refrigeration.

Throughout this book, a word or phrase used for the first time that may not be familiar to all readers will be in blue font type. These words and phrases will also be included in the glossary.

TEMPERATURE RANGES OF REFRIGERATION

The following is a list of the space temperatures of the more common ranges of refrigeration discussed in this book:

- 75°F, AC (comfort cooling)
- 55°F, high-temperature refrigeration
- 35°F, medium-temperature refrigeration
- −10°F, low-temperature refrigeration
- −25°F, extra-low-temperature refrigeration

Most of the examples in the next few chapters are concerned with medium- and low-temperature applications. Medium-temperature walk-in refrigerators usually operate at a range of 35°F to 37°F, whereas reach-in refrigerators run at slightly higher temperatures, from 38°F to 40°F. Walk-in freezers normally run at −10°F, and reach-in freezers operate at about 0°F.

The difference between the temperatures in walk-ins and reach-ins is mainly due to how the box is used and how the equipment is designed. The lower temperatures of walk-ins allow them to keep large amounts of product fresh for relatively longer periods of time.

Reach-ins, on the other hand, are used more for convenience. Because they are smaller than walk-ins, reach-ins can be located closer to where they are needed. A reach-in is usually restocked from a walk-in at least once a day. Therefore, the slightly higher storage temperature of a reach-in is acceptable because the product is in the box for a relatively shorter period of time.

THE REFRIGERATION CYCLE

Figure 1-1 is an illustration of a very simple AC system showing a compressor and an expansion valve; cylindrical tanks represent the condenser and evaporator. The pressures and temperatures represent those of a standard-efficiency R22 AC system on a 95°F day.

The compressor develops a pressure of 278 pounds per square inch gauge (psig) and discharges superheated vapor at 175°F. The vapor drops to 165°F when it enters the condenser and continues to be cooled by the air around the cylinder. When the vapor temperature falls to 125°F, the gases start to condense into droplets of liquid—the vapor has reached its condensing temperature, which is the saturation temperature of R22 refrigerant at a pressure of 278 psig (refer to the pressure/temperature [P/T] chart in the appendix). The condensing continues at 125°F until all the vapor turns to liquid at the bottom of the tank. Additional cooling, called subcooling, of that liquid by the 95°F ambient air reduces the temperature of the liquid to 115°F as it leaves the bottom of the condenser. By the time the liquid enters the expansion valve, it has been further subcooled to 105°F.

During this process, the pressure in the high side of the system, between the compressor outlet and the expansion valve inlet, remains constant at 278 psig. However, the refrigerant changes temperatures when the hot discharge gas cools, then subcools, as it flows through the condenser. The significance of these different temperatures is important to understanding the refrigeration process.

Low-Pressure Side of the System

As the 278-psig liquid refrigerant goes through the thermostatic expansion valve (TEV), its pressure drops to 69 psig. The 209-psig decrease in pressure, from one side of the valve to the other, is accompanied by a decrease in temperature. The TEV acts like a garden-hose nozzle, changing the solid stream of liquid from the condenser to a spray mixture of vapor and liquid refrigerant droplets. Small droplets are more easily

FIGURE 1-1 **Simple R22 AC system.** *Courtesy of Refrigeration Training Services.*

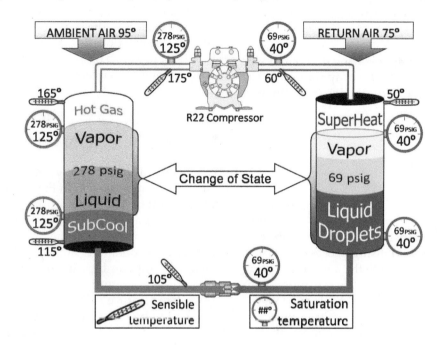

boiled off in the evaporator than a solid stream of liquid is. The R22 refrigerant boils, or evaporates, at 40°F when its pressure is reduced to 69 psig (refer to the P/T chart in the appendix).

The heat from the 75°F air blowing across the tank is absorbed into the refrigerant, causing the refrigerant droplets to boil. The refrigerant temperature remains at 40°F until all of it has vaporized. Only then will its temperature rise as it absorbs more heat from the surrounding air. By the time the suction vapor leaves the tank, the refrigerant temperature will be raised to 50°F. The temperature of the refrigerant above its 40°F boiling point (saturation temperature) is called superheat.

How Is Heat Absorbed into the Evaporator?

Starting at the TEV, a fog of liquid droplets is sprayed into the tank. The warm air blowing over the evaporator tank is cooled as its heat is absorbed into the boiling refrigerant. Much more heat is absorbed in the refrigerant as it boils off than is absorbed before or after it has boiled. Boiling, or the change of state from a liquid to a vapor, absorbs heat without a change in temperature. Strange as it may sound, this temperature change cannot be measured with a thermometer. Almost all the refrigerating effect achieved in the evaporator is accomplished as the refrigerant boils. The type of heat absorbed during the evaporation process is called latent heat. The ability to remove tremendous amounts of heat in a small area makes it possible for

manufacturers to design refrigeration systems small enough to be used in both homes and businesses.

When the refrigerant is fully vaporized, it is totally saturated with all the latent heat it can absorb. The 40°F saturated vapor can raise its temperature only by absorbing sensible heat. This sensible heat can be measured with a thermometer, and any temperature rise above the refrigerant's saturation temperature is called superheat.

How Does the Condenser Get Rid of the Heat Absorbed by the Evaporator?

To reject the heat absorbed by the evaporator, as shown in Figure 1-1, the cool suction vapor must be raised to a temperature higher than the 95°F outside air. In Figure 1-1 the refrigerant temperature is increased to 125°F. The 30°F difference between the condensing temperature and the outdoor air is great enough to easily transfer heat from the hot condenser to the warm outdoor air.

NOTE: **The greater the difference in temperature between two substances, the faster the transfer of heat from one to the other.**

Compressing the 69-psig suction vapor to 278 psig increases its boiling point from 40°F to 125°F (see the P/T chart in the appendix). Raised to 125°F, the vapor from the evaporator releases latent heat to the cooler ambient air as the refrigerant condenses to a liquid.

NOTE: **The difference between the *condensing temperature* of a refrigerant and the *ambient temperature* is called the** condenser split.

125°F condensing temperature −95°F ambient = 30°F condenser split

In fact, the discharge vapor leaving the compressor is above 125°F. In addition to the evaporator's latent heat, the discharge vapor also contains the following sensible heat:

- Evaporator superheat
- Suction line superheat
- Compressor motor heat
- Heat of compression

In Figure 1-1, the 175°F hot gas leaving the compressor must de-superheat, or get rid of its superheat, before it can start condensing at its saturation temperature of 125°F. The condensing process continues at 125°F, rejecting latent heat into the ambient. Cooling the fully condensed liquid below its saturation temperature is called subcooling. To calculate subcooling, determine the condensing temperature from the head pressure and then subtract the temperature of the liquid line leaving the condenser.

EXAMPLE: **2**

The head pressure is 278 psig, and the temperature of the liquid line at the condenser outlet is measured at 115°F. Therefore, 125°F condensing temperature −115°F liquid line temperature = 10°F subcooling.

The liquid travels out of the condenser to the TEV, where the process starts again. This cycle removes heat from where it is not wanted (cooled space) and rejects it somewhere else (outdoors). This is the basic definition of the refrigeration process.

This section on the basic refrigeration cycle is nothing new for most readers. However, the review is still important to form a basis for much of what is discussed in the following chapters.

COMPARING COMMERCIAL REFRIGERATION WITH AC

What is the difference between the medium-temperature refrigeration system in Figure 1-2 and the AC system in Figure 1-1?

The pressures and temperatures on the condenser side are the same because they both use R22 refrigerant and reject evaporator heat into 95°F ambient air. However, the return air blowing over the evaporator in the medium-temperature system is only 35°F. Because the space temperature is lower, the evaporator temperature had to be lowered. In Figure 1-2, the evaporator was lowered to 25°F by reducing the pressure of the R22 refrigerant to 49 psig.

Therefore, a refrigeration system metering device drops the evaporator pressure and temperature to a level lower than that achieved by a metering device designed for an AC system. Similarly, a refrigeration compressor should be capable of increasing the lower evaporator pressure up to a level high enough to reject the heat into 95°F ambient air.

Figure 1-3 is a more elaborate diagram of an AC system, with labels to identify what is happening in the refrigerant circuit.

FIGURE 1-2 **Simple R22 refrigeration system.** *Courtesy of Refrigeration Training Services.*

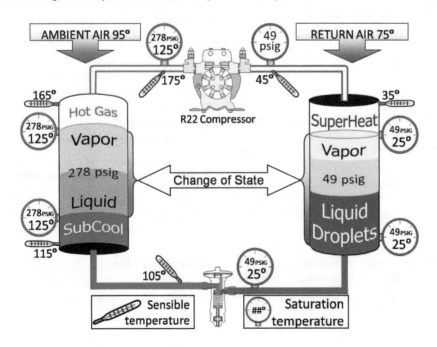

FIGURE 1-3 **Basic R22 AC system with fixed metering device.** *Courtesy of Refrigeration Training Services.*

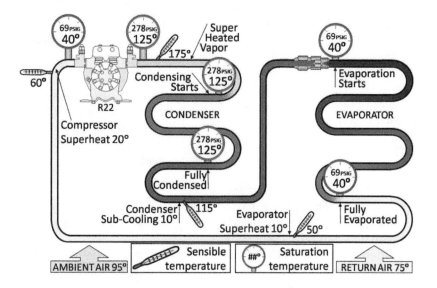

Figure 1-4 is similar to Figure 1-3, except that it shows the different temperatures and low side pressures of a typical walk-in refrigerator using R22.

NEWER REFRIGERANTS IN COMMERCIAL REFRIGERATION

Many experienced AC technicians diagnose standard R22 AC units based on approximate pressure readings. For instance, they may say that the suction pressure should be about 65 psig to 75 psig if the return temperature is 75°F. Or, when checking head pressure on a 95°F day, the technician may estimate that it should be around 260 psig to 290 psig. Although this technique is not as precise as using a P/T chart, these technicians

seem to be satisfied with the results of this method of service diagnostic.

Similarly, before 1990, some commercial refrigeration technicians also used approximations for three refrigerants: R12, R22, and R502. When the Clean Air Act of 1990 came into effect, R12 and R502 were replaced with more than 10 different refrigerants designed to maintain the same temperatures, but at different pressures. As a result, refrigeration technicians had to relearn how to use system temperatures, rather than pressures. Surprisingly, this method proved to be easier and better than working with pressures alone.

Therefore, this book concentrates on using system temperatures so that the reader will better understand all refrigeration systems, no matter what refrigerant is used.

FIGURE 1-4 **Basic R22 walk-in refrigerator with TEV and receiver.** *Courtesy of Refrigeration Training Services.*

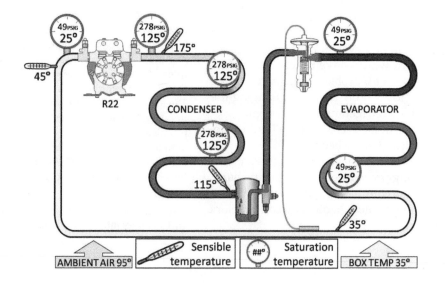

EXAMPLE: 3

The condensing and evaporator temperatures in Figure 1-4 will be the same regardless of the refrigerant the system uses: R22, R12, R502, R134a, or R404A.

Currently, the following refrigerants are being used in new equipment:

- R404A—walk-in refrigerators and freezers
- R134a and R404A—reach-in refrigerators
- R404A—reach-in freezers
- R404A, R407A, R407C, and R744 (CO2)—parallel rack systems in supermarkets

R134a is a single chemical refrigerant that boils and condenses at a single pressure and temperature. However, 400 series refrigerants, called *zeotropes,* are blends of different refrigerants. The different refrigerant components change state at different pressures. This property is known as *glide* and has been a source of both concern and myths for years. The following are four of these concerns and reasons they are not as great a problem as originally feared:

1. There is flashing off in the liquid line of at least one component of the refrigerant.
   - In practice, limited flashing of a single component does not seem to greatly affect the overall operation of the system.
2. There is no way to accurately measure superheat and subcooling.
   - On P/T charts, the refrigerant temperatures designated as dew point are used to calculate superheat, and bubble point temperatures are used to calculate subcooling.
   - Most current P/T charts have removed the dew point and bubble point and now have a single temperature reached by simply averaging. Although not as accurate, the single temperature at least removes some of the confusion most techs have when figuring out superheat and subcooling.
3. Even a small refrigerant leak will change the chemical composition of the remaining refrigerant in the system.
   - Most refrigerant and system manufacturers say that the system would have to lose most of its charge for this to be a problem. One or two small leaks can be "topped off" without changing the properties of the refrigerant in the system.
4. Refrigerant blends must be charged in a liquid state. However, technicians have been taught that liquid in the suction line can damage the compressor.
   - Proper charging procedures ensure that no harm will come to the compressor. This process is covered more fully in later chapters on compressors and charging.

NOTE: Technicians should be familiar with Environmental Protection Agency (EPA) regulations on how many times a system can be topped off. Also, they should be aware of who keeps track of the leaks and the amount of refrigerant, and who is ultimately responsible for compliance with the laws.

Some applications require accurate measurement of superheat, such as supermarket cases and very-low-temperature systems. In those situations, it is important the technician is aware of the refrigerant being used, its bubble point, and how to take those factors into consideration when measuring superheat.

THE FOUR BASIC COMPONENTS OF A REFRIGERATION SYSTEM

Chapters 2 through 5 cover in detail each of the four basic components of a refrigeration system. The following is a brief overview of the part each component plays in the refrigeration cycle (Figure 1-5).

Evaporator

Warm air from the space is blown over the evaporator. Heat from the air is absorbed into the refrigerant as it boils within the evaporator tubing. The heat remains in the refrigerant, which flows to another area and is ejected.

Sometimes when a technician is not able to obtain the exact information needed to solve a problem, he must rely on past experience with similar equipment under similar conditions. The technician has subconsciously determined approximate values for certain conditions. Although they may not be easy to put into words, every technician has developed certain rules of thumb that he uses to diagnose equipment problems.

The New Dictionary of Cultural Literacy defines the term *rules of thumb* as "a practical principle that comes from the wisdom of experience and is usually but not always valid."

Most experienced service and installation technicians use rules of thumb every day on the job. When used in this book, they will be referred to as *Technician's Rules of Thumb* (TROT). The acronym TROT is easy to remember if one thinks of a horse breaking into a trot when it wants to move faster. Likewise, a technician can both learn and work faster by using some rules of thumb. There will be a special notation in the text when TROT are relevant. In addition, the appendix has a complete list of all the TROT used in this book.

NOTE: Factory specifications and guidelines always take precedence over TROT. These rules of thumb should be used only when factory information is not available.

FIGURE 1-5 **The four basic components of a refrigeration system.** *Courtesy of Refrigeration Training Services.*

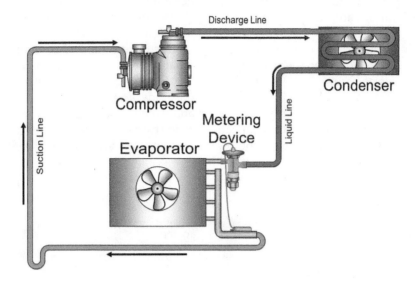

Condenser

The condenser is a mirror image of the evaporator. However, instead of absorbing heat, it rejects heat. There is tremendous heat transfer as the refrigerant changes state. Latent heat is released as the vapor condenses into a liquid within the condenser.

Compressor

The heat in the refrigerant can be removed only if it is exposed to relatively cooler ambient temperatures. Since the outside air can be 95°F or higher, the refrigerant temperature must be raised much higher. The compressor can raise the refrigerant temperature by raising its pressure. Therefore, the hotter it gets outside, the higher the compressor pressures become.

Metering Device

The metering device, either expansion valve or capillary tube, reduces the liquid pressure by forcing it through a nozzle or small opening. Lowering the pressure of the refrigerant allows it to boil at a lower temperature. To make the refrigerant boil more easily, the metering device changes the stream of liquid into a dense fog of liquid droplets before it enters the evaporator.

SUMMARY

Most of the commercial refrigeration applications in this book operate at a range of 35°F to 40°F for medium-temperature units and 0°F to −10°F for low-temperature units.

Because most readers are familiar with the pressures and temperatures of R22 used in AC units, many illustrations of commercial refrigeration are presented using R22. Hopefully, this will make it easier to understand the differences and similarities between the two types of applications.

The more popular refrigerants currently used in new commercial refrigeration are R134a and R404A. However, as of July 2015, the EPA issued regulations for phasing out R134a, R404A, R407, R422, R507, and many others for both new equipment and retrofits. Following are the guidelines for terminating the use of R134a and R404A in new food refrigeration equipment on January 1 of the years noted:

- Food refrigeration for supermarkets   2017
- Remote condensing units   2018
- Self-contained medium-temp under 2,200 Btuh   2019
- Self-contained medium-temp over 2,200 Btuh   2020
- Self-contained low-temp   2020

The use of blended refrigerants has required technicians to learn new skills but has not been the problem the industry once feared. The learning curve for using even more new refrigerants brought on by the latest EPA regulations should not be that difficult.

There are four basic components of the refrigeration system. The evaporator, condenser, compressor, and metering device are covered in greater detail in the next four chapters. A thorough understanding of these components is necessary before moving on to the many accessories, operating controls, and safety controls covered in later chapters.

Technicians should always follow factory specifications and recommendations. However, there are times when a Technician's Rules of Thumb (TROT) can help speed up the diagnostic process.

In commercial refrigeration, as in any service business, time is money. The quicker a technician can diagnose a problem, the more efficient she is. Better efficiency means more success for the technician and her company. Just as important to success is a technician's positive attitude as a result of working at something she enjoys. The goal of this book is to help the reader become a better technician, one who enjoys what she does and makes a good living doing it.

There are some very talented female technicians in this trade, and many more are needed. In this book, male and female pronouns will be used interchangeably to describe technicians. Not only is this fairer, but it is less cumbersome than to keep using *him/her*, *he/she*, and so on.

REVIEW QUESTIONS

1. What is considered the "normal" box temperature of a walk-in refrigerator?

   a. 35°F to 37°F
   b. 38°F to 40°F
   c. −10°F
   d. 0°F

2. What is considered the "normal" box temperature of a reach-in refrigerator?

   a. 35°F to 37°F
   b. 38°F to 40°F
   c. −10°F
   d. 0°F

3. What is considered the "normal" box temperature of a walk-in freezer?

   a. 35°F to 37°F
   b. 38°F to 40°F
   c. −10°F
   d. 0°F

4. What is considered the "normal" box temperature of a reach-in freezer?

   a. 35°F to 37°F
   b. 38°F to 40°F
   c. −10°F
   d. 0°F

5. Why is the temperature of walk-ins usually lower than the temperature of reach-ins?

   a. Walk-ins are designed for long-term storage and lower temperatures.
   b. Reach-ins are designed for long-term storage and lower temperatures.
   c. Reach-ins are small and cannot hold low temperatures.

Refer to Figures 1-1 to 1-4 for questions 6–11:

6. What is the pressure throughout the high side of the system?

   a. 229 psig
   b. 278 psig
   c. 49 psig
   d. 69 psig

7. At what temperature does the refrigerant condense?

   a. 125°F
   b. 115°F
   c. 100°F
   d. 95°F

8. What is the temperature of the subcooled liquid in the condenser?

   a. 125°F
   b. 115°F
   c. 95°F
   d. 10°F

9. What is the pressure drop across the TEV?

   a. 229 psig
   b. 278 psig
   c. 49 psig
   d. 69 psig

10. R22 refrigerant at 49 psig boils at what temperature?

   a. 75°F
   b. 35°F
   c. 25°F
   d. 10°F

11. What is the temperature of the superheated vapor at the outlet of the evaporator?

    a. 75°F
    b. 35°F
    c. 25°F
    d. 10°F

12. How can the latent heat in a cold suction line be transferred to higher outdoor air temperatures?

    a. Add heat to the suction vapor until it is higher than the ambient
    b. Compress the suction vapor until its condensing temperature is higher than the ambient
    c. Cool the discharge vapor until it condenses just above the ambient

13. What is the primary difference between the AC system in Figure 1-1 and the refrigeration system in Figure 1-2?

    a. The AC unit has to raise the head pressure higher.
    b. The refrigeration unit has to lower the evaporator temperature.
    c. The AC unit cannot use a TEV.

14. Which refrigerant is used most in recently installed walk-in refrigerators?

    a. R12
    b. R502
    c. R404A
    d. R134a

15. Which refrigerant is used most in recently installed reach-in refrigerators?

    a. R12
    b. R502
    c. R123
    d. R134a

16. Which refrigerant is used most in recently installed walk-in freezers and reach-in freezers?

    a. R12
    b. R502
    c. R404A
    d. R134a

17. Why do zeotropic blends have temperature glide?

    a. The component refrigerants boil off at different rates.
    b. The refrigerants are so new they have not had time to stabilize.

18. Under what circumstances should a technician use TROT?

    a. When the technician is too rushed to look up the correct information
    b. When the factory information is not available
    c. When the technician has not had enough training

OBJECTIVES

After completing this chapter, you should be able to

- Describe the functions of an evaporator
- Explain evaporator temperatures and temperature difference
- Describe evaporator types and styles
- Explain latent heat, sensible heat, and superheat
- Explain the relationship of coil temperature difference and humidity
- Describe evaporator defrost methods

CHAPTER OVERVIEW

This chapter begins with an explanation of what an evaporator does and how it is designed to accomplish its functions. The concepts of evaporator temperature and temperature difference (TD) are introduced to help explain the function and performance of an evaporator.

The practice of superheat measurement is shown as the only reliable means of determining whether an evaporator has too much or too little refrigerant. Understanding these key concepts and conditions is essential to solving the complex system troubleshooting problems in later chapters.

Humidity is discussed as a function of TD, and moisture plays an important role in most commercial refrigeration applications. Because refrigeration evaporators operate below freezing point, the practice of defrosting both medium-temperature and low-temperature evaporators is covered in detail.

FUNCTIONS OF THE EVAPORATOR

The primary function of an evaporator is to absorb heat from the refrigerated space. The secondary function is to remove, or maintain, humidity in the space.

Evaporator Temperature

Evaporator temperature is the term used to describe the temperature of the refrigerant *inside* the evaporator tubing. It is not practical to drill a hole in the tubing to insert a thermometer. An easier and more accurate method is to take the suction pressure and compare it to a pressure/temperature (P/T) chart. For instance, if the suction pressure is 50 psig on an R22 walk-in cooler, according to the P/T chart, the temperature of the coil is about 26°F (see Figure 2-1).

Temperature Difference

The term temperature difference, or TD, is used in this book to stand for the difference between the evaporator temperature and the temperature of the refrigerated space. For instance, if the evaporator temperature of a walk-in refrigerator is 25°F and the box temperature is 35°F, then the TD will be 10°F (35°F − 25°F = 10°F). The reasons for using evaporator temperature, instead of the leaving air temperature of the evaporator, are ease and accuracy of measurement. With walk-ins and reach-ins, it is often difficult to measure discharge air

accurately because of the turbulence of the air passing through the fan blades. However, the inlet air is always at the box temperature, and the evaporator temperature is easily determined from the suction pressure.

The air conditioning (AC) industry has always used *ΔT (delta T)* as the difference between supply and return air temperatures. For AC applications, it is easy to put temperature probes in the ductwork where the air is thoroughly mixed.

In the supermarket industry, there are a number of ways to measure temperatures to determine the appropriate TD or ΔT. For instance, one could use the temperature of the air discharged at the supply grille or "honeycomb," return air at the inlet grille, product or case temperature at the shelf area, and evaporator temperature from the suction pressure along with a P/T chart to determine the saturation temperature.

The important thing is for technicians to ensure they are using the same definition when discussing TD. In any discussion of temperature difference between two or more parties, they should clarify what they mean when using TD and ΔT.

Refrigerant inside the Evaporator

Refrigerant boils inside the evaporator, absorbing latent heat. The temperature at which the refrigerant boils is referred to by several different terms:

- Evaporator temperature
- Suction temperature
- Saturation temperature
- Saturated suction temperature (SST)

FIGURE 2-1   **Cross-section of a Heatcraft evaporator.** *Photo by Dick Wirz.*

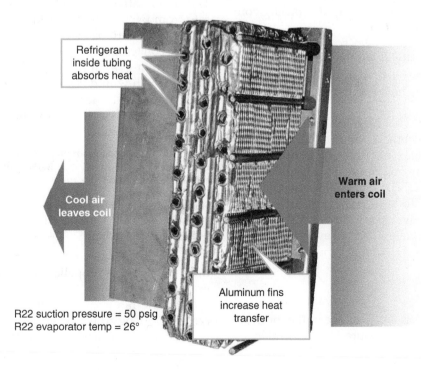

Although these terms sound different, they all essentially mean the same thing.

As the temperature of the air over the evaporator tubing increases, so does the temperature of the refrigerant inside the evaporator tubing. The rise in temperature causes the refrigerant to boil more rapidly. The more vigorous the boiling, the more the suction pressure rises. Therefore, if hot product is put into a reach-in, walk-in, or refrigerated case, the temperatures and pressures will rise accordingly. Conversely, as the temperature is lowered or the load falls due to blocked coils or fan problems, the pressures and temperatures will also fall.

This relationship of refrigerant temperature to box temperature is an important concept to understand to properly troubleshoot refrigeration systems.

What Happens on Start-up?

Hot pull-down refers to the start-up of a new system, a freezer after defrost, or when there is warm product in the box. This means there is a tremendous load on the system initially, and it will not have proper temperatures and pressures until it is near its intended design conditions or normal operating conditions. However, the abnormal system temperatures and pressures observed during a hot pull-down do give an indication of how excessive load affects a system. Consider what happens in the evaporator at the start-up of a warm walk-in refrigerator.

The expansion valve feeds refrigerant by fully opening 100 percent, but it is not enough to fill the evaporator. Therefore, the superheat (the difference between the evaporator temperature and the temperature of the refrigerant leaving the evaporator) is high. The suction pressure is also high because the warm air in the box boils off the refrigerant rapidly, raising both its pressure and its temperature.

As the box temperature falls, so do the pressures and the superheat. Finally, when the box is close to its design temperature, the suction pressure is equivalent to about 10°F below the box, and the superheat is also close to 10°F.

We have just seen what high load does to evaporator conditions. If a technician finds abnormally high pressures and superheat on a service call, he should first consider the load in the box before adjusting the thermal expansion valve (TEV), adding refrigerant, or condemning components. Technicians have been known to adjust an expansion valve open to lower superheat on initial start-up, only to have a flooded evaporator when the box temperature reaches design conditions.

Humidity and TD

Humidity is measured as a percentage of the moisture air can hold at a certain temperature. Strictly speaking, it should be called relative humidity (RH); however, in this book, we will simply use the term *humidity*. Raise the air temperature, and it can hold more moisture. Lower the air temperature enough, and the air will start to release its moisture when it reaches its *dew point*. When the temperature of outdoor air drops below its dew point, rain is likely. Or early on a summer's morning when the ground is colder than the dew point of the air above it, dew forms on the grass. Another example is when the steam from a hot shower condenses on the cooler surface of the bathroom mirror. In refrigeration, when air inside a refrigerated space drops below its dew point, it condenses on cool surfaces, namely the evaporator coil.

All refrigeration removes humidity from a space, some more than others. The process of pulling humidity out of a space is called dehumidifying. During refrigeration, an evaporator operates below the dew point temperature of the air passing over it. Therefore, the moisture in the air condenses on the cold tubing and evaporator fins, collects in the drain pan, and flows outside through the condensate drain line. The air leaving the evaporator now has less moisture, or lower humidity, than when it entered.

Air Conditioning TD and Humidity

The TD between the air entering the coil and the evaporator temperature has a lot to do with the amount of moisture condensed out of the air. For instance, on an AC coil, the TD is about 35°F (75°F return air temperature −40°F evaporator temperature). From psychrometric charts, it can be determined that the air in the space will have a humidity of about 50 percent, which is desirable in a living space for comfort cooling. This low relative humidity allows moisture to evaporate from peoples' skin, making the room seem both cool and comfortable. If the evaporator had only a 25°F TD, the coil temperature would be 50°F instead of 40°F. The warmer evaporator would not remove as much water, which would result in higher space humidity. With 50 percent relative humidity as a benchmark, the higher the room humidity, the less comfortable the occupants will be.

NOTE: AC technicians are more familiar with the Δ*T*. This is the difference between the air handler's return air temperature and that of the supply air. The supply air is usually measured in the trunk line after it mixes well in the supply air plenum. In the previous example, a properly operating AC unit would have about a 20°F Δ*T* (75°F return air temperature −55°F supply air temperature = 20°F Δ*T*). The evaporator temperature would be about 40°F.

Unfortunately for refrigeration technicians, the air discharged by the propeller fans of refrigeration evaporators has different temperatures at several places around the fan area. The difficulty of determining accurate leaving air temperatures has resulted in the use of coil temperature as a better method of measurement.

Walk-in TD and Humidity

In a walk-in cooler, such as in Figure 2-2(b), large coil surface and high air volume combine to produce a TD of only about 10°F (35°F box temperature −25°F coil temperature). At a box temperature of 35°F with an evaporator TD of 10°F, the humidity in a walk-in refrigerator is about 85 percent. This high humidity is very good for foods such as meat and produce that need to be stored in a moist location to prevent them from drying out. However, if the product in the walk-in is covered or contained, such as packaged meats or bottled beverages, then the equipment could be designed for a higher TD of 15°F to 20°F. Refer to Chapter 11, "Walk-in Refrigerators and Freezers," for how equipment matching affects TD.

Reach-in TD and Humidity

Reach-in refrigerators have a limited amount of interior space (see Figure 2-2(a)). Therefore, the evaporator must be as small as possible, yet capable of maintaining adequate temperatures for medium-temperature applications. The smaller evaporator still maintains about 38°F to 40°F, but at a higher TD of about 20°F. A TD this high results in a lower humidity of approximately 65 percent. Although unwrapped product dries out faster in a reach-in than in a walk-in, a reach-in is used primarily for short-term storage. Reach-ins are designed for convenience, where it is desirable to have refrigerated product easily accessible for customers or kitchen staff.

The amount of coil surface and the quantity of airflow can affect box humidity. For instance, high space humidity can be gained by using an evaporator with a large coil surface and greater British thermal units per hour (Btuh) capacity than its condensing unit. A lower fan speed can prevent damage to product that is sensitive to airflow. A combination of these two factors is used in low-velocity coils. These evaporators are often used in meat rooms, deli cases, and florist coolers (see Figure 2-3).

FIGURE 2-2 **A reach-in (a) and a walk-in (b).** *Courtesy of Master-Bilt Products.*

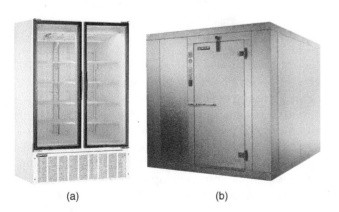

(a)   (b)

EVAPORATOR TD AND HUMIDITY

AC = **35°F** TD at **50** percent

AC: **75°F** return temperature **−40°F** evaporator temperature = **35°F** TD

Reach-ins = **20°F** TD at **65** percent

Reach-in: **40°F** box temperature **−20°F** evaporator temperature = **20°F** TD

Walk-ins = **10°F** TD at **85** percent

Walk-in: **35°F** box temperature **−25°F** evaporator temperature = **10°F** TD

High-humidity boxes = **8°F** TD at **90**+ percent (flower boxes, fresh meat, and deli cases)

NOTE: **As a rule, the lower the evaporator TD, the higher the humidity.**

These *rules of thumb* will be used for troubleshooting in a later section. However, to prevent misapplying these general statements, following are a few conditions and exceptions to keep in mind.

First, the AC ratings are in TD rather than ΔT. Also, AC is included in the list as a source of comparison to commercial refrigeration for those readers who are already familiar with ΔT and comfort cooling. The values given are based on "standard" units of 10 SEER (seasonal energy efficiency ratio) and below. The high-efficiency units have higher coil temperatures, which raises sensible cooling capacities to increase their ratings. However, this lowers the TD and raises indoor humidity, an issue that will not be covered in this textbook.

Although the TDs of reach-ins can vary between 15°F and 30°F, most of them operate in about a 20°F TD. Walk-ins range from 8°F to 20°F TD, and for most of the examples and troubleshooting in this book, 10°F TD will be used (see Figures 2-4 and 2-5).

These TD values can be used on both freezers and refrigerators. However, the humidity ratings are not applicable to freezers because the food is usually packaged.

Using TROT for TD and Troubleshooting

If a walk-in cooler is running at 35°F box temperature, what should the evaporator temperature be? According to TROT, it should be 10°F lower than the box temperature, or 25°F (35°F box temperature −10°F TD = 25°F evaporator temperature). To verify the evaporator temperature, the technician should check the suction pressure as close as possible to the evaporator. If the system contains the refrigerant R22, the suction pressure will be about 49 psig. If it contains R404A, the pressure will be about 62 psig. If the suction pressure is taken at the

FIGURE 2-3   **Low-velocity, high-humidity evaporator coil.** *Courtesy of Heatcraft Worldwide Refrigeration.*

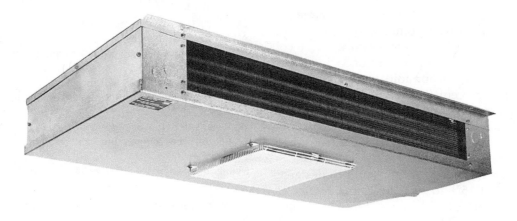

FIGURE 2-4   **Reach-in evaporator.** *Courtesy of Heatcraft Worldwide Refrigeration.*

FIGURE 2-5   **Walk-in evaporator.** *Courtesy of Heatcraft Worldwide Refrigeration.*

FIGURE 2-6   **The most common evaporator construction: copper tubing and aluminum fins.** *Photo by Sharon Rounds.*

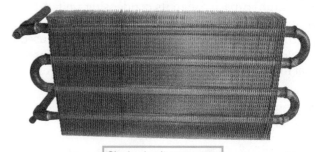

Single circuit evaporator

compressor inlet, say 50 feet away, there may be a pressure reading of about 2 pounds lower than at the evaporator. This pressure drop in the suction line is normal for a correctly installed remote system. A larger pressure drop would indicate a problem with pipe sizing or some restriction in the suction line.

Evaporator Materials

Manufacturers use different materials for their evaporators based on design conditions of heat transfer, cost, and corrosion resistance. Copper gives the best heat transfer but at a relatively high cost. Aluminum is the next best choice for heat transfer and is cheaper. To reach an effective balance of cost-to-heat transfer, most forced air evaporators are made of copper tubing

with aluminum fins. The aluminum fins add surface area for optimum heat transfer (see Figure 2-6).

Manufacturers occasionally try to lower the cost of their equipment by replacing the copper tube evaporators with aluminum tubing. Eventually, the evaporators develop small leaks that are very difficult to repair. For this reason, aluminum evaporators are not popular with service technicians.

Reach-in sandwich and pizza preparation units, in which tomatoes and vinegar are often stored, can develop small pinhole leaks in the evaporator copper tubing from the corrosive effects of the acids in these foods. Therefore, manufacturers offer an optional epoxy coating on these coils to prevent the acids from attacking the copper.

Ice machine manufacturers are concerned about the minerals in water collecting on their evaporators. The greater the mineral buildup, the lower the heat transfer. Most ice machines, such as Manitowoc, use nickel-coated copper evaporators on which the ice is formed. Hoshizaki ice machines have had great success with stainless steel evaporators, even though stainless steel is not as efficient in its heat transfer and costs more than copper. Chapter 12 discusses ice machines more fully.

In water-cooled condensers, mineral buildup causes problems with heat transfer, water flow, and corrosion. Alloys of copper and nickel (cuprous nickel) have been very effective in reducing these problems. However, there is no evidence of their use in evaporators yet.

Heat Exchange Efficiency

Liquid is the most efficient heat exchange "medium." Because of its density, liquid provides better heat transfer than vapor. For example, heat the end of a 7/8-inch piece of copper pipe until it is cherry red. What will cool it faster, blowing 70°F air over it or pouring 70°F water over it? The water, of course, will cool it much quicker because the dense water absorbs heat faster than air.

Liquid-to-liquid heat exchange is very efficient. One example is a chiller that has a float-type metering device that allows liquid to fill, or flood, the evaporator. The secondary refrigerant of water, glycol, or brine is circulated through the tubing covered by liquid refrigerant. The flooded evaporator is cooled as the surface of the liquid boils off and the vapor is pulled into the compressor.

In most of the evaporators covered in this book, the only liquid medium is inside the coil tubing. As air from the space passes over the evaporator coils, the liquid refrigerant boils off as it absorbs heat from the air. This boiling greatly increases heat transfer, as it causes the refrigerant to absorb latent heat.

To illustrate how heat transfer is increased by boiling, consider the concept of heating water taught in the first classes of basic refrigeration. It only takes 1 Btuh of heat to raise the temperature of 1 pound of water from 211°F to 212°F. However, it requires the addition of 970 Btuh to make 212°F water change state to a vapor at 212°F. The point is that the water must absorb nearly a thousand times more heat (latent heat) to boil than it does to simply change temperature. The heat contained in the air passing over evaporator tubing is efficiently absorbed into the refrigerant inside the tubing as the liquid refrigerant boils into a vapor.

If the liquid entering the evaporator is converted to small droplets, the boiling process is easier to accomplish. The metering device (TEV, capillary tube, or fixed nozzle) changes, or "expands," the incoming stream of liquid refrigerant into droplets. This is where the term *DX*, or direct expansion coil, comes from.

Flow Effect on Heat Exchange

The speed of air moving across an evaporator or the rate of liquid flow through tubing can change the rate of heat exchange. Readers familiar with AC are well aware of the effects air volume has on the system. In a standard AC system, the evaporator absorbs about 12,000 Btuh with approximately 400 cubic feet per minute (cfm) of airflow. If the air is too slow, or too fast, it will not cool and dehumidify properly.

In refrigeration, it is unusual to find ductwork or multispeed fans. The volume of airflow is designed into the fan-coil unit from the manufacturer. However, it is important to understand the effects on system pressures and temperatures if a fan motor has the wrong fan speed or burns out, if the fan blade turns in the wrong direction, or if the evaporator is dirty or iced up.

As airflow decreases, there is less heat for the refrigerant to absorb. With less heat, the refrigerant does not boil as much. Therefore, the suction pressure drops, as does the coil temperature. The lower evaporator temperature produces even more frost. Eventually, the entire evaporator will be covered with frost.

The frost acts as an insulator between the warm air in the box and the cold refrigerant in the evaporator. Therefore, because of the reduced heat transfer, the box temperature increases, and the thermostat (tstat) keeps the compressor running. Eventually, the frost will turn to ice, allowing the box temperature to rise high enough to cause food to spoil.

TYPES OF EVAPORATORS

Early in the evolution of refrigeration, the first evaporators were long sections of refrigeration pipe mounted near the ceiling of a refrigerated room. The warm room air rises to the ceiling, the cold pipes lower its temperature, and as the cold air falls, it cools the space. A natural and gentle circulation of air continues through the process of natural convection. This method of refrigeration produces a high-humidity environment perfectly suited to fresh meat and produce. The addition of aluminum fins on the copper pipe greatly increases the heat exchange and allows a smaller amount of tubing to accomplish the same refrigeration effect. These evaporators are known as finned tube convection coils, or gravity coils, and are still in use today, especially in full-service meat and deli cases in super-markets. An interesting application is to preserve live crabs for the seafood industry. Crabs can survive outside water as long as they have a cool environment with very little air movement (see Figure 2-7).

FIGURE 2-7   **Gravity coil or convection coil.** *Photo by Dick Wirz.*

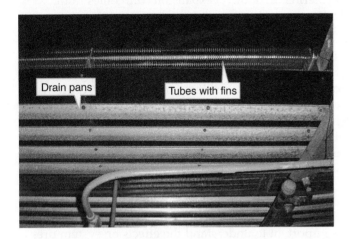

More innovation came by adding a fan to force air across the coil surfaces. This further increased evaporator efficiency while greatly reducing the size of the evaporator. In fact, coils could be made small enough to fit inside a cabinet. Thus, the commercial reach-in refrigerator was born (see Figure 2-8).

On the early large evaporators, the single-pipe design resulted in a significant pressure drop by the time the refrigerant left the coil. Because a drop in pressure causes a drop in temperature of saturated refrigerant, the second half of the coil would be much colder than the first. This arrangement caused uneven coil temperatures

FIGURE 2-8   **Fan-coil unit.** *Courtesy of Heatcraft Worldwide Refrigeration.*

and excessive coil frost. To solve this problem, the large evaporators were redesigned into smaller evaporator sections, or circuits. A multicircuit evaporator is like stacking a group of single coils on top of each other and feeding them with one expansion valve. The outlet of each circuit is tapped into the side of a straight pipe, or header. This header becomes the suction line that returns the evaporator's superheated vapor to the compressor. Using multiple circuits is more efficient than a long single circuit and helps keep pressure drop and coil frosting to a minimum (see Figure 2-9).

FIGURE 2-9   **A multicircuit evaporator coil.** *Photo by Dick Wirz.*

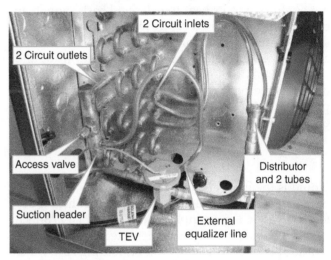

Stamped evaporators utilize two sheets of metal, with passages stamped into the metal where the refrigerant can flow. The extended surface of the plates provides good heat transfer between the refrigerant and other liquids. This method is used by some ice machine manufacturers and is one of the preferred methods for cooling the water in large commercial fish tanks (see Figure 2-10).

FIGURE 2-10   **Stamped or "plate-type" evaporator.** *Compliments of Parker Hannifin–Sporlan Division.*

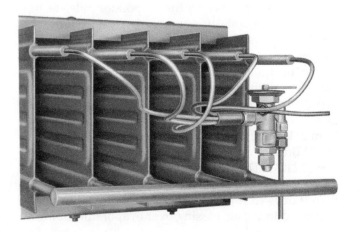

EVAPORATOR OPERATION

It is important that technicians understand what happens inside an evaporator and how different evaporators affect different conditions. Following is a sample of the evaporator type and proper operating conditions a technician could expect in a medium-temperature walk-in refrigerator.

A standard walk-in refrigerator designed for a 35°F box temperature would use a fan-coil unit designed for a 10°F TD. The evaporator temperature, or temperature of the refrigerant inside the evaporator, would be 10°F below the box temperature, or 25°F.

The most accurate measurement of evaporator temperature is obtained by using the suction pressure. In this application, an R22 system at 25°F suction would have a pressure of 49 psig, or 62 psig for R404A. To verify, refer to the P/T chart in the appendix.

The metering device changes the entering liquid to a dense fog of liquid droplets. During the same process, the high-pressure liquid is lowered to what is called the evaporator pressure, or suction pressure. This pressure relates to the evaporator temperature (use the P/T chart). During evaporation, the refrigerant remains the same temperature (its saturation temperature) throughout the coil until all droplets of liquid are vaporized or totally saturated.

When the refrigerant nears the end of the evaporator, the fully saturated vapor can only absorb sensible heat. Although absorbing sensible heat does not contribute much to the overall refrigeration effect, being able to measure sensible heat with a thermometer is very important. Every 1°F the suction vapor is above its saturation temperature is 1°F of superheat.

MEASURING SUPERHEAT

Measuring the amount of superheat allows the technician to determine if all refrigerant has boiled off in the evaporator and to measure approximately the efficiency of coil operation. It can also be a measurement of the margin of safety against flooding the compressor with liquid refrigerant.

The four steps in calculating superheat are as follows:

1. Measure the suction pressure with gauges.
2. From the pressure, determine evaporator temperature using a P/T chart.
3. Measure the temperature of the suction line at the outlet of the evaporator.
4. Subtract the evaporator temperature from the suction line temperature.

Because evaporator temperature is determined from suction pressure, accuracy depends on how physically close to the evaporator the pressure reading is taken. An excellent reading can be taken by installing a tee in the TEV external equalizer line or at the outlet of the evaporator.

Here is an example of a superheat calculation on a 35°F walk-in refrigerator (see Figure 2-11):

35°F (at the coil outlet or TEV bulb) −25°F (evaporator temperature) = 10°F superheat.

The amount of evaporator superheat designed into a system varies from as much as 15°F for some AC systems to as low as 3°F for most ice machines. As a rule of thumb, 10°F is usually adequate for commercial refrigeration systems. However, for optimum efficiency, the technician should check with the equipment manufacturer.

The technician should know which type of superheat measurement is recommended by the factory. An evaporator manufacturer might specify that most of its refrigeration evaporators are designed with a 10°F superheat and that it is calculated by using the suction line temperature at the evaporator outlet. However, a compressor manufacturer may recommend a 20°F or 30°F compressor superheat. This is the total of the superheat in the evaporator plus the sensible heat picked up in the suction line. The compressor superheat, or total superheat, calculation is determined by taking the suction line temperature 6 inches from the compressor inlet, not at the outlet of the evaporator (see Figure 2-11). Compressor manufacturers want to make sure there is enough superheat at the compressor inlets to ensure that the compressors will not be damaged by liquid flooding from the evaporator.

Flooding and Starving Evaporators

A flooding evaporator is not boiling off enough of its refrigerant to prevent liquid from leaving the evaporator. Refrigerant that has not totally boiled off cannot pick up any sensible heat. A flooded system has no superheat. However, flooding is sometimes used to describe an evaporator that has a superheat much lower than normal.

EXAMPLE: 1

A system with a normal superheat of 10°F is considered to be flooding if its superheat is below 5°F.

A starving evaporator is one in which the refrigerant is boiling off too soon. The refrigerant does not fill the evaporator sufficiently; therefore, it picks up more sensible heat than normal. A starving evaporator has high superheat.

EXAMPLE: 2

A system with a normal superheat of 10°F is considered to be starving if its superheat is above 20°F.

FIGURE 2-11 **An example of pressures and temperatures on the low side of a refrigeration system.** *Courtesy of Refrigeration Training Services.*

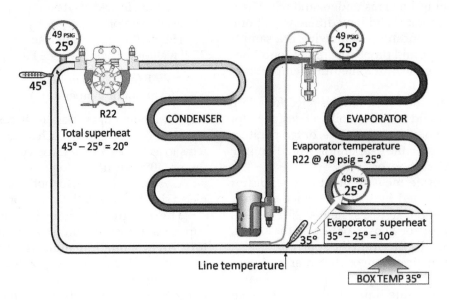

Hot Pull-Down

If the box or space temperature is far above its normal or there is a heavy product load, the process of bringing down the temperature is called hot pull-down. During hot pull-down, none of the pressure/temperature relationships previously discussed seems to hold true. The metering device, whether it is capillary tube or TEV, feeds as much refrigerant as possible to the evaporator. However, the liquid refrigerant quickly boils off because of the high heat-load condition, resulting in a starving evaporator with high superheat.

Although in a hot pull-down the system is out of balance, it is advisable to wait until pressures and temperatures settle down and get closer to the system's usual, or *design*, conditions before determining things like TD and superheat.

According to TROT, superheat should be measured only when the system is within 5°F of design conditions. For instance, a walk-in box designed for 35°F operation will not be even close to its normal superheat until it has dropped to at least 40°F. Even then, it will probably be feeding more refrigerant than it does when it is nearer to 35°F, but at least it would be fairly close to the desired 10°F superheat.

Evaporator Problems

Evaporators have two main problems:

1. Air flow problems: dirty or iced evaporator, fan motor, or blade problems
2. Refrigerant problems: too much refrigerant or not enough, metering device problems, or distributor problems

Troubleshooting these problems is fully covered in later chapters. However, because both medium- and low-temperature refrigeration units develop frost, which can cause airflow problems, this is a good place to discuss how to get rid of frost.

Medium-Temperature Evaporator Air Defrost

Medium-temperature commercial refrigeration usually operates at box temperatures between 34°F to 40°F with evaporator temperatures about 15°F to 25°F. Because the coil temperature is well below freezing, it is normal for frost to build up on the evaporator fins during the "on cycle" when the compressor is running.

When the tstat (thermostat) is satisfied, the compressor shuts off, and the evaporator fans continue to circulate air from the space through the coil fins. Because the space and product temperatures are above freezing, the evaporator warms up, and the frost melts. This process, called random (or off-cycle) defrost, occurs each time the tstat turns off the compressor.

To help ensure enough defrost time, commercial refrigeration tstats are designed with a relatively wide temperature differential. Differential refers to the difference between a control's cut-in and cut-out. The cut-in of a tstat is the setting at which the contacts close when there is a rise in box temperature. The cut-out is the setting at which the contacts open, stopping the refrigerating cycle. Tstats that sense box temperature usually have a differential of 4°F or 5°F, wide enough to air defrost the evaporator coil.

Planned Air Defrost

Sometimes the standard off cycle is not long enough to defrost the evaporator. This problem usually occurs when the box temperature is maintained between 34°F and 36°F. The evaporator can also develop too much frost when the box receives very heavy usage from warm product, bad door gaskets, or just excessive door openings.

Technicians have to plan when, and for how long, to shut off the compressor to clear frost from the coils. For this reason, it is called a *planned* defrost and uses a time clock to shut off the compressor long enough to accomplish the necessary air defrosting.

Normally, defrosts are scheduled for times the box is not in use. For example, technicians usually set the clock to go into defrost at 2:00 in the morning for an hour or two. This gives the coil enough time to melt the frost accumulated during the day. The product temperature in the box may rise by a few degrees, but not enough to cause food spoilage. When the defrost period is complete, the compressor quickly restores the product to its original temperature.

Below 34°F box temperature, additional heat must be used to accomplish complete defrost. For example, meat cutters prefer meat at 28°F because at that temperature, it is firm and easier to cut. Therefore, a meat box will have a medium-temperature condensing unit, a medium-temperature expansion valve, and a freezer coil with electric heaters to accomplish defrosting. Usually, it requires only one or two short defrosts every 24 hours.

The Paragon time clock in Figure 2-12 simply opens and closes a set of switches according to the setting of the "trippers." The clock opens a set of contacts and turns off the refrigeration when the black tripper reaches 2:00 A.M. The silver tripper closes the contacts and turns the refrigeration back on at 4:00 A.M.

Figure 2-13 shows the basic wiring of a walk-in refrigerator that uses a tstat and pump-down solenoid. When the tstat is satisfied, it breaks the circuit to the solenoid coil, allowing the solenoid plunger to drop, which stops the flow of liquid to the expansion valve. The compressor runs until the suction pressure drops or "pumps down" to the cut-out setting on the low-pressure control. The evaporator fans run continuously, circulating the air in the box.

When the temperature in the space rises, the tstat energizes the solenoid coil. The valve plunger lifts, allowing refrigerant to flow through the TEV into the evaporator and back to the compressor. When the suction pressure rises, the low-pressure control closes, and the compressor starts.

There are several reasons for using a pump-down solenoid with a remote condensing unit:

1. When the solenoid closes, the compressor pulls refrigerant out of the low side of the system and stores it in the receiver before shutting down. This prevents refrigerant migration during the off cycle.
2. By removing all refrigerant from the suction line, it eliminates the possibility of refrigerant vapor condensing to a liquid and slugging the compressor on start-up.
3. Compressor starting is easier from an unloaded condition.
4. Control wires are not needed between the tstat in the refrigerated space and the remote condensing unit.

When it is time for a planned defrost, contacts 2 and 3 open (see Figure 2-14). The magnetic coil on the solenoid valve is de-energized, stopping the flow of liquid refrigerant to the evaporator. The compressor pumps down and shuts off on low pressure. The fans run constantly, melting any frost as the evaporator temperature rises to the temperature of the refrigerator.

In Figure 2-15, the wiring is the same as in Figures 2-13 and 2-14, except for the addition of a two-pole switch to shut off the fan. Some customers want to shut off the fans while they are inside the cooler. In addition, the electrical inspector may require a switch to act as a disconnect for servicing the evaporator fan motor.

FIGURE 2-12 **Paragon 4000 series defrost clock for planned defrost.** *Photo by Dick Wirz.*

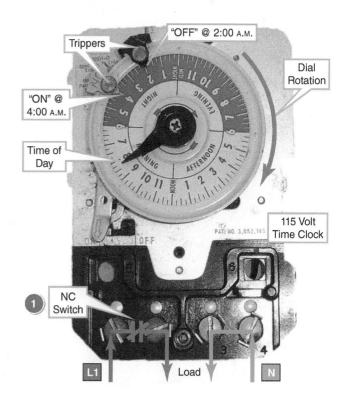

FIGURE 2-13   **Basic wiring of 115-volt evaporator on a walk-in refrigerator.** *Courtesy of Refrigeration Training Services.*

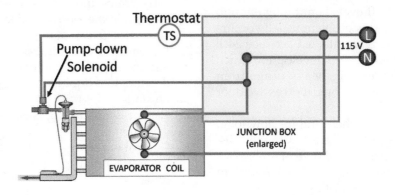

FIGURE 2-14   **Defrost clock used for planned defrost.** *Courtesy of Refrigeration Training Services.*

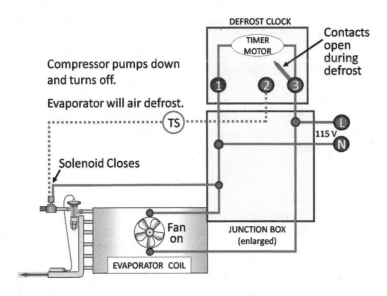

FIGURE 2-15   **Evaporator fan switch wiring.** *Courtesy of Refrigeration Training Services.*

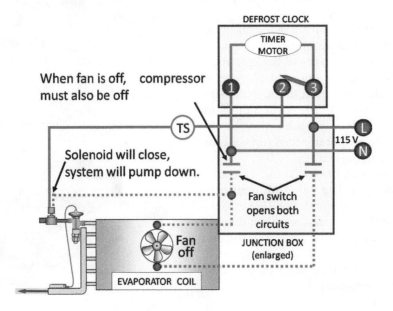

Whatever the reason, it is important to make sure the switch cuts power to the pump-down solenoid so that the compressor is off whenever the fan is off. If not, the compressor will continue running, freezing the coil and causing refrigerant floodback.

When to Defrost

Following is a list of TROT for refrigerated space temperatures when the tstat should be able to perform random or off-cycle defrost, when a clock for planned defrost is needed, and when heat must be added to defrost the evaporator:

> Box temperature 37°F (and above) = **Off Cycle** (no clock needed)
> Box temperature 35°F = **Planned** (clock only)
> Box temperature 33°F (and below) = **Heat** (and clock)

Most medium-temperature refrigeration systems automatically defrost during the off cycle as long as the box temperature is 37°F or higher and the product is the same temperature.

However, when the box temperature is lowered to 35°F, there is usually too much frost to melt during the off cycle. Therefore, a clock is installed to force the compressor to stay off long enough to defrost the coil.

If the box temperature is 33°F or below, there is no way the box is warm enough to air defrost. Therefore, supplemental electric strip heat or hot gas is needed for defrost.

Low-Temperature Evaporator Defrost

Low-temperature evaporators require a time clock, controls, and a source of heat to melt their normal accumulation of frost. Also, because frost is produced easily when the coil temperature is below 0°F, the fin spacing must be wide enough to prevent frost bridging between fins for at least 4 to 6 hours of normal operation.

AC coils may have fin spacing of 15 fins per inch (fpi), whereas medium-temperature units operate at 10 fpi. However, freezer coil fin spacing should be no more than 7 fpi. The wider fin spacing slows the bridging of frost and eventual frost buildup (see Figure 2-16).

The most common defrost system for walk-in and reach-in freezer evaporators is electric resistance strip heaters. They are usually attached to coil fins at the air inlet side of the evaporator. The heaters and all the controls are preassembled at the factory, making it easier for the installers.

Hot gas defrost is primarily used in supermarket applications and ice machines because it is quick and efficient. During defrost, hot gas enters the evaporator

FIGURE 2-16 **Measuring evaporator fin spacing.** *Photo by Dick Wirz.*

downstream of the TEV. The hot refrigerant warms the tubing inside the evaporator, which is more effective than electric heaters that warm the fins outside the tubing. Installing a hot gas defrost system can be very labor intensive due to additional piping and valves. However, hot gas is more efficient because the compressor can generate the same amount of Btuh as electric heaters but with less electrical energy.

Defrost Operation of Freezers

The number of defrosts per day and the maximum length of defrost time depend on the conditions of operation and the location of the equipment.

EXAMPLE: 3

In the Washington, D.C., area, the high heat and humidity during the summer normally require four defrosts in a 24-hour period.

Under design conditions, an electric defrost will last for only about 15 to 20 minutes. However, occasional excessive frost will cause longer defrost time. If the defrost is too long, the clock has an adjustable fail-safe setting that will automatically take the system out of defrost and put it back into refrigeration mode. The setting of the fail-safe varies depending on the climate where the system is located.

EXAMPLE: 4

Most technicians in the Washington, D.C., area set the fail-safe to 45 minutes.

Too many defrosts can be as bad as too few. Even in very humid climates, more than six defrosts a day may indicate a system problem that needs to be addressed. With too many defrosts, there is not enough time to properly freeze the product. Also, shorter defrosts may not fully defrost the evaporator. To verify complete defrosting, a technician should use a flashlight to check all sections of the evaporator. Remember, *defrost heaters defrost frost, not ice.* Once coil frost turns to ice, it has to be defrosted manually.

In hot, dry regions like Phoenix, Arizona, there is little humidity to turn to frost on freezer coils. Therefore, it is possible to use as few as two defrosts a day.

The freezer defrost sequence is as follows:

1. Defrost is initiated, or started, according to the time setting on the defrost clock.
2. The compressor and the evaporator fan(s) stop.
3. The electric heater (or hot gas) starts warming the coil.
4. When the coil reaches a temperature high enough to have melted all the frost, a temperature sensor, called the defrost termination switch, takes the

coil out of defrost and returns it to refrigeration operation.
5. The compressor starts, but the fan(s) stay off. The refrigerant now cooling the coil removes defrost heat and refreezes any remaining water droplets from defrost.
6. When the evaporator drops to approximately 25°F, the fan delay switch closes. The fans start and continue circulating air until the next defrost cycle.

NOTE: Usually the defrost termination and fan delay are combined into a single three-wire control designated by the letters *DTFD* (defrost termination/fan delay) (see Figure 2-17). The control has a common wire and two sets of contacts. One of the contacts is closed when warm and the other set is closed when cold. The DTFD will be discussed in more detail later in this chapter.

FIGURE 2-17 **Electric defrost heaters and controls on a freezer evaporator.** *Photo by Dick Wirz.*

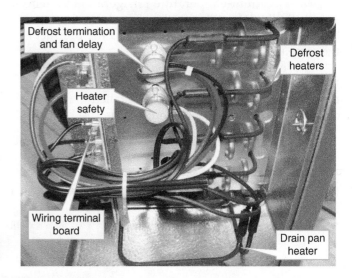

The defrost cycle is also important for proper oil return. When the compressor is off, oil migrates to the coldest spot in the system, which, in a freezer, is usually the evaporator. During the freeze cycle, the cold oil thickens as it travels through the coil and tends to become trapped in the evaporator. Therefore, the heat during defrost warms the oil sufficiently to return it to the compressor on start-up.

Freezer Defrost Clocks

Figure 2-18 is one of the more common 208 to 230 volt defrost clocks, the Paragon 8145-20. The face of the clock has the time of day on the outer ring. The black areas denote night from 6:00 P.M. to 6:00 A.M.

The screws on the face are defrost trip pins. The clock rotates slowly and continuously. When the tripper gets to the "TIME" pointer, the system is mechanically put into defrost. The clock in Figure 2-18 is set for four defrosts per day (6:00 A.M., noon, 6:00 P.M., and midnight).

The fail-safe setting is set on the center dial. The system is supposed to come out of defrost in response to the defrost termination temperature sensor in the evaporator. If it does not switch to refrigeration mode by the time the fail-safe time is reached, the clock will mechanically shift the contacts out of defrost and back into the freeze cycle. The clock in Figure 2-18 is set for a 45-minute fail-safe.

Setting the proper time of day, or manually turning the system into defrost, is accomplished by turning the center of the dial counterclockwise.

Freeze Cycle

On the clock in Figure 2-19, contacts 2 and 4 are closed, sending power to terminal 4 on the evaporator. Terminal 4 is connected to the freezer tstat solenoid and the evaporator fan.

NOTE: **N is used as the common wire.**

FIGURE 2-18   **Paragon defrost time clock Model 8145-20.** *Photo by Dick Wirz.*

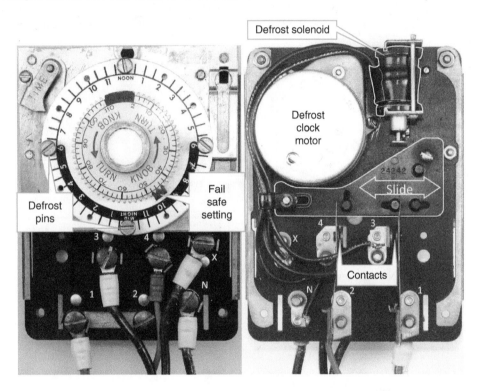

FIGURE 2-19   **Defrost clock in freeze cycle.** *Courtesy of Refrigeration Training Services.*

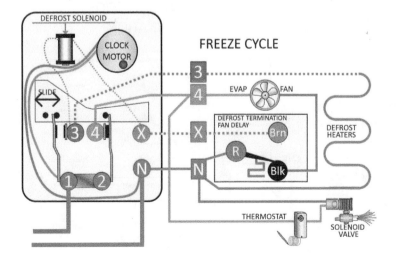

Defrost Cycle

Defrost is time-initiated by the defrost clock. Following is the sequence of operation:

1. On the clock in Figure 2-20, contacts 2 and 4 are opened. This shuts off power to the tstat, solenoid, and fan.
2. On the clock, contacts 1 and 3 are closed. This sends power to terminal 3 on the evaporator, energizing the defrost heater. When 3 on the clock is energized, it also sends power to one side of the defrost termination solenoid shown in the upper left corner of Figure 2-20. In an actual defrost timer, the solenoid is located behind the front panel of the clock, next to the clock motor (see Figure 2-18).

Defrost Termination

When the heaters warm up the evaporator to about **55°F**, the defrost termination contacts close between **R** and **Brn** (brown) on the **DTFD** control (see Figure 2-21). This allows power to flow through the common wire from **N** through **X** and to the other side of the defrost solenoid. When energized, the defrost solenoid coil pulls a lever that moves the slide bar to the right. This mechanically changes the switch positions. Contacts 1 and 3 will open, and contacts 2 and 4 will close (see Figure 2-22).

The defrost clock runs continuously when the system is in freeze or defrost. If the defrost termination control does not bring the system out of defrost, the fail-safe switch will force the system back into freeze mode.

FIGURE 2-20 **Defrost clock in defrost cycle.** *Courtesy of Refrigeration Training Services.*

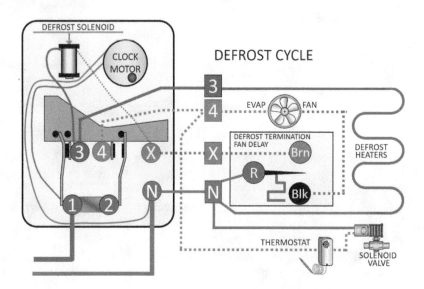

FIGURE 2-21 **Defrost clock at defrost termination.** *Courtesy of Refrigeration Training Services.*

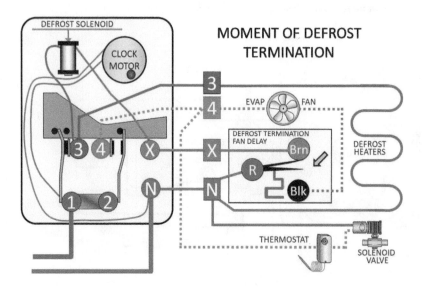

FIGURE 2-22   **Defrost clock returns to freeze cycle**. *Courtesy of Refrigeration Training Services.*

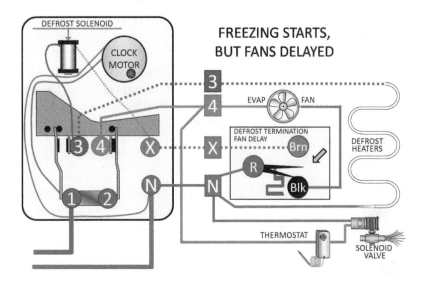

Return to Freeze Cycle

The closing of clock contacts 2 and 4 (see Figure 2-22) sends power to terminal 4 on the evaporator. This energizes the solenoid valve and one side of the evaporator fan.

Fan Delay

The evaporator fan stays off at the beginning of the freeze cycle because the fan delay contacts of the **DTFD** control are open between **R** and **Blk** (black) (see Figure 2-22). They will close when the evaporator cools down to **25**°F, delaying the fan until the evaporator is below freezing (see Figure 2-19). This ensures the heat from defrost is removed from the evaporator and the few droplets of water left after defrost are refrozen onto the evaporator fins. The fan delay control prevents both defrost heat and water droplets from being blown out into the box when the fan restarts. One indication of a bad fan delay would be icicles on the box ceiling and ice on the fan blades because closed contacts would allow the fan to start as soon as the unit is back in the freeze cycle.

Check this type of DTFD control by putting it into another freezer until the temperature of the control is down to about 0°F. Then take it out and use an ohmmeter to measure continuity between the common wire and the wire that will go to the fan. Below 25°F, this circuit should be closed. As the control warms up, the contacts will open. At about 55°F, the contacts between the common wire and the wire that goes to the X terminal should close (see Figure 2-23).

When a warm freezer is starting up, the fans may seem to take a very long time to come on. In addition, the fans may cycle on and off a few times until the box

FIGURE 2-23   **Three-wire defrost termination and fan delay.**
Photo by Dick Wirz.

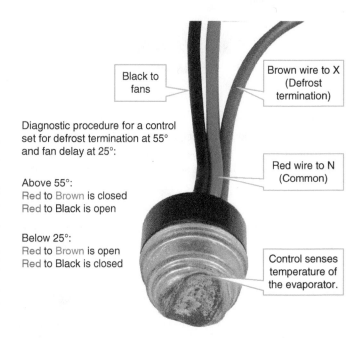

Diagnostic procedure for a control set for defrost termination at 55° and fan delay at 25°:

Above 55°:
Red to Brown is closed
Red to Black is open

Below 25°:
Red to Brown is open
Red to Black is closed

temperature drops below 25°F. Some technicians jump out the control during start up (see Figure 2-24).

Adjustable Defrost Termination

Some units have an adjustable defrost termination control mounted on the front of the evaporator coil, with the sensing bulb located in the evaporator compartment (Figure 2-25). The adjustable range to end defrost is between 60°F and 75°F depending on the control model. Although the defrost termination has some adjustment, the fan delay is usually fixed at

FIGURE 2-24   **Exploded view of DTFD.** *Photo by Dick Wirz.*

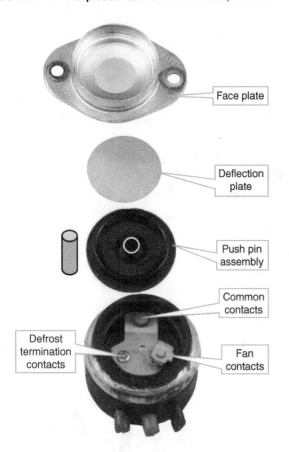

Face plate

Deflection plate

Push pin assembly

Common contacts

Defrost termination contacts

Fan contacts

about 25°F. However, the fan delay has an adjustment screw behind the control cover next to the duration adjustment. Each clockwise turn of the screw increases the fan delay by 3°F. The screw should not be adjusted

more than four turns. Making this adjustment also increases the defrost temperature setting by a similar amount.

Heater Safety

The heater safety control will break the circuit to the defrost heaters and pan heater if the evaporator overheats. The factory setting for most walk-ins is 70°F cutout and 40°F cut-in.

Electronic Time Clocks

In the 1990s, a company called Grasslin came out with an electronic version of the old electro-mechanical time clocks (Figure 2-26). It was more accurate and easier to adjust, had battery backup available, and—most importantly—one clock could replace most of the older clocks. To make it easy to replace the Paragon with the Grasslin, the new company used a numbering system similar to Paragon. The idea caught on, and Paragon (later called Invensys) finally joined the movement about 5 years later with their own version (Figure 2-27).

Demand Defrost

In standard low-temperature refrigeration systems, the defrost clocks are set for a definite number of defrosts per day. Based only on time, the clock forces the unit to defrost whether the evaporator has a substantial frost buildup or not. In the more advanced electronic-based refrigeration control systems, one of the many features is defrosting the coil only when needed. Because defrost is based on demand, it has been referred

FIGURE 2-25   **Adjustable defrost termination control.** *Photo by Dick Wirz.*

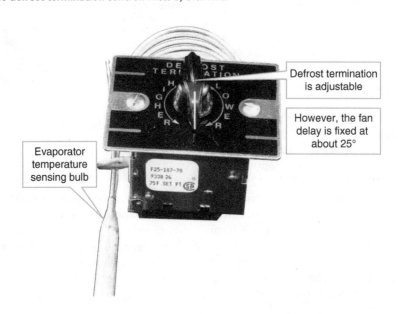

Defrost termination is adjustable

However, the fan delay is fixed at about 25°

Evaporator temperature sensing bulb

FIGURE 2-26   **Grasslin electronic defrost time clock.** *Photo by Dick Wirz.*

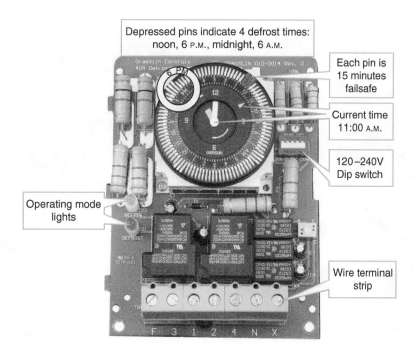

FIGURE 2-27   **Paragon electronic defrost time clock.** *Photo by Dick Wirz.*

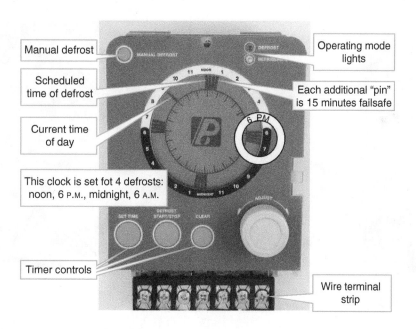

to as demand defrost. Based on electronic inputs of pressure (using transducers; see Figure 2-28), temperature (using thermistors; see Figure 2-29), run cycle times (the amount of time for each cycle from the start of one cycle to the start of the next cycle), and other data, the controller determines when and if defrost is necessary. Heatcraft claims its factory-installed Beacon

II™ (Figure 2-30) and aftermarket Smart Defrost Kit™ (Figure 2-31) can easily reduce defrost requirements by 40 percent. Fewer defrosts mean less energy consumption for defrosting as well as for the refrigeration needed to remove the heat added to the coil during defrost. In addition, box and product temperatures are more stable.

FIGURE 2-28   **Pressure transducer.** *Photo by Dick Wirz.*

FIGURE 2-29   **Thermistors for temperature inputs.**
Photo by Dick Wirz.

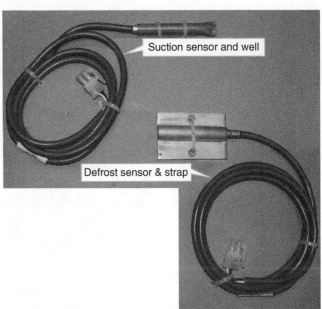

FIGURE 2-30   **Beacon controller mounted in evaporator.**
Courtesy of Heatcraft Worldwide Refrigeration.

FIGURE 2-31   **Beacon II™ remote Smart Controller.**
Photo by Dick Wirz.

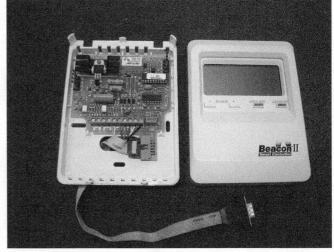

SUMMARY

The evaporator's function is to absorb heat from the refrigerated space. There is a difference in the evaporator temperature for different applications from AC to very-low-temperature freezing.

TD is a function of the difference between air entering the evaporator and the temperature of the refrigerant inside the evaporator tubing. The greater the TD, the more moisture is removed from the air by the evaporator, thus lowering the relative humidity.

There are several types of evaporators, but the most common is the fan coil made of copper tubing with aluminum fins.

Three terms used to describe evaporator heat are the following:

- **Latent heat** is the heat absorbed in the evaporator by changing the state of the refrigerant, without a change in the temperature of the refrigerant.
- **Sensible heat** is the heat absorbed in the evaporator after all refrigerant has been evaporated. There is a change in refrigerant temperature as sensible heat is absorbed.
- **Superheat** is a measurement of the sensible heat after all the refrigerant has boiled off. It is an

indication of the efficiency of the coil and whether the evaporator is starving or flooding.

Evaporators must have some type of defrost if the coil temperature is below freezing. In commercial refrigeration, all coils develop frost. They may only need some time while the compressor is off for the air in the box to defrost them, or they may need supplemental electric or hot gas to remove the frost. Electronic defrost clocks, temperature thermistors, pressure transducers, and electronic controllers have made refrigeration operation more precise and energy efficient.

REVIEW QUESTIONS

1. What is considered the "evaporator temperature"?

   a. The temperature of the refrigerant inside the evaporator tubing
   b. The temperature of the air entering the evaporator
   c. The temperature of the air leaving the evaporator

2. How do you calculate evaporator temperature from the suction pressure?

   a. Use a thermometer to check the air temperature
   b. Use a P/T chart
   c. Use an incline manometer to measure the pressure difference

3. How is evaporator TD calculated for a walk-in cooler?

   a. Subtract the entering air temperature from the leaving air
   b. Find the head pressure and add the box temperature
   c. Subtract the SST from the box temperature

4. What is "hot pull-down"?

   a. When the evaporator is subject to higher temperatures and loads than under normal operating conditions
   b. The defrost period of a freezer
   c. The head pressure a system experiences during pump-down

5. During hot pull-down, what is the evaporator experiencing?

   a. Flooding from the TEV being wide open
   b. Starving because the refrigerant is boiling off quickly
   c. Excessive frost because the evaporator is very cold

6. What is the approximate humidity in a walk-in refrigerator with a 10°F TD evaporator?

   a. 50 percent
   b. 65 percent
   c. 85 percent

7. As TD increases, how does it affect box humidity?

   a. Increases humidity
   b. Decreases humidity

8. If airflow across an evaporator is decreased, what effect does it have on evaporator temperatures and suction pressure?

   a. Evaporator temperature increases, and suction pressure increases
   b. Evaporator temperature decreases, but suction pressure increases
   c. Evaporator temperature and suction pressure both decrease

9. Why are multiple circuits used on larger evaporator coils?

   a. They provide less pressure drop than a long single-circuit evaporator
   b. They cost less to produce than a single-circuit evaporator
   c. They allow use of multiple TEVs on one evaporator

10. If the suction pressure of an R22 unit is 55 psig, what is the approximate evaporator temperature?

    a. 25°F
    b. 30°F
    c. 35°F

11. If the suction pressure of an R404A unit is 21 psig, what is the approximate evaporator temperature?

    a. −14°F
    b. −4°F
    c. +14°F

12. How do you determine the evaporator superheat of a refrigeration system?

    a. Subtract head pressure from suction pressure
    b. Subtract evaporator temperature from the temperature of the suction line at the expansion valve bulb
    c. Add the evaporator temperature to the suction line temperature

13. A −10°F walk-in freezer using R404A has a suction pressure of 15 psig. The suction line temperature at the TEV bulb is −20°F. What is the evaporator superheat?

    a. 2°F
    b. 10°F
    c. 20°F

14. Based on the superheat in the previous question, is the evaporator normal, flooding, or starving?

    a. Flooding
    b. Normal
    c. Starving

15. Within how many degrees of the design box temperature can you check evaporator superheat?

    a. Any temperature is good for checking superheat
    b. 25°F
    c. 5°F

16. What are the two main categories of evaporator problems?

17. Why is it necessary for the fin spacing in a freezer evaporator to be wide?

    a. Frost buildup will not occur as fast if the fin spacing is wide
    b. It helps to move more air because more space means less resistance
    c. Defrost heat needs to get through the fin openings

18. What is the basic sequence of operation at the beginning of a freezer defrost?

    a. Compressor shuts off, evaporator fans run, and heaters come on
    b. Evaporator fans shut off, compressor shuts off, and heaters come on
    c. Fans and heaters are delayed until the compressor cycles off

19. Which freezer defrost system is more efficient, hot gas or electric? Why?

20. If a walk-in is operating at a box temperature of 35°F, what type of defrost should it have?

    a. Heated defrost: a defrost clock and electric heat or hot gas
    b. Planned defrost: a defrost clock only
    c. Random defrost: no clock or heaters, just off-cycle defrost with tstat

21. What is an effective way to check a DTFD control?

    a. Clip the common wire on the control and see if it goes into defrost
    b. Replace it with a new one from the factory and see if doing so corrects the problem
    c. Freeze the control in another freezer then measure the continuity of the contacts with an ohmmeter

CONDENSERS

3

CHAPTER OVERVIEW

This chapter begins with an explanation of what a condenser does and how it is designed to accomplish its functions. The concepts of condensing temperature and condenser split are introduced to help explain the performance of a condenser.

The practice of subcooling measurement is shown as a reliable means of determining condenser performance and system charge. Understanding these key concepts and conditions is essential to solving the complex system troubleshooting problems in later chapters.

Different types of condensers are described, as well as their operation and proper maintenance. Because most commercial refrigeration remote condensers have to operate in cold ambient conditions, head pressure controls are covered in detail.

OBJECTIVES

After completing this chapter, you should be able to
- Describe the functions of a condenser
- Explain how a condenser operates
- Describe condensing temperature and condenser split
- Describe condenser types and applications
- Describe low-ambient controls
- Explain condenser troubleshooting and maintenance procedures

FUNCTIONS OF THE CONDENSER

Condensers are the mirror image of the evaporator. Whereas the function of the evaporator is to absorb heat from the refrigerated space, the condenser must reject that heat outside the refrigerated space.

In addition to evaporator heat and suction line superheat, the condenser must also reject the heat of compression and motor heat picked up by the suction vapor on its way through the compressor. This additional heat can be as much as one-third more than that absorbed by the evaporator. For example, a 36,000-Btuh-per-hour system must reject about 48,000 Btuh of heat. To accomplish this, there must be more effective condensing coil surface than evaporator surface. Airflow through condensers is also an important factor. AC air handlers move about 400 cubic feet per minute per ton (cfm/ton), whereas condenser fans move about 1,000 cfm/ton. *Most condenser coils are designed to have airflow across them of about 1,000 cfm/ton of refrigeration. This is two and a half times the airflow of most AC evaporators, which is 400 cfm/ton.*

CONDENSER OPERATION

Those new to the trade may find it difficult to understand how the system could absorb so much heat into the evaporator yet still have a cold suction line. This strange condition is due to the fact that the heat absorbed is almost entirely latent heat. This type of heat does not change the temperature of the cool suction vapor. Latent heat is trapped inside the refrigerant vapor as the liquid droplets boil off in the evaporator. The temperature of this vapor remains the same throughout the evaporator. After the refrigerant has become fully saturated with latent heat, any additional heat (superheat) is sensible heat and is measurable with a thermometer.

Before the evaporator can reuse the refrigerant, two things must happen in the condenser:

1. The heat in the suction vapor must be removed.
2. The vapor must change to liquid before entering the metering device.

When a vapor condenses, it releases a tremendous amount of latent heat as a result of its change of state from a vapor to a liquid. In theory, the 35°F vapor in the suction line could condense if it were recooled sufficiently. However, in reality, the suction vapor is almost always exposed to temperatures higher than itself; therefore, it is nearly impossible to discharge the heat contained in the cool suction vapor into an ambient that is warmer.

Assuming an outdoor temperature of 95°F, how can a 35°F vapor be condensed back into a liquid?

Somehow, the temperature of this vapor must be raised high enough that 95°F ambient air is cool enough to condense the vapor.

According to the laws of thermodynamics, the temperature of a vapor can be increased by raising its pressure. This process is accomplished in refrigeration with the help of a compressor. The mechanical operation of compression is covered in detail in Chapter 4, "Compressors." This chapter concentrates on the pressure–temperature relationship of compression as it aids the process of heat removal, eventually returning suction vapor to liquid.

EXAMPLE: 1

R22 vapor at 25°F and 49 psig is compressed until it reaches 185 psig. Based on the pressure/temperature (P/T) chart, the higher-pressure-vapor temperature is now 96°F. Theoretically, if the vapor is cooled by one degree, down to 95°F, it will no longer be totally saturated and should begin condensing back into a liquid.

At only one degree difference between the vapor and the ambient air, the cooling process is fairly slow. In Example 2, the difference between the temperature of the warm vapor in the condenser and the temperature of the ambient air is increased to 30°F. The greater the temperature difference between two substances, the faster the transfer of heat.

EXAMPLE: 2

R22 vapor at 25°F and 49 psig is compressed until it reaches 278 psig. At this pressure, the saturated vapor is 125°F and will condense much faster when cooled by 95°F ambient air.

The two preceding examples were theoretical and did not take into account the addition of several other sources of sensible heat that are part of an actual compression process. However, the intent is to show that increasing pressure increases heat and that temperature difference affects the rate of heat transfer.

THREE PHASES OF THE CONDENSER

Following are three phases of the condenser:

1. De-superheat
2. Condense
3. Subcool

De-superheat Phase

The discharge gas, or hot gas, leaving the compressor contains much more heat than it did when it entered the compressor and is at a much higher temperature.

As stated earlier, the suction vapor picks up motor heat and the heat of compression as it undergoes the compression process. The hot gas is called superheated vapor. Just like in the evaporator, if the temperature of a vapor is above its saturation temperature, it is superheated.

Refer to Figure 3-1. The temperature of the hot gas leaving the compressor is 175°F, but the condensing temperature of R22 at 278 psig is only 125°F. The vapor must be cooled by 50°F in the first part of the condenser before condensation can begin.

Condensing Phase

The condensing temperature is the temperature at which a vapor returns to a liquid state.

Condensing is a process, not an action that happens all at once. Condensing starts when the condensing temperature is reached and continues throughout most of the condensing coil until the vapor has fully condensed into a liquid. At this point, the refrigerant is a saturated liquid at a temperature corresponding to condensing pressure listed on a P/T chart.

Subcooling Phase

Cooling the saturated liquid below the condensing temperature is known as *subcooling*. The amount of subcooling is determined by subtracting the temperature of the liquid leaving the condenser from the condensing temperature.

Refer to Figure 3-1. The condensing temperature (and saturation temperature) of R22 at 278 psig is 125°F. The liquid line temperature at the condenser outlet is 115°F.

- 125°F condensing temperature − 115°F liquid line temperature = 10°F subcooling

Why check for subcooling? By knowing the amount of subcooling, a technician can tell a lot about the system she is troubleshooting. For instance:

- No subcooling (0°F) means there is not enough refrigerant in the system to condense into a liquid.
- Low subcooling (less than 5°F) may mean there is flashing of refrigerant before it reaches the metering device.
- High subcooling (over 20°F) may mean there is too much refrigerant in the system.

NOTE: **Although many systems operate properly between 5°F and 20°F subcooling, it is dangerous to make a diagnosis based entirely on measurements outside this range. Other factors should be considered, and they are covered in Chapter 7, "Refrigeration System Troubleshooting."**

⚠ C A U T I O N

If you cannot measure any subcooling at the condenser outlet, but you are fairly certain the unit has enough refrigerant, then check the temperature of the liquid line leaving the receiver. Some manufacturers (Heatcraft, for one) use the condensing unit receiver as the subcooler rather than adding tubing to the condenser coil.

FIGURE 3-1   **Phases of a condenser.** *Courtesy of Refrigeration Training Services.*

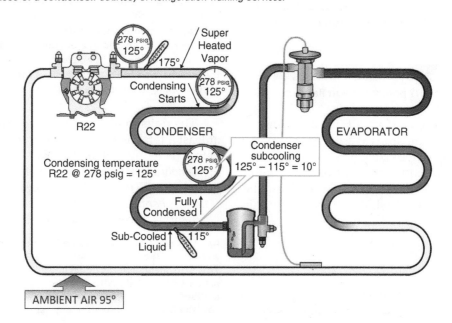

FLASH GAS

Flash gas is liquid refrigerant boiling off into vapor. Flashing of saturated liquid occurs if its temperature rises or if its pressure is lowered.

EXAMPLE: 5

Refer to Figure 3-1. Liquid R22 at 278 psig is saturated at 125°F. It will flash if

1. Its temperature is increased by one degree to 126°F.
2. Its pressure is lowered to 277 psig.

Pressure drop is the primary cause of flash gas and is a desirable event when the refrigerant passes through the metering device. The refrigerant drops from the liquid line temperature to the evaporator temperature as it flashes off.

However, flashing in the liquid line before the refrigerant enters the metering device can cause the device to behave erratically. This flash gas can sometimes be seen as bubbles in the sight glass, a good reason to install the sight glass just before the thermostatic expansion valve (TEV).

NOTE: Bubbles in the sight glass caused by low refrigerant charge or pressure drop may be corrected simply by adding refrigerant. However, bubbles will form in the sight glass during low load conditions and may occur when using blended refrigerants. With these conditions, adding refrigerant will cause more problems than it will solve.

Using Subcooling to Prevent Flash Gas

If the liquid in the previous example had a temperature drop equivalent to the pressure drop, then the liquid would have remained totally saturated, or in a liquid state. Therefore, subcooling a liquid can prevent it from flashing from a pressure drop. The amount of subcooling determines how much pressure drop the liquid can endure before it flashes into a vapor.

EXAMPLE: 6

If R22 at 278 psig drops to 277 psig, it will start flashing at 125°F.

If the liquid is subcooled to 120°F, it will still be in a liquid state if the pressure drops to 260 psig.

Proper liquid line pipe sizing is important in preventing a decrease in liquid pressure on long piping runs. Pressure drop is the result of friction between the inside of the pipe and the fluid flow. If the pipe size is

SUBCOOLING
Subcooling in most refrigeration systems is 10°F.

too small or the length of run is too long, the pressure drop will increase. This is similar to rolling friction on the wheels of a car. If a car is coasting on level ground, the friction will slow the car, using up its forward momentum.

Vertical lift, or the opposing pressure of the weight of liquid being pushed straight up, results in a large pressure drop. A 3/8-inch vertical liquid line will experience a pressure drop of about 1/2 psig per foot of rise.

EXAMPLE: 7

A walk-in refrigerator is located in the third floor cafeteria. Its condensing unit is located on the ground 30 feet below. The 3/8-inch liquid line will experience a 1/2-psig pressure drop per foot of vertical rise. The 15-pound pressure drop (30 feet × 0.5 psig/ft = 15 psig) on an R22 system will require about 4°F subcooling to prevent the liquid refrigerant from flashing before it enters the ceiling space (Figure 3-2).

Following are the calculations for the 4°F subcooling needed to prevent flashing in the liquid line riser piping in Figure 3-2.

According to the P/T chart, R22 at 278 psig will flash above its saturation temperature of 125°F. The 15 psig pressure drop from the liquid line riser will reduce the condensing pressure from 278 psig to 263 psig. The saturated temperature of R22 at 263 psig is about 121°F. Therefore, the condensing unit needs to subcool the liquid at least 4°F (125°F − 121°F) to prevent the liquid from flashing before it enters the ceiling space.

EXAMPLE: 8

In the same example in Figure 3-2, the liquid line turns horizontally into the attic. There is no pressure drop, but now the liquid line is exposed to heat in the attic space. The heat in the attic raises the liquid temperature an additional 5°F. Liquid will flash off as soon as the temperature rises above its condensing, or saturation, temperature. Therefore, to prevent flashing, there will have to be one degree of subcooling for every degree of heat added to the liquid above its condensing temperature. If the attic heat adds 5°F to the heat of the liquid, it will need 5°F of subcooling to prevent flashing.

The units in the previous two examples will need at least a total of 9°F of subcooling to prevent flashing before the liquid enters the TEV. *To increase the subcooling, simply add refrigerant. Most split systems are designed to have about 10°F to 15°F of subcooling when properly charged. Detailed charging procedures will be covered in Chapter 9, "Retrofitting, Recovery, Evacuation, and Charging."*

NOTE: A sight glass and liquid line pressure tap near the TEV would help the technician make sure there is no flashing at the TEV. Also, insulating the liquid line in the ceiling space may help prevent some of the attic heat from entering the liquid line. In addition, a heat exchanger in the suction line near the TEV will cool the liquid line and return any vapor to a liquid.

FIGURE 3-2 **Subcooling is required for pressure drop and added heat.** *Courtesy of Refrigeration Training Services.*

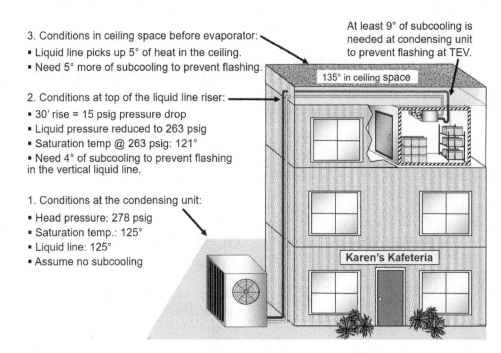

At least 9° of subcooling is needed at condensing unit to prevent flashing at TEV.

3. Conditions in ceiling space before evaporator:
- Liquid line picks up 5° of heat in the ceiling.
- Need 5° more of subcooling to prevent flashing.

2. Conditions at top of the liquid line riser:
- 30' rise = 15 psig pressure drop
- Liquid pressure reduced to 263 psig
- Saturation temp @ 263 psig: 121°
- Need 4° of subcooling to prevent flashing in the vertical liquid line.

1. Conditions at the condensing unit:
- Head pressure: 278 psig
- Saturation temp.: 125°
- Liquid line: 125°
- Assume no subcooling

135° in ceiling space

Karen's Kafeteria

FIGURE 3-3 **Different condenser splits for different applications.** *Courtesy of Master-Bilt Products, Russell Heat Transfer Products Group, Heatcraft Worldwide Refrigeration, and York.*

Master Bilt medium temp 30° split

Russell freezer 25° split

Heatcraft remote 10° to 30° split

York high efficiency 20° split

 T . R . O . T

CONDENSER SPLIT

Most CSs are designed for 30°F at a 95°F ambient.

This applies to standard medium-temperature refrigeration condensing units and AC systems 10 seer and lower. Manufacturers have determined this split is a good balance between the cost of production and the cost of operation. However, high efficiency units and freezers will have a CS closer to 20. See examples in Figure 3-3

CONDENSER SPLIT

Condenser split (CS) is the temperature difference between a unit's condensing temperature and the temperature of the ambient air entering the condenser.

EXAMPLE: 9

If the condensing temperature of the refrigerant is 125°F and the ambient is 95°F, the CS is 30°F (125°F − 95°F = 30°F).

The system manufacturer determines the CS of a unit at a given ambient temperature. For example, the higher the split, the more rapid the heat transfer, and the smaller the condenser coil required.

NOTE: **The CS will vary slightly with ambient temperatures and evaporator loads.**

EXAMPLE: 10

Consider a unit designed for a CS of 30°F at 95°F ambient:

- At a 70°F ambient the CS could drop to about 25°F. A condenser is more efficient at temperatures below its maximum design conditions.
- Low evaporator loads do not require as much CS. The less heat absorbed by the evaporator, the less heat there is to be rejected by the condenser, and the lower the CS needed to reject that heat.
- High evaporator loads (such as during hot pull-down) will require higher CS to reject the additional heat picked up in the evaporator.

A lower split means a more energy-efficient system. The compressor does not have to work as hard (using less power) because the condensing temperature and pressure are lower.

Manufacturers can lower the CS by simply making the condensers larger. However, there are two reasons that all units do not have large condensers:

1. The cost of the unit is higher because of increased production costs.
2. Larger condensers require more space, which limits where the units can be located.

Fortunately, cost and size are not always factors. Following are several instances where the cost of a larger condenser is justified:

- High-efficiency AC units

   The 12 SEER units have a CS of 20°F at a 95°F ambient.

- Freezer condensers

   Freezer compressors are exposed to high compression ratios because they operate at very low suction pressures. This ratio can be lowered if the CS is lowered. Most low-temperature compressors have a 20°F to 25°F CS. (Compression ratios are discussed in detail in Chapter 4, "Compressors.")

- Remote condensers

   Manufacturers of remote condensers justify the cost of larger condensers because lower CSs can lower the total horsepower requirements of a system. This reduces not only the initial cost of the total equipment package but also the customer's operating costs. The manufacturers of remote condensers offer a range of optional CSs as low as 10°F. However, the majority of remote condensers are the standard 30°F split.

EXAMPLE: 11

A supermarket may require a compressor rack with three 40-hp compressors if the CS is 30°F. Using a larger condenser can lower the split to 10°F. The lower CS may allow the same box load to be handled by only three 30-horsepower compressors. This is a saving of 25 percent in equipment costs and a similar saving in operation. (NOTE: This is just an example; actual figures vary.)

CLEANING AND MAINTAINING AIR-COOLED CONDENSERS

A dirty condenser prevents the rejection of heat from the condenser and results in higher condensing pressures and temperatures. A condenser may look clean because the fins do not have dust or dirt on them. However, the only way a technician can be sure a condenser is clean all the way through is to clean it himself. The technician should use a pump sprayer to soak the condenser with a good cleaning solution. On outdoor units, use as much water as possible to flush out the condenser. If possible, use a hose connected to a water faucet or hose bib to get the water pressure and volume needed for a good cleaning. Cooking grease buildup on condenser fins may require hot water, or even steam cleaning, to remove the deposits.

Clean the condenser fan blades to make sure they are moving the proper air quantities. Dirty fan blades move less air, raising the condensing pressures.

Air-cooled condensers inside the customer's building, especially kitchens, present special problems. A technician must try not to interfere with the customer's operation, make a mess in her work area, or damage her equipment with chemicals.

The technician should place towels around the base of the condensing unit to prevent water and cleaner from dripping down the side of the reach-in.

 R E A L I T Y   C H E C K   1

Some equipment is so inaccessible that it must be moved outside for proper cleaning. If the unit cannot be moved, you may need to cut the condensing unit loose and take it outside. All Environmental Protection Agency (EPA) guidelines for recovery must be observed.

Some cleaners can streak the box's metal surface, especially aluminum. The technician should be prepared to go through a lot of towels and rags.

Small portable steam cleaning machines can also be used to clean condensers. The high-temperature steam is very effective in melting accumulated cooking grease, yet it uses very little water. In addition, the use of a steam cleaner cuts down on the need for harsh cleaning chemicals.

GOOD BUSINESS TIP: Proper cleaning can be very time consuming and expensive. Give your customer a quote first; they may choose to try cleaning the units themselves. After their first attempt, most will be happy to pay you to clean their equipment the next time.

GOOD BUSINESS TIP: When the customer is aware of the expense involved in cleaning, she may be more receptive to a quote on a maintenance contract. Periodic maintenance is much easier and less expensive than major cleanings as necessary. Also, a breakdown due to a dirty condenser has the added expense of lost product and the customer's inconvenience of working without the piece of refrigeration.

Coil Cleaners

Use an alkaline (not acid) cleaner that will handle both the dirt and the grease found on most refrigeration equipment condensers. Two popular choices are Triple "D" by DiversiTech and CalClean by Nu-Calgon (Figure 3-4). These cleaners are

concentrates and must be properly mixed at a ratio of 1 part cleaner to 6 parts water. The undiluted cleaner does not clean well. It would be like trying to clean your car with car wash soap without mixing it with a bucket of water first.

Another important tip is to allow enough time for the cleaner to work before rinsing; usually about 5 to 10 minutes.

Do not use foaming cleaners. The foaming action is hydrogen gas released as the chemical reacts with the aluminum, not with dirt or grease. They do not clean better than nonfoaming cleaners, they eat away the aluminum fins, and they etch the aluminum so that dirt and grease collects faster than it would on the original smooth aluminum surface.

Acid cleaners are good only for removing salt water spray or drywall dust residue. Acid will not clean grease off a condenser coil.

Always wear eye protection and gloves when using chemicals of any kind. Use a mask or breathing apparatus if the cleaner you are using gives off any fumes.

Maintenance Inspections

Once the equipment is thoroughly cleaned and the customer is on a periodic maintenance schedule, the technician should install a filter on every indoor condensing unit. This will make inspections easier and

FIGURE 3-4 **CalClean and Triple "D" coil cleaners.** *Photos by Dick Wirz.*

FIGURE 3-5 **Filters on condensers.** *Photos by Dick Wirz.*

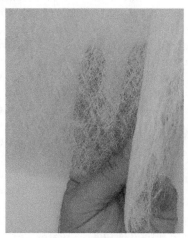

Metal mesh condenser air filter
on ice machine

1" fiberglass material allows
good airflow

Fiberglass air filter on reach-in
condenser

faster because the filters are changed without having to reclean each condenser (Figure 3-5).

NOTE: **Condenser fans cannot handle much resistance to airflow. They will slow down, causing the head pressure to rise. To prevent this, use the inexpensive type of fiberglass filter material found in 1-inch disposable furnace filters. It is sold by the roll and can be cut to the size needed.**

This filter material has wide spaces between the fibers to allow air to pass but will collect a lot of dust and grease. Hang it on the back of the condenser using two "S-hooks" fashioned from a piece of wire coat hanger.

Make sure the inspection intervals are frequent enough to prevent a heavy buildup on the filters. Some customers may require monthly inspections of filters on condensers in one section of their business, like the cooking area or a loading dock, whereas the rest of the unit filters may need to be changed only every three months.

Ice machine manufacturers have had fewer service problems since they began installing filters on their condensers. They usually use a thin, but dense, mesh of plastic or aluminum. To overcome the air resistance of these types of filters, the manufacturers had to upgrade to larger condenser fan motors.

LOW-AMBIENT CONTROLS FOR AIR-COOLED CONDENSERS

Most standard condensing units are designed to operate satisfactorily in ambient conditions between 60°F and 100°F. Below 60°F, the condensing pressure is too low for the metering device to operate properly. Standard expansion valves require a minimum pressure drop between the high-pressure liquid entering the valve and the low-pressure fluid leaving. When the inlet pressure is too low, the valve does not throttle properly, and erratic operation results. This causes the valve to hunt for its equilibrium point, starving the evaporator one minute and flooding it the next.

EXAMPLE: 12

R22 head pressure is at 168 psig; condensing temperature is 90°F. The evaporator temperature of a medium-temperature walk-in is 25°F at 49 psig. The pressure drop across the TEV is approximately 119 psig (168 pisg − 49 psig = 119 psig). Therefore, a pressure drop less than 119 psig may cause erratic TEV operation.

The solution is to keep the head pressure above minimum, usually equivalent to a 90°F condensing temperature. For example, if the unit uses R22, the minimum pressure allowed would be 168 psig. The following sections describe three of the most common methods for keeping the head pressure up in low-ambient conditions:

• Cycling the condenser fans
• Using dampers to control the airflow through the condenser
• Flooding the condenser with refrigerant to decrease its effective condensing surface

Fan Cycle Controls

Fan cycle controls can help manage head pressure by monitoring one of two conditions, the discharge pressure or the ambient temperature (Figure 3-6). Most fan

FIGURE 3-6 **Fan cycle for low-ambient conditions.** *Courtesy of Refrigeration Training Services.*

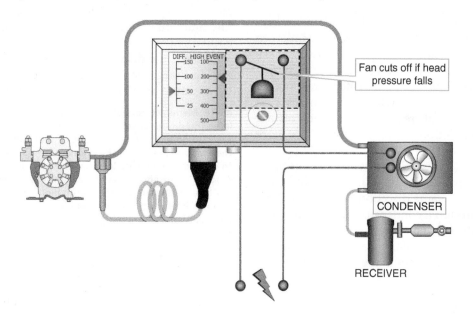

cycle controls stop the condenser fan when the head pressure drops to a minimum pressure cut-out setting. Without the fan pulling cool air through the condenser, the system cannot readily reject heat. As a result, the head pressure rises. The fan starts when the pressure reaches the control's higher cut-in setting.

EXAMPLE: 13

An R22 condensing unit is operating in an ambient of 50°F. When the head pressure drops to the equivalent of 90°F condensing temperature (168 psig for R22), the fan shuts off. This is called the control's *cut-out pressure*.

The *cut-in pressure* is usually set about 40 psig above the cut-out, or 208 psig (168 psig + 40 psig = 208 psig) in this example. The condenser fan cut-out is equivalent to an ambient of 60°F, and the cut-in is equivalent to a 74°F ambient. The unit will be operating in an average ambient condition of about 67°F, midway between 60°F and 74°F. The pressures will remain high enough for proper operation of the metering device.

The differential (difference between cut-in and cut-out) should be between 30 psig and 50 psig. If it is lower than 30 psig, the fan motor will short cycle, resulting in premature motor failure. If the cut-in is greater than a 50 psig differential, the fan will be off too long. Long off cycles cause wide swings in head pressure, which can result in erratic operation of the expansion valve.

On remote condensers with multiple fans, the manufacturer may use a combination of temperature and pressure fan controls. For instance, the first fan may be set to shut off at 70°F ambient, the second at 60°F ambient, and the last fan closest to the condenser outlet may be operating on condensing temperature. Check with the manufacturer for its recommendations.

Limitations to Using Fan Cycle Head Pressure Controls

On condensing units with air-cooled compressors, the condenser fan has the important function of providing the airflow needed to cool the air-cooled compressor motor. When the fan is cycled off, the lack of air movement could cause the compressor's motor to overheat, even in very cold weather.

Copeland does not recommend fan cycle controls on its single-fan condensing units equipped with an air-cooled compressor. However, if the unit has multiple condenser fans, the control can cycle all but one of the fans. The remaining fan ensures compressor motor cooling.

Fan cycle controls are most effective in the southern United States where winter temperatures are normally above freezing.

Fan Speed Controls

Variable frequency drive (VFD) and *electrically commutated motors (ECM)* allow condenser motors to modulate motor speed in order to maintain a more consistent head pressure. Unlike standard fan cycle controls that vary the head pressure 30 psig to 50 psig, motors that vary their speed can keep the head pressure nearly constant. Although more expensive, these fan motors are gaining acceptance because they increase system performance and efficiency, which lowers operating costs.

Air Dampers

Shutters, or dampers, are occasionally used on large remote condensers to keep head pressure up. Similar to fan cycling, dampers limit the airflow across the condenser surface. The dampers have multiple positions and close in stages. Some damper systems respond to the head pressure of the unit, whereas others respond to ambient air temperatures. Fan cycle controls are often used in conjunction with the dampers for more complete head pressure control.

Dampers are also used to prevent low-ambient conditions based on prevailing winds, or winds that are blowing from the same direction most of the time. Wind blowing through a condenser can drop the condensing temperature, even if the fan has cycled off. A good rule to follow when installing air-cooled condensers is to place them 90 degrees to the prevailing wind.

CONDENSER FLOODING

Condenser flooding is like overcharging a unit; it increases the head pressure. Manufacturers of units with condenser flooding use a type of head pressure regulating (HPR) valve to restrict the liquid leaving the condenser. Backing up in the condenser, the liquid refrigerant fills condenser tubing, leaving less space for the discharge gas from the compressor to condense. As a result, the head pressure is elevated to a minimum pressure based on the HPR valve setting. The temperatures and pressures in the condenser rise as if there is warm ambient air going through the condenser.

Most HPR valves are actually three-port valves that close down the liquid leaving the condenser during low-ambient conditions to increase the head pressure equivalent to a 90°F condensing temperature. At the same time, the discharge pressure port opens to bypass hot discharge vapor to maintain a warm refrigerant temperature in the receiver (see Figures 3-7 and 3-8).

NOTE: Manufacturers of flooded condenser systems have calculated the proper charging procedures to ensure adequate refrigerant to flood the condenser during low-ambient conditions. If the unit does not have the information, the technician will have to contact the factory in order to properly charge the system. Also, the unit should have a receiver large enough to contain the total refrigerant charge during high-ambient conditions. See Chapter 9 for more information on charging flooded condensers and standard remote refrigeration systems.

Troubleshooting HPR Valves

To properly troubleshoot an HPR valve (Figure 3-9), the technician must know the following:

- Pressure setting of the HPR valve
- Current head pressure
- Liquid line temperature
- Ambient air temperature entering the condenser

As in all troubleshooting, the technician must determine what is currently happening in the system and compare it to how the system is supposed to be operating. Some technicians believe it would be nice if all refrigeration tubing were clear glass so they could see the flow of refrigerant in all the parts of the unit. However, it is actually better to rely on gauges and thermometers

FIGURE 3-7 **Cutaway of an HPR valve.** *Courtesy of Refrigeration Training Services.*

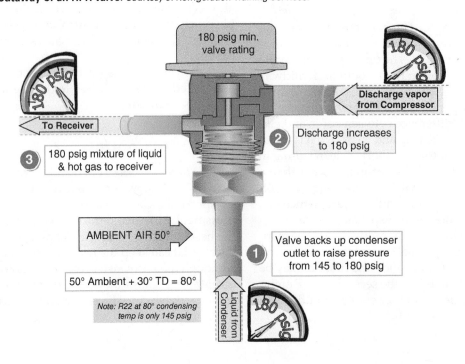

180 psig min. valve rating

180 psig

180 psig

Discharge vapor from Compressor

To Receiver

③ 180 psig mixture of liquid & hot gas to receiver

② Discharge increases to 180 psig

AMBIENT AIR 50°

① Valve backs up condenser outlet to raise pressure from 145 to 180 psig

50° Ambient + 30° TD = 80°

Note: R22 at 80° condensing temp is only 145 psig

Liquid from Condenser

180 psig

FIGURE 3-8   **HPR valve in a refrigeration system.** *Courtesy of Refrigeration Training Services.*

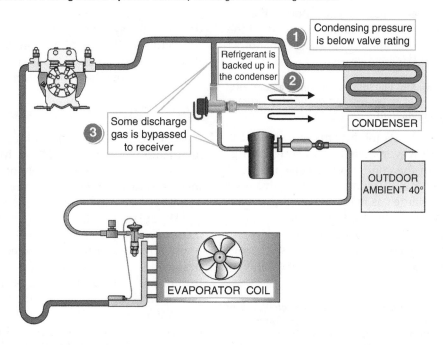

FIGURE 3-9   **A diagnostic chart for HPR valve problems.** *Courtesy of Refrigeration Training Services.*

DIAGNOSIS OF HEAD PRESSURE REGULATING VALVES (HPR's)			

INFORMATION NEEDED:

HEAD PRESSURE **(HP)** _____
LIQUID LINE TEMPERATURE **(LLT)** _____
PRESSURE RATING OF HPR VALVE _____
AIR TEMP. ENTERING THE CONDENSER _____

NOTES	APPLICATION	REFRIGERANT	RATING
	High pressure valves	R502, R404A, HP81, & R22	180 psig = 90° condensing temp
	Low pressure valves	R12, R134a, MP39, R414B	100 psig = 90° condensing temp

Summer (ABOVE 60°)	Winter (BELOW 60°)	REASON	DIAGNOSIS
HP - NORMAL **LLT - NORMAL**	**HP - LOW** **LLT - COLD**	HPR IS **NOT** BACKING UP REFRIGERANT IN CONDENSER. NORMAL FLOW THROUGH THE CONDENSER. NO HOT GAS BYPASSING.	HPR STUCK OPEN OR HEAD HAS LOST CHARGE. REPLACE HPR
HP - HIGH **LLT - HOT**	**HP - HIGH** **LLT - WARM/HOT**	HPR TOTALLY BLOCKING CONDENSER OUTLET. NO LIQUID LEAVING CONDENSER. HOT GAS REPLACES LIQUID IN LL.	HPR STUCK CLOSED. REPLACE HPR
HP - LOW **LLT - HOT**	**HP - LOW** **LLT - WARM/HOT**	NOT ENOUGH GAS FOR GOOD HP, HOT GAS REPLACES LIQUID IN LL. *Note: a remote unit that requires 18 pounds of refrigerant for low ambient conditions may operate properly on only 8 pounds in warm weather.*	LOW CHARGE ADD REFRIGERANT *Not charged properly for low ambient conditions or there is a refrigerant leak.*

than to trust one's eyes. Correct temperature and pressure determination is the most accurate method of diagnosing a refrigeration system.

When the HPR is working properly, the head pressure is equal to the valve setting, and the liquid line is warm (about 90°F).

Following are two HPR problems during low-ambient operation:

- If the head pressure is low and the liquid line is cold, the HPR valve port at the condenser outlet is stuck open. It is not backing up refrigerant in the condenser or bypassing hot gas.

- If the head pressure is high and the liquid line is warm or hot, it means the HPR valve port at the condenser outlet is stuck closed. It has closed off the outlet to the condenser and is bypassing only hot gas into the liquid line.

NOTE: A low charge can fool you into condemning a good HPR valve. If the liquid line is warm (or hot) but the head pressure is low, the unit is low on refrigerant.

The valve is doing its job by shutting down the condenser outlet. However, there is not enough refrigerant backing up in the condenser to raise the head pressure. No liquid is coming through the valve inlet to mix with the discharge gas in the bypass line. Therefore, it is just hot gas entering the liquid line.

A technician who observes these symptoms can verify his diagnosis by adding a few pounds of refrigerant. If the pressures return to normal, the technician is justified in his assumptions.

NOTE: This situation often occurs during the first cold days of the fall or winter. If the unit was originally charged during the summer, it is not uncommon for the technician to have forgotten to add enough refrigerant for condenser flooding during cold weather operation.

FLOATING HEAD PRESSURE

Floating head pressure describes the practice of allowing a system's head pressure to drop as the ambient drops. The main reason for the use of low-ambient controls is to keep the liquid pressure high enough for conventional expansion valves to operate properly. However, several manufacturers have introduced balanced port expansion valves that operate correctly even as the liquid line pressure drops. Russell Coil Company uses these valves on its Sierra remote refrigeration systems installed in many convenience stores. Russell has designed the system to operate efficiently and correctly down to a 30°F ambient without low-ambient controls. Below that temperature, condenser fan cycling is usually the only low-ambient control necessary.

Water-Cooled Condensers

Air-cooled units rely on ambient air to transfer the heat from inside the condenser to the air outside the air conditioned space. The rate of heat transfer can be increased in several ways, including moving a greater quantity of air through the condenser.

Water is denser than air; therefore, it can absorb heat more efficiently than air. Water-cooled condensers are physically smaller than air-cooled condensers and are designed to operate at a consistent condensing temperature of only about 105°F. The only time an air-cooled unit operates at this temperature is on a mild 75°F day.

Water-cooled units are used indoors where the ambient air is hot or contains flour, dust, or grease, like in a commercial kitchen. Another application is when the condensing unit is located in an unventilated area. Ice machines situated in small vending areas are often water cooled. If they were air cooled, they would recirculate their own air and eventually go out on high head pressure.

The most popular water-cooled condensers for commercial refrigeration are:

- Tube-in-tube condensers like the coiled water-cooled condensers used in small condensing units and ice machines (Figure 3-10)
- Shell-and-tube condensers like the condenser-receivers used on larger condensing units and centrifugal chillers

FIGURE 3-10 **Water-cooled condensing unit with tube-in-tube condenser.** *Courtesy of Russell Heat Transfer Products Group.*

Tube-in-tube condensers have the water circulating through an inner pipe, which is surrounded by an outer pipe that contains the refrigerant. In this configuration, the refrigerant can be cooled by both the inner water tube and the ambient air in contact with the exterior surface of the refrigerant pipe. (See Figure 3-11 for an interior view of this type of condenser.) A tube-in-tube condenser is designed so the incoming water enters where the subcooled condensed liquid leaves the condenser. This counterflow configuration maintains a relatively constant rate of heat exchange throughout the entire condenser. In addition, if cold entering water comes in contact with the hot gas line, the thermal shock of the great temperature difference between the two fluids can cause metal fatigue and cracking. Furthermore, the greater the temperature difference between the two fluids, the more mineral buildup at that point in the system.

Tube-in-tube condensers are most often in a coil configuration and are relatively inexpensive. Another version is the flange type that has straight runs of stacked tube-in-tubes with removable flanges or plates on either end for cleaning.

Shell-and-tube type condensers utilize a large tank, or shell, for the purpose of condensing and holding the refrigerant. The hot gas entering the shell comes in contact with cool water pipes, or tubes, that run through the tank. Another way to describe the configuration is that it looks like a liquid receiver with interior water tubes. In fact, some technicians call these units condenser/receivers. The end plates can be removed for cleaning the water tubes (see Figures 3-12 and 3-15).

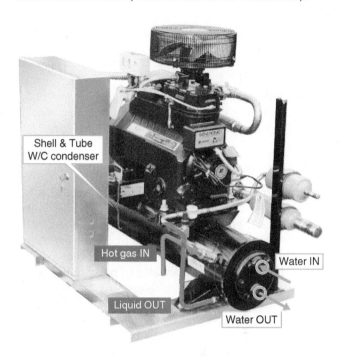

FIGURE 3-12 **Water-cooled condensing unit with shell-and-tube condenser.** *Courtesy of Russell Heat Transfer Products Group.*

Water-Regulating Valves

Most water-cooled units use water-regulating valves (WRV) to modulate the water flow entering the condenser in response to head pressure (see Figure 3-13). When the compressor is off, spring pressure pushes down on the valve's internal flange to shut off the flow of water into the condenser. When the compressor starts, the head pressure rises, pushing up on the bellows, which puts opening pressure on the valve. The two pressures equalize to allow only enough water into the condenser to maintain the head pressure for which the valve is adjusted (see Figure 3-13). When adjusting the WRV, use a gage wrench. Turning the shaft with a pair of pliers

FIGURE 3-11 **Flange-type condenser.** *Photo by Dick Wirz.*

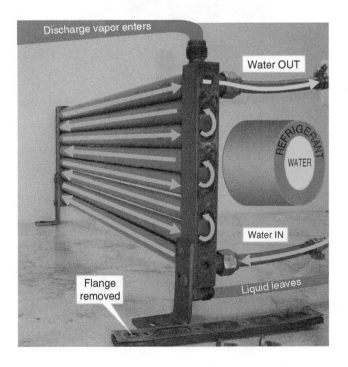

FIGURE 3-13 **Water-regulating valve.** *Photo by Dick Wirz.*

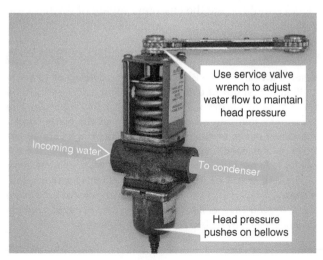

T . R . O . T

The cost of installing a cooling tower is usually justified when the combined water-cooled equipment capacities reach about 24,000 Btuh to 36,000 Btuh.

or a screwdriver often results in damaging the end of the adjuster. Turning the shaft clockwise (looking down on the end of the adjuster) increases the head pressure.

Cooling Towers

Cooling towers used in refrigeration are the same as those used for comfort cooling. However, it is important to remember that towers must be chemically treated to lower the scale buildup in condenser tubes. Also, they require some type of constant bleed-off or "blow-down" of the sump water to prevent excessive mineral concentration in that reservoir.

Operations large enough for cooling towers usually employ professional water treatment companies to maintain tower water quality. Without proper tower maintenance the efficiency of the refrigeration systems will decrease as the cost of operation and service rises. Cooling towers recirculate their water at the rate of approximately 3 gallons per minute (gpm) per ton of refrigeration (12,000 Btuh) at an 85°F sump water temperature.

Wastewater Systems

Wastewater systems are water-cooled units that use tap water to cool the condenser and then drain the waste water. This method uses about 1.5 gpm per ton at an incoming water temperature of 75°F.

NOTE: **The flow rate of water depends on the temperature of the entering water. The colder the incoming water, the less the flow of water needed to condense the refrigerant. The higher the temperature of incoming water, the greater the flow of water needed for condensing.**

When suggesting a water-cooled system, the technician should make the customer aware of the increase that will occur in her water and sewer bill.

EXAMPLE: 14

Assume a 1-ton (12,000-Btuh) unit runs an average of 16 hours (960 minutes) out of every 24 hours. The water use will be 1,440 gallons per day (1.5 gpm × 1 ton × 960 minutes). The customer's monthly water bill will reflect the water-cooled condenser usage of 43,200 gallons. If the charges are $3 per 1,000 gallons used, the water-cooled unit's portion of the bill will be $130 per month, or $1,560 annually. Multiple units, dirty condensers, or leaking HPR valves can make this figure rise considerably. *Although the installed cost of water-cooled units using waste water is cheaper initially, a cooling tower or even a remote air-cooled unit may be less expensive in the long run.*

NOTE: **Check with the local building inspector before installing a wastewater system. Many municipalities do not allow this type of refrigeration system, primarily due to its water usage but also because of the added load on the waste treatment facilities.**

Service and Maintenance of Water-Cooled Equipment

Whether it is personal health or refrigeration equipment, periodic inspection and maintenance makes more sense than waiting until something goes wrong (Figures 3-14 and 3-15). Following are some guidelines:

1. What is the temperature of the water leaving the condenser?

   Hint: Take the temperature of the leaving water or strap and insulate an electronic thermometer probe to the condenser drain line. The pipe temperature should be very close to the actual water temperature.

2. What is the condensing temperature?

   Hint: Use gauges to determine the head pressure. Use a P/T chart to determine the condensing temperature.

FIGURE 3-14 **Mineral buildup in tube-in-tube condenser.** *Photo by Dick Wirz.*

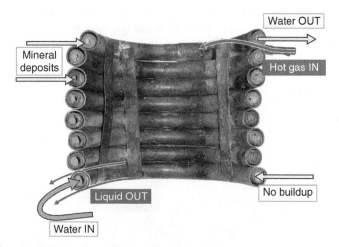

FIGURE 3-15 **Mineral buildup in shell-and-tube condenser.** *Photo by Dick Wirz.*

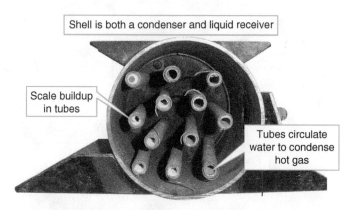

3. What is the flow rate of the leaving water?

   Hint: Use a measurable container, like a 1-gallon bucket. Put it under the condenser outlet and determine how long it takes to fill. For instance, if it takes 30 seconds to fill a 1-gallon bucket, the flow rate is 2 gpm.

If the condenser is piped into a larger drain pipe, cut the drain line from the water-cooled condenser that you are checking. Measure the flow as previously described. Install a union or compression coupling not only to reconnect the main drain but to make it easy to check the flow again during the next inspection.

Proper Operating Conditions of a Water-Cooled Unit on a Wastewater System

The CS of most water-cooled condensers is designed for 30°F above an average entering water temperature of 75°F. Therefore, the condensing temperature should be about 105°F (75°F + 30°F = 105°F). The *leaving* water temperature of a water cooled unit should be *10°F below* the condensing temperature. In this case, it should be 95°F (105°F − 10°F = 95°F).

Following is why it is so important to check the water flow of the condenser outlet: The WRV is on the inlet water line of the condenser (see Figure 3-13). It regulates the flow of water to maintain the proper head pressure. When the condenser tubes become coated with a mineral buildup, the head pressure starts to rise. As the head pressure increases, the regulating valve opens wider, allowing more water to flow through the tubing to bring the head pressure back down to its original setting. If the technician checked only the head pressure, he would not realize the tubing was already starting to be coated with minerals.

EXAMPLE: 15

A technician checks the flow rate of a 12,000-Btuh unit and finds the condensing temperature at 95°F but a flow rate of 2 gpm instead of the normal 1.5 gpm. The technician would know he is starting to have a problem. He can schedule a convenient time to clean the unit and will probably have a relatively easy job of it.

If the technician depends only on seeing high head gauge pressure to signal the need for cleaning, it will be too late. The water-regulating valve keeps opening more and more to allow greater water flow to compensate for the insulating effect of the mineral buildup. Only when the water valve is fully open and the scale is thick enough to prevent proper heat transfer will high head pressure show up, probably tripping the high-pressure reset. At this point, the condenser has developed a very thick coating of minerals, making it more difficult to clean. In addition, the technician has an emergency situation on his hands.

CLEANING WATER-COOLED CONDENSERS

As mentioned earlier, the ends on a flange-type condenser can be removed for cleaning. Special brushes and drill-powered rods can be used to clean the minerals from the water tubing.

Some technicians first try removing the mineral deposits from flanged condensers with chemicals. This may prevent having to disassemble the condenser and clean it with rods and brushes. The next paragraph explains chemical cleaning of water-cooled condensers (see Figure 3-16).

FIGURE 3-16 **Acid cleaning illustration on a water-cooled condenser.** *Photo by Dick Wirz.*

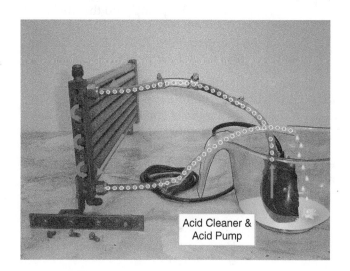

Acid Cleaner & Acid Pump

 REALITY CHECK 2

Planning in advance is very important. The technician should have flange gaskets, spare bolts, and a bolt extractor before starting the work. Occasionally, the bolts snap off in the process of removing the flange. The gaskets usually need to be replaced before reinstalling the flanges.

 REALITY CHECK 3

Cleaning water-cooled condensers is easy if the scale deposits on the tubes are not too thick. However, cleaning condensers that have a heavy buildup is not always successful. The cleaner may not be able to remove enough of the minerals to lower the head pressure. In some cases, the cleaning can dislodge chunks of minerals, resulting in a complete blockage of the condenser tubing.

The only way to clean coil-type condensers is to circulate a strong cleaning solution through them that is specifically made for this purpose. The refrigeration supply house that has the cleaner should also be able to furnish the small epoxy-coated submersible pump needed just for this task. Following is a brief description of the cleaning procedure:

1. Cut the water inlet and outlet pipes.
2. Install unions in the cut pipes. This allows an easy means of hooking up the hoses for cleaning the condenser as well as reconnecting the piping upon completion of the cleaning process.
3. Attach a hose from the outlet of the acid pump to the inlet of the condenser.
4. Run a hose from the outlet of the condenser back to the bucket.
5. Mix the proper amount of cleaning solution in the bucket then put the pump into the bucket of cleaner.
6. Turn on the pump and circulate the acid solution until the condenser is clean.

Follow the directions that come with the cleaner to determine when the condenser is clean. Or simply reconnect the condenser's inlet water and check the head pressure along with the outlet water flow rate.

For units in very bad condition, it may be advisable to give the customer two quotes. One quote is to attempt acid cleaning of the existing unit. The other

⚠ C A U T I O N

Safety Note: Some cleaning solutions are caustic acid that can burn skin and even cause blindness. Make sure to use thick rubber gloves and eye protection. Also, cleaning solutions can damage the condenser if they are circulated too long (for instance, overnight). Always read and follow the manufacturer's instructions.

---

quote includes an additional charge to replace the condenser if cleaning is unsuccessful. The customer may just decide to replace the condenser and be sure of the results, and the quote.

BUSINESS TIP: **It is best to provide customers with alternatives, whenever possible. If there is more than one way to perform a service, you should give your customer the appropriate information and then let the customer make the final decision.**

Microchannel Technology

The fin and tube condenser coils may be a thing of the past. *Microchannel* is the term used for the type of condenser and evaporator coils being introduced in some units by Trane, York, Heatcraft, and other manufacturers (see Figure 3-17). The new configuration produces smaller coils with higher efficiency. The design is similar to that currently being used by the auto industry in radiators.

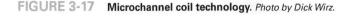

FIGURE 3-17 **Microchannel coil technology.** *Photo by Dick Wirz.*

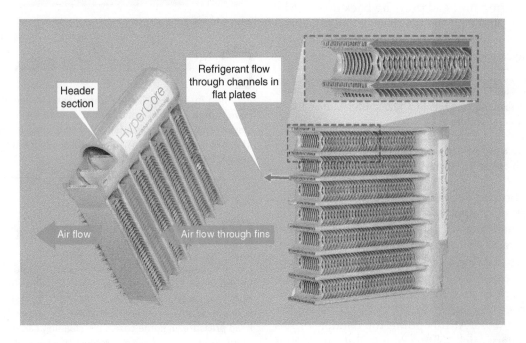

SUMMARY

Refrigerant boils in the evaporator as it absorbs heat. The compressor pumps the refrigerant and raises its pressure and temperature before discharging it to the condenser. The condenser rejects the heat and condenses the vapor back into a liquid.

Being able to measure condenser subcooling helps the technician determine if all refrigerant vapor has condensed into a liquid. Subcooling is also beneficial in preventing flash gas in the liquid line.

Low-ambient conditions require controls to maintain a minimum of 90°F condensing temperature in air-cooled condensers. On water-cooled condensers, the water-regulating valve automatically adjusts the water flow to maintain the proper head pressure.

The best way to be sure a condenser is clean is to clean it. There are proper procedures for cleaning both air- and water-cooled condensers. A consistent maintenance program is beneficial to the business of both the customer and the servicing technician.

REVIEW QUESTIONS

1. What is the primary function of the condenser?

   a. Absorb heat from the refrigerated space
   b. Reject heat from the refrigerated space

2. Why does the suction vapor have to be increased in temperature before it can be condensed?

   a. So the condensing temperature is above the ambient
   b. So the ambient temperature will be warmer than the refrigerant

3. How can the temperature of refrigerant be increased without adding excessive heat?

   a. Raise the temperature going through the condenser
   b. Lower the pressure of the refrigerant
   c. Raise the pressure of the refrigerant

4. What are the three phases of the condenser?

   a. Subcool, supercool, and condense
   b. De-superheat, condense, and subcool
   c. Superheat, condense, and subcool

5. Where does the superheat in discharge gas come from?

   a. Suction vapor superheat
   b. Compressor motor heat
   c. Heat of compression
   d. All of the above

6. When does condensing start?

   a. When the discharge gas is cooled to its condensing temperature
   b. As soon as the discharge gas enters the condenser
   c. After the refrigerant leaves the condenser

7. What is subcooling, and what does it indicate?

8. After the vapor has been condensed to a liquid, what are two causes for the liquid to flash back into vapor before it reaches the metering device?

   a. A rise in temperature with a rise in pressure
   b. A fall in temperature with a fall in pressure
   c. A rise in temperature or a fall in pressure

9. If a 3/8-inch liquid line has a vertical lift of 40 feet, how much pressure drop is there?

   a. 10 psig
   b. 20 psig
   c. 30 psig
   d. 40 psig

10. Based on the pressure drop in the question above and a design condensing temperature of 125°F, for R22, what is the minimum subcooling required?

    a. 2°F
    b. 5°F
    c. 12°F
    d. 14°F

11. What is CS?

    a. The difference between ambient air and condenser discharge air
    b. The difference between suction temperature and condensing temperature
    c. The difference between ambient air and the condensing temperature

12. If the CS of a unit is 30°F, what should the condensing temperature be with an ambient air temperature of 70°F entering the condenser?

    a. 80°F
    b. 100°F
    c. 125°F
    d. 40°F

13. Based on the previous question, if the system is using R22, what should the head pressure be?

    a. 100 psig
    b. 180 psig
    c. 196 psig
    d. 278 psig

14. What, or who, determines the CS of a unit?

    a. The condensing unit manufacturer
    b. The installing technician
    c. The customer

15. What are the approximate CSs of the following units and why?

    a. Standard medium temperature refrigeration condensing unit?
    b. Commercial freezer?
    c. Remote refrigeration condenser?

16. How does a technician make sure a condenser is clean?

    a. Ask the customer.
    b. Look for dirt sticking out of the fins.
    c. The technician thoroughly cleans it himself.

17. Why is it necessary to make sure the condenser fan blades are clean?

    a. Dirty blades provide less airflow than clean ones.
    b. It will not look right if the condenser is clean and the blades are not.
    c. The customer is paying for a good cleaning.

18. The term *low ambient* usually describes ambient temperatures below __ degrees?

    a. 0°F
    b. 32°F
    c. 40°F
    d. 60°F

19. What is the minimum condensing temperature for standard refrigeration units?

    a. 60°F
    b. 90°F
    c. 105°F
    d. 125°F

20. What are three common methods of head pressure control during low ambient?

    a. HPR valves, CPR valves, and dampers
    b. Dampers, fan cycle controls, and condenser flooding
    c. Floating head pressure, HPR valves, and WRVs

21. Fan cycle controls can respond to what two conditions?

    a. Water flow and airflow
    b. Head pressure and suction pressure
    c. Ambient air temperature and head pressure

22. What problems can occur if the condenser fan stays off too long?

    a. None; the head pressure needs to rise anyway.
    b. Air-cooled compressor motor overheating and erratic TEV operation result.
    c. The TEV will starve the evaporator, and the head pressure will drop.

23. What problem occurs when the condenser fan short cycles?

    a. The fan motor can be damaged.
    b. The thermostat in the box will not function properly.
    c. Nothing; it is good to keep the pressure swings close together.

24. What region of the United States is best suited for fan cycle controls?

    a. North
    b. South
    c. East
    d. West

25. How do prevailing winds affect a condenser's operation?

    a. They lower head pressure even when the fan stops.
    b. They keep the compressor cool during the fan's off cycle.
    c. They plug up the condenser with dirt.

26. How does condenser flooding maintain head pressure in low ambients?

    a. The refrigerant takes up space in the condenser tubing, forcing head pressure up.
    b. It lowers the head pressure until it reaches optimum pressures.
    c. It makes the TEV flood refrigerant through the evaporator.

27. What four things must a technician know before diagnosing problems in an HPR valve?

    a. Manufacturer, model number, serial number, and color
    b. Ambient temperature, head pressure, liquid line temperature, and valve rating
    c. Head pressure, discharge pressure, subcooling, and valve size

28. What is floating head pressure?

    a. Head pressure of condensing units used on ships
    b. Allowing the head pressure to drop along with the ambient
    c. Using balanced-port TEVs to feed evaporator to keep head pressure up

29. What does tube-in-tube mean when referring to water-cooled condensers?

    a. The water pipe is surrounded by the refrigerant tubing.
    b. The shell is a refrigerant tank with water tubes running through it.
    c. The condenser is a big water tube with small refrigerant tubes running through it.

30. What is a shell-and-tube water-cooled condenser?

    a. The water pipe is surrounded by the refrigerant tubing.
    b. The condenser is a big tube with water tubes running through it.
    c. The condenser is a big water tube with small refrigerant tubes running through it.

31. Wastewater condensers use how much water per minute at an incoming water temperatures of 75°F?

    a. 1.0 gpm/HP
    b. 1.5 gpm/ton
    c. 3.0 gpm/HP
    d. 3.5 gpm/ton

32. If the compressor in the previous question is rated at 6,000 Btuh, what should be the leaving water flow rate in gallons per minute?

    a. 0.5 gpm
    b. 0.75 gpm
    c. 1.5 gpm
    d. 3.0 gpm

33. What is the approximate condensing temperature of a water-cooled condenser?

    a. 75°F
    b. 90°F
    c. 105°F
    d. 125°F

34. The water leaving a water-cooled condenser should be how many degrees below the condensing temperature?

    a. 5°F
    b. 10°F
    c. 15°F
    d. 20°F

35. If the head pressure is correct, do you need to measure the flow rate of the outlet water? Justify your answer.

36. If the head pressure is high and the flow rate is high, what is the condition of the water-cooled condenser?

    a. There is too much mineral buildup in the condenser.
    b. The water-regulating valve needs to be adjusted.
    c. The unit has a refrigerant leak.

37. Is cleaning a water-cooled condenser always successful? Justify your answer.

OBJECTIVES

After completing this chapter, you should be able to

- Describe the functions of a compressor
- Depict how reciprocating compressors operate
- Explain the refrigeration system's effect on compressor operation
- Describe compression ratios and their importance
- Explain the similarities and differences between hermetic and semihermetic compressors
- Detail the importance of compressor lubrication
- Describe the causes for compressor problems and their cures

CHAPTER OVERVIEW

This chapter concentrates on reciprocating compressors, the most common type of compressor used in commercial refrigeration. To a lesser degree, scroll compressors will also be covered because they have proven their worth in AC and are now being used in commercial refrigeration.

The goal of this chapter is to explain how reciprocating compressors operate and what conditions in the refrigeration system affect their operation. In addition, the final section provides an opportunity to put this information to good use by analyzing examples of compressor problems and how they can be corrected.

FUNCTIONS OF A COMPRESSOR

The compressor can be considered the "heart" of a refrigeration unit because it pumps refrigerant through the system. In addition, it raises the temperature of the returning suction vapor to a higher temperature before discharging it into the condenser. As discussed in Chapter 3, "Condensers," the discharge gas can be condensed only if its saturation temperature is hotter than the air or water cooling the condenser.

The refrigerant vapor temperature is increased by compressing it to a higher pressure. However, in the process of going through the compressor, the vapor picks up sensible heat from two sources:

1. Motor heat, or the heat generated by the compressor motor
2. Heat of compression, or the heat generated during the compression process

This additional sensible heat, plus suction-line sensible heat, takes the form of superheated vapor. It must be rejected in the first part of the condenser before the vapor can cool enough to condense.

How a Reciprocating Compressor Operates

The following descriptions of the reciprocating compressor cylinder sequence are illustrated in Figure 4-1. Starting at Step A, the piston is at the top of its travel. The suction vapor will enter the cylinder from the left and the discharge vapor will leave on the right.

In Step B, as the crankshaft turns, the piston is lowered in its cylinder and the pressure within the cylinder drops. When the cylinder pressure is less than the suction-line pressure, the suction reed valve bends down and allows the cylinder to fill with suction vapor. The action of the piston moving down is called the intake stroke or suction stroke because the vapor from the evaporator is sucked into the compressor cylinder.

Step C is the bottom of the suction stroke. The pressures in the cylinder and above the suction reed valve are equal. The suction valve springs closed.

In Step D, when the piston starts to rise, it compresses the vapor in the cylinder to a smaller space, increasing the vapor's pressure. This is called the compression stroke. When the pressure in the cylinder is above the pressure in the condenser, the discharge reed valve is pushed open. The high-pressure *discharge gas* is discharged into the condenser.

At the top of the compression stroke, Step E, the pressure in the cylinder equalizes to the pressure in the condenser. This allows the reed valve to spring closed. When the discharge valve is closed, there remains a small area of high-pressure gas between the top of the piston and the bottom of the valve (at the arrow). This space is known as clearance volume.

In Step F, the piston starts its downward suction stroke. The high pressure in the clearance volume re-expands until it is below the suction pressure. Only then can the suction valve open.

The higher the pressure in the clearance volume, the more the piston has to travel downward before it can take on suction vapor. The lower the piston travels

FIGURE 4-1   **Reciprocating compressor operation.** *Courtesy of Refrigeration Training Services.*

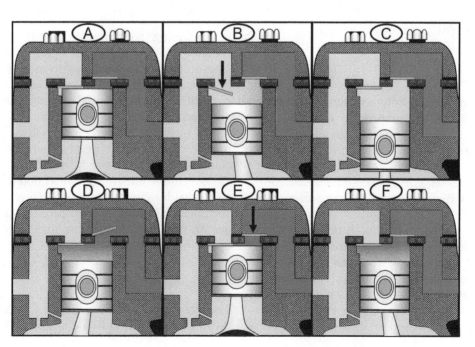

in the cylinder before the suction valve opens, the less volume of suction vapor it can pull in. This adversely affects the volumetric efficiency of the compressor.

Volumetric efficiency is a term used to describe how well a compressor is operating as compared with its capacity at design conditions. Compressor capacity can be affected by variations in suction pressure and discharge pressure, as well as by the condition of its valves and rings. Volumetric efficiency is discussed further in Chapter 7, "Refrigeration System Troubleshooting."

Compression Ratio

In refrigeration, the compression ratio is the relationship of the head pressure to the suction pressure in pounds per square inch absolute (psia). To determine psia, just add 14.7 pounds to the gauge pressure in pounds per square inch gauge (psig).

EXAMPLE: 1

The compression ratio of an R22 walk-in refrigerator system at design conditions of 25°F evaporator and 125°F condensing temperature is calculated as follows:

1. The low-side pressure at 25°F is 49 psig + 14.7 = 64 psia.
2. The high-side pressure at 125°F is 278 psig + 14.7 = 293 psia.
3. The compression ratio at design conditions is 293/64 = 4.6, which is expressed as 4.6-to-1, or 4.6:1.

In other words, the suction pressure is increased by 4.6 times to reach the head pressure. Compressors are designed to operate efficiently within certain compression ratios. Following are the compression ratios of two other R22 applications:

- Air conditioning (AC) compressor at 40°F evaporator and 125°F condensing = 3.5:1 ratio
- Freezer compressor at −20°F evaporator and 125°F condensing = 11.8:1

Based on the difference between these two ratios, it is easy to see why an AC compressor would not operate very well in a freezer application. In addition, the higher the compression ratio, the higher the heat of compression; therefore, the hotter the discharge vapor leaving the compressor. For this reason, freezer compressors have very high discharge temperatures even under normal conditions.

In Chapter 3, the condenser split (CS) of a standard medium-temperature refrigerator was said to be about 30°F. However, standard freezers have a CS between 20°F and 25°F, and high-efficiency freezer condensers are now being produced with a CS as low as 10°F. The unit's designed CS is important to know when diagnosing a unit based on its condenser split.

Manufacturers design their freezers with a lower CS to lower the compression ratio and the compressor discharge temperature. To illustrate, assume an R404A reach-in freezer operates at a 0°F box temperature with an evaporator temperature of −20°F. At a 30°F CS, the condensing temperature at a 95°F ambient would be 125°F. The compression ratio is calculated as follows:

$$\frac{(333\,\text{psig} + 14.7)}{16\,\text{psig} + 14.7} = \frac{348\,\text{psia}}{31\,\text{psia}} = 11.2{:}1\,\text{ratio}$$

If the unit were manufactured with a lower CS of 20°F, the condensing temperature would be lowered to 115°F. Therefore, the compression ratio would also be lower:

$$\frac{(292\,\text{psig} + 14.7)}{16\,\text{psig} + 14.7} = \frac{307\,\text{psia}}{31\,\text{psia}} = 9.9{:}1\,\text{ratio}$$

Keeping the compressor discharge temperature as low as possible will extend the useful life of a freezer. High discharge temperatures cause discharge valve failure and lubrication problems due to oil breakdown.

It is common knowledge that high ambients, dirty condensers, and fan problems cause high condensing temperatures, which in turn cause high discharge temperatures. However, increased hot gas temperatures can also come from running a unit down to a lower box temperature than it was designed for. Using the previous example, the R404A reach-in freezer designed for 0°F box temperature has a compression ratio of 9.9:1 at an evaporator temperature of −20°F. Following is an illustration of how the compression ratio would change if the box temperature were reduced by only 10°F:

$$\frac{(292\,\text{psig} + 14.7)}{10\,\text{psig} + 14.7} = \frac{307\,\text{psia}}{25\,\text{psia}} = 12.3{:}1\,\text{ratio}$$

Lowering the thermostat (tstat) by 10°F caused the suction pressure to drop by 6 psig. This small drop increased the compression ratio more than 20 percent. In comparison, it would have taken a 60 psig increase in head pressure to equal the same increase in compression ratio.

Diagnosing compressor failure problems on a freezer, or any other type of system, should always include causes of high head pressure. However, the previous example illustrates why a good technician should also consider damage caused by low-load conditions. Technicians should look for causes like a frozen evaporator, starving evaporator, and lower-than-normal operating temperature.

Hermetic Compressors

The word *hermetic* comes from a Greek word meaning "secret" and "sealed in a container." It describes this style of compressor well because the entire motor and piston assembly is mounted on springs inside a metal shell, then welded shut (Figure 4-2). Other terms used to describe this compressor are *welded* and *tin-can*.

The suction vapor enters the shell, cools the motor as it passes over it, and then is pulled into the cylinder. Cooling for the compressor motor comes entirely from the suction vapor. Therefore, it is important to return enough dense vapor refrigerant to properly cool the compressor motor windings. The engineering terminology for this is mass flow rate and is given in terms of pounds per hour (lbs/hr) of refrigerant.

In other words, the lower the pressure, the lower the density of the refrigerant. Technically speaking, the low-pressure vapor has fewer atoms, or molecules, per cubic inch to carry away the heat energy from the motor. Therefore, low-pressure vapor is less effective at transferring the heat away from the motor's surface by convection.

EXAMPLE: 2

If a pipe was just brazed, what would cool it quicker: blowing 70°F air over it or spraying a fine mist of 70°F water on it? The mist would transfer heat quickly because the water droplets are denser than air. Also, as the water boils off, it absorbs large amounts of latent heat.

Compressor manufacturers insist that suction pressures be kept high enough to provide proper cooling. They know very well that slightly warmer dense vapor at a higher pressure will actually cool their suction-cooled compressor motors better than colder, less dense refrigerant vapor at a lower pressure.

If an R22 compressor is designed for a 25°F saturated suction temperature, the compressor motor will be adequately cooled with a suction pressure of 49 psig. What would happen if the metering device was restricted or the system was running low on refrigerant? The suction pressure would drop, correct? If the suction pressure dropped to 28 psig, the suction temperature would drop to 5°F. Although the temperature is colder, the lower-pressure vapor does not properly cool the motor, possibly causing it to overheat.

Semihermetic Compressors

This type of compressor is also called a *serviceable hermetic*, *cast-iron*, or *bolted* compressor. Unlike the hermetic compressor that is welded shut, the semihermetic compressor is bolted together, which allows limited serviceability. The motor is sealed inside, yet the valve plate assembly is exposed. Two serviceable operations can be performed on this compressor: valve repairs and oil replacement. In addition, on most semihermetics of 5 horsepower (hp) and above, the oil pump can be replaced, although it seldom fails. Another important feature is the possibility to tear down or disassemble this type of compressor when it fails in order to determine what caused the failure. Few semihermetic compressors are completely rebuilt on the job. However, defective compressors are recycled by the factory, where the compressor housing is refitted with all new internal parts and the entire assembly is resold as a new compressor.

There are two types of semihermetic compressors: air cooled and suction cooled. In a suction cooled

FIGURE 4-2   **Hermetic compressor in welded shell.** *Courtesy of Copeland Corporation.*

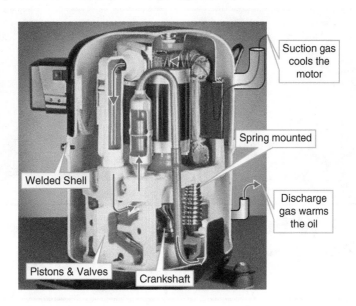

compressor, the suction vapor enters the compressor near the motor and cools the compressor motor before entering the cylinder (see Figure 4-3).

In the air-cooled compressor, the suction valve is mounted close to the valve plate and cylinder (see Figure 4-4). The suction vapor goes directly into the cylinder without cooling the compressor motor. Cooling of the motor is by air blown across the compressor body, usually from the condenser fan. On water-cooled condensing units, water tubing is wrapped around the compressor body to cool the motor.

Because the condenser fan provides the motor cooling on air-cooled compressors, it is easier to understand why fan cycling controls are not recommended on single-fan motor condensing units.

The smaller semihermetic compressors from 1/4 hp to about 3 hp are air cooled. Most semihermetic compressors of 5 hp and above are suction cooled.

FIGURE 4-3   **Semihermetic suction-cooled compressor.** *Courtesy of Copeland Corporation.*

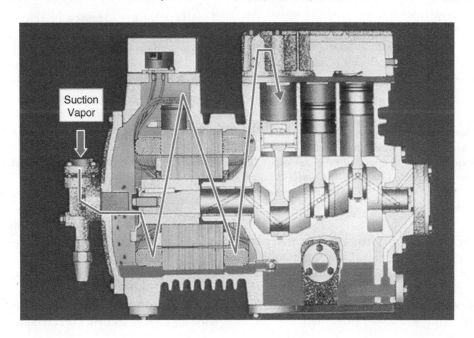

FIGURE 4-4   **Semihermetic air-cooled compressor and water-cooled tubing wrap.** *Courtesy of Copeland Corporation.*

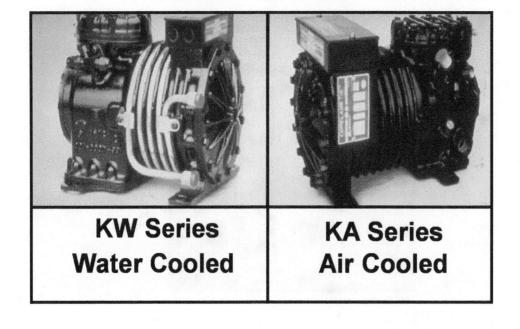

When in doubt, the technician should check the placement of the suction service valve on the compressor. If it is on the motor end of the compressor, the unit is suction cooled (see Figure 4-3). If the suction valve is on the side of the body, near the compressor head, it is air cooled (see Figure 4-4).

Additional cooling for the compressor heads may be required on some low-temperature units. The very low suction pressures in these compressors create high compression ratios and high temperatures in the compressor head. The high discharge temperatures created require separate fans mounted above the compressor heads to prevent overheating of the discharge valves and pistons (see Figure 4-5).

FIGURE 4-5 **Copeland low-temp compressor with head fan and oil cooler.** *Photo by Jerry Meyer, Hussmann.*

COMPRESSOR LUBRICATION

Every metal surface in a compressor that has a metal part moving against it must have adequate oil between the two surfaces. This may sound obvious, but it is surprising how many technicians fail to grasp the importance of this point. When a compressor motor is turning at 1,725 or 3,450 revolutions per minute (rpm), it takes only a few seconds to destroy a compressor when metal hits metal.

Compressors splash, pump, and sling oil inside the compressor to adequately lubricate all bearing surfaces. In the process, some of the oil becomes part of the suction vapor. As the suction vapor is compressed and discharged into the condenser, the oil moves with the refrigerant. The oil travels through the entire system until it returns to the compressor crankcase and resumes its lubrication job. Therefore, maintaining sufficient pressures and velocities in the tubing is very important. The system piping must be correctly sized, sloped, and p-trapped to ensure proper oil return. Piping practices are covered in more detail in Chapter 11, "Walk-in Refrigerators and Freezers."

Compressor Problems and Failures

Compressor problems associated with electricity are covered in Chapter 8, "Compressor Motor Controls." Although electrical problems can prevent a compressor from running, they are not the primary cause of compressor failure. Even the electrical problems associated with compressor failure are often caused by refrigeration problems.

For instance, moisture in a system causes acids to form in the oil. The acids attack the insulation on the windings and can cause a short circuit in the motor. In addition, floodback can cause enough damage to the crankshaft bearings that the motor's rotor will drop onto the stator and cause a motor burnout. There is more on this situation later in the chapter.

Mechanical problems, on the other hand, often result in damage so severe that the compressor must be replaced. Therefore, it is important to understand what is considered improper operation and to correct it before it causes compressor damage.

Most compressor manufacturers would agree with this statement: "Compressors don't die, they're murdered!" Based on that premise, it would be a reasonable assumption that technicians should perform an "autopsy" to find out what caused the demise of the compressor.

Manufacturers of semihermetic compressors encourage technicians to tear down a failed compressor, even when it is still under warranty. Determining the cause of failure is necessary to prevent a repeat problem with the replacement compressor.

Some distributors and manufacturers have "teardown seminars" where they allow technicians to take apart a variety of failed compressors. The seminar leader helps the technicians recognize the type and form of the damages to help them diagnose the cause of the failure. Participating in one of these seminars is one of the most beneficial educational experiences a technician can have.

The following sections cover the most common mechanical problems that result in compressor failures. After each description of the problem is a listing of the possible causes. Next is a list of what to look for if the compressor is torn down for inspection. Finally, there are suggested cures to solve the compressor's problem.

The primary problems associated with mechanical compressor failure are

- Lubrication
- Flooding and slugging
- Flooded starts
- Overheating

Lubrication Problems

Too Much Oil. This can cause compressor noise and vibration when the compressor crankshaft hits the oil surface. Although this situation is rare, a good technician should be aware of its symptoms.

Causes

- There is too much oil from multiple compressor changes.
- If the compressor has been replaced several times, the original problem may still be active. The additional oil from each replacement compressor just adds another problem.
- Someone added too much oil.

On a freezer system, the oil tends to slow down and collect in the very cold evaporator. During defrost, the evaporator warms, as does the oil inside it. When the compressor starts, the oil returns. If oil was added during the freeze cycle, when the oil level in the sight glass is lowest, there may be too much oil in the crankcase after defrost.

NOTE: Copeland maintains that the proper level in an oil sight glass is not necessarily at the halfway point, but anywhere that you can see it. If the level is visible, but at the bottom of the sight glass, don't add oil. Carlyle states that its 6D compressor oil should be between 1/4 and 3/4 of the sight glass. Its 6E and 6CC should be from 1/8 to 3/8.

Most manufacturers recommend that after starting up a new installation, it should be necessary to add oil only once. If more oil is needed, there may be a piping problem.

What to look for Signs of slugging but with oil accumulated in the piston area.

Cures

- Drain out oil and correct the cause of multiple compressor changes.
- Make sure the piping is correctly sized and sloped, and that p-traps are installed if necessary.
- Use an oil separator if the system has a very long run to the condenser. However, check with the system manufacturer first.

Too Little Oil. This can cause the compressor to seize because of the friction between metal parts.

Causes

- There is low refrigerant charge or low load. Low mass flow of the refrigerant will not return oil.
- Improper installation of equipment and piping can prevent proper oil return.
- There is inadequate freezer defrost or a frozen evaporator.
- Short cycling of the compressor pumps oil out of the crankcase, but it does not run long enough to pull it back.

What to look for

- All rods and bearings worn or scored
- Crankshaft uniformly scored
- Worn pistons rings and cylinders (Figure 4-6)
- Little or no oil in crankcase

FIGURE 4-6 **Piston worn from lack of lubrication.** *Courtesy of Copeland Corporation.*

Cures

- Correct the refrigerant charge or the reason for low-load conditions.
- Correct installation and piping problems (more on this in Chapter 11).
- Make sure the evaporator is clear after defrost.
- Correct the cause of compressor short cycling.

Flooding

Flooding refrigerant, or floodback, causes lubrication problems.

- Air-cooled compressors (Figure 4-7)—flooding refrigerant washes oil from the cylinder walls during the suction stroke. Without lubrication, the piston rubs against the cylinder. The softer metal of

FIGURE 4-7 **Copeland air-cooled compressor oil system and cutaway view.** *Courtesy of Emerson.*

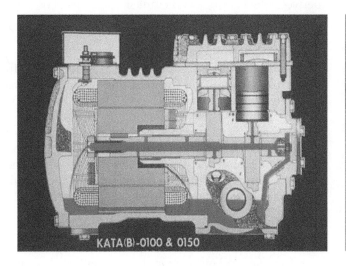

the piston wears away, opening a gap between the piston and the cylinder wall. During the compression stroke, the refrigerant leaks past, or *blows by,* the piston and is pushed into the crankcase.

The result is an *inefficient* compressor. The compressor cannot compress the vapor in the cylinder, so the head pressure drops. The suction vapor is pumped into the crankcase, which raises the suction pressure.

- Suction-cooled compressors (Figure 4-8)—flooding refrigerant dilutes the crankcase oil that lubricates the bearings, crankshaft, and rods. The refrigerant boils off quickly as it is pumped between these parts, leaving insufficient oil to properly lubricate them.

Causes

- Flooding is caused by too much refrigerant entering the compressor.

What to look for on an air-cooled compressor

- Worn pistons and cylinders
- No evidence of overheating

What to look for on a suction-cooled compressor

- Center and rear bearings worn or seized
- Dragging rotor, shorted stator
- Progressively scored crankshaft
- Worn or broken piston rods

FIGURE 4-8 **Suction-cooled compressor crankshaft bearings worn by flooding.** *Courtesy of Copeland Corporation.*

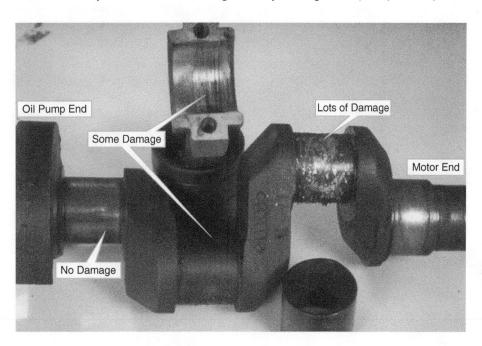

Cures

- Check for proper superheat, even on fixed orifice metering devices.

At design conditions, all systems should have some superheat. This verifies there is no liquid refrigerant returning to the compressor. Most compressor manufacturers recommend at least 20°F superheat at the compressor.

Fixed metering systems are critically charged. This means the manufacturer has specified an exact amount of refrigerant to be in the unit. Hermetic compressors, used in most critically charged systems, can tolerate a limited amount of liquid refrigerant in the crankcase oil. By factory design, even at extreme conditions, there will not be enough refrigerant in a critically charged system to harm the compressor. The only way floodback can damage this type of system is if the unit was overcharged.

Thermal expansion valve (TEV) systems flood if the valve-sensing bulb is not properly strapped to the suction line. The valve operates as if the suction line is warm and opens fully to flood the evaporator to cool the line. A TEV will also flood if the valve is stuck open by debris or ice.

Slugging

Slugging is a severe form of refrigerant flooding. Whereas flooding will eventually damage the compressor, slugging causes almost immediate compressor failure.

On suction-cooled compressors, the effect of slugging is a more rapid process of displacing oil with refrigerant. Damage from metal-to-metal contact between the bearings and crankshaft occurs soon after the liquid enters the compressor crankcase.

On air-cooled compressors, the refrigerant enters almost directly into the reed valve and cylinder area. The slug of liquid hitting the suction reed valve will break it (Figure 4-9). Or the head gasket will be blown out between the low side and the high side. Either type of damage results in an inefficient compressor because the discharge gas is pumped back into the suction line.

If enough liquid enters the cylinder on the suction stroke, the piston will not be able to compress it. The piston will stop, but the crankshaft does not. The result is a broken connecting rod, or even a broken crankshaft.

Causes

- Slugging usually occurs on startup from the refrigerant and oil lying in the suction line close to the compressor.
- There is severe overcharge in a critically charged system using a fixed metering device.
- The TEV is stuck open or the TEV was oversized.

What to look for

- Broken reeds, rods, or crankshaft
- Loose or broken discharge valve backer bolts
- Blown head gaskets

Cures

- Use the same corrections as for refrigerant flooding.
- Check the refrigerant piping for sags or traps close to the compressor that may cause oil or refrigerant to collect during the off cycle.

FIGURE 4-9 **Broken reeds from slugging.** *Courtesy of Copeland Corporation.*

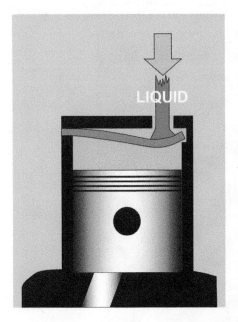
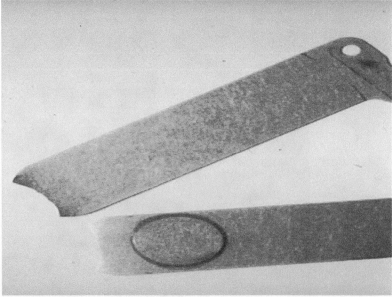

Oil slugging is a piping problem that cannot be detected by checking superheat. If the inspection of a slug-damaged air-cooled compressor reveals oil lying on top of the piston, oil slugging was probably the cause of failure.

All remote refrigeration units should have a liquid-line pump-down solenoid. This will shut down the supply of refrigerant and oil when the tstat is satisfied. By the time the low-pressure control shuts off the compressor, all refrigerant and most of the oil should be pumped out of the suction line. On startup, there is little chance of slugging, unless the piping is installed incorrectly.

Flooded Starts

When the compressor is off, any refrigerant remaining in the suction line will migrate to the coldest location in the system. If the compressor is located outdoors on a cold day, the refrigerant will travel to the compressor. Some refrigerants, like R22, have an affinity for, or are attracted to, refrigerant oil. Therefore, the refrigerant will travel as a vapor to the crankcase. Once in the crankcase, the refrigerant changes to a liquid and settles under the oil because the liquid refrigerant is heavier (denser) than oil.

When the compressor starts, the refrigerant almost explodes into a foaming mixture of refrigerant and oil. This mixture causes excessive bearing wear on the crankshaft only where it splashes the most. As the refrigerant boils off, there is not enough oil remaining to properly lubricate the surfaces. If there is an excessive amount of refrigerant in the crankcase, it could reach the cylinder, causing refrigerant slugging during the compression stroke.

Causes

- Refrigerant is migrating from the suction line to the crankcase oil.

What to look for

- Worn or scored rods or bearings
- Rods broken or seized
- Erratic bearing wear pattern on crankshaft (Figure 4-10)

Cures

- Crankcase heaters warm the oil to prevent the refrigerant from turning to a liquid. However, a crankcase heater has little effect if the outdoor temperature is near freezing.
- A pump-down solenoid ensures there is no refrigerant in the suction line during the off cycle. It is very effective in preventing refrigerant from migrating to the crankcase. Solenoids are covered more fully in Chapter 6, "Controls and Accessories."

FIGURE 4-10 **Erratic bearing wear from flooded start.**
Courtesy of Copeland Corporation.

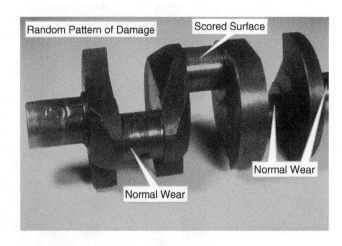

NOTE: **Although crankcase heaters drive refrigerant out of the compressor oil, the danger of flooding may not be completely solved. When the vaporized refrigerant leaves the oil, it may condense back into liquid refrigerant when it comes in contact with a cold suction line. When the compressor starts, the liquid laying in the suction line will flood the compressor. This is one reason a pump-down cycle is more effective at preventing flooded starts than a crankcase heater is.**

Lubrication Problems Disguised as a Motor Burnout

The compressor motor has a stationary part called the stator that contains the insulated motor windings. The rotor, which rotates within the stator, is attached to the compressor crankshaft. Lack of lubrication can damage the bearings that support the compressor crankshaft (Figure 4-11). If the crankshaft drops as little as 1/16 inch, the rotor will hit the motor stator, shorting out the motor windings. The compressor failure, complete with burned oil and black carbon, will look like a bad electrical problem. Many technicians are fooled into diagnosing an electrical problem that is actually a mechanical failure.

Overheating

Overheating can occur in the motor windings and also in the discharge valve area of a compressor.

Causes for high motor heat

- Suction-cooled compressor—an overheated motor occurs when there is inadequate cooling from suction vapor.
- Air-cooled compressor—the motor windings overheat due to improper cooling on the outside of the compressor body.

What to look for

- Stator spot burn from metal debris

FIGURE 4-11   **Motor burnout caused by bearing failure due to flooding.** *Courtesy of Copeland Corporation.*

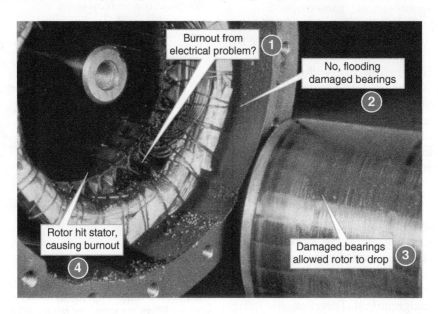

Cures

- Suction-cooled compressors—the suction vapor is the coolant for the compressor motor, so keep the suction pressure up.
- Both suction- and air-cooled compressors—check superheat at the suction line 6 inches from the compressor. Most compressor manufacturers want at least 20°F superheat at the compressor to prevent flooding. However, if the superheat rises above 50°F, overheating problems can occur.

EXAMPLE: 3

Assume an acceptable 10°F superheat at the evaporator of a refrigerated case with a 50-foot suction-line run. The coil is clear, and the box temperature is within design conditions. If the superheat at the compressor is 60°F, assume the additional 50°F of heat is being picked up after leaving the evaporator. The problem concerns the suction line; maybe some of the pipe insulation is missing, or it is going through a hot area.

EXAMPLE: 4

In the previous example, assume the superheat at the outlet of the evaporator is 40°F and at the compressor it is 60°F. The primary problem is a starving evaporator because that is where most of the superheat (30°F) is being generated. The 20°F suction line superheat is not excessive.

Causes for high discharge heat

- High heat loads in the box, and additional heat added to the suction line, are passed directly to the compressor. The higher the heat entering the compressor, the higher is the discharge temperatures leaving the compressor.

- High compression ratios will also increase discharge gas temperature. Imagine a compressor that normally has a 4:1 ratio. If the suction pressure drops or the condensing pressure rises, the ratio becomes higher. For example, assume the compression ratio has increased to 6:1. The compressor motor has to work harder, and hotter, and the heat of compression is greater.

For Copeland reciprocating compressors, the maximum allowable discharge line temperature, measured 6 inches from the compressor, is 225°F. On Carlyle mineral oil compressors, it is 275°F and 250°F for polyol ester (POE). Above that, the high heat in the cylinder will cause the oil to break down. There will not be sufficient lubrication between the piston and the cylinder.

What to look for

- Worn pistons, rings, and cylinders
- Discolored valve plates
- Burned discharge reed valves

Cures

- Stress the importance, to the customer, of not overloading the box with warm product.
- Look for evidence of low-load conditions: frozen evaporator, starving evaporator, or operating the box at too low a temperature.
- Properly insulate suction lines.
- Check compression ratios against what they should be for design conditions.

Blow-By

Blow-by is a description of what happens during the compression stroke. High discharge pressure pushes the cylinder walls down past the piston rings and into the crankcase. The image in our mind of high-pressure vapor blowing by the piston rings is how it came to be called blow-by.

Most compressors have a little blow-by. It may be less than 1 psig when the compressor is new, but as it ages, the pressure may go above 3 psig. Copeland suction-cooled compressors have a vent valve between the motor section and the crankshaft area of the crankcase. If the crankcase section is up to 3 psig higher than the suction side in the motor compartment, the vent valve dumps the excess crankcase pressure into the suction side to equalize the pressure between the two compartments. This is probably the reason for the "rule of thumb" of 3 psig many technicians use as the highest allowable blow-by pressure.

If the blow-by goes above 3 psig, then the crankcase will be overpressurized, preventing oil from the motor compartment from entering the crankshaft area of the crankcase. If the compressor continues to run it will soon deplete the oil in that part of the crankcase causing the oil failure control to trip. Let's try to visualize how that happens by looking at Figure 4-12, a cutaway of a semihermetic suction-cooled Copeland compressor. Among other things, the figure shows the two compartments, one for the motor and the other for the crankcase. Unfortunately, it does not show the exact location of the pressure vent valve but does show an oil check valve, which will be discussed next.

The suction vapor and oil return to the compressor through the suction service valve, shown on the left side of the picture. The oil drops to the bottom of the motor compartment, where it drains through a valve to the crankcase or crankshaft area on the right half of the compressor. Note the small oil check valve in the middle section at the oil level. As long as the pressures are the same in both sections, the oil check valve will allow oil to flow freely from the motor compartment to the crankcase.

However, if the crankcase section on the right is pressurized above the pressure in the motor compartment from blow-by, then the oil check valve will close. The result is too much oil in the motor compartment but not enough in the crankcase. This causes one of two things to happen—either the compressor goes out on motor overload or the oil safety switch trips.

To overload the motor, the oil builds up in the motor compartment and drenches the motor windings. This causes drag on the motor, which raises amperage. The high amperage causes the overload to trip. While the overload is cooling down, the pressures equalize between the motor compartment and the crankcase, which allows the oil check valve to open. By the time the compressor restarts, there is oil in both sections, but the compressor will soon experience the same problem and go out on motor overload.

To trip the oil safety switch (also called the oil failure switch), the crankcase section has to run out of oil. Looking at Figure 4-12 again and also at Figure 4-13, try to visualize the pressure building in the crankcase area from blow-by. The oil pump, shown on the

FIGURE 4-12 **Copeland semihermetic suction-cooled compressor.** *Courtesy of Emerson.*

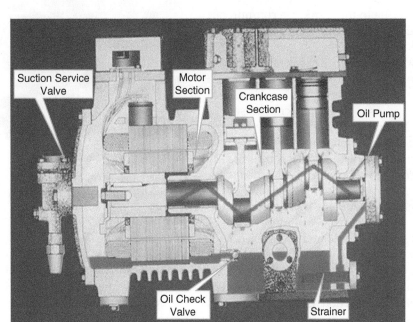

FIGURE 4-13 **Copeland compressor cutaway.** *Courtesy of Emerson.*

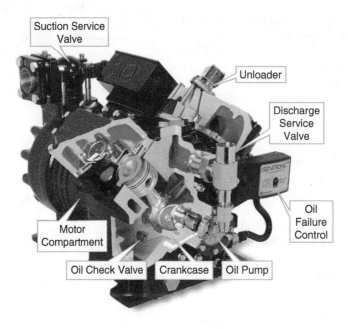

far right of the picture, continues to pull oil from the crankcase up its dip tube and pushes oil down the crankshaft and up the connecting rods. Eventually, the crankcase runs out of oil, resulting in the oil safety tripping and shutting the compressor off. While the compressor is off, the pressures in the two sections equalize. This allows oil to flow from the motor compartment to the crankcase. It is important to understand how blow-by can cause the oil safety to trip. Because when a service technician shows up to reset the oil safety and start the compressor, he will see there is oil in the sight glass and will not realize that the oil safety tripped due to the crankcase running out of oil. This is why it is important to understand and be able to check for blow-by.

To diagnose blow-by, take the crankcase pressure inside the crankcase and compare it to the suction pressure taken at the suction service valve. Copeland does not provide an access valve to the crankcase, but there is usually a 1/4-inch MPT (male pipe thread) plug where an access port can be installed, as in Figure 4-14. For example, if the suction pressure is 50 psig and the crankcase pressure is 55 psig, then there is 5 psig of blow-by.

If the difference is not evident by comparing the two pressures, but you still suspect blow-by, there is one more method some technicians have used successfully. While the compressor is running, throttle the suction service valve closed about halfway. Often at this point there is little difference between the two pressures. However, as the suction service valve closes down more and more, the suction pressure starts dropping. If the crankcase pressure does not fall with the

FIGURE 4-14 **Access valve for checking crankcase pressure.** *Photo by Jess Lukin.*

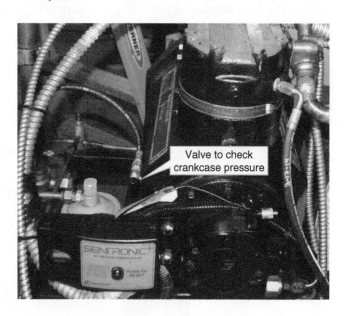

Valve to check crankcase pressure

suction, but remains steady, then this shows the crankcase is being pressurized by blow-by.

Diagnostic Chart

Figure 4-15 is a chart to help diagnose semihermetic compressor failures after doing a compressor teardown. The idea is to circle all the Xs in each row of the symptom that applies. Next, total all the Xs circled in each column. The column with the most circled Xs is the cause of the failure.

FIGURE 4-15 **Semihermetic compressor diagnostic chart.** *Courtesy of Refrigeration Training Services.*

DIAGNOSING SEMI-HERMETIC COMPRESSOR FAILURES
© 2004 Dick Wirz of **Refrigeration Training Services, LLC**

STEPS: (1) Circle ALL the Xs in the rows which apply
(2) Total each column of circled Xs
(3) Column with the most Xs is the problem

AIR COOLED COMPRESSORS:

		Flood back	Flooded Start	Slugging	High Disch. Temp.	Loss of Oil
VALVE PLATE	NO EVIDENCE OF OVERHEATING	X				
	DISCOLORED FROM HEAT				X	
VALVE REEDS	BURNED				X	
	BROKEN			X		
HEAD GASKETS - BLOWN				X		
PISTONS & CYLINDERS - WORN		X			X	
PISTON RODS - BROKEN			X	X		X
PISTON RODS & BEARINGS - *ALL* WORN OR SCORED						X
PISTON ROD CAP BOLTS - BROKEN OR LOOSE				X		
CRANKSHAFT	BROKEN			X		
	UNIFORMLY SCORED					X
	ERRATIC WEAR PATTERN		X			
STATOR SPOT BURN FROM METAL DEBRIS					X	
LITTLE OR NO OIL IN CRANKCASE						X
TOTALS:						
HIGHEST TOTAL IS CAUSE OF FAILURE		Flood back	Flooded Start	Slugging	High Disch. Temp.	Loss of Oil

SUCTION COOLED COMPRESSORS:

		Flood back	Flooded Start	Slugging	High Disch. Temp.	Loss of Oil
VALVE PLATE - DISCOLORED FROM HEAT					X	
VALVE REEDS - BURNED					X	
HEAD GASKETS - BLOWN				X		
PISTONS & CYLINDERS - WORN					X	
PISTON RODS - BROKEN		X	X			X
PISTON RODS & BEARINGS - *ALL* WORN OR SCORED						X
PISTON ROD CAP BOLTS - BROKEN OR LOOSE				X		
CRANKSHAFT	BROKEN			X		
	UNIFORMLY SCORED					X
	PROGRESSIVELY SCORED	X				
	ERRATIC WEAR PATTERN		X			
CENTER & REAR BEARINGS WORN OR SEIZED		X				
STATOR SHORTED, DRAGGING ROTOR		X				
STATOR SPOT BURN FROM METAL DEBRIS					X	
LITTLE OR NO OIL IN CRANKCASE						X
TOTALS:						
HIGHEST TOTAL IS CAUSE OF FAILURE		Flood back	Flooded Start	Slugging	High Disch. Temp.	Loss of Oil

CORRECTIVE ACTION:

FLOODBACK - Liquid refrigerant returning **during the running cycle** washes oil off the metal surfaces.
    (1) Maintain proper evaporator & compressor superheat.
    (2) Correct abnormally low load conditions.
    (3) Install accumulator to stop uncontrolled liquid return.

FLOODED STARTS - Refrigerant vapor migrates to the crankcase **during the off cycle**, diluting the oil on startup.
    (1) Install pumpdown solenoid.
    (2) Check crankcase heater operation.

SLUGGING - Attempt to compress liquid. Extreme floodback in air cooled compr.; severe flooded start in suct. cooled.
    See all of above corrective actions for Floodback & Flooded starts.

HIGH DISCHARGE TEMPS - High temperatures in compressor head and cylinders causes oil to lose lubrication.
    (1) Correct abnormally low load conditions.
    (2) Correct high discharge pressure conditions.
    (3) Insulate suction lines.
    (4) Provide proper compressor cooling.

LOSS OF OIL - When oil is not returned to the crankcase there is uniform wearing of all load bearing surfaces.
    (1) Check oil failure control, if applicable.
    (2) Check system refrigerant charge.
    (3) Correct abnormally low load conditions or short cycling.
    (4) Check for incorrect pipe sizing and / or oil traps.
    (5) Check for inadequate defrosts.

NOTE: This chart was developed by the author and is based on general information available from the Copeland Corporation. It is intended to be used as a training tool to help simplify the diagnosis of most mechanical failures of semihermetic compressors. It is not intended to represent the views of, nor is it sanctioned by, the Copeland Corporation.

The first six symptoms under air-cooled compressors and the first five under suction-cooled compressors can be checked by simply removing the compressor head. If this does not reveal enough symptoms to make a good diagnosis, the technician will need to inspect the crankcase. Disassembling the crankshaft takes a little more effort, but it could be well worth the time if it helps determine the reason for the compressor failure.

NOTE: Inefficient compressor diagnostics caused by valve problems, piston blow-by, blown gaskets, and so on will be covered in Chapter 7.

Compressor Capacity Control

The load on a compressor varies based on ambient conditions as well as the load on the evaporator. As the temperature of refrigerated product drops, the suction pressure also comes down. When the tstat (thermostat) is satisfied, the compressor will cycle off. This on−off control is the simplest type of capacity control. If the load on the compressor is low, the unit may cycle too often, which causes increased wear and greater operating cost.

Every time a compressor starts, it draws lock rotor amperage (LRA). LRA is the amperage a motor draws for a split second each time it starts. LRA is 3 to 5 times the motor's normal running amperage. Each time a motor starts, the amperage goes from 0 to LRA and back down to normal running amps in about 1.5 to 3 seconds. Large compressors draw very heavy starting current, which can cause voltage drops in power to the rest of the building. In addition to the burst of heat generated in the motor winding on every start, the large inrush of current also increases the cost of power for the customer. If that were not bad enough, starting puts a mechanical strain on the compressor parts, and it takes time for the oil pressure to build up enough to properly lubricate the compressor. If humidity control is an issue, a compressor that does not run long enough will not remove humidity from the space it is conditioning. Most manufacturers agree that a compressor that has more than six cycles per hour is considered to be short cycling.

Additional problems will occur if the load drops too far. The decrease in suction pressure causes an increase in compression ratio, which causes more heat in the compressor head. Also, the lower mass flow of the refrigerant not only reduces the cooling effect of the refrigerant on a suction-cooled motor but may prevent proper oil return to the crankcase.

There are several ways to control compressor capacity in order to prevent the compressor from short cycling and also to keep the returning suction pressure up. One method is to divert hot gas to the low side of the system to increase the load on the compressor. The additional load prevents low suction pressure, produces longer run times, and ensures good oil return. Hot gas bypass is especially beneficial in low-temperature applications where low suction pressure produces very high compression ratios and high heat in the compressor head. Adding refrigerant to the suction vapor increases its density, raises its pressure, cools the compressor motor, lowers the compression ratio, and reduces the temperature of the discharge vapor in the compressor head. Bypassed refrigerant can come from the hot discharge gas, as shown in Figure 4-16. This figure shows it entering the low side just before the evaporator, but hot gas can be injected into the low-pressure side of the system in several places. For all its benefits, bypassing hot gas does not reduce energy usage significantly. Hot gas bypass valves will be explained in greater detail in Chapter 6.

Combining multiple compressors for a single load or group of loads is another type of capacity control. One (or more) compressor cycles off when the load decreases and comes back on line when the load increases (see Figure 4-17). Although this arrangement, called parallel rack systems, provides great capacity control and power savings, it has a high initial expense. Multiple compressor systems are covered more thoroughly in Chapter 10, "Supermarket Refrigeration."

Another type of capacity control is unloading the compressor, which means preventing the compressor from performing to its rated capacity. One of the most popular methods is blocking the suction vapor from entering some, but not all, of the cylinders in the compressor. Without suction vapor, there is no compression of the refrigerant, which not only reduces the compressor capacity but, because there is less work performed, also lowers the amperage; the longer run time means no short cycling. Following are examples to help illustrate the basics of how unloaders operate.

NOTE: Unloaders and their associated controls are quite sophisticated and differ from one manufacturer to the other. Consult the valve or compressor manufacturer to find out how its particular controls operate and are adjusted. Hopefully the following explanations and pictures will make it easier to understand the concepts of both unloading and capacity control.

Assume there are three walk-ins, and each requires a 10-hp condensing unit. Instead of using three 10-hp systems, the contractor has determined that it will cost much less to use a single 30 hp unit to run all three walk-ins. As long as all three walk-ins are calling for cooling at the same time, the compressor will be fully loaded, and the suction pressure will be basically the same in all three evaporators.

FIGURE 4-16 **Hot Gas Bypass to evaporator.** *Adaption of Sporlan drawing by Refrigeration Training Services.*

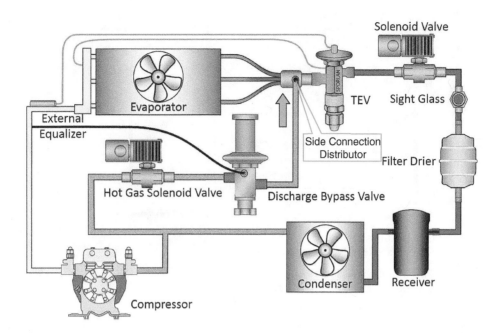

FIGURE 4-17 **Parallel Rack systems.** *Photo by Jess Lukin.*

When the first walk-in is down to temperature, its thermostat closes a liquid line (pump-down) solenoid valve, which stops the flow of refrigerant to its evaporator. The load on the compressor is now decreased by one-third. In other words, the 30-hp compressor has a 20-hp load because the evaporators in only two of the walk-ins are absorbing heat. Because the compressor is now too large for the load, the suction pressure will drop, and the unit may short cycle.

When the second walk-in is satisfied, the compressor load is decreased by another third. The 30-hp condensing unit is now handling only 10 hp worth of a load. This will cause erratic behavior, including short cycling, overheating, and poor oil return, which will almost certainly cause compressor damage.

The 30-hp compressor in Figure 4-18 has six cylinders and three heads (two cylinders per head). To illustrate capacity control, we will assume two of the

FIGURE 4-18   **Three walk-ins on a single compressor.** *Courtesy of Refrigeration Training Services.*

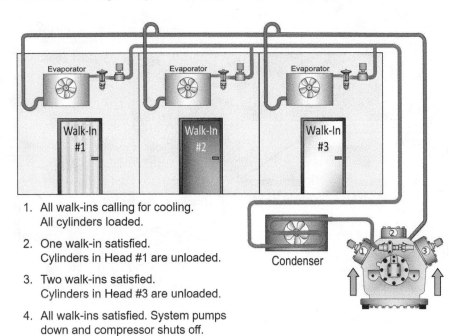

1. All walk-ins calling for cooling. All cylinders loaded.

2. One walk-in satisfied. Cylinders in Head #1 are unloaded.

3. Two walk-ins satisfied. Cylinders in Head #3 are unloaded.

4. All walk-ins satisfied. System pumps down and compressor shuts off.

standard compressor heads have been replaced with cylinder heads, which have unloaders installed in them. When the first evaporator is satisfied, the resulting drop in suction pressure will cause the first unloader to be activated. If the first unloader is still in operation when the second evaporator is satisfied, the drop in suction pressure will activate the second unloader to block off the suction vapor to the second set of cylinders. If all three evaporators are satisfied at the same time, the suction pressure will drop low enough for the low-pressure control to shut off the compressor.

NOTE: **A separate low-pressure control is required for each unloader. An additional pressure switch functions as both an operating control and a safety control. When the system goes into a pump-down, the pressure control shuts off the compressor. If the unit loses too much refrigerant, it will also shut off the compressor.**

Unloaders can be either fully mechanical or electrically operated. Either way, the unloader will block off the suction vapor to specific cylinders when the suction pressure to the compressor drops below an acceptable pressure. When the load increases, the suction pressure also increases. The rise in pressure will result in the unloaders opening the port to the suction valves. This allows the cylinders to once again start pumping refrigerant at full capacity.

The unloader control has a set point or cut-in pressure at which it will unload. In addition, there is an adjustable differential that determines at what pressure the control will cut out, or go back to its normal open or fully loaded operation.

For example, imagine there are two unloaders on a six-cylinder compressor. Each unloader will handle two cylinders. Under normal operation, assume the suction pressure is 62 psig. The first unloader is set to activate at 50 psig with a 20-psig differential (it will go back to full load at 70 psig). The second unloader is set to activate at 40 psig with a differential of 25 psig.

If all three medium-temperature walk-ins in Figure 4-18 require refrigeration, the 30-hp, R404A compressor would have a suction pressure of about 62 psig (25°F evaporator temperature). If one walk-in is satisfied and pumps down, there is only the load of two remaining evaporators on the compressor. Because the 30-hp compressor is still at full capacity but with only the load of two evaporators, the suction pressure will drop below 50 psig. At this point, the first pair of cylinders will unload, and the compressor capacity will drop by 33 percent. With the compressor capacity matched to the load of the two remaining evaporators, the suction pressure will return to the original 62 psig. When the walk-in comes back online, the suction pressure will rise to 70 psig, causing the unloader to open and bringing the system back to 100 percent capacity (see Figures 4-19 through 4-22).

NOTE: **These drawings of unloader operations have been modified to make this difficult concept a little easier to understand. Actual Carlyle and Copeland compressor unloaders differ slightly from these drawings.**

Assume the first walk-in is satisfied. Unloader 1 will unload two cylinders. The compressor will be

FIGURE 4-19   **Normal conditions: cylinder loaded for full capacity (Carlyle style).** *Courtesy of Refrigeration Training Services.*

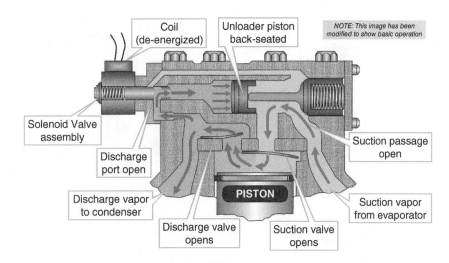

FIGURE 4-20   **Low load condition: cylinder unloaded for reduced capacity (Carlyle style).** *Courtesy of Refrigeration Training Services.*

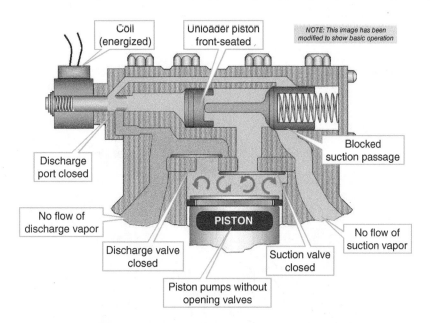

running at 67 percent capacity. When a second walk-in also becomes satisfied, there is 20 hp worth of capacity running the one remaining walk-in that requires only 10 hp. Because the compressor capacity is twice the evaporator load, the suction pressure will quickly drop to the cut-out of 40 psig. The pressure control will energize unloader 2, and two more cylinders will unload to decrease the compressor capacity to 33 percent. The suction will rise up to the original 62 psig because the compressor capacity is now matched to the load of the last remaining evaporator.

There are as many different applications as there are types of unloaders. Just remember that unloading a compressor means preventing it from doing what is was designed to do, namely pull in suction vapor from the evaporator, compress it, and discharge it to the condenser. By blocking the suction vapor to the cylinders, the capacity of the compressor is adjusted to the load of the evaporators. As a result, the compressor is protected from high compression ratios, draws less current during unloading, does not short cycle, and has adequate oil return; in addition, the evaporator temperatures are stabilized for optimum temperature and humidity control.

Capacity Control Considerations

Along with the benefits of capacity control, there are a few factors to consider. The expansion valves used should be able to operate properly at the lowest capacity conditions. Most standard TEVs will maintain control down to 50 percent of their rated capacity, while

FIGURE 4-21   **Normal conditions: cylinder loaded for full capacity (Copeland style).** *Courtesy of Refrigeration Training Services.*

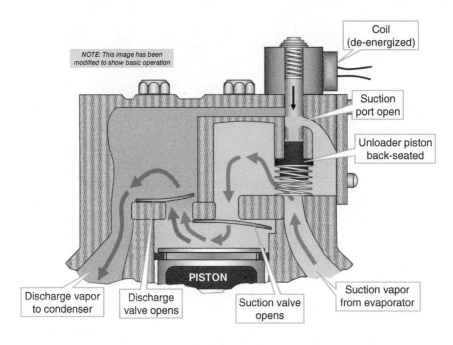

FIGURE 4-22   **Low load condition: cylinder unloaded for reduced capacity (Copeland style).** *Courtesy of Refrigeration Training Services.*

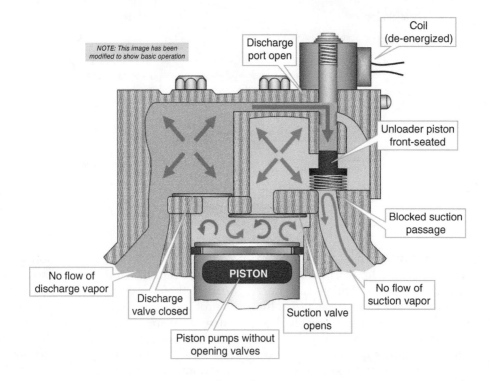

balanced port TEVs will operate down to 25 percent of their capacity. Electronic expansion valves have an even greater range of capacity control, some as low as 10 percent.

Suction-line piping is also important on systems with unloaders to make sure there is adequate oil return during unloaded conditions. If the evaporator requires a vertical riser, a two-pipe system is required. Suction vapor and oil will flow only in the smaller of the two pipes during low-load conditions. However, during full load, both pipes will be in use. This topic will be covered thoroughly in the piping section of Chapter 11.

Low-Temperature Applications

The primary problem with compressor operation at low suction pressures is that the compressor's efficiency decreases as its compression ratio increases. One reason for the loss of efficiency has to do with volumetric efficiency, which is the ratio of discharge vapor actually pumped out of the cylinders compared to the compressor's displacement. After the piston has completed its compression stroke, the small amount of discharge vapor left in a cylinder must re-expand during the piston's downstroke (see Figure 4-1, steps B and F). Only after the vapor has decreased below the pressure in the suction line will the suction valve open to allow more vapor into the cylinder. The lower the suction pressure, the more that the residual gas expands and the less is the space available for fresh vapor to enter. Therefore, as the compression ratio increases, the compressor pumps a smaller volume of refrigerant, which is a reduction in the mass flow rate.

Another reason for decreased efficiency is the increase in vapor temperature due to an increase in the heat of compression. High compression ratios cause the cylinder walls to become hotter. The heat increases the temperature of the incoming vapor, expanding it into a lighter vapor. As a result, the compressor is pumping a less-dense vapor, which means it is moving a decreased weight of refrigerant for every cycle of the piston. This reduces its capacity to absorb heat in the evaporator, which reduces the refrigeration effect, or the ability of the refrigerant to move heat.

In addition to inefficiency, high cylinder heat from high compression ratios can also damage the compressor. Oil prevents metal-to-metal contact in the compressor. However, at about 310°F, oil will start to vaporize, which causes ring and piston wear. At 350°F, oil will break down or burn, causing contaminants and even faster damage to the compressor. Although it is not possible for a technician to accurately measure the temperature inside a compressor cylinder, the manufacturers recommend checking for overheating by taking the temperature of the discharge line 6 inches away from the compressor. The temperature inside the compressor head is about 75°F hotter than the discharge line. Based on this information, Copeland has stated that for maximum safe operation, the discharge-line temperature should remain below 225°F. At that line temperature, the cylinder temperature would be about 300°F, just below the oil's vaporizing temperature. Carlyle sets a high limit of 275°F for their compressors using mineral oil and 250°F for those using POE.

Manufacturers recommend a number of methods to help prevent compressor problems such as overheating. One is to install a hot gas bypass system with a de-superheating expansion valve, which injects refrigerant into the suction line to keep suction pressure up and lower its temperature at the same time (see Figure 4-23). This method lowers the compression ratio and prevents compressor overheating. A de-superheating valve is for special applications and will be discussed more in the section on hot gas bypass valves in Chapter 6.

FIGURE 4-23 **De-superheating valve with hot gas bypass.** *Adaption of Sporlan drawing by Refrigeration Training Services.*

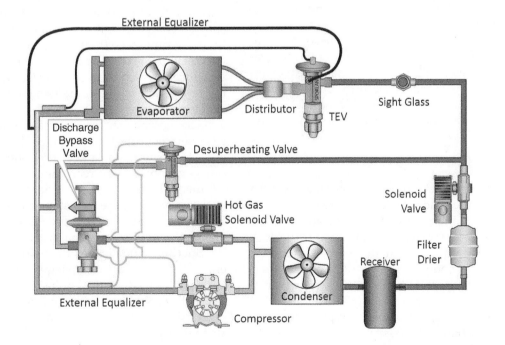

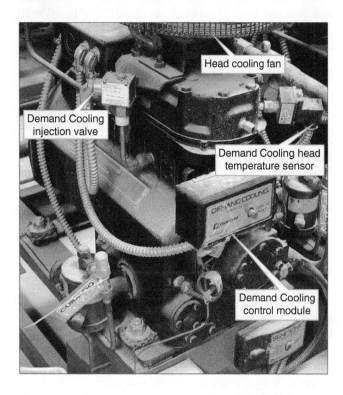

FIGURE 4-24 **Picture of low-temperature Copeland compressor with Demand Cooling™ and oil cooler.** *Photo by Jerry Meyer, Hussman.*

Another way to prevent discharge heat problems in low-temperature compressors is to install head cooling fans. Most manufacturers install these fans on their compressors that operate at evaporator temperatures below 0°F.

An even more direct and effective method is called demand cooling. Copeland's Demand Cooling™ is used on their R22 Discus low-temperature compressors. Saturated refrigerant is injected into the suction cavity when the compressor's internal head temperature reaches 292°F. Injection continues until the temperature is reduced to 282°F. If the temperature remains above 310°F for one minute, the control shuts down the compressor on a manual reset. See Figure 4-24 for examples of both head cooling fans and demand cooling.

Two-Stage Compound Systems

As explained in the previous paragraphs, the lower the temperature of the space being refrigerated, the lower the system's suction pressure and temperature. As the suction pressure decreases, the volumetric efficiency and compressor cooling also decrease. A standard low-temperature compressor has no problem operating at −20°F evaporator temperatures. However, at "extra-low" temperatures between −30°F and −80°F, a single-stage compressor can experience problems associated with very high compression ratios.

Figure 4-25 is an example of how much discharge temperatures rise when suction temperature falls from an AC application down to an extra-low-temperature application. The illustration is a greatly simplified *R22* pressure−enthalpy chart. This chart plots the theoretical discharge temperatures for three evaporator temperatures: +40°F, −20°F, and −40°F. In all three cases, the condensing temperature will be 120°F, and vapor entering the compressors will be at 60°F.

NOTE: **A 60°F return vapor temperature means the 40°F vapor from an AC evaporator has picked up 20°F of superheat. However, to have a 60°F return temperature from a −40°F evaporator requires a superheat of 100°F. Although this scenario is a little unrealistic, using a common 60°F return temperature for all the evaporators makes the results of the example easier to plot (and more impressive).**

In our example, a standard AC at 40°F evaporator and 120°F condensing will have discharge gas leaving the cylinder at about 175°F. As the evaporator temperature falls to −20°F, the discharge temperature rises to 280°F. When the evaporator temperature drops to −40°F, the discharge temperature is 320°F, and the oil is already vaporizing as it loses its lubricating ability.

NOTE: **The accepted method of calculating compression ratios is with absolute pressures (psia), which are used in the charts in Figures 4-25 and 4-27. For gauge pressure in psig, subtract 14.7 pounds.**

The use of a single-stage compressor to go from a suction temperature of −40°F to a condensing temperature of 120°F subjects the compressor to a very high (18:1) compression ratio. A solution to the problems encountered when a single compressor operates in extra-low temperatures is the use of compound compression, as is illustrated in Figure 4-26. Suction vapor entering the first stage (low stage) is compressed but only to a level partway to the final condensing pressure. Discharge from the first stage becomes the suction of the second stage (high stage), which compresses it the rest of the way. The compression ratio of each stage is reduced, which increases efficiency; and by cooling the refrigerant between the two stages, the compression temperatures are decreased.

The "interstage," or the area of intermediate pressure between the first-stage discharge and the second-stage suction, is a key component to compound refrigeration. As you read the following example of the theory behind two-stage compound systems, refer to the drawing of the two-stage compound system in Figure 4-26 and the pressure−enthalpy diagram in Figure 4-27 of two-stage compression.

Suction vapor leaves the −40°F evaporator and enters the first stage (low stage) compressor at about 60°F and 15 psia. Instead of having to be compressed to 120°F condensing, it only has to reach the pressure of 65 psia, which is the suction pressure of the second stage (high stage). The temperature of the first-stage

FIGURE 4-25 **Pressure–enthalpy diagram to illustrate how discharge temperature can increase as suction temperature decreases.** *Courtesy of Refrigeration Training Services.*

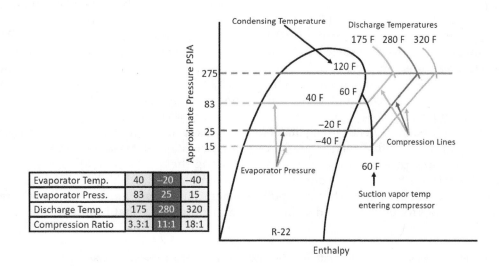

Evaporator Temp.	40	−20	−40
Evaporator Press.	83	25	15
Discharge Temp.	175	280	320
Compression Ratio	3.3:1	11:1	18:1

FIGURE 4-26 **A basic two-stage compound system for extra-low-temperature operation discharge temperatures in extra-low-temperature applications.** *Courtesy of Refrigeration Training Services.*

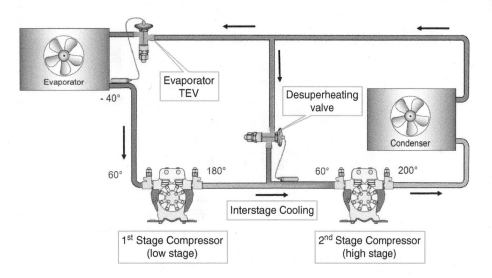

FIGURE 4-27 **Pressure–enthalpy diagram illustrating how two-stage compression can lower.** *Courtesy of Refrigeration Training Services.*

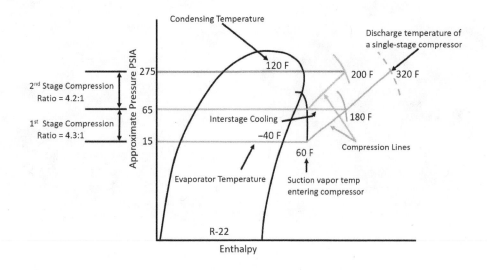

discharge vapor is 180°F, which is much too high for the suction of the second-stage compressor. A de-superheating expansion valve injects cool vapor into the interstage between the low and high stages. This refrigerating effect reduces the 180°F discharge vapor to 60°F suction vapor before it enters the high stage. The high-stage-65-psia suction vapor is compressed to 275-psia discharge pressure at 200°F before it is released to the condenser, where it is changed to a liquid so the process can start all over again.

In Figure 4-28, imagine a three-head (two cylinders under each head) compressor with a suction pressure of 15 psia. Because the low-pressure vapor has very little density, the low-stage section of the compressor requires the four cylinders under the two heads to pull in enough refrigerant vapor for the two cylinders in the

one head high-stage section to compress. The discharge from the two heads in stage 1 is sucked into the motor crankcase, which becomes the interstage. Before the discharge vapor enters the motor area, it is cooled by the de-superheating expansion valve. The vapor then enters the single head of the second stage and is compressed to 275-psia vapor at a condensing temperature of 120°F. See Figure 4-29 for two 2-stage compressors; one has three heads, and the other has a single head divided into two sections.

There are many controls and accessories required for two-stage extra-low-temperature compressors. Among them are solenoid valves, accumulators, oil separators, and subcoolers. There are also many considerations of piping, insulation, superheat, subcooling, and the like that have not been addressed in the

FIGURE 4-28 **Example of an extra-low-temperature system using a two-stage compressor.** *Courtesy of Refrigeration Training Services.*

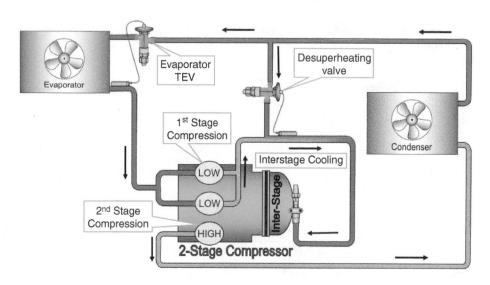

FIGURE 4-29 **Copeland 2-stage compressors.** *Courtesy of Emerson.*

previous examples. However, it is hoped that the illustrations and explanations are enough to help the reader begin to understand not only compound and two-stage refrigeration but the necessity of keeping a compressor's operation in line with its design specifications.

Scroll Compressors

Although the first scroll compressor was patented in 1905, it was not until the late 1980s that they were commercially produced for AC. It took Copeland nearly 10 years to sell a total of a million scrolls, but they now produce more than a million each year. Although the AC industry was slow to embrace the new technology, almost all residential manufacturers currently use scroll compressors. Reciprocating compressors are still dominant in the commercial refrigeration industry; however, scrolls are being incorporated into more applications every year, from single systems to parallel rack systems (Figures 4-30 and 4-31).

FIGURE 4-30 **Scroll compressor design.** *Courtesy of Refrigeration Training Services.*

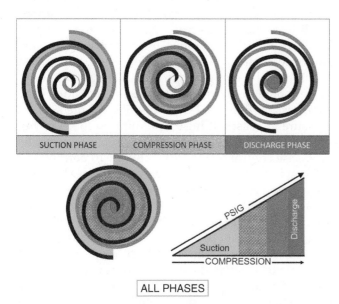

Instead of using a piston-driven reciprocating motion to compress vapor, the scroll compressor uses two precisely machined parts that look like the end of a rolled up paper or "scroll." In this type of compressor, the top scroll section is fixed, and the bottom scroll is moved in an orbital motion as the motor crankshaft rotates. Suction vapor enters the first open chambers formed by the outside edges of the two scrolls. As the orbiting scroll moves within the stationary scroll, a series of pockets or sealed moving compression chambers are formed. As the moving scroll continues to revolve, the vapor is compressed into smaller and smaller sections of the scroll. Eventually, the vapor is forced into the smallest section at the center

FIGURE 4-31 **Upper and lower scrolls.** *Photo by Dick Wirz.*

of the scroll, where it reaches its highest pressure as it is discharged to the condenser.

High efficiency is the most important advantage of scroll compressors over reciprocating compressors. The scroll compression process is nearly 100 percent volumetrically efficient in pumping refrigerant.

The reciprocating compressor basic design requires some space between the piston and the head or valve plate. This inevitably leaves a small amount of discharge gas in the cylinder. This small amount of remaining high-pressure vapor from the previous compression cycle must be re-expanded and reduced from the discharge pressure to below the suction pressure before the suction reed valve will open to admit more suction vapor for compression. The higher the compression ratio, the further the piston must drop before it opens the suction reed to allow more suction vapor into the cylinder to be compressed. The reduction in the usable capacity of cylinder space (volumetric efficiency) depends on the suction and discharge pressures. The higher the discharge pressure and the lower the suction pressure, the greater the inefficiency. This is a major consideration in low-temperature applications.

Although the capacity of scroll compressors is also reduced as compression ratios increase, it is not as much as reciprocating compressors. For this reason, it is likely that the use of scroll compressors will increase in commercial refrigeration applications.

As commercial refrigeration technicians encounter more scrolls in an industry that used to be dominated by reciprocating compressors, they should not be apprehensive about the difference in technology. Instead, they should continue to apply the same basic principles of good refrigeration installation and service practices to scrolls that they have been using on reciprocating compressors. Although most of the evaporator and condensing temperatures of scroll units are similar to those of reciprocating compressors, the technician should always refer to the manufacturer's literature to determine if the scroll unit is operating properly under its current conditions.

However, one difference in scroll operation is true and easily checked: Three-phase scroll compressors must be turning in the proper direction. On the initial start-up, if the compressor is rotating in the wrong direction, it will make a knocking noise and will not be pumping. If the suction pressure does not drop and the discharge pressure does not rise to proper levels, any two of the compressor power leads should be reversed and then the pressures rechecked. Reverse rotation will not damage the compressor, but after several minutes, the internal protector will trip.

Like reciprocating compressors, scroll compressors sometimes require some kind of capacity control and protection from high compression ratio operation. Copeland uses an injection system similar to its Demand Cooling™ to prevent overheating and high discharge temperatures in low-temperature applications.

The Copeland Scroll Ultra Tech™ has two bypass ports in the base of the scroll plates, which is used for capacity control. Under low-load conditions, an electric solenoid opens the ports, effectively reducing the displacement of the scroll by one-third. The compressor will operate at either 67 percent capacity or 100 percent capacity depending on the load requirements. This compressor is currently being used only in AC applications (see Figures 4-32 and 4-33).

FIGURE 4-32 **Copeland Ultra Tech™ at 67 percent part load capacity.** *Adaption of Copeland drawing by Refrigeration Training Services.*

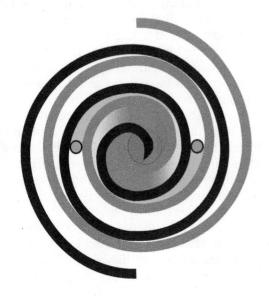

FIGURE 4-33 **Copeland Ultra Tech™ upper scroll will solenoid.** *Photo by Dick Wirz.*

Copeland's Digital™ Scroll capacity control is achieved by axially (vertically) separating the upper scroll from the lower scroll. During the time the two parts are slightly separated, there is no compression and only about 10 percent power usage. Separation is achieved by energizing a solenoid valve that bypasses some discharge gas to the suction side of the compressor (Figures 4-34 and 4-35). As the suction pressure increases, it raises the upper scroll. Varying the amount of time the scrolls are separated results in precise control of capacity between 10 percent and 100 percent. For example, during a 20-second period, the solenoid can be de-energized for 16 seconds (100% capacity), then energized for 4 seconds (0% capacity). The resulting average capacity for that 20-second period is 80 percent (16 seconds ÷ 20 seconds × 100). This on−off operation of the solenoid valve will continue as long as the compressor is running and the load requires the capacity reduction. Because the discharge vapor is introduced to the suction side for such a relatively short period of time, there is no chance of overheating the compressor. The Digital™ Scroll is currently being used in medium- and low-temperature commercial refrigeration applications where reciprocating compressors with unloaders were previously used.

Variable Frequency Drives

Another form of capacity control is modulating the speed of the compressor motor. Variable frequency drive (VFD) technology provides the ability to change the speed of a motor (and its capacity) by electronically

FIGURE 4-34 **Copeland Digital™ Scroll capacity control.** *Adaption of Copeland drawing by Refrigeration Training Services.*

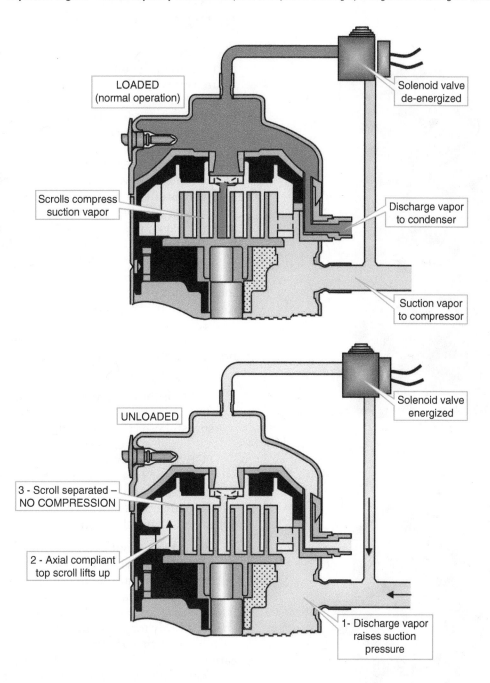

varying the frequency of the motor. For instance, the speed of a 60-cycle motor that normally runs at 1,750 rpm will decrease to about 1,400 rpm at 50 cycles. Not only does this drop the capacity of the motor, but it also reduces its power consumption as well. VFD motors are currently being used in blowers, pumps, and compressors (see Figure 4-36).

In the past, VFD technology was very expensive and usually reserved for only larger motor loads.

However, over time, the cost of this technology has continued to decrease, making it available to smaller motor applications. In commercial refrigeration, the benefit of varying the evaporator fan speed is a more consistent evaporator discharge air temperature, which can greatly extend the shelf life of refrigerated products. Varying the condenser fan will help maintain consistent condensing temperatures, especially in low-ambient conditions.

FIGURE 4-35 **Copeland Digital™ Scroll.** *Courtesy of Emerson.*

FIGURE 4-36 **Copeland Variable Frequency Drive (VFD) Scroll.** *Courtesy of Emerson.*

READING MODEL AND SERIAL NUMBERS

The model numbers on compressors give important information. The only trouble for service techs is to figure out what they mean. Each manufacturer has its own numbering system. One way to find out what the numbers stand for is to call the local supply house or distributor that carries that brand of equipment. Another way is to search the Internet and find the manufacturer's website. Usually, these websites have instructions on how to read the compressor numbers, and there is often a search box where you can enter the compressor number and find out almost everything about it. Also, manufacturers have handbooks that have that information in them. If you do not have one, you can buy one on the Internet, download a copy for yourself, or just read an electronic version. They also include component information for that compressor, like the part number and rating of capacitors and relays.

Following are the basics of how the Copeland compressor numbering system works:

WWW − XXX − YY − ZZZ
WWW is the compressor family and application.
XXX is the capacity in Btuh. Note some compressors still use HP: 0100 is 1 hp; 1000 is 10 hp.
YY is the model variation and oil type. E is for POE oil but nothing for mineral oil.
ZZZ is for type of starting, overload protection, voltage, and phase.
The three numbers after this are the B/M (bill of material), which doesn't usually concern us.

Be very careful about assuming the capacity rating on the nameplate is what the compressor will be putting out for all applications in which it can be used. Its actual output or capacity is based on the evaporator temperature for which it is designed. A compressor approved only for air conditioning may very well have a Btuh rating that accurately describes that one application at a 40°F. However, if it is approved for more than one application, the rating may reflect the capacity of only one of its applications. For instance, a compressor rated for both medium- and low-temperature refrigeration may have a nameplate showing a rating of 24,000 Btuh at an evaporator temperature of 25°F. But when the same compressor is used for low temperatures at an evaporator temperature of only −20°F, its rating is only 12,000 Btuh. In Figure 4-39, the compressor's nominal rating of 170,000 Btuh is quite different from the three applications for which it is approved.

The serial numbers for Copeland are pretty easy: XX-Y-ZZZZZ.

XX is the last two digits of the year it was made.
Y is the month; A is January, B is February, C is March, etc.
ZZZZZ is the numeric sequence of numbers assigned to it when it was manufactured.
For instance, serial number 00B12345 was built in February 2000; serial number 95F67899 was built in June 1995.

See Figures 4-37–4-39 for examples of Copeland model numbers for their "C" style hermetic, scroll, and Discus compressors.

Figure 4-40 is a chart for Tecumseh compressor number codes. The company does not have a wide range of compressor types like Copeland does. Therefore, the coding for its hermetic compressors is fairly easy to figure out and is similar to Copeland's. Tecumseh's code for the serial number uses the same alphabetic range for the months (A = January, B = February, C = March, etc.). However, the following two numbers are the day of the month. The two numbers after that are the last two digits of the year. Then there is a space and about six digits for the numbered sequence assigned to that compressor at the factory. For instance, C0106 123456 would be March 1, 2006, and compressor number 123456.

FIGURE 4-37 **Model numbers of a CR24K7-PFV Copeland Hermetic Compressor.** *Courtesy of Emerson.*

C – Compressor Family Series
R – Application: R is R22 High Temp, A/C, & Heat Pump
              S is R404A Medium Temp Refrigeration
              F is R404A Low Temp Refrigeration
24 – Nominal BTUH capacity to two significant digits
K – Capacity Multiplier K = 1,000
7 – Model Variation
P – Motor Type: P is Permanent Split Capacitor
             T is Three Phase
F – Motor Protection: F is an internal overload
V – Voltage: V is 208/230-1
         C is 208/230-3
         D is 460-3

CR24K7PFV is for high temp refrigeration, A/C, or Heat Pump using R22 and has a capacity of 24,000 Btuh. It requires a run capacitor, has an internal overload, and is 208/230V single phase.

FIGURE 4-38 **Model number of a ZS21KAE–TFD Copeland Scroll Compressor.** *Courtesy of Emerson.*

Z – Compressor Family Series – Scrolls are always "Z"
S – Application: R is R22 High Temp, A/C, & Heat Pump
              S is R404A Medium Temp Refrigeration
              F is R404A Low Temp Refrigeration
21 – Nominal BTUH capacity to two significant digits
K – Capacity Multiplier K = 1,000
7 – Model Variation
E – Ester Oil (POE)
T – Motor Type: P is Permanent Split Capacitor
              T is Three Phase
F – Motor Protection: F is an internal overload
D – Voltage: V is 208/230-1
          C is 208/230-3
          D is 460-3

ZS21KAE-TFD is a scroll compressor for medium temp refrigeration using R404A. It's capacity is 21,000 Btuh. It has POE oil. It is 460V three phase with an internal overload.

FIGURE 4-39   **Model 3DS3R17ME-TFC Copeland Semi-Hermetic Discus Compressor.** *Courtesy of Emerson.*

3 – Compressor Family Series
D – Discus compressor
S3 – Valve plate and model variation
R – Application: R is R22 High Temp, A/C, & Heat Pump
     S is R404A Medium Temp Refrigeration
     F is R404A Low Temp Refrigeration
17 – Nominal BTUH capacity to two significant digits
M – Capacity Multiplier K = 1,000, M = 10,000
E – Ester Oil (POE)
T – Motor Type: P is Permanent Split Capacitor
     T is Three Phase
F – Motor Protection: F is an internal overload
C – Voltage: V is 208/230-1
     C is 208/230-3
     D is 460-3

3DS3R17ME-TFC A Discus high temp or A/C using R22. It has POE oil. It is 208/230V three phase with an internal overload. It's nominal capacity is 170,000 Btuh. Checking the Copeland website it is rated for A/C at 194,000 Btuh using R22, but only 116,000 Btuh using R404A in a medium temp refrigeration application.

FIGURE 4-40   **Compressor model number chart.** *Courtesy of Refrigeration Training Services.*

Compressor Model Number Codes for **AJA7494ZXD**

AJ	A	7	4	94	Z	XD
Family	Revision	Application	# of Digits in Rated BTU Capacity	First Two Digits of Rated Capacity	Refrigerant	Voltage

Family	Revision
AE	A
AG	B
AH	C
AJ	etc...
AK	
AN	
AV	
AW	
AZ	

Example:
4 digits in total BTU.
First two are 94.
BTU rating is 9,400.

Voltage Codes

XA = 115-60-1
XD = 208-230-60-1
XF = 208-230-60-3

Evap Temp	Rating Point	Motor Starting
1. Low	-10°F	Normal
2. Low	-10°F	High
3. High	+45°F	Normal
4. High	+45°F	High
5. A/C	+45°F	Normal
6. Med	+20°F	Normal
7. Med	+20°F	High
8. A/C	+49°F	Normal

Refrigerants

A=R12
B=R410A
C=R407C
E=R22
J=R502
Y=R134a
Z=R404A/R507

SUMMARY

The compressor can be described as the "heart" of the system because it pumps refrigerant throughout the system. The reciprocating compressor piston sucks in the suction vapor on the downstroke and then compresses it on the upstroke. This process increases the suction pressure and temperature so the condenser can return the refrigerant vapor to a liquid.

The rate at which a compressor increases suction vapor to discharge vapor can be expressed as its compression ratio. Compressor damage can be the result of operating at compression ratios outside those designed for the unit.

Hermetic compressors are suction cooled and welded in a steel shell. Semihermetic compressors are more expensive than hermetic compressors, but valve repairs can be made on this cast-iron compressor.

The primary cause of compressor failure is lack of lubrication. The other problems contributing to the failure of a compressor are flooding, flooded starts, and overheating. These conditions directly or indirectly cause lubrication problems.

Slugging is an example of having too much of a good thing: refrigerant or oil. Compressors are vapor pumps; they cannot compress liquid. Slugging damage is quick and drastic, resulting in broken rods, pistons, and crankshafts.

Manufacturers of semihermetic compressors encourage technicians to tear down a failed compressor, even when it is under warranty. Determining the cause of failure is necessary to prevent a repeat problem with the replacement compressor.

Cycling a compressor on and off can be hard on the compressor. Capacity control can extend the run time, protect the compressor from damage under low-load conditions, and reduce energy requirements as well as increase efficiency. Some of the methods of capacity control are hot gas bypass, multiple compressors, and unloaders.

In low-temperature applications, the primary problem is that as the suction pressures decrease, so does the compressor's efficiency. Increased cylinder heat, less motor cooling, and reduced mass flow rates of refrigerants are encountered as compression ratios increase.

Extra-low-temperature applications require the use of compound two-stage compression systems. Either two separate compressors or a single compressor with two stages operate well when refrigeration temperatures are −30°F to −80°F.

Scroll compressors used in commercial refrigeration have an advantage over reciprocating compressors because scrolls have nearly 100 percent volumetric efficiency. Scrolls do not have the re-expansion of vapor caused by the clearance volume inherent in piston-type compressors.

The use of VFD technology on large motors is helping to increase compressor efficiency. By varying the motor's frequency, the compressor speed and capacity can be controlled.

REVIEW QUESTIONS

1. The discharged vapor picks up which of the following two additional sources of sensible heat as the suction vapor is processed through the compressor?

   a. Motor heat and the heat of compression
   b. Evaporator superheat and motor heat
   c. Motor heat and condensing heat

2. What is clearance volume, and how can it affect compressor efficiency?

3. What is volumetric efficiency?

   a. The size of the cylinder in relation to the clearance volume
   b. How well a compressor is operating compared with its design capacity
   c. The head pressure divided by the suction pressure

4. If an R404A commercial freezer has design conditions of −20°F evaporator temperature and 125°F condensing temperature, what is the compression ratio?

   a. 14.1:1
   b. 12.3:1
   c. 11.2:1
   d. 10.5:1

5. In the example above, what is the compression ratio if the customer lowers the box temperature so that the evaporator is operating at −30°F with the same head pressure?

   a. 14.1:1
   b. 12.3:1
   c. 11.2:1
   d. 10.5:1

6. Assume the above freezer evaporator temperature remains at −20°F, but the condenser is designed for a 25°F condenser split and the condensing temperature drops to 120°F. What is the compression ratio?

   a. 14.1:1
   b. 12.3:1
   c. 11.2:1
   d. 10.5:1

7. Which situation will damage the above freezer compressor quicker, lowering the suction pressure by 5 psig or raising the head pressure by 20 psig, and why?

   a. Raising the head pressure because it raises the compression ratio more than lowering the suction pressure

b. Lowering the suction pressure because it raises the compression ratio more than raising the head pressure

c. Either will damage the compressor because they will increase the compression ratio about the same

8. In the above question, what kind of damage will occur to the compressor?

a. Slugging
b. Flooding
c. Overheating

9. Does lowering the suction pressure increase or decrease motor cooling in a suction-cooled compressor?

a. It decreases motor cooling, causing the motor to overheat.
b. It increases motor cooling because vapor is at a lower temperature.

10. What repairs can be made on the job to a semihermetic compressor?

a. Motor stator replacement only
b. Motor rotor replacement only
c. Crankcase oil change and valve plate replacement

11. What is the difference between an air-cooled compressor and a suction-cooled compressor?

12. What are the causes of too much oil in a compressor?

a. The factory put in too much oil.
b. There were multiple compressor replacements, or another technician added oil.
c. The customer was told to add oil every month.

13. What are the causes of too little oil in a compressor?

a. Low refrigerant charge or low load
b. Piping problems preventing oil return
c. Frozen evaporator or starving evaporator
d. Compressor short cycling
e. All of the above

14. What would a teardown show if there was lack of oil?

a. Burned valves
b. Oil on top of pistons
c. All rods and bearings worn or scored

15. What causes flooding and slugging?

a. Too much refrigerant returning to the compressor
b. Too little refrigerant in the system
c. Compressor running too slow to pump vapor

16. With the compressor running, which symptom is the most positive indicator of flooding or slugging at the compressor?

a. Sweating suction line
b. High superheat
c. Low to no superheat
d. Compressor noise

17. If the suction valves are broken, what is the cause?

a. Flooding
b. Slugging
c. Flooded start

18. What would cause scoring of the cylinder walls and omit blow-by?

a. Flooding
b. Slugging
c. Flooded start

19. What is the difference between flooding and slugging?

a. Slugging is a little too much refrigerant; flooding is a lot of liquid.
b. Flooding is a little too much refrigerant; slugging is a lot of liquid.
c. Both cause overheating by washing oil off pistons.

20. What is the primary cause of a flooded start?

a. Refrigerant migration
b. Floodback
c. Compressor short cycling

21. How can flooded starts be prevented?

a. Adjust superheat and subcooling
b. Install a crankcase heater or pump-down solenoid
c. Install a crankcase pressure regulating valve

22. On a teardown, what would indicate a flooded start?

a. Erratic wear pattern on the crankshaft
b. Progressive wear on the crankshaft
c. Broken valves and a blown head gasket

23. What causes high motor heat on an air-cooled compressor?

a. Inadequate airflow from the condenser fan
b. Low-mass flow of suction vapor to cool the compressor motor
c. Broken discharge valves

24. What causes high motor heat on a suction-cooled compressor?

a. Inadequate airflow from the condenser fan
b. Low-mass flow of suction vapor to cool the compressor motor
c. Broken discharge valves

25. What is compressor capacity control?

    a. Preventing the customer from loading up the walk-in
    b. Raising the head pressure to prevent loading up the compressor
    c. Controlling the amount of work a compressor can perform

26. Which of the following would be considered short cycling?

    a. Six cycles per hour
    b. Four cycles per hour
    c. Two cycles per hour

27. Which of the following is considered compressor unloading?

    a. Cycling the compressor on and off
    b. Blocking the discharge valves to reduce capacity
    c. Preventing suction vapor from being compressed

28. What is usually used to start the unloading process?

    a. A drop in suction pressure
    b. A rise in suction pressure
    c. Either a rise or a fall in head pressure

29. As the compression ratio of a compressor is increased . . .

    a. Less heat is produced in the head of the compressor.
    b. More refrigerant is pumped through the system.
    c. Less refrigerant is pumped through the system.

30. What is the maximum discharge-line temperature for Copeland compressors?

    a. 200°F
    b. 225°F
    c. 275°F

31. Two-stage compound systems are used in what application?

    a. Low temperature (0°F to −20°F)
    b. Extra-low temperature (−30°F to −80°F)
    c. Ultra-low temperature (−100°F to −150°F)

32. What is the interstage on a two-stage system?

    a. Where the discharge of one enters the suction of the other
    b. Where the suction of one enters the discharge of the other
    c. Where refrigerant is added to increase the suction pressure

33. Why are scrolls considered more efficient then reciprocating compressors?

    a. They run at lower suction pressures.
    b. They put out more British thermal units per hour (Btuh) for every pound of refrigerant.
    c. They have better volumetric efficiency.

34. How does VFD vary compressor capacity?

    a. It changes motor speed by increasing motor resistance.
    b. It frequently unloads the suction valves to vary capacity.
    c. It modulates the frequency of the electrical power.

35. What method is used to check for blow-by?

    a. Compare crankcase pressure to suction pressure
    b. Calculate compression ratio, then divide by the suction pressure
    c. Subtract pressure at the suction service valve from the head pressure

36. What is the maximum allowable blow-by pressure?

    a. 9 psig
    b. 6 psig
    c. 3 psig

37. What is the date code for Copeland serial number 15F00123?

    a. December 2000
    b. January 2005
    c. June 2015

38. What is the date code for Tecumseh serial number L2015 123456?

    a. December 20, 2015
    b. January 23, 2014
    c. June 12, 2005

39. Does the compressor nameplate give the actual Btuh capacity of the compressor?

    a. Not always; it depends on the condenser it is connected to.
    b. Not always; the compressor capacity is based on the evaporator temperature.
    c. Yes; the nameplate rating is always an accurate indicator of its output.

OBJECTIVES

After completing this chapter, you should be able to

- Understand thermostatic expansion valves and capillary tubes
- Explain how they meter refrigerant to the evaporator
- Detail how to size them for proper operation
- Describe how system pressures and temperatures affect them
- Depict how thermostatic expansion valves maintain evaporator superheat and capillary tubes do not
- Describe metering device problems and troubleshooting

CHAPTER OVERVIEW

If liquid refrigerant were allowed to flow directly into the evaporator, some cooling would occur, but not all of the refrigerant would vaporize. The remaining liquid would flood the compressor. In this chapter, you learn how refrigerant is metered into the evaporator to get the most efficient heat absorption without damaging the compressor.

Thermostatic expansion valves and capillary tubes are the primary metering devices in commercial refrigeration. Automatic expansion valves have limited, but specialized, applications.

FUNCTIONS OF A METERING DEVICE

The primary function of a metering device is to provide refrigerant to the evaporator in a condition essential for efficient heat absorption. To accomplish this, the device must change a full column of liquid refrigerant in the liquid line into tiny droplets of refrigerant that are easily evaporated, or vaporized, in the evaporator. The liquid is pushed through an orifice, or opening. It is very similar to what happens when water passes through a water nozzle on a garden hose: The full stream of water in the hose comes out in a fine spray (Figures 5-1 through 5-3).

FIGURE 5-1 **Sporlan TEV cutaway view.** *Compliments of Parker Hannifin–Sporlan Division.*

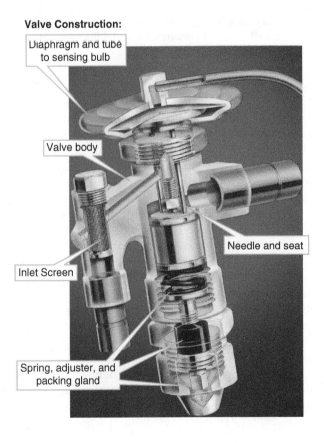

Valve Construction:

FIGURE 5-2 **Danfoss TEV.** *Photo by Dick Wirz.*

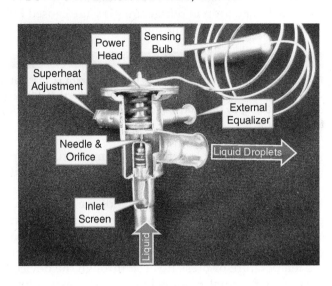

FIGURE 5-3 **Cap tube.** *Photo by Dick Wirz.*

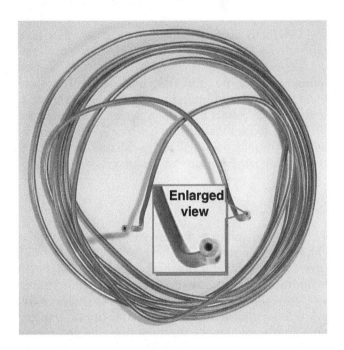

Ideally, most of the evaporator should be filled with the refrigerant droplets, fully vaporizing before they reach the evaporator outlet. However, the heat loads on the evaporator occasionally vary. As a result, the metering device may sometimes starve or underfeed the evaporator. Other times, it may flood or overfeed the evaporator. Starving the evaporator decreases cooling efficiency; flooding can actually damage the compressor.

The metering device not only changes the liquid to a spray but also lowers the temperature of the refrigerant entering the evaporator. As long as the saturation temperature of the refrigerant is below the space temperature, the warmer air in the box causes the refrigerant inside the evaporator tubing to boil into a vapor. The boiling refrigerant absorbs heat, which, in turn, removes heat from the space. As the heat is removed, the space temperature falls.

Reducing refrigerant temperature is accomplished partially by the drop in refrigerant pressure caused by its being forced through the metering device.

EXAMPLE: 1

The saturation temperature of liquid R22 refrigerant is 100°F at 196 psig (refer to a pressure/temperature [P/T] chart). If it is reduced to 49 psig by the metering device, its temperature drops to about 25°F.

The other factor in refrigerant temperature drop is a result of some of the refrigerant flashing off as it enters the evaporator. This flashing is called *adiabatic expansion*. The process uses some of the refrigerant passing through the metering device to cool the rest of the refrigerant. It does nothing for the refrigeration effect and should be kept to a minimum. The higher the temperature or pressure drop through the device, the more refrigerant is needed for adiabatic expansion. Depending on conditions, about 25 percent to 33 percent of the liquid entering the evaporator is flashed off, leaving the remaining amount of refrigerant to absorb latent heat from the space as it flows through the coil tubing.

This inefficiency in the system increases as the head pressure rises and sends hotter liquid to the metering device. As a result, more refrigerant is used to lower the temperature of the remaining refrigerant. Therefore, there is less refrigerant left to absorb heat in the evaporator, which results in a rise in the space temperature. This situation is illustrated in Chapter 7, "Refrigeration System Troubleshooting," when the problems of high condensing temperatures are discussed.

SUPERHEAT

The primary job of a thermostatic expansion valve (TEV) is to maintain superheat. In so doing, the TEV provides the proper amount of refrigerant to the evaporator; at the same time, it protects the compressor from floodback. Thermostatic expansion valve systems store a relatively large amount of refrigerant. To prevent flooding, there must be some *margin of safety* or cushion. This is achieved by using a thermostatically controlled metering device that adjusts itself to fill most of, but not all, the coil with refrigerant. Technicians can measure this margin of safety by simply calculating the superheat. Although Chapter 2, "Evaporators," discusses superheat at length, a short review is beneficial for understanding the process of metering device operation and troubleshooting.

Superheat is the amount of sensible heat absorbed in the refrigerant after all the droplets of liquid refrigerant have completely boiled off to a vapor. During the boiling process, the refrigerant absorbs large amounts of latent heat, or the heat of vaporization. When the refrigerant has absorbed all the latent heat possible at that pressure and temperature, the only heat it can absorb is sensible heat. Although picking up sensible heat does little for the total cooling process, it does provide an accurate means of measurement.

EXAMPLE: 2

A walk-in refrigerator has an evaporator temperature of 25°F. Measuring the temperature of the suction line at the outlet of the coil is the only other information necessary to calculate superheat:

- **If the outlet temperature is 35°F, the superheat is 10°F (35°F − 25°F).**
- **If the outlet temperature is 45°F, the superheat is 20°F (45°F − 25°F).**
- **If the outlet temperature is 25°F, the superheat is 0°F (25°F − 25°F).**

Based on superheat measurements, a technician can determine how efficient the evaporator is and if the compressor is in danger of experiencing floodback.

What Is the Proper Superheat?

Superheat adjustment is based on the operating characteristics of the system and the evaporator (Figure 5-4). Air conditioning (AC) units are susceptible to airflow problems and dirty filters that can cause refrigerant to flood the compressor. High superheat settings provide a good margin of safety for the compressor.

Freezers need to use as much of their evaporator as possible to remove heat from the cold space. Low superheats indicate a more efficient coil.

Medium-temperature refrigeration operates best somewhere between the two other systems.

Knowing how to properly calculate the superheat is just as important as knowing what the superheat should be. Following are three suggestions:

1. Use accurate gauges and electronic thermometers.

Pocket thermometers, even when strapped to the pipe and insulated, are not accurate enough for taking superheat measurements. Electronic thermometers

T . R . O T

EVAPORATOR SUPERHEATS
1. 15°F for AC
2. 10°F for medium-temperature refrigeration
3. 5°F for low-temperature refrigeration

NOTE: These rules of thumb only apply when you have no other information to refer to. Whenever possible, follow the recommendations of the equipment manufacturer.

are much more accurate. Several brands even have an accessory with the thermistor as part of a quick-release pipe clamp. This option costs more than $100, but the extra investment will pay for itself quickly. It makes measuring superheat easier, faster, and more accurate, which means better diagnosis and fewer callbacks.

2. Take pressure readings close to where the temperature readings are taken.

The suction-line temperature should be taken at the expansion valve bulb on the outlet of the evaporator. Ideally, the suction pressure reading should also be taken at, or near, this point. To check suction pressure at the coil outlet, some technicians use a flared Schrader tee in the external equalizer line or install a pressure access fitting in the suction line at the coil outlet.

FIGURE 5-4 **Taking superheat at the evaporator of a walk-in freezer.** *Photo by Dick Wirz.*

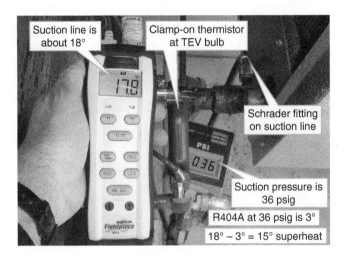

If the compressor is only 5 feet away from the evaporator outlet, use the pressure reading at the suction service valve. However, if the compressor is far from the evaporator and has a suction-line accumulator or has many elbows in the line, the pressure drop could make a difference. In this situation, using the pressure reading at a distant compressor suction service valve and then taking the temperature of the suction line back at the evaporator would not give an accurate superheat calculation.

Sometimes, it is just not cost effective for a technician to take the time to install a pressure tap at the evaporator outlet or on the external equalizer line. Therefore, it is necessary to use an alternative calculation to estimate the evaporator pressure. If the piping and accessories in the suction line appear to be sized properly, there should be a pressure drop of not more

than 2 psig. Adding 2 psig to the pressure reading at the suction service valve should give a fairly accurate evaporator pressure reading. A technician can use this pressure and the suction-line temperature at the evaporator outlet for calculating evaporator superheat.

3. Take readings within 5°F of the design conditions.

It is best to take superheat readings at the system's design temperature, or at least within 5°F of it. If the walk-in box is designed for 35°F, do not measure superheat until the box is down to at least 40°F.

Never take superheat at start-up of a warm system. This condition is considered a hot pull-down. The valve is feeding as much refrigerant as it can, but it is quickly boiled off by the high heat load in the box. The superheat is very high, and the coil is starved. Before checking superheat, let the unit run until the box is near its design temperature.

Low load is just as bad as high load. If the box is more than 5°F below its design temperature or if the equipment is operating in low-ambient conditions without head pressure control, the expansion valve will lose control. It will hunt, overfeeding then underfeeding, trying to maintain superheat. Correct the low-load situation before trying to adjust the valve.

HOW A TEV OPERATES

A TEV is influenced by the following forces:

1. Diaphragm pressure from the sensing bulb—opens the valve
2. Evaporator pressure—closes the valve
3. Spring pressure—adjustable closing pressure

NOTE: There is actually a fourth pressure, the force of liquid refrigerant entering the valve. However, it does not become a factor unless the liquid pressure is far outside the design conditions, either very high or very low. These conditions are discussed more in Chapter 7.

In Figures 5-5 and 5-6, it is important to note the only opening force on the TEV is the pressure exerted downward on the diaphragm. This pressure comes from the boiling of refrigerant inside the TEV bulb. The bulb is affected by the temperature of the suction line to which it is attached. In Figure 5-6, the suction-line temperature at the sensing bulb is 35°F. Assume the expansion valve bulb contains R22 to match the system. The equivalent pressure of R22 at 35°F is 62 psig. This is the opening pressure exerted downward on the diaphragm.

The evaporator pressure is a closing force on the valve. As the pressure pushes up on the diaphragm, the needle rises into the seat, restricting the flow of liquid refrigerant. The evaporator temperature in the drawing is 25°F. The evaporator pressure equivalent is 49 psig for R22.

FIGURE 5-5   **Simplified Sporlan diagram of three pressures acting on a TEV.** *Illustrated by Irene Wirz, RTS.*

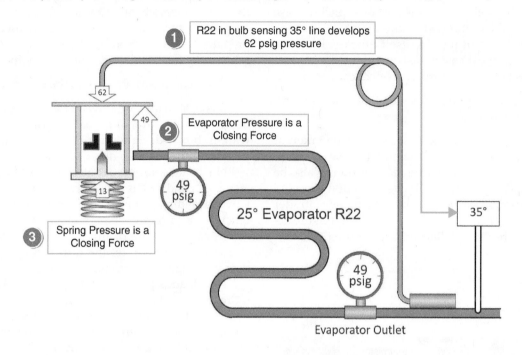

FIGURE 5-6   **Three pressures acting on a TEV.** *Courtesy of Refrigeration Training Services.*

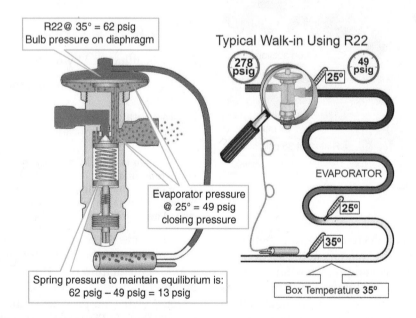

If the opening pressure is 62 psig and the closing pressure is only 49 psig, the valve will stay open, allowing too much refrigerant to flow through the evaporator. To prevent flooding, the adjustable spring provides another closing pressure against the valve needle. The spring exerts a force of 13 psig (62 psig bulb – 49 psig evaporator). The opening pressure is now equal to the closing pressure, which means the valve is in equilibrium. When a TEV is in equilibrium, the flow rate of refrigerant into the evaporator is balanced to the amount of heat the evaporator is absorbing, plus some superheat.

TEV BODY STYLES

In commercial refrigeration, most TEVs use either flare- or sweat-piping connections (Figure 5-7). Sweat connections are preferred by manufacturers that install the TEV on the evaporator before it leaves the factory.

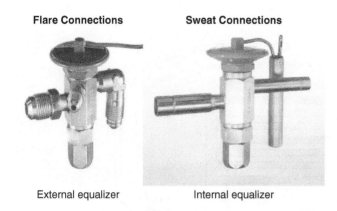

Flare Connections   **Sweat Connections**

External equalizer   Internal equalizer

On the assembly line, it is fairly easy to braze the valves. Sweat connections, unlike flare connections, do not leak from the rigors of shipping and handling.

When installing or replacing a sweat-type valve, it is very important not to overheat the valve during the brazing process. To prevent the valve from becoming too hot, wrap a damp rag around the valve body and direct the flame away from the valve. Semicircular brazing tips have been used primarily on factory assembly lines and are now available to technicians at local supply houses. These tips surround the joint with multiple flames, allowing faster brazing with less heat reaching the valve body.

EQUALIZED VALVES, INTERNAL AND EXTERNAL

Internally equalized TEVs use the pressure of the refrigerant leaving the valve as the primary closing force of the valve. This force pushes up on the bottom of the valve diaphragm, "equalizing" the opening pressure that the sensing bulb exerts on the top of the diaphragm. There is only a 2-psig pressure drop through most single-circuit evaporators; therefore, the evaporator outlet pressure is very close to the pressure leaving the valve and entering the evaporator inlet. As a result, the pressure at the valve outlet is an adequate indicator of the evaporator pressure and temperature. Internally equalized TEVs are used on small, single-circuit evaporators. Reach-in boxes and ice machines use this type of valve.

Externally equalized valves are used on larger, multicircuited evaporators like those installed in walk-in boxes. How does a technician know when an externally equalized TEV is needed? He should check for distributors or feeder tubes between the TEV connection and the tubes on the evaporator. Only multicircuited coils use distributors. The evaporator manufacturer usually installs a 1/4-inch equalizer tube near the evaporator outlet.

Most multicircuited evaporators have a 35-psig pressure drop in their distributors. Therefore, the evaporator pressure is about 35 psig lower than the pressure leaving the TEV. The higher pressure at the TEV outlet provides too much closing force on the valve diaphragm. In order for the valve to feed enough refrigerant to all the circuits in the evaporator, the closing force on the valve diaphragm would have to be the lower pressure at the evaporator's outlet as it enters the suction line.

As the name implies, an externally equalized TEV uses pressures external to the valve, not the internal pressure developed at the valve outlet. The external equalizer line provides a path between the space under the TEV diaphragm and the coil outlet. The pressure at the evaporator outlet, not the valve outlet, becomes the closing force under the valve diaphragm.

To help show the importance of an externally equalized TEV, Figure 5-8 illustrates how installing an internally equalized valve on a multicircuit evaporator causes evaporator starving and high superheat. The equalizer line is shown pinched off, as there is no connection for it on the internally equalized TEV.

NOTE: In Figures 5-8 and 5-9, there are two pressures omitted in an effort to simplify the illustration and the calculations: the adjustment spring pressure and the 2-psig pressure drop in the evaporator.

In Figure 5-8, the valve's closing pressure under the diaphragm would be the TEV outlet pressure of 49 psig. However, the pressure drop in the distributors is 35 psig. Therefore, the evaporator outlet pressure would only be 14 psig (49 psig – 35 psig). At that pressure, R22 is only –14°F. At such a low temperature, the medium-temperature TEV sensing bulb would not develop enough pressure on top of the diaphragm to open the valve. Therefore, the evaporator would starve.

Now refer to Figure 5-9. An externally equalized TEV has been installed with the external equalizer line connected to it. The valve's closing pressure is now the outlet pressure of the evaporator. To maintain the pressure of 49 psig at the evaporator outlet, the TEV opens wide enough to allow 84 psig pressure to leave the TEV. The 35-psig pressure drop in the distributors results in a 49-psig (84 psig – 35 psig) evaporator. The higher pressure at the outlet also means a higher temperature. Now that the suction line is warmer, the pressure rises in the sensing bulb. The bulb pressure is high enough to provide an opening force on the top of the diaphragm that is great enough to bring the feeding of the TEV to a point of equilibrium. The evaporator will operate at its required capacity, and the evaporator superheat will be determined by the valve's adjustable spring.

FIGURE 5-8   **Internally equalized TEV on a multicircuit evaporator.** *Courtesy of Refrigeration Training Services.*

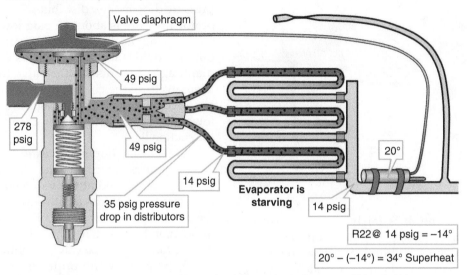

Valve senses evaporator high inlet pressure.
Bulb senses low outlet temperature.

Valve diaphragm

49 psig

278 psig

49 psig

14 psig

35 psig pressure drop in distributors

Evaporator is starving

14 psig

20°

R22 @ 14 psig = −14°

20° − (−14°) = 34° Superheat

Note: For this example, the normal 2 psig pressure drop in the evaporator is not part of this calculation

FIGURE 5-9   **Externally equalized TEV on a multicircuit evaporator.** *Courtesy of Refrigeration Training Services.*

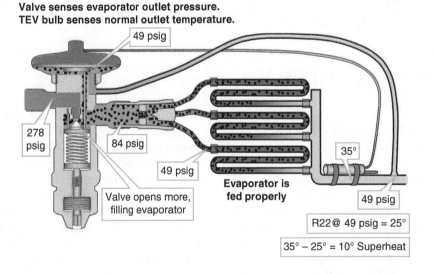

Valve senses evaporator outlet pressure.
TEV bulb senses normal outlet temperature.

49 psig

278 psig

84 psig

49 psig

Valve opens more, filling evaporator

Evaporator is fed properly

35°

49 psig

R22 @ 49 psig = 25°

35° − 25° = 10° Superheat

The illustrations show the external equalizer line installed in the suction line downstream of the TEV sensing bulb (Figure 5-10). In many valve manufacturers' literature, this is the recommended location. The reasoning of the valve company is that if a small amount of liquid should somehow travel down the external tubing, it will not adversely affect the sensing bulb. However, evaporator manufacturers usually install the external equalizer line into the suction header of the evaporator. This location is upstream of the bulb location.

The question is: Should the service technician change the location of the factory-installed external line? At least one major valve manufacturer sees no problem with using the evaporator manufacturer's external line, claiming its new TEVs are more reliable than older valves. Therefore, there is little chance of liquid entering the external equalizer line.

FIGURE 5-10 **External equalizer line tapped into suction header.** *Photo by Dick Wirz.*

ADJUSTING SUPERHEAT

A word of caution: When diagnosing a TEV problem, adjusting the superheat should be the last choice, not the first. A new valve is set by the factory for a superheat that should be satisfactory for the application and usually does not need further adjustment. Also, expansion valves do not get out of adjustment during normal operation. If an adjustment is required, care should be taken not to exert excessive force on the adjuster at the end of either the forward or back-seated position. This can damage the valve needle and also cause the valve to leak refrigerant.

If the initial adjustment does not result in a change of superheat, the technician should return the adjuster to its original position before checking other possible causes of superheat problems. Finally, adjusting a TEV should be performed only by technicians who fully understand expansion valve operation.

Most TEVs have a superheat adjustment stem on the bottom of the valve. Turning the adjuster into the valve (clockwise) will increase the spring tension. More spring pressure closes the valve, reducing the amount of refrigerant to the evaporator and increasing the superheat.

According to most textbooks, and valve manufacturers, the proper way to adjust an expansion valve is to make a one-quarter turn of the stem, wait 15 minutes, recheck the superheat, and then make another one-quarter turn as necessary. *Each full turn of the adjustment stem will change the superheat from as little as 1°F superheat to as much as 5°F, depending on the make and type of valve. On most Sporlan valves, each complete turn of the stem will adjust the superheat 4°F.* Assume we are adjusting a Sporlan TEV in the following example.

EXAMPLE: 3

EXAMPLE: 3

If the superheat needs to be changed by 6°F, make one and a half turns of the adjuster and then wait 15 minutes to make sure it settles out correctly.

If the valve is completely out of alignment, it is wise to return the adjustment to the factory setting, which is close to the middle of the adjustment stem. Go through the following steps to find the center position of the valve stem:

1. Turn the adjustment stem in clockwise until it stops. Be gentle when making this adjustment, as the needle can be easily damaged.
2. Count the turns as you back the stem out counter-clockwise until it stops.
3. Finally, turn the adjustment stem in clockwise half the total number of turns.

It is now at the middle of its adjustment range, which is the factory setting. Recheck the superheat after the unit has been in operation long enough to come down to design conditions.

NOTE: Once adjusted, most valves do not need to be readjusted. Although it is a good idea to check the superheat, it is not a good idea to adjust the valve without first checking at least the following conditions:

- Make sure the system is within 5°F of design conditions.
- If the superheat is high, determine first if the valve is starving because of a lack of refrigerant. It could be anything from low refrigerant to a restricted filter drier or clogged strainer in the valve inlet.
- If the superheat is low, check the bulb to make sure it is on a clean surface, strapped properly, and in the correct position on the suction line.

TEV BULB PLACEMENT

The textbooks and valve manufacturers have the following guidelines for TEV bulb placement on the suction line (Figure 5-11):

- For a pipe size of 7/8-inch outer diameter (OD) and larger, place the bulb at a 45-degree angle from center. View the end of the pipe as if it were a clock face. Install the bulb at 8 o'clock or 4 o'clock.

Fasten the bulb to the side of the suction line.

Placing the TEV bulb on a vertical suction line is not recommended. The oil traveling up the entire inner surface of a vertical suction riser prevents the bulb from sensing true suction temperature.

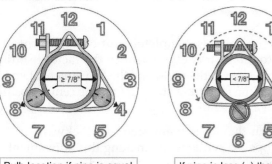

Bulb location if pipe is equal to or greater (≥) than 7/8" place bulb at 8 or 4 o'clock.

If pipe is less (<) than 7/8" place bulb anywhere from 8 to 4 o'clock.

REALITY CHECK 1

Sometimes the vertical suction line is the only place available to mount the TEV bulb. Fortunately, the thin layer of oil spread out around the vertical section of pipe will not greatly affect the ability of the bulb to sense the suction vapor temperature. If you must mount it vertically, common practice is to position the sensing bulb with the cap tube coming out of the top (see Figure 5-12).

The reason for this position is that it is the coolest part of the pipe. The top is the warmest part, and the bottom is where the oil collects. The oil acts as an insulator, which prevents the bulb from sensing the proper suction-line temperatures.

• For a pipe size of 5/8-inch OD and smaller, put the bulb anywhere, except on the bottom of the suction line.

As in the large pipes, the oil flows in the lowest section of the smaller pipes, acting as an insulator between the temperature of the tubing and the refrigerant temperature (Figure 5-12). Therefore, the bulb should never be installed on the bottom of the tubing.

Most service technicians find that the top of the tubing is the easiest place to clamp the sensing bulb. Because the temperature of the refrigerant is fairly even throughout smaller suction lines, locating the bulb on the top is as good a place as any.

Some technicians find it confusing when the factory recommendations state bulb locations of 8 o'clock and 4 o'clock, 10 o'clock and 2 o'clock, or 9 o'clock and 3 o'clock. It is easier to remember that the mid-position on the side of the suction line works quite well for any size pipe up to about 1 5/8-inch OD suction line.

Always install the TEV bulb on a clean, flat horizontal section of pipe. Never attempt to strap the bulb on a pipe joint or elbow. No matter how tight you strap it to the uneven surface, there will be gaps in the contact area between the bulb and pipe (see Figure 5-13).

Most TEV manufacturers supply brass straps and bolts to fasten the bulb to the suction line. The assembly is adequate but can be difficult to install properly. Some technicians use stainless steel radiator clamps that have a screw-type adjuster. These clamps make the job of bulb mounting much easier. Also, the screw clamps make a service technician's job less difficult when checking for corrosion or ice between the bulb and suction line. It is important to tighten the clamps enough to make a snug fit of the bulb to the pipe.

FIGURE 5-12 **Sensing bulb on a vertical line.** *Courtesy of Refrigeration Training Services.*

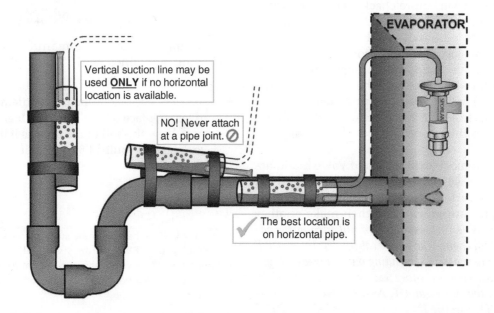

Vertical suction line may be used **ONLY** if no horizontal location is available.

NO! Never attach at a pipe joint. 🚫

The best location is on horizontal pipe. ✓

EVAPORATOR

FIGURE 5-13 **Examples of bulb on pipe joint and use of radiator clamps.** *Photo by Dick Wirz.*

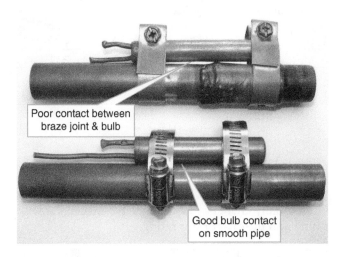

Poor contact between braze joint & bulb

Good bulb contact on smooth pipe

However, overtightening can damage or deform the sensing bulb. Initially, there was some concern that the stainless steel straps on copper pipe would cause galvanic action, resulting in corrosion. This has not proved to be the case. In fact, at least one manufacturer is now using stainless steel straps. *If the bulb is deformed too much, it could affect the pressure on the valve diaphragm, which would change the TEV's ability to maintain the proper superheat.*

Insulating the bulb and pipe assembly is recommended to ensure that the bulb senses suction-line temperature, not the air around the suction line. Use 3/8- or 1/2-inch-thick closed-cell pipe insulation that does not absorb moisture.

HOW THE SYSTEM AFFECTS TEVs

Thermostatic expansion valves adjust the flow rate of refrigerant into the evaporator based on suction-line temperature and its relationship to the evaporator pressure and temperature. TEVs respond well to the load in the box by allowing more refrigerant to enter when the bulb senses a high heat load. It decreases the amount of refrigerant as the evaporator temperature drops. Also, TEVs are not adversely affected by high-ambient conditions at the condenser. The valve can handle very high inlet pressures without allowing refrigerant to flood the evaporator.

Thermostatic expansion valves perform a balancing act between the temperature of the liquid entering the valve and the pressure drop from condensing pressure to evaporator pressure. As the pressure difference increases, so does the valve capacity. However, as the liquid line temperature increases, the capacity of the valve decreases. The effect of one offsets the

effect of the other, so the valve capacity remains almost constant.

The following examples are intended to clarify general concepts of TEVs, rather than deal with exact figures for specific valves. The values in the chart (Figure 5-14) closely approximate pressures and temperatures of a 35°F walk-in refrigerator with a 25°F evaporator temperature. The ambient temperatures affecting the head pressure cover a range from 50°F to 100°F outdoor-ambient conditions. The condenser has a 30°F condenser split (CS), and the refrigerant is R22.

Actual factory correction factors are used to illustrate increased and decreased valve capacity. Correction factors are a percentage of operating capacity compared to the valve's capacity at design conditions. If the valve is operating at design conditions, the factor is 1.00 because it is operating at 100 percent of its design capacity. A correction factor of 0.90 means it is operating at only 90 percent of its capacity; if 1.10, the valve is operating at 110 percent of its capacity.

NOTE: **Numbers in the table are approximations and are rounded off to simplify the examples.**

Correction factors are primarily used by factory application engineers when sizing TEVs for systems. However, technicians should have some understanding of how much valve operation is affected by pressure drop through the valve and the liquid temperature entering the valve.

The chart in Figure 5-14 illustrates how higher condenser loads affect expansion valves. As head pressure increases, the pressure drop across the valve also increases. The result is higher valve capacity because of the greater refrigerant flow rate.

At 100°F ambient, the pressure drop across the valve is 250 psig. The higher pressure entering the valve has the ability to force more refrigerant into the evaporator. The additional refrigerant can increase evaporator capacity by 40 percent above design and is given a correction factor of 1.40. However, the higher temperature of the liquid decreases capacity because more refrigerant must be flashed off when entering the evaporator to drop the refrigerant temperature from 130°F to the 25°F evaporator temperature. The correction factor is 0.80, which means the valve loses 20 percent from design conditions. By multiplying the increase factor of 1.40 by the decrease factor of 0.80, the result is an overall correction factor of 1.10, or 10 percent over the design conditions.

The 10 percent capacity increase does not mean the evaporator will necessarily have that much more refrigerant fed into it. It is similar to reserve capacity that is available if needed by an increase in the evaporator loading. Remember, the TEV bulb senses suction-line temperature and maintains superheat. Even with the

FIGURE 5-14   **TEV correction factors balance valve capacity.** *Adapted from Sporlan 201 catalog.*

Outdoor Ambient (°F) A	Condensing Temperature (°F) B	Head Pressure (psig) C	Evaporator Pressure (psig) D	Pressure Drop (psig) (C – D)	Correction Factor for Pressure Drop E	Correction Factor for Liquid Temperature F	Overall Correction Factor (E × F)
100	130	300	50	250	1.40	0.80	1.10
70	100	200	50	150	1.10	1.00	1.10
50	80	150	50	100	0.90	1.15	1.05

increased inlet pressure on the valve, it will not allow any more refrigerant through the valve unless the bulb pressure senses increased superheat from increased evaporator load. Only then will the bulb increase pressure on the diaphragm and open the valve.

TEVs are capable of preventing very high pressures from forcing refrigerant into the evaporator. For this reason, high ambients, dirty condensers, and slight overcharges of refrigerant have less effect on the evaporator pressure, or the superheat, than might be expected. This is discussed more in Chapter 7 on system troubleshooting.

TEV SIZING

The most popular method for sizing expansion valves is to match the TEV capacity to the evaporator capacity. However, this is correct only if the evaporator and the condenser capacity are the same. Following are the basic steps for choosing the correct expansion valve:

1. Calculate the total system British thermal units per hour (Btuh) capacity based on the combination of the evaporator and the condensing unit.
2. Choose the correct valve based on all three of the following:
    A. Total system capacity
    B. Evaporator temperature
    C. Liquid temperature

When the evaporator and condenser are mismatched (have different capacities), then some research is required to determine the total system capacity and the actual evaporator temperature. This will be discussed in detail in Chapter 11, "Walk-in Refrigerators and Freezers."

TEVs are rated in tons at 12,000 Btuh per ton.

EXAMPLE: 4

A system rated at 6,000 Btuh would require a 1/2-ton valve rated at 6,000 Btuh *at its normal operating evaporator temperature and liquid temperature.*

It is best to use the charts furnished by the valve manufacturer to properly size TEVs. An oversized valve can be just as bad as an undersized valve. Occasionally, there is no exact match between the valve rating and the evaporator rating.

EXAMPLE: 5

An R22 walk-in evaporator has a capacity of 9,000 Btuh. The only valves available for the particular refrigerant and application are 6,000 Btuh (1/2 ton) and 12,000 Btuh (1 ton). Which would you choose, and why?

In Figure 5-15, the choice of two valves is in bold letters (FVE-1/2 and FVE-1). The 1-ton valve is recommended because the smaller valve may starve the evaporator. Also, most valve manufacturers agree that if the coil rating is between two valves, the larger valve will still maintain the proper superheat without flooding.

If the evaporator system in the previous example is rated at 7,000 Btuh, it would be best to check the valve charts. The correction factor of the 1/2-ton valve under normal conditions may very well be 1.10. This would probably place the valve close enough to the capacity of the 7,000-Btuh evaporator (6,000 × 1.10 = 6,600 Btuh).

Even an 8,000-Btuh evaporator, with a suction-line heat exchanger, may have enough of a liquid line temperature correction factor to allow using the 1/2-ton valve. The benefits of heat exchangers are discussed more fully in Chapter 6, "Controls and Accessories."

It is important to understand the effect evaporator temperature has on the capacity of a TEV. This is especially critical in low-temperature applications. For example, a freezer system that has a capacity of 36,000 Btuh would require a TEV with a capacity of 3 tons. In Figure 5-16, the chart shows that a valve rated at 3 tons at a +20°F evaporator temperature will provide only 2.10 tons of capacity at a +20°F evaporator temperature. A larger valve would be required. Based on the chart, a 4-ton valve will have a capacity of almost 3 tons (2.94 tons) at a −20°F evaporator temperature.

FIGURE 5-15 **Sporlan Type F TEV selection chart.** *Compliments of Parker Hannifin–Sporlan Division.*

REFRIGERANT (Sporlan Code)	TYPE F		TYPE EF		NORMAL CAPACITY Tons of Refrigeration	Thermostatic Charges Available	CONNECTIONS - Inches SAE Flare / ODF Solder Blue figures are standard and will be furnished unless otherwise specified	
	SAE Flare		ODF Solder					
	Internal Equalizer Flare	External Equalizer Flare	Internal Equalizer Solder	External Equalizer Solder			INLET	OUTLET
22 (V) 407C (N) 407A (V)	FV-1/5	FVE-1/5	EFV-1/5	EFVE-1/5	1/5	C Z ZP40	1/4 OR 3/8	3/8 OR 1/2
	FV-1/3	FVE-1/3	EFV-1/3	EFVE-1/3	1/3			
	FV-1/2	FVE-1/2	EFV-1/2	EFVE-1/2	1/2			
	FV-1	FVE-1	EFV-1	EFVE-1	1			
	FV-1-1/2	FVE-1-1/2	EFV-1-1/2	EFVE-1-1/2	1-1/2			
	-	FVE-2	-	EFVE-2	2		3/8	1/2
	FV-2-1/2	-	EFV-2-1/2	-	2-1/2			
	-	FVE-3	-	EFVE-3	3			
134a (J) 12 (F) 401A (X) 409A (F)	FJ-1/8	FJE-1/8	EFJ-1/8	EFJE-1/8	1/8	C	1/4 OR 3/8	3/8 OR 1/2
	FJ-1/6	FJE-1/6	EFJ-1/6	EFJE-1/6	1/6			
	FJ-1/4	FJE-1/4	EFJ-1/4	EFJE-1/4	1/4			
	FJ-1/2	FJE-1/2	EFJ-1/2	EFJE-1/2	1/2			
	FJ-1	FJE-1	EFJ-1	EFJE-1	1			
	FJ-1-1/2	FJE-1-1/2	EFJ-1-1/2	EFJE-1-1/2	1-1/2		3/8	1/2
	-	FJE-2	-	EFJE-2	2			
404A (S) 502 (R) 408A (R)	FS-1/8	FSE-1/8	EFS-1/8	EFSE-1/8	1/8	C Z ZP	1/4 OR 3/8	3/8 OR 1/2
	FS-1/6	FSE-1/6	EFS-1/6	EFSE-1/6	1/6			
	FS-1/4	FSE-1/4	EFS-1/4	EFSE-1/4	1/4			
	FS-1/2	FSE-1/2	EFS-1/2	EFSE-1/2	1/2			
	FS-1	FSE-1	EFS-1	EFSE-1	1			
	FS-1-1/2	FSE-1-1/2	EFS-1-1/2	EFSE-1-1/2	1-1/2		3/8	1/2
	-	FSE-2	-	EFSE-2	2			

SPORLAN Type F Valve Selection Chart

FIGURE 5-16 **TEV capacity is reduced as evaporation temperature decreases.** *Adapted by RTS from Sporlan 201 catalog.*

THERMOSTATIC EXPANSION VALVE CAPACITIES				
Commercial Refrigeration Applications R404A REFRIGERANT				
Valve Types	Nominal Rated Capacity	Tonnage at Evaporator Temperature		
		20°	−10°	−20°
C-S	3 Ton	3	2.45	2.1
C-S	4 Ton	4.28	3.42	2.94

READING AN EXPANSION VALVE

Thermostatic expansion valve manufacturers usually have a numbering system on the valve that identifies the type of valve, what refrigerant it is designed for, and what application (type of system) it is designed for.

Although the valve coding can be a little confusing at times, the valve in Figure 5-17 is typical of most Sporlan valves. The first letter (F) is the body style, the second is the refrigerant, an E as the third letter means it is externally equalized, and the number one means it is rated at 1-ton capacity.

Note on connections: No letter before the body style means the valve has flare connections. If there is the letter S before the body style, the valve has short sweat connections. However, the letter E before the body style means extended (long extension) sweat connections. The type F valves in the chart in Figure 5-15 are available in either flare or extended sweat connections.

Note on equalized valves: The letter E after the refrigerant code means externally equalized. However, if there is no letter between the refrigerant and the tonnage, the valve is internally equalized (refer to the chart in Figure 5-15).

FIGURE 5-17 **FIGURE 5-17** **Reading a Sporlan TEV.** *Photo by Dick Wirz.*

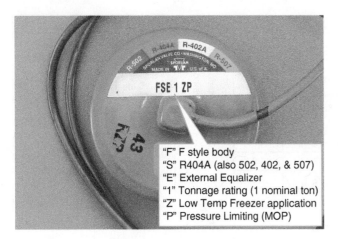

FSE 1 ZP

"F" F style body
"S" R404A (also 502, 402, & 507)
"E" External Equalizer
"1" Tonnage rating (1 nominal ton)
"Z" Low Temp Freezer application
"P" Pressure Limiting (MOP)

After the tonnage, rating is the application: GA for AC, C for commercial medium temperature, and Z for freezer. The P on this valve stands for "pressure limiting," which means the valve will limit the amount of refrigerant entering the evaporator to a maximum pressure of approximately 40 psig. This feature is used primarily on low-temperature TEVs to prevent the compressor from being overloaded during a hot pull-down. For example, an R404A system operating at −20°F evaporator will have a low suction pressure of 16 psig (refer to P/T chart in the appendix) during the refrigeration cycle. When it goes into defrost, the compressor turns off, the defrost heaters come on, the evaporator warms up to about 55°F, and the pressure inside the evaporator rises to almost 115 psig. When defrost ends and the compressor starts, not only will it have trouble starting against 115 psig vapor, but a standard TEV will be wide open, feeding refrigerant to the evaporator, which increases the load on the compressor. The result is high starting amperage, which often causes the compressor to shut off on overload.

To prevent this, pressure-limiting expansions valves limit the maximum suction pressure going to the compressor. For example, during the hot pull-down after defrost, an R404A low-temperature externally equalized pressure-limiting TEV will open only enough to maintain a maximum of 40 psig in the evaporator. The compressor can handle 40 psig, so the compressor starts and runs without high amps. As the evaporator cools and the pressure drops below 40 psig, the valve will open 100 percent to fully feed the evaporator until it reaches its designed superheat. This pull-down period is relatively short, so the pressure-limiting TEV will be operating like a standard TEV in only a few minutes.

The pressure-limiting function is also known as *maximum operating pressure* or MOP. This is another way to express the ability of the TEV to restrict refrigerant flow to the evaporator in order to limit the rise of suction pressure.

At this time, it will be useful to explain why low-temperature compressors have trouble with high suction pressure when they try to start and when they are running. It is definitely not because they have less powerful motors or because they are not built as strong as higher-pressure compressors. The reason is that low-temperature compressors are designed to run at low suction pressure. Low-pressure refrigerant is much less dense than higher-pressure refrigerant; therefore, the piston and cylinder must take a larger volume of vapor on each stroke than a medium- or high-temperature compressor. This can be accomplished by a larger cylinder, longer stroke, or both. If the pressure goes too high and the cylinder is filled with a denser refrigerant, the power needed to compress it will cause high motor amps.

In our previous example, the compressor that was running normally at 16 psig had to try to start at 115 psig, which is over 7 times its normal design. Compare that to an R22 air conditioning unit that has a normal operating evaporator temperature of about 40°F with a pressure of about 70 psig. If the pressure was increased 7 times like our freezer example, the AC compressor would be under about 490 psig on the suction side. I'm sure anyone familiar with AC will immediately realize that a 490-psig suction pressure will cause high amps and an overloaded compressor.

As you progress through this book, try to keep in mind that low-temperature compressors will have start and run problems anytime the suction pressures rise too far above their normal range.

TROUBLESHOOTING TEVs

For an expansion valve to work correctly, there are three main criteria:

1. The valve must be sized properly.
2. The sensing bulb must be strapped to the suction line correctly.
3. There must be a *full column of liquid* (all liquid, no flashing) to the valve.

Flooding

Whenever there is no measurable superheat, the evaporator is flooding. It could be that the TEV is feeding more refrigerant into the evaporator than heat in the box can vaporize. One of the primary causes of TEV flooding is ice in the valve. Any moisture that gets through the filter drier can freeze in the valve. If ice holds the valve open, the valve will flood the evaporator. To check for this problem, shut the unit off. Then use a hot rag or hot air gun to heat the valve; never use a torch. After the valve has warmed enough to melt the ice, restart the compressor. The superheat should return to normal when the system is operating within its normal temperature range.

Ice problems in a TEV are usually caused from water released by a filter drier that is at its maximum moisture-holding capacity. Replacing the filter drier should solve the moisture problem. However, for a more comprehensive remedy, the technician could consider recovering the refrigerant, evacuating the system, and recharging with new refrigerant in addition to replacing the filter drier.

While waiting for the ice to melt, check the TEV sensing bulb. Make sure it is tightly attached to a clean section of suction line. If in doubt, remove the straps, clean the pipe and bulb, then reattach and insulate the bulb. A TEV bulb that is not sensing the suction line is responding to a warmer ambient temperature. If this is the case, the valve will be feeding more refrigerant than it should.

If the bulb is not the problem, and there does not seem to be any ice, reset the superheat adjustment to the factory setting (as described earlier). If the flooding continues, replace the valve.

Starving

High superheat is an indication that the evaporator is starving. It could be the TEV is not allowing enough refrigerant into the evaporator. As a result, the suction pressure will be low or may even be in a vacuum. The head pressure will be slightly low because the condenser does not have much heat to process. Once again, ice in the valve may be the problem. This time, it may be restricting the flow of refrigerant. The solution is to warm the valve. A rapid rise in the suction pressure indicates that the ice has melted. Replace the filter drier, restart the compressor, and check the superheat.

A loose sensing bulb will not make the valve shut down. However, the valve will slam shut if the sensing bulb loses its charge. Without bulb pressure on top of the valve diaphragm, there is no force to move the valve needle from its seat. The result is no refrigerant flow. Check for a cracked or a broken capillary tube between the valve head and bulb. Equipment vibration is a common cause of tubing damage.

There is one last possibility for no refrigerant flow: The strainer on the valve inlet may be clogged. Some sweat valves have a screen below a bolt, like the one in Figure 5-18. The screens in flare valves are part of the inlet flare fitting. It is worthwhile to check the inlet screens because doing so may prevent the time and expense incurred in needlessly replacing the valve. If the strainer is clogged, it is a good idea to replace the filter drier.

FIGURE 5-18 **Sporlan TEV inlet strainers.** *Photo by Dick Wirz.*

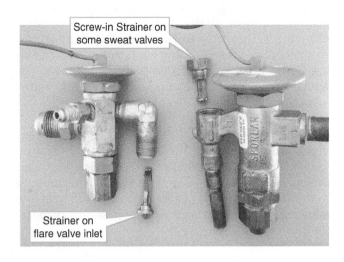

Screw-in Strainer on some sweat valves

Strainer on flare valve inlet

If there is still no flow through the valve, then adjusting the superheat will not do much good. The solution is probably replacing the TEV.

TEV Hunting

If the superheat is fluctuating from low or no superheat to high superheat every 10 seconds to 15 seconds, the valve is **hunting**. The TEV is looking for its balance point, or equilibrium, where it will be able to maintain proper evaporator temperature and superheat.

Make sure the valve is correctly sized and there is 100 percent liquid being supplied to the valve. The factory recommendation is to open the valve by backing the adjustment stem counterclockwise about one turn. The higher volume of refrigerant through the valve will probably help it to settle down. Adjust the superheat as necessary.

Balanced Port Expansion Valves

Hunting can also be caused by fluctuation of the head pressure. When the condenser fan is cycled on and off in low-ambient conditions, the head pressure rises and falls as much as 50 psig. As the pressures go up and down, so does the pressure of refrigerant on the inlet of the TEV. As a result, a conventional valve may feed erratically, which can cause flooding as well as starving of the evaporator.

A balanced port expansion valve uses a balancing force to steady the feeding of the valve under conditions such as low- and high-ambient and fluctuating conditions encountered with condenser fan cycle controls. During low condensing pressures, a standard TEV will have a tendency to starve because there is not enough pressure drop across the valve to feed properly. In a balanced port valve, even low-pressure liquid entering the valve will tend to push down on the flattened head of the needle valve, allowing refrigerant to flow into the evaporator. To prevent overfeeding the evaporator, the incoming liquid also pushes up on the valve diaphragm, which has a closing or balancing effect on the valve (see Figure 5-19).

In very hot ambients, the high inlet pressure will try to push open the needle valve and flood the evaporator. To prevent this, there is an opposite force on the underside of the diaphragm, which will balance the opening effect on the needle valve. The pressure from the bulb strapped to the suction line will therefore remain as the primary opening force based on evaporator loading (see Figure 5-20).

A chart for balanced port valves is similar to a chart for standard expansion valves. The primary difference is that the tonnage rating is a range (e.g., 3/4 thru 1 1/2 tons) rather than a specific number (e.g., 1 ton). This has the advantage of allowing more flexibility when choosing the right size TEV for the job (see Figure 5-21). Although balanced port valves are more stable over a wide range of operating conditions, they still need to be sized properly for the application. You must know the operating range (evaporator temperature), tonnage, and the refrigerant used. From these three pieces of information, you should be able to determine the proper valve to use. For a more complete understanding of choosing a TEV, consult the valve manufacturer's sizing chart. For Sporlan, go to their website at www.sporlanonline.com and search for Bulletin 10-10.

FIGURE 5-19   **Standard and Balanced Port TEVs in low-ambient conditions.** *Courtesy of Refrigeration Training Services.*

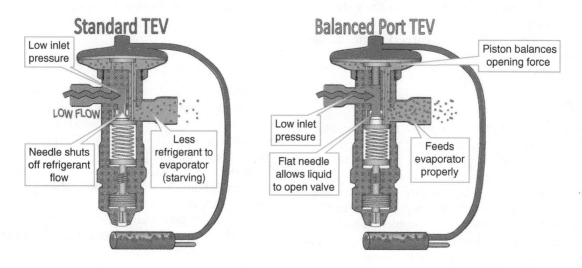

FIGURE 5-20   **Standard and Balanced Port TEVs in high-ambient conditions.** *Courtesy of Refrigeration Training Services.*

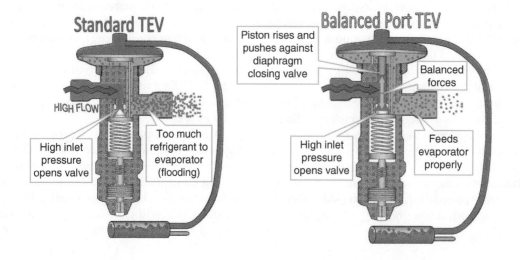

FIGURE 5-21 **Balanced Port TEVs have a capacity range.** *Compliments of Parker Hannifin–Sporlan Division.*

REFRIGERANT (Sporlan Code)	TYPE		Port Size	NORMAL CAPACITY Tons of Refrigeration	Thermostatic Charges Available	CONNECTIONS - Inches Blue figures are standard and will be furnished unless otherwise specified		
	SAE Flare					SAE Flare		
	Internal Equalizer	External Equalizer				INLET	OUTLET	External Equalizer
22 (V) 407C (N) 407A (V)	BFV-AAA	BFVE-AAA	AAA	1/8 thru 1/3	C Z ZP40	1/4 OR 3/8 1/4 OR 3/8 3/8		1/4
	BFV-AA	BFVE-AA	AA	1/2 thru 2/3				
	BFV-A	BFVE-A	A	3/4 thru 1-1/2				
	BFV-B	BFVE-B	B	1-3/4 thru 3				
	BFV-C	BFVE-C	C	3-1/4 thru 5-1/2				
134a (J) 12 (F) 401A (X) 409A (F)	BFJ-AAA	BFJE-AAA	AAA	1/8 thru 1/5	C	1/4 OR 3/8 1/4 OR 3/8 3/8	3/8 or 1/2	1/4
	BFJ-AA	BFJE-AA	AA	1/4 thru 1/3				
	BFJ-A	BFJE-A	A	1/2 thru 1				
	BFJ-B	BFJE-B	B	1-1/4 thru 1-3/4				
	BFJ-C	BFJE-C	C	2 thru 3				
404A (S) 502 (R) 408A (R)	BFS-AAA	BFSE-AAA	AAA	1/8 thru 1/5	C Z ZP	1/4 OR 3/8 1/4 OR 3/8 3/8		
	BFS-AA	BFSE-AA	AA	1/4 thru 1/3				
	BFS-A	BFSE-A	A	1/2 thru 1				
	BFS-B	BFSE-B	B	1-1/4 thru 2				
	BFS-C	BFSE-C	C	2-1/4 thru 3				

SPORLAN BALANCED PORT TYPE F VALVE SELECTION CHART

CAPILLARY TUBE METERING DEVICES

Capillary tubes, or cap tubes, are simply small copper pipes. They meter refrigerant by changing the high-pressure liquid entering one end of the pipe to a spray of liquid at a lower pressure by the time it leaves the other end. Cap tubes come in various inside diameters (IDs) and lengths. They are used in small commercial refrigeration equipment like reach-in boxes.

The use of cap tube systems is very cost effective. Cap tubes are inexpensive and have no moving parts. Unlike TEVs, the system pressures equalize quickly so the compressor is not starting under a load. Therefore, compressors on these systems do not need start capacitors to help them start. TEVs require a receiver and extra refrigerant; cap tube systems do not.

Systems using a cap tube work fine as long as they are used only for storing already refrigerated product. Cap tube systems do not respond well to box load changes.

A restaurant cook has a 40°F reach-in refrigerator located next to the grill area in which he works. Meat patties from the 35°F walk-in box in the back kitchen are placed in the reach-in just before dinner. The reach-in box only has to keep the product cold

until it is removed for cooking. Therefore, a cap tube refrigerator has little problem maintaining the product temperature because load on the evaporator is not heavy.

How a Cap Tube Works

The friction of fluids (refrigerant) traveling through tubing can result in a pressure drop. Cap tubes use this principle to change high-pressure liquid refrigerant from the liquid line to a lower-pressure spray of refrigerant that boils easily in the evaporator. The longer the tube or the smaller the ID of the tube, the lower the pressure will be when the refrigerant exits the pipe and enters the evaporator.

Capillary tube sizing is critical, especially in the tubing diameter. A difference of 0.005 inch (five-thousandths of an inch), from 0.026 inch to 0.031 inch, can double the flow rate of refrigerant to the evaporator.

However, cap tubes depend on their length as well as their diameter to determine their total restriction, or flow rate. Manufacturers of cap tubes have determined what combination of tubing ID and length is needed to accomplish a given amount of refrigeration for systems of various sizes. Figure 5-22 is a sizing chart for cap tubes. The tubing size and length are based on the condensing unit size, refrigerant, and the evaporator temperature.

FIGURE 5-22 **Cap tube sizing chart.** *Adapted from J/B Industries.*

H.P.	REF.	CAPILLARY TUBE SIZING CHART (R-12 AND R-22) NORMAL EVAPORATING TEMPERATURE DEGREES F							
		−10 to +5		+5 to +20		+20 to +35		+35 to +50	
1/5	R-12	8 Ft.	TC-31	8 Ft	TC-36	10 Ft.	TC-42	6 Ft.	TC-42
1/4	R-22	12 Ft.	TC-36	6 Ft	TC-36	8-1/2 Ft.	TC-42	6 Ft.	TC-49
1/4	R-12	10 Ft.	TC-36	6 Ft	TC-36	8 Ft.	TC-42	6 Ft.	TC-49
1/3	R-22	10 Ft.	TC-36 →	6 Ft	TC-36	11 Ft.	TC-49		
1/3	R-12	12 Ft.	TC-42	6 Ft	TC-42	9 Ft.	TC-49	6 Ft.	TC-54
1/2	R-22	6 Ft.	TC-36	9 Ft	TC-42	7-1/2 Ft.	TC-54	10 Ft.	TC-64
1/2	R-12	11 Ft.	TC-54	9 Ft	TC-49				
3/4	R-22	11 Ft.	TC-54	9 Ft	TC-54				
3/4	R-12	7-1/2 Ft.	TC-54	12 Ft	TC-70	10 Ft.	TC-80		
1	R-22	10 Ft.	TC-64	12 Ft	TC-70				
1	R-12	10 Ft.	TC-70	11 Ft	TC-54	7-1/2 Ft.	TC-54 (2 pcs)		
1-1/2	R-22	7-1/2 Ft.	TC-54 (2 pcs)	7-1/2 Ft	TC-54 (2 pcs)	8 Ft.	TC-64 (2 pcs)		
1-1/2	R-12			9 Ft	TC-64 (2 pcs)	10 Ft.	TC-80 (2 pcs)		
2	R-22			10 Ft	TC-70 (2 pcs)	9 Ft.	TC-75 (2 pcs)		
2	R-12	10 Ft.	TC-70 (2 pcs)	9 Ft	TC-75 (2 pcs)	10 Ft.	TC-85 (2 pcs)		
3	R-22			10 Ft	TC-70 (3 pcs)	9 Ft.	TC-75 (3 pcs)		
3	R-12	10 Ft.	TC-70 (2 pcs)	8 Ft	TC-64 (4 pcs)	10 Ft.	TC-80 (4 pcs)		
4	R-22			10 Ft	TC-70 (4 pcs)	9 Ft.	TC-75 (4 pcs)		
4	R-12			10 Ft	TC-70 (5 pcs)	9 Ft.	TC-75 (5 pcs)		
5	R-12			10 Ft	TC-80 (5 pcs)	9 Ft.	TC-85 (5 pcs)		

EXAMPLE: 7

A 1/3-hp R22 condensing unit with a 15°F evaporator would use 6 feet of TC36 (0.036 inch) cap tube.

When replacing a cap tube, sometimes the only tubing available has a different ID than the original. Figure 5-23 shows how to determine what size and length of tubing will work in place of the original. The chart uses multipliers to adjust the length of the replacement tubing relative to the original tubing in order to achieve the same pressure drop as the original.

Assume the replacement cap tube has a larger ID than the original cap tube. The replacement tubing will have to be longer than the original to achieve the same pressure drop. Multiply the original tubing length by the correction factor to determine the length of the replacement tubing.

EXAMPLE: 8

In Figure 5-23, the original cap tube was 0.040 inch ID and 9 feet long. However, the replacement is larger at 0.042 inch ID. Therefore, the replacement must be longer than the original.

- **Per the chart, multiply the original tubing length (9 feet) by a factor of 1.25.**
- **1.25 × 9 feet = 11.25 feet is the length needed for 0.042-inch ID tubing.**

If the ID of the replacement cap tube is smaller than the original, the replacement will have to be shorter. The smaller ID replacement has more pressure drop than the original tubing, so it does not need to be as long.

EXAMPLE: 9

In Figure 5-23, the original cap tube was 0.0400 inch ID and 9 feet long. However, the replacement is smaller at 0.036 inch ID. Therefore, the replacement must be shorter than the original.

- **Per the chart, multiply the original tubing length (9 feet) by a factor of 0.62.**
- **0.62 × 9 feet = 5.5 feet is the length needed for 0.036-inch ID tubing.**

How Cap Tubes Respond to System Conditions

Capillary tubes meter the amount of refrigerant entering the evaporator based on the pressure exerted on the liquid entering the tube from the condenser. If the load on the evaporator is fairly constant, the higher the head pressure, the more refrigerant is fed into the evaporator.

FIGURE 5-23 **Cap tube length conversion chart.** *Adapted from J/B Industries.*

CAPILLARY TUBE LENGTH CONVERSION CHART

This conversion chart enables the user to translate the recommended length of tube diameter into sizes stocked by J/B. In using the chart it is recommended that conversions be made using factors within the unshaded area.

TO USE CHART

(1) Locate recommended cap tube I.D. in left hand column.

(2) Read across and find conversion factor under copper cap tube size.

(3) Multiply the given length of the recommended cap tube by the conversion factor.

(4) The resultant length (Min. 5 feet/Max. 16 feet) of copper cap tube will give the same flow characteristics as the original recommended cap tube.

EXAMPLE

Recommended cap tube: 9'-.040 I.D.

Locate .040 in left hand column and reading across gives the following conversion factor: No. TC-36 (.62) and TC-42 (1.25).

Multiplying the recommended cap tube length of 9' by the conversion factor gives the following results:

5-1/2' (5'6") TC-36 and 11-1/4' (11'3") TC-42. Either of these two cap tubes will give the same results as the original cap tube of 9'-.040 I.D.

Part No.	TC-26	TC-31	TC-36	TC-42	TC-44	TC-49	TC-50	TC-54	TC-55
Tube I.D.	.026	.031	.036	.042	.044	.049	.050	.054	.055
.024	1.44								
.025	1.20								
.026	1.00	2.24							
.028	.72	1.59							
.030	.52	1.16							
.031	.45	1.00	2.00						
.032		.86	1.75						
.033		.75	1.54						
.034		.65	1.35						
.035		.58	1.16	2.31					
.036		.50	1.00	2.10					
.037		.45	.90	1.79					
.038		.39	.80	1.59	1.9?				
.039		.35	.71	1.41	1.75				
.040		.31	.62	1.25	1.55	2.51			
.041		.28	.56	1.12	1.38	2.26	2.50		
.042		.25	.50	1.00	1.24	2.03	2.23		
.043		.23	.45	.87	1.11	1.83	1.98		
.044		.20	.39	.81	1.00	1.62	1.79		
.045			.35	.73	.90	1.47	1.60	2.32	
.046			.32	.67	.82	1.34	1.47	2.08	2.27
.047				.59	.74	1.20	1.31	1.89	2.06

9' × .62 = 5½'   9' × 1.25 = 11¼'

Capillary tubes do not respond well to load changes on the evaporator. The limited amount of refrigerant being fed into the evaporator will quickly boil off if there is a high heat load in the box. The evaporator will be starved, and the superheat will be high. Therefore, it takes a cap tube system longer than a TEV system to pull down the box temperature.

Manufacturers of cap tube systems determine how much refrigerant must be in the unit for proper operation, without damaging the compressor. The specific amount of refrigerant is known as the critical charge. If the unit is overcharged, it could flood back to the compressor during both low evaporator load conditions and high condenser loading. If the unit is undercharged, it will not refrigerate properly and may even cause the compressor to overheat.

T . R . O . T

REPLACEMENT CAP TUBES

Larger must be longer; smaller must be shorter.

TROUBLESHOOTING CAP TUBES

A starving evaporator, as evidenced by high superheat, can be caused by high evaporator loads. Heat in the box rapidly boils off the limited amount of refrigerant the cap tube can supply the evaporator. The suction pressure will be lower than normal because the compressor is pulling hard on a small amount of refrigerant. The head pressure will be slightly high because the condenser is processing more heat load than usual.

To help the situation, remove some of the hot product from the box, add something cold to the box like ice, or just wait for the temperature to come down. The following example illustrates the slow pull-down of a cap tube system with a high box load.

EXAMPLE: 10

A brand new reach-in display unit is delivered. The customer fills the unit with warm drinks. The refrigerator could take nearly 24 hours to bring the drinks down to 40°F. There is nothing wrong with the reach-in, it is just that cap tube systems are designed to keep refrigerated product cold, not to pull down warm product quickly.

The primary service problem with a cap tube is a tubing restriction. The ID of the tubing is so small that it is easily plugged. The main culprit is the filter drier.

Most small driers for cap tubes use a beaded desiccant or filter material. Sometimes, the beads become loose and rub against each other, and small particles flake off (Figure 5-24). The fine desiccant powder eventually stops up the cap tube. The symptoms include low suction pressure (maybe even in a vacuum) and lower-than-normal head pressure.

FIGURE 5-24 **Cap tube and drier.** *Courtesy of Refrigeration Training Services.*

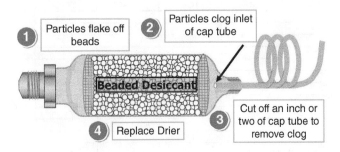

① Particles flake off beads
② Particles clog inlet of cap tube
Beaded Desiccant
④ Replace Drier
③ Cut off an inch or two of cap tube to remove clog

Fortunately, the blockage is usually just at the tubing inlet; therefore, it is not necessary to replace the entire cap tube. The recommended repair is as follows:

1. Recover the existing refrigerant.
2. Cut out the existing filter drier.
3. Cut off the first inch or two of cap tube.
4. Install a new drier.
5. Evacuate and weigh in the proper amount of new refrigerant.

NOTE: **The loss of a couple of inches of cap tube is not enough to adversely affect system performance.**

SERVICE TIP: **After separating the cap tube from the filter drier, insert a long piece of picture hanging wire to check for blockage in the cap tube (see Figure 5-25).**

FIGURE 5-25 **Using picture hanging wire to check for blockage in cap tube.** *Photo by Jess Lukin.*

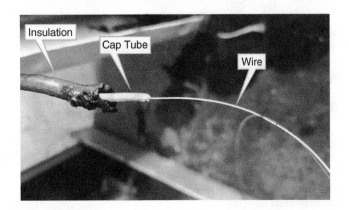

Insulation
Cap Tube
Wire

It is important that cap tubes are cut properly to retain their full ID. Do not use a tubing cutter; it will actually reduce the size of the tubing. The approved method is to score or notch a line across the tubing with the sharp edge of a file. With the notch on the top of the pipe, place your thumbs under it and your fingers around the pipe. Pull down as if snapping a pencil between your two hands. The tubing will break apart, and the ID will remain its full size.

Be careful not to use too much brazing or soldering material when reinstalling the cap tube. The tubing of the filter drier is usually only 1/4 inch and is easily filled with solder, which can also plug the end of the cap tube. The recommended procedure is to insert the cap tube about 2 inches into the end of the drier but not far enough to touch the inside strainer of the filter drier. Jamming the cap tube into the screen may also block the flow of refrigerant into the end of the cap tube. The next step is to hold the cap tube tight to one side of the tubing while crimping the remaining portion of the 1/4-inch tubing. Finally, apply only enough brazing material to seal the crimped opening.

Whereas evaporator starving in a cap tube system is normally due to the load on the evaporator, flooding is a result of a high load on the condenser. The rate of refrigerant fed through a cap tube is directly proportional to the liquid pressure on the inlet. If the head pressure rises, for any reason, more refrigerant will be forced through the tube into the evaporator.

However, floodback is limited by the amount of refrigerant in the system. The manufacturer's critical charge of cap tube units is partially determined by how much liquid refrigerant the compressor can handle without experiencing damage.

SERVICE TIP: **When servicing cap tube systems or any system with a critical charge, it is recommended that gauges not be used to check the operation. Every time the gauges are removed, a little refrigerant is lost from the system. Critically charged systems will not operate correctly if they are undercharged by as little as 10 percent of their rated charge.**

Instead, look for overloading with warm product, bad door gaskets, or very heavy use with frequent door openings. Also, check for a dirty condenser, dirty or iced evaporator, or bad fan motor. Ask if someone recently added refrigerant or serviced the unit.

Only when absolutely necessary should gauges be used on the system. If the pressures are a little off, do not just add the refrigerant. Recover the existing refrigerant, pull a good vacuum, and weigh in the proper charge.

If the low side of the system is in a vacuum and the head pressure is slightly low for the ambient conditions, the problem is most likely a restricted cap tube. Remove the refrigerant and filter drier. Cut off the first inch or two of the cap tube. Install a new filter drier, evacuate, and weigh in the proper charge of new refrigerant. Start the unit and allow it time to come down to temperature.

NOTE: Make sure the thermostat (tstat) is set properly; customers usually turn down the tstat at the first sign of trouble.

By this time, every repairable possibility has been checked. If the unit still does not come down to temperature, the problem is most likely an inefficient compressor.

Problems and solutions related to cap tube refrigeration units are discussed in more detail in Chapter 7.

AUTOMATIC EXPANSION VALVES

Automatic expansion valves (AEVs) are specialized metering devices used in applications in which there is a fairly constant load. This type of device is basically a pressure regulating valve that once set, maintains a constant evaporator temperature. It is not adversely affected by high condenser loads like a cap tube is. Unlike a TEV, an AEV does not try to maintain evaporator superheat. This is important when the unit has a problem because the evaporator temperature dipped too low (see Figure 5-26).

FIGURE 5-26   **AEV.** *Photo by Dick Wirz.*

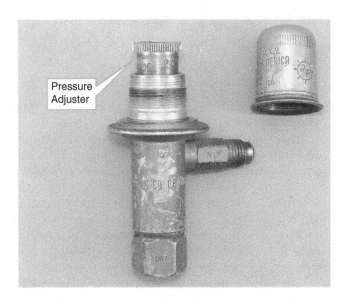

Pressure Adjuster

Automatic expansion valves operate just the opposite of TEVs. An increase in the load on an evaporator causes a rise in evaporator pressure and suction-line temperature. If the evaporator used a TEV, it would open, adding more refrigerant to the evaporator to lower its temperature. An AEV, on the other hand, will actually shut down the flow of refrigerant and slightly starve the evaporator. Although this does not bring

down the load quickly, it prevents the evaporator from becoming too cold.

Automatic expansion valves are used on drinking water coolers to maintain a cold water temperature. However, this type of metering device will keep the evaporator above freezing, so there is little chance of the cold water tank bursting.

Another use of AEVs is on soft-serve ice cream freezers and frozen drink (slush) dispensers. The temperature of the product must be kept within one or two degrees of design to dispense properly.

ELECTRIC EXPANSION VALVES

In current designs, the electronics controlling the valve are separate from the valve itself. Therefore, the correct term to describe the valves would be *electronically controlled electric valves*. To keep things simple, we will instead use the term *electric valve* in this book.

The primary functions of a TEV and an electric expansion valve (EEV) are basically the same: to maintain the proper superheat in the evaporator. If the evaporator has the proper superheat, then the technician can be reasonably sure it is properly fed with refrigerant and there is little chance of floodback to the compressor. Both valves maintain superheat by sensing evaporator pressure and suction-line temperature. The standard TEV in use is a self-contained unit that utilizes pressure and temperature to meter the proper amount of refrigerant to the evaporator.

On the other hand, to perform the same functions, the EEV needs electrical power to operate and a separate controller to send signals to the valve. After receiving information from the sensors, the controller determines how the valve needs to operate. A small electric motor in the top of the valve opens and closes the valve port. The electronic technology required makes the EEV cost more than the TEV, but the EEV operates more precisely, which improves system efficiency and lowers the entire system's operating costs. On larger systems, the higher initial cost of EEVs has a relatively quick payback because of the savings in energy.

Electric valves can perform many different tasks based on the software programmed into their controller and the type of needle or plunger in the valve. Following is a partial list of some of the electric valves now being used in refrigeration systems:

- Expansion valves
- Hot gas bypass
- Evaporator pressure regulators
- Crankcase pressure regulators
- Head pressure regulators
- Heat reclaim three-way valves

Regardless of the valve's task, the electric motor portion of all these valves is very similar. Although there have been several types of motors used in electric valves, the most common is currently the step motor (refer to Figures 5-27, 5-28, and 5-29). Traditional induction motors used in fans, pumps, and compressors rotate as long as power is applied to them. However, step motors rotate only in a defined arc or distance, then they stop. Each time power is applied then removed, the motor moves a fixed amount, or step, before it stops. The benefit of small steps is that the valve is able to control the flow of refrigerant very precisely. Whereas small electric valves can use a direct drive mechanism to open and close the port (Figure 5-28), larger valves must increase

FIGURE 5-27 **Sporlan EEV.** *Adapted from Sporlan by RTS.*

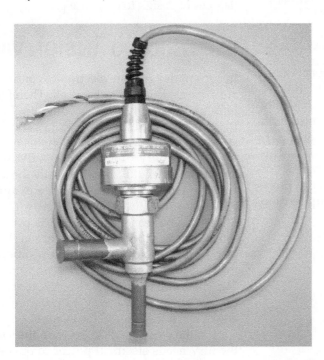

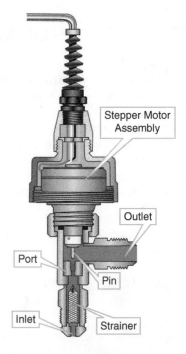

FIGURE 5-28 **Alco EEV cutaway view.** *Photo by Dick Wirz.*

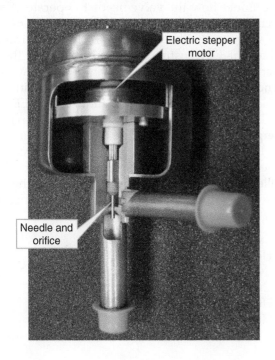

the motor's output force by using a simple gear train (see Figure 5-29).

The controller that tells the EEV what to do uses solid-state switches called transistors (Figure 5-30). The term *solid state* indicates they are fabricated from a solid chip of silicon and have no moving parts. They act as switches or relays by using a small electrical signal to turn a large signal on and off. Inside the controller is a microprocessor, also known as a computer, which sends signals to the transistors to make the step motor open or close the valve in very precise increments.

The valve, controller, microprocessor, and wiring are known as "hardware." The "software" is the set of instructions, or algorithms, needed to make the valve perform a function. Extensive testing is required to make sure the algorithms make the valve operate correctly. Due to the expense of this testing, superheat algorithms are held as secrets (proprietary information) by most control manufacturers. Algorithms are based on "if–then" logic; for example, "if superheat rises 1°F, then open the EEV two steps."

To determine superheat, the controller must know the saturated suction temperature and the temperature

FIGURE 5-29   **Sporlan uses gears to increase force in larger valves.** *Adapted from Sporlan by RTS.*

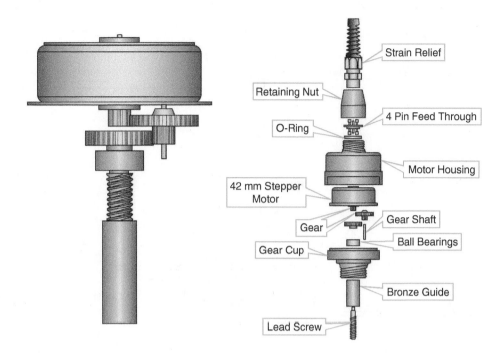

FIGURE 5-30   **Sporlan controller for electric valves.** *Compliments of Parker Hannifin–Sporlan Division.*

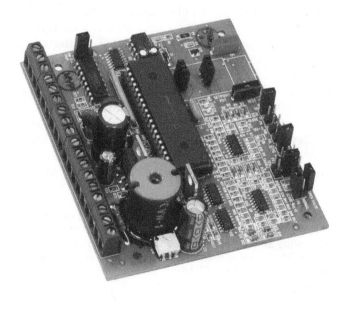

FIGURE 5-31   **Pressure transducer (left) and thermistor (right).** *Photo by Dick Wirz.*

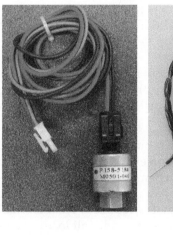

of the suction line. The controller receives this information in a recognizable electronic form from a pressure transducer and a thermistor (Figure 5-31).

A pressure transducer is a three-wire device. Two wires supply power, and the third wire sends the output signal to the controller. The temperature sensor, or thermistor, is a solid-state device that changes its electrical resistance in response to a change in temperature. The controller is able to determine temperature based on the resistance encountered by the electrical signal it sends through the thermistor.

Because the valves are electronically controlled, they have the ability to very accurately monitor and control pressure, temperature, superheat, and many other system requirements. EEVs used in medium-temperature cases and walk-ins are credited with lower frost accumulation by maintaining a more consistent evaporator temperature than traditional TEVs (Figure 5-32).

In addition, EEVs can shut off completely, which eliminates the need for a pump-down solenoid. The valve also prevents leakage between the high and low sides during

FIGURE 5-32 **Sporlan diagram of EEV and controller, with inputs of pressure and temperature.** *Compliments of Parker Hannifin–Sporlan Division.*

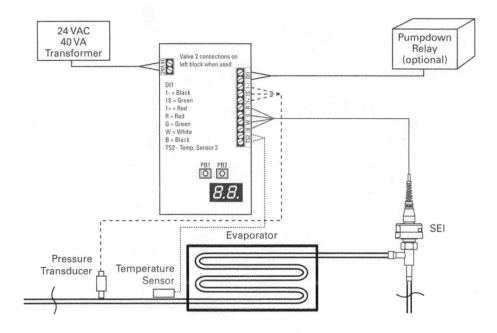

the off cycle, which prevents the two pressures from equalizing. The benefit is that the system will begin refrigerating faster after start-up for two important reasons:

1. The compressor quickly brings the head pressure up to operating conditions because the discharge pressure did not drop very much after the compressor's previous cycle.
2. The compressor does not have to spend time pulling down the suction pressure to its operating pressure because the suction did not fall that much after the previous cycle.

When used with multiple capacity compressors, the valves maximize system efficiency while matching refrigerant flow rate to changing compressor capacities. For instance, when an unloader is utilized, the capacity of the compressor is reduced. The EEV will match the compressor capacity under all loading and unloading conditions.

Troubleshooting EEVs is not that difficult, even though they employ a high degree of technical

hardware and software. When a system is not operating correctly, technicians tend to condemn the part of the system they do not understand. This is especially true when it comes to electronic components. Like the control boards in ACs and ice machines, EEVs respond only to the signals supplied by their controllers. Before condemning an EEV, the technician must make sure it is receiving the proper signal from the controller. Most modern controllers have built-in diagnostic capabilities, and manufacturers provide detailed instructions on how to check both the controller and the valves. Sporlan even has a SMA-12 Step Motor Actuator to help diagnose systems by confirming or proving the correct operation of their valve's step motor.

NOTE: EEVs can maintain very low superheats. Theoretically, they could maintain an evaporator superheat of 1°F. However, our ability as technicians to properly check those superheats, even with electronic thermometers, is often inaccurate by several degrees. Therefore, EEV manufacturers recommend a minimum superheat of 5°F.

SUMMARY

Metering devices provide refrigerant to the evaporator in order to absorb heat from the box. Both the pressure and temperature of entering liquid refrigerant must drop tremendously to cool the evaporator. Cap tubes do an adequate job of metering refrigerant in smaller commercial units like reach-in boxes. TEVs have the ability to handle more evaporator load and are

always used on larger equipment with multicircuited evaporators.

Proper sizing of the metering device is critical to overall system operation. Cap tubes perform much differently than TEVs. Troubleshooting metering devices requires a thorough understanding of how they are sized, how they operate, how the system affects them, as well as their limitations. Electronic expansion valves are becoming more common in commercial refrigeration systems. Although their cost is greater than conventional TEVs, the savings in energy from their efficient operation provides a quick payback on the initial investment in this new technology.

REVIEW QUESTIONS

1. What is the primary function of all metering devices?

   a. To provide refrigerant to the evaporator in order to absorb heat
   b. To prevent floodback
   c. To maintain superheat

2. What are two ways the metering device accomplishes its primary function?

   a. Prevents floodback and evaporator flooding
   b. Changes incoming liquid to a spray and lowers its temperature
   c. Increases the evaporator pressure and superheat

3. What are two ways the metering device lowers the temperature of refrigerant as it enters the evaporator?

   a. Subcooling and de-superheating
   b. Drop in pressure and increase in superheat
   c. Drop in pressure and adiabatic expansion

4. Why is superheat important in a TEV system?

   a. It acts as a margin of safety to prevent compressor damage due to flooding.
   b. It keeps the temperature of the evaporator up so it does not freeze.
   c. It allows the evaporator to operate at its maximum efficiency.

5. The suction pressure at the outlet of the evaporator is 65 psig on an R404 system, and the temperature of the suction line at the bulb is 35°F. What is the superheat?

   a. 6°F
   b. 8°F
   c. 10°F
   d. 12°F

6. According to TROT, what is the approximate evaporator superheat setting for a medium-temperature walk-in box?

   a. 5°F
   b. 10°F
   c. 15°F
   d. 20°F

7. According to TROT, what is the approximate evaporator superheat setting for a walk-in freezer?

   a. 5°F
   b. 10°F
   c. 15°F
   d. 20°F

8. Can you use a pocket thermometer to calculate superheat? Why or why not?

   a. Yes, pocket thermometers are very accurate.
   b. No, they are not accurate enough for measuring line temperature.

9. If the compressor is 60 feet from the evaporator, can you use the suction pressure at the compressor suction valve to calculate evaporator superheat? Why or why not?

   a. Yes, because the suction pressure is the same throughout the low side.
   b. Probably, if you allow for a 2-psig pressure drop in the suction line.
   c. No, because there would be no way of knowing what the actual evaporator pressure is.

10. When is the best time to take superheat readings?

    a. When the space temperature is within 5°F of design temperature
    b. On the initial start-up of the system
    c. When the system has just shut down

11. What are the three primary forces on a TEV?

    a. Head pressure, suction pressure, and valve pressure
    b. Bulb pressure, evaporator pressure, and spring pressure
    c. Bulb pressure, liquid line pressure, and evaporator pressure

12. What is the only opening force of a TEV?

    a. The spring pressure
    b. The evaporator pressure
    c. The sensing bulb pressure

13. What is the prime consideration when installing a sweat-type TEV?

    a. Not to overheat the valve
    b. Not to allow moisture into the valve
    c. Not to allow oxidation into the valve

14. What type of evaporator will need an externally equalized valve?

    a. Single-circuit evaporator without equalizer tube
    b. Multicircuit evaporator with equalizer tube

15. Why are externally equalized valves needed on some evaporators?

    a. Because of the high pressure drop in the distributors
    b. Because of the number of coil fins
    c. Because of the suction header at the evaporator outlet

16. When adjusting a Sporlan TEV for 2°F more super-heat, how many turns of the adjustment stem are needed and in what direction?

    a. Counterclockwise (out) one-half turn
    b. Clockwise (in) one-half turn
    c. Clockwise (in) one turn

17. How is the factory setting of the superheat adjust-ment determined on a Sporlan expansion valve?

    a. Find the midpoint of the superheat adjustment stem
    b. Adjust the stem all the way in until it stops
    c. Adjust the stem all the way out until it stops

18. According to valve manufacturers, where should the TEV sensing bulb be mounted on a horizontal suction line that is 7/8 inch in diameter or larger?

    a. On the top of the suction line
    b. On the bottom of the suction line
    c. At a 4 o'clock or 8 o'clock position

19. Based on the previous question, where would you mount the bulb if the pipe size was 5/8 inch OD or under?

    a. At the bottom of the suction line
    b. Anywhere except the bottom of the suction line

20. What happens to the capacity of a TEV as the head pressure decreases?

    a. The capacity increases because of more pressure drop across the valve.
    b. The capacity decreases because of less pressure drop across the valve.

21. What happens to the capacity of a TEV as the liquid temperature decreases?

    a. The capacity increases due to less adiabatic expansion needed.
    b. The capacity decreases due to less refrigerant flow.

22. An 18,000-Btuh evaporator would require what size TEV? Give the answer in both Btuh and tonnage.

    a. 18,000 Btuh, 1.5 ton
    b. 12,000 Btuh, 1 ton
    c. 12,000 Btuh, 1.5 ton

23. What are the three criteria needed for a TEV to work correctly?

    a. Valve sized correctly, bulb installed correctly, and complete liquid to the valve
    b. Valve sized correctly, bulb tight to the pipe, and high pressure to the valve
    c. Valve sized correctly, bulb mounted on pipe bottom, and pressure to the valve

24. How do you know if a TEV is flooding?

    a. Low superheat
    b. High superheat
    c. No superheat

25. What are the first three things to check if a TEV is flooding?

    a. Ice in valve, head pressure, and superheat
    b. Head pressure, suction pressure, and superheat
    c. Ice in valve, bulb sensing properly, and superheat

26. How do you know if a TEV is starving?

    a. Low superheat
    b. High superheat
    c. No superheat

27. What four things would you check if a TEV is starving?

    a. Ice in the valve, bulb charge, strainer blockage, and superheat
    b. Head pressure, suction pressure, strainer, and superheat
    c. Ice in the valve, bulb location, suction pressure, and superheat

28. How do you know if a TEV is hunting?

    a. Low superheat
    b. High superheat
    c. Superheat fluctuates

29. What three things would you check if a TEV is hunting?

    a. Valve sizing, liquid to the valve, and if opening the valve stops the hunting
    b. Valve sizing, head pressure, and bulb location
    c. Valve sizing, superheat, and mid-position of the valve

30. How do cap tubes meter refrigerant?

    a. Lower evaporator pressure to the superheat setting
    b. Lower liquid pressure to evaporator pressure
    c. Spray liquid into the evaporator

31. If a replacement cap tube has a larger ID than the original, should it be longer or shorter than the original cap tube?

    a. Longer
    b. Shorter

32. A convenience store customer just bought a reach-in refrigerator with a cap tube system. He wants it repaired because the warm bottled beer he loaded into it after lunch is not cold by his busy period at 5:00 P.M. What would you tell him and why?

33. You are servicing a warm reach-in with a cap tube system. The customer says it runs, but at a temperature 10°F to 20°F higher than normal, and the product entering the box is from a properly operating walk-in box. How would you check the box without putting your gauges on it?

34. Upon putting gauges on the unit in the previous question, you find the low side in a vacuum and the head pressure slightly lower than normal for the ambient conditions. What would you suspect as the problem, and how would you correct it?

35. What is the primary function of a TEV?

    a. Maintain the proper evaporator superheat
    b. Fill the evaporator completely with liquid refrigerant
    c. Feed enough refrigerant to prevent compressor problems

36. What are the main reasons for using an expensive EEV over a standard TEV?

    a. Customers want units with better technology
    b. Better system efficiency and control
    c. EEVs are more profitable

37. What is an algorithm?

    a. An electrical theory used to develop EEVs
    b. A mathematical function that provides parameters for a TEV to operate
    c. A set of instructions in the microprocessor that controls an electric valve

38. What is a pressure transducer?

    a. A safety control to prevent low pressure
    b. A pressure-sensing device to monitor
    c. A pressure-sensing device that sends information to the controller

39. What is a thermistor?

    a. A safety control to prevent low temperature
    b. A safety control to prevent high temperature
    c. A device that changes resistance in response to temperature changes

40. What is the benefit of preventing a system's pressures from equalizing during the off cycle?

    a. The system reaches its rated capacity sooner after start-up
    b. The evaporator is less likely to flood the compressor
    c. The compressor will draw less starting current

41. What is the minimum superheat recommended by EEV manufacturers?

    a. 1°F superheat
    b. 5°F superheat
    c. 10°F superheat

6

CONTROLS AND ACCESSORIES

OBJECTIVES

After completing this chapter, you should be able to

- Describe temperature controls
- Explain the operation of service valves
- Describe solenoid valves for pump-down and hot gas bypass
- Explain the operation of hot gas bypass valves
- Describe crankcase pressure regulators and evaporator pressure regulators
- Describe low-pressure and high-pressure controls
- Describe oil separators and oil safety controls
- Explain the operation of receivers and accumulators
- Explain the operation of filter driers and sight glasses
- Describe heat exchangers and vibration eliminators

CHAPTER OVERVIEW

The last four chapters covered the four basic components of a refrigeration system: evaporator, condenser, compressor, and metering device. This chapter covers the controls that manage the system operation and the accessories that improve system performance.

There are many controls and accessories available for refrigeration systems. However, the main emphasis of this chapter is on the controls technicians typically encounter in food service equipment, up to about 7.5 hp. On larger equipment, the component size may be a little bigger and seem more complicated, but the operating characteristics are similar to what is covered here.

TEMPERATURE CONTROLS

The primary controller for temperature is a thermostat (tstat). In response to temperature, an electrical circuit is switched on or off, causing the compressor to stop or start. Commercial refrigeration tstats are line voltage and use a gas- or liquid-filled sensing bulb to monitor temperature. The pressure from the bulb is transmitted through a cap tube to the diaphragm in the control in response to one of the following temperatures:

- Return air or box temperature
- Supply air temperature
- Evaporator temperature
- Suction-line temperature

The majority of tstats monitor the return air, which is the average box temperature. Tstats come in many temperature-sensing ranges and temperature differentials (TDs) (the difference between cut-in and cut-out). When sensing air temperature, the differential is about 3°F to 5°F. This fairly wide temperature swing prevents compressor short-cycling and also allows most medium-temperature refrigerators to self-defrost during the off cycle.

EXAMPLE: 1

Assume a refrigerator is supposed to maintain a box temperature of 40°F. When the temperature in the box rises to 40°F, the tstat starts the compressor. The tstat has a 5°F differential, or temperature swing. When the space temperature drops to 35°F (40°F – 5°F differential = 35°F), the tstat shuts the compressor off. The evaporator fan(s) run continuously, circulating the air throughout the box. When the air temperature rises to 40°F, the tstat starts the compressor, and the cycle is repeated (see Figure 6-1).

FIGURE 6-1 **Air sensing thermostat.** *Photo by Dick Wirz.*

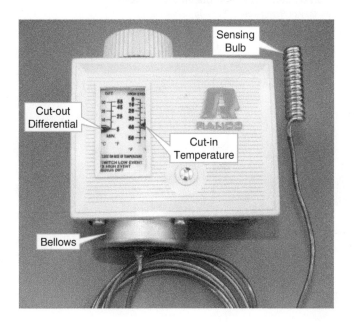

Some medium-temperature boxes have a problem with excessive coil frosting. This condition is often due to heavy product loads or low evaporator temperatures. Reach-in refrigerator evaporators have about a 15°F to 20°F TD. That means a 40°F box has coil temperatures of 25°F to 20°F. At these temperatures, frost accumulates when the compressor is running (see Figure 6-2).

FIGURE 6-2 **Diagram of the relationship between box temperature and evaporator temperature in a refrigerator with a 20°F TD.** *Courtesy of Refrigeration Training Services.*

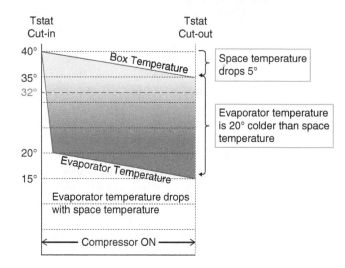

To ensure the evaporator is cleared of frost during the off cycle, some reach-in manufacturers place the tstat sensor into the evaporator fins or strap it to the suction line. In these positions, the tstat reacts to the coil temperature rather than the air temperature in the space. If the tstat is set to cut in at 40°F, it will do so when the evaporator temperature is up to 40°F. Usually, this is the same as the air temperature that is blowing through the evaporator fins, but with one important difference: Any frost that accumulated on the evaporator during the refrigeration cycle must be melted before the coil temperature will reach 40°F. This just about guarantees the evaporator will be free of frost before it turns the compressor on, thereby providing the self-defrosting feature of this type of tstat use. Figure 6-3 is a picture of a reach-in refrigerator tstat with its bulb inserted inside the evaporator to sense coil temperature rather than the box temperature. The picture was taken with the ceiling-mounted evaporator pan dropped to expose the coil, tstat, and wiring.

An evaporator temperature–sensing tstat must have a wide enough difference between its cut-in and cut-out to properly monitor evaporator temperature, usually a 15°F to 20°F differential for

FIGURE 6-3 **Reach-in refrigerator with coil-sensing thermostat.** *Photo by Dick Wirz.*

Tstat bulb sensing coil temperature

Thermostat

A refrigerator is packed with warm product, and the box temperature shoots up to 60°F. The compressor starts, but long before the box temperature drops to 40°F, the coil frosts, blocking airflow and heat transfer. Because the frost is insulating the coil from the warm box temperature, the evaporator temperature continues to fall. An air-sensing control would keep the compressor running, adding to the frost problems. However, if the sensing bulb is mounted in the evaporator fins, the tstat will shut off the compressor when the evaporator drops to the control's cut-out temperature. While the compressor is off, the fans continue to circulate the warm box air over the evaporator, melting the frost. When the coil is defrosted, its temperature reaches the 40°F cut-in, and then the compressor starts again. This cycle continues until both the box temperature and the product finally come down to the 40°F design temperature.

reach-ins, which matches reach-in TDs. Assume the air temperature in a reach-in swings from 40°F when the compressor starts down to 35°F when the compressor shuts off. While the box temperature is 40°F, the evaporator temperature is about 20°F (assuming a 20°F evaporator TD). Just before the temperature control stops the compressor when the box temperature drops to 35°F, the evaporator temperature would be running at about 15°F (35°F – 20°F TD = 15°F). In Figure 6-4, the tstat would have a cut-in at 40°F and a cut-out at 15°F.

The following is an example of how a tstat with its sensing bulb mounted in the evaporator can prevent frost problems better than a tstat sensing box air temperature.

If the tstat were only sensing box temperature, it would keep the compressor running even though the coil was frosted. The fan motors would be adding heat to the box, while the evaporator, insulated by frost, would be forming ice on the coil tubing from the inside out.

Walk-in refrigerators can also be kept free of frost in the same manner by using a tstat with a remote sensing bulb (see Figure 6-5).

The tstat's large sensing bulb is installed in the middle section of the evaporator, about 3 inches from the top. The installer embeds the sensing bulb between the coil fins at a slightly downward angle. A screwdriver can be used to gently spread the fins to make room for the bulb. The bulb should not touch the coil tubing, only the coil fins.

Most remote bulb tstats have their cut-out temperature as the primary feature on the front of the dial. On

FIGURE 6-4 **Illustration of how the evaporator defrosts during the off cycle before the thermostat will start the compressor.** *Courtesy of Refrigeration Training Services.*

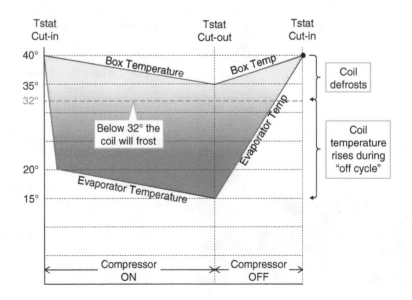

FIGURE 6-5   **Remote bulb thermostat set for wide differential when bulb is sensing coil temperature.** *Photo by Dick Wirz.*

FIGURE 6-6   **Electronic thermostat with remote bulb.** *Photo by Dick Wirz.*

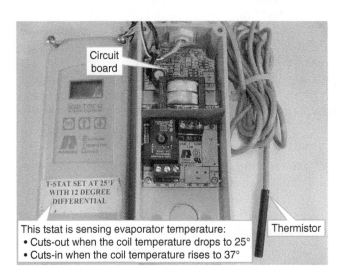

FIGURE 6-7   **Ranco constant cut-in thermostat.** *Photo courtesy of Ranco® by Robertshaw®.*

temperature while having a very effective self-defrost feature in every operating cycle.

COMPRESSOR SERVICE VALVES

Compressor service valves are used for much more than just checking system pressures. Technicians also use compressor service valves for checking compressor reed valves and isolating the compressor. Figure 6-8 shows a larger-than-normal view of a suction service valve for illustration purposes. The position of the stem in this figure is called *back-seated* because the back of the valve stem flange is against the seat.

All valves have a packing around the stem to prevent refrigerant leaks. Some valves have a packing nut

FIGURE 6-8   **Suction service valve.** *Courtesy of Refrigeration Training Services.*

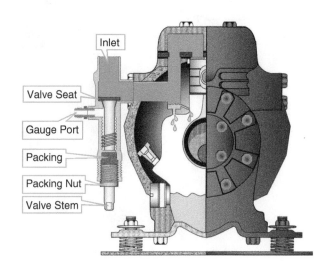

the tstat in Figure 6-5, the large pointer is the cut-out, the small pointer is the cut-in, and the area between the two points is the differential. A system with a 10°F TD would probably have the cut-out approximately 10°F to 15°F lower than the cut-in.

Figure 6-6 is an electronic version of the remote bulb tstat. It is more accurate than the mechanical models. It also has a wider differential of up to 30°F, which allows it to be used in many more applications.

For reach-in refrigerators, the coil-sensing tstats are much smaller than those used for walk-ins. Also, some manufacturers use what is called a constant cut-in tstat (Figure 6-7). This means that the cut-in is fixed, but the cut-out differential is adjustable to match the evaporator TD of the refrigerator in which it is used. The advantages of using this type of temperature control include maintaining a fixed, nonadjustable box that should be loosened a quarter turn before turning the valve stem and then tightened afterward. The cover for the valve stem is not shown, but it is called a *dust cap*. It keeps the valve stem clean. If dirt or rust collects on the valve stem, it will damage the packing as the stem is turned in and out. Eventually, this can allow refrigerant to leak past the packing whenever the valve is cracked open or front-seated. The leaking should stop when the valve is back-seated.

NOTE: One way to check for an inefficient compressor is to front-seat the suction service valve and pull the compressor into a vacuum. A rise in crankcase pressure may be an indication of bad reed valves inside the compressor. A leaking service valve would also cause the pressure to rise, giving a false diagnosis of leaking reed valves.

In Figure 6-9(a), the valve is *cracked* or *mid-seated*. By turning the valve clockwise, the valve stem rotates inward, causing the back of the valve flange to move away from its seat. System pressure rushes into the area around the stem and out the gauge port to the pressure gauge.

Figure 6-9(b) shows the suction service valve front-seated. The front of the valve flange is seated, which completely blocks off the suction line to the compressor. In this position, you can perform the following four troubleshooting procedures:

1. Check the condition of reed valves.

Run the compressor until it pulls a vacuum in the crankcase. Turn the compressor off and watch the suction gauge. If the pressure starts rising, the suction valves are leaking through, or the discharge vapor is leaking past the piston and cylinder wall.

2. Check for high back-pressure overloading.

Compressor overloading can occur on a low-temperature unit during the initial startup after installation. It can also happen in a freezer when it comes out of defrost. In both the cases, the compressor starts, runs for a short period, and then goes out on overload. The high back pressure (suction pressure) from a warm evaporator (hot pull-down) is more than the compressor can handle.

However, by starting the compressor with the suction valve front-seated, the compressor is starting unloaded. With the compressor running, open the suction service valve one turn at a time as you check the compressor's amperage. Stop opening when the amperage reaches the compressor's rated load amps (RLA), as listed on the compressor's electrical rating plate. As the box cools, the suction pressure and amperage will fall. Continue opening the valve, while monitoring RLA, until the valve is fully open. If the box cools down and if the compressor cycles off on the tstat and then restarts, the compressor is probably going to be just fine.

A compressor needs some help if it continues to trip on overload when it is starting under high suction pressure. A crankcase pressure regulator (CPR) or a pressure-limiting thermal expansion valve (TEV) will probably solve the problem. Both of these pressure limiters are described in a later section of this chapter.

3. Isolate the suction side of the compressor for oil changes.

If the suction valve is front-seated, all the refrigerant in the low-pressure (LP) part of the system is on the

FIGURE 6-9 **Service valve cracked or mid-seated (a) and front-seated (b).** *Courtesy of Refrigeration Training Services.*

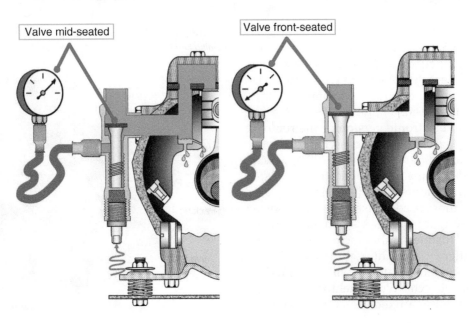

Valve mid-seated | Valve front-seated

other side of the valve flange. The small amount of remaining refrigerant between the service valve and the cylinder can be vented. The oil can now be changed, if the compressor is a semihermetic. Oil changes are discussed more fully in Chapter 9, "Retrofitting, Recovery, Evacuation, and Charging."

4. Isolate both sides for valve plate inspection and compressor replacement.

In Figure 6-10, the suction service is front-seated. If the discharge service valve is also front-seated and the pressure in the head is bled off, the compressor is completely isolated from both sides of the system.

With the compressor isolated, the valve plate can be removed for inspection or replacement.

FIGURE 6-10 **Isolated compressor.** *Courtesy of Refrigeration Training Services.*

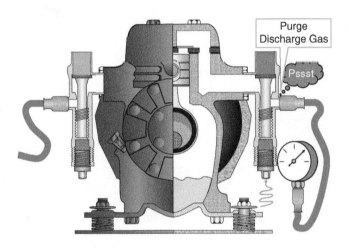

Purge Discharge Gas

Pssst

⚠ C A U T I O N

Safety Note: Never run the compressor with the discharge service valve front-seated. It will immediately blow the valve plate gasket. Or worse, it could pump dangerously high pressure into your gauge hose if it is connected to the service valve.

⚠ C A U T I O N

Safety Note: Make sure all the pressure is bled off the compressor before unbolting the compressor head. It takes only a few pounds of pressure to blow the head into the air.

The isolated compressor can be replaced without removing the refrigerant from the rest of the system. The following are the basics for this type of compressor replacement:

1. Unbolt the service valves from the compressor.
2. Remove the old compressor.
3. Set the new compressor in place.
4. Rebolt the service valves.

5. Pull a vacuum on the new compressor.
6. Back-seat the service valves.
7. You're ready to start it up.

SOLENOID VALVES FOR PUMP-DOWN AND HOT GAS BYPASS

In commercial refrigeration, valves with a solenoid (an electrical magnetic coil) are used to stop and start the flow of refrigerant (Figure 6-11). When the power to the solenoid is off, an iron plunger holds the round, disc-shaped seat closed inside the brass body. This blocks the flow of refrigerant. When the electrical coil is energized, the magnetic field inside the coil lifts the plunger. The refrigerant pressure pushes the seat up, and the refrigerant flows through the valve.

As stated in Chapter 4, "Compressors," the most effective way to prevent refrigerant migration during the off cycle is to use a pump-down solenoid. The valve stops the flow of liquid, and the compressor pulls all remaining refrigerant out of the evaporator and the suction line. A pump-down solenoid is located in the liquid line, either at the condensing unit or near the evaporator (Figure 6-12). The following is the sequence of operation for a pump-down solenoid (refer to Figures 6-13 and 6-14):

FIGURE 6-11 **Solenoid valve.** *Compliments of Parker Hannifin–Sporlan Division.*

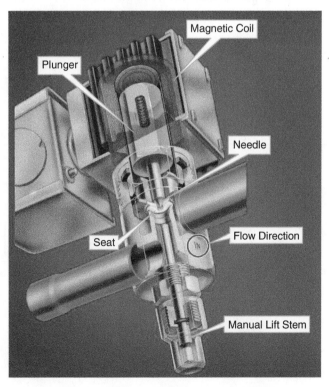

FIGURE 6-12   **Solenoid valve for pump-down.** *Photo by Dick Wirz.*

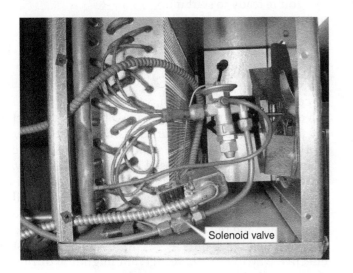

Solenoid valve

FIGURE 6-13   **Diagram of pump-down control: automatic or continuous.** *Courtesy of Refrigeration Training Services.*

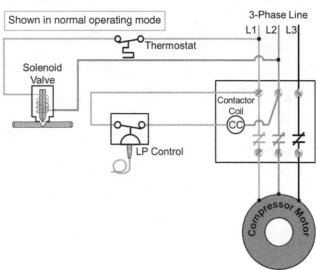

1. When the tstat is satisfied (i.e., has reached its temperature setting), the power to the solenoid is interrupted.
2. The plunger drops, and the needle and seat block the flow of liquid.
3. The compressor continues to run, pumping refrigerant out of the evaporator and suction lines and into the high side of the system.
4. When the suction pressure drops to the LP control cut-out, the compressor stops. At this point, only LP vapor remains between the solenoid valve and the inlet of the compressor.

5. When the tstat calls for refrigeration, the solenoid is energized.
6. The plunger is lifted up into the solenoid coil, opening the valve.
7. The refrigerant flows through the valve, through the evaporator, and into the suction line.
8. The suction pressure rises.
9. When the pressure reaches the LP control cut-in, the compressor starts.
10. The system is back in the refrigeration mode.

FIGURE 6-14   **Thermostat satisfied, pumps down, cycles off.** *Courtesy of Refrigeration Training Services.*

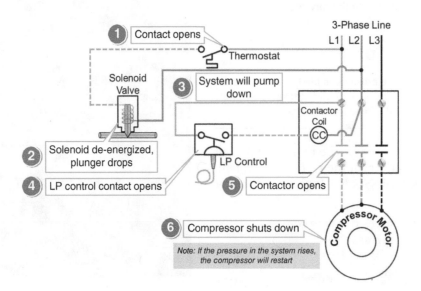

The pump-down method just described is called an *automatic* pump-down. This method is uncomplicated and effective; however, it is not without concerns. On a long piping run after pump-down, there may be enough refrigerant left to boil off and restart the compressor for a short cycle. According to most manufactures, this is acceptable as long as it is two cycles or less. The manufacturers believe it is better to have a couple of short cycles than to set the LP control so low (like below 15 psig for medium temperature or below 5 psig for low temperature) that the compressor experiences high compression ratios each time it cycles off on the pump-down.

Another concern is that once pumped down, the system pressures may equalize if there is any leakage through the solenoid valve or compressor valves. If the crankcase pressure builds up to the LP cut-in point, the compressor will start for another pump-down cycle. The compressor will pull down the pressure quickly, but the leaking valves will cause it to continually short-cycle until the tstat calls for cooling again. Obviously, if a technician observes this type of short cycling, she will investigate and determine the reason for it.

Solenoid valves can be installed on vertical lines but work better when installed on horizontal lines. They must be in the upright position if installed on the horizontal line. Solenoid valves installed upside down will not close fully.

Also, the valve must be installed in such a way that the refrigerant flow is in the proper direction. To ensure proper installation, an arrow—or the word *in*—is stamped into the valve body. If a solenoid valve is installed backward, the liquid pressure will not allow the valve to fully close. As a result, the system will not pump down, the compressor will not shut off, and the box will continue to drop in temperature.

Solenoid valves, and all valves for that matter, should be sized based on tonnage or capacity, not by pipe size. Sporlan was informed of problems with solenoids not fully closing on its E6 solenoid valves with manual lift stems and 3/8-inch pipe connections. The intent of the application was to use the manual lift stem to open the valve during evacuation without having to put power to the solenoid. This was a good idea as long as the valve capacity was not oversized for the system capacity. E6 valves have a minimum rating of over 3 tons (36,000 Btuh), which would have about a 5-hp medium-temperature compressor or a 10-hp low-temperature compressor. If the system was rated at 12,000 or 20,000 Btuh, the valve's capacity would be so oversized that it might not fully close during pump down. This situation could be very frustrating for the technician and a big waste of time until the true cause was finally diagnosed.

NOTE: Electronic expansion valves (EEVs) can be programmed to shut down so completely that they can be used to pump the system down as well as meter the refrigerant.

HOT GAS BYPASS VALVES

Hot gas bypass valves are modulating control valves that bypass enough discharge gas to the low side of the system to maintain a minimum suction pressure at the compressor.

EXAMPLE: 3

A single compressor is feeding three units. Each unit has its own tstat controlling a pump-down solenoid. If two units are shut down, the remaining evaporator may not be able to return enough refrigerant for sufficient compressor cooling.

A hot gas bypass valve would correct this problem by adding refrigerant to the suction line. A control opens the valve if the suction pressure or temperature falls below the control's setting.

Instead of entering the suction line, the discharge gas could be connected to the evaporator inlet. In this installation, the hot gas would not only prevent compressor overheating but also prevent the evaporator from becoming too cold (see Figure 6-15).

FIGURE 6-15 **Hot gas bypass to the evaporator.** *Adaption from a Sporlan diagram.*

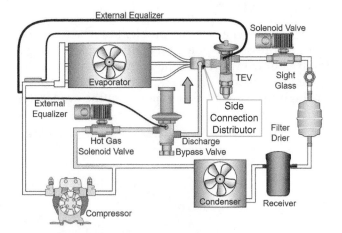

If the evaporator is far away from the compressor, or if there are multiple evaporators, the hot gas is usually returned directly into the suction line at the condensing unit. The problem with this method is that the hot gas may greatly increase the temperature of the suction vapor as it enters the compressor. This condition will cause overheating of the compressor. A solution is to cool the hot gas before it enters the suction line by mixing it with cooler refrigerant from a de-superheating expansion valve. The sensing bulb of the de-superheating valve is attached to the suction line in order for it to monitor the temperature of the vapor entering the compressor. This special valve is basically a standard expansion valve with a very high superheat setting (see Figure 6-16).

FIGURE 6-16 **Hot gas bypass to suction line with de-superheating valve.** *Adaption from a Sporlan diagram.*

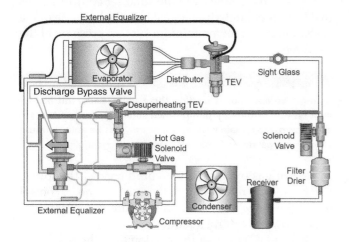

Whatever the method used, the primary result is the same: to prevent low load conditions from causing compressor overheating because of high compression ratios and short-cycling the compressor.

CRANKCASE PRESSURE REGULATORS

Earlier in this chapter, the problem of low-temperature compressors tripping on their overload during a hot pull-down was mentioned. A CPR can be used to solve this problem. This valve automatically throttles the high-pressure suction vapor from the evaporator

until the compressor can handle the load. The CPR valve should be installed in the suction line close to the compressor service valve (Figure 6-17).

Low-temperature compressors operate near their RLA as they pump LP, low-density suction vapor during the freeze cycle. However, after defrost, the suction vapor entering the cylinder is high-pressure, high-density vapor, which often overloads the compressor.

<table>
<tr><td align="center">EXAMPLE: 4</td></tr>
</table>

A –10°F reach-in ice cream freezer has an evaporator temperature of –30°F. The suction pressures of this R404A system range from 24 psig (at –10°F), when the tstat cuts in, to below 10 psig (at –30°F), when the tstat is satisfied. After defrost, the evaporator has been heated to about 50°F, which raises the suction pressure to over 100 psig on startup. The starting pressure after defrost is more than four times greater than when it starts during the normal freeze cycle (100 psig ÷ 24 psig ≈ 4 psig).

To put this into perspective, assume an air-conditioning (AC) system that starts at a 75°F evaporator temperature. The initial suction pressure of R22 at the evaporator would be about 132 psig until the valve starts throttling down. What if the suction pressure were four times its normal suction pressure, or 528 psig (132 psig × 4 psig)? Even an AC compressor would have a problem starting with suction pressure that high.

CPRs prevent overloading by restricting the flow of suction vapor if the outlet pressure of the valve (on the crankcase side) increases. In accordance with how this valve responds to pressures, Sporlan refers to this type of valve as a CRO (closes on rise of outlet) valve.

FIGURE 6-17 **Crankcase pressure regulator valves.** *Compliments of Parker Hannifin–Sporlan Division.*

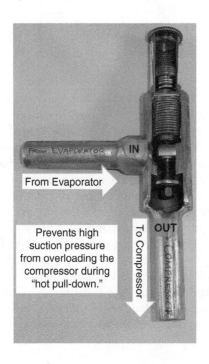

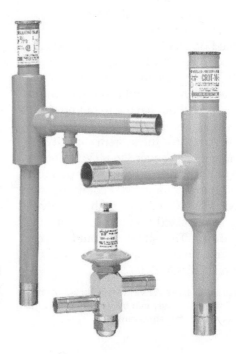

EXAMPLE: 5

If the compressor can handle only 30 psig on startup, the valve can be adjusted to allow a maximum of 30 psig to enter the crankcase, even if the suction pressure entering the valve is 100 psig.

REALITY CHECK 1

There is more than one way to prevent compressor overloading.

If the compressor can handle up to 40 psig, a pressure-limiting expansion valve will work. A Sporlan low-temperature valve that keeps suction pressure down to about 38 psig uses the designation "ZP40"; Z for freezer, P for pressure limiting, and 40 for approximate maximum pressure. At anything above 40 psig, the valve is shutting down to restrict pressure. But when pressure drops below 38 psig, the valve is wide open.

Do not install a pressure-limiting valve, or a CPR, unless it is absolutely necessary. Restricting the suction pressure during a hot pull-down means it will take longer for the evaporator temperature to come down after each start.

If the compressor is having a hard time on the initial startup after installation, the crankcase pressure can be lowered manually to keep the compressor overload from opening. As described earlier, the suction service valve can be front-seated on startup and opened slowly as the suction pressure falls and the amperage stabilizes.

Another way to lower amperage on initial startup is to lightly spray the condenser with water. Sometimes, even a small hand-pump sprayer is enough to keep the

compressor from going off on overload. Once the box has come down to temperature, the unit may not need this type of help again, even after defrost. However, if it does, you can then decide which valve to use.

The following is a procedure to determine the maximum pressure the compressor can handle. Knowing the maximum pressure can help in determining whether a ZP (low-temperature, pressure-limiting) expansion valve can handle the problem or whether a CPR valve is needed.

1. Front-seat the suction service valve.
2. Start the compressor when the evaporator is warm.
3. Open (back-seat) the suction service valve slowly and watch the compressor amperage.
4. When the compressor is drawing RLA, note the suction pressure.
5. This is the maximum pressure the compressor can handle and is also the CPR valve setting.

EVAPORATOR PRESSURE REGULATORS

The evaporator pressure regulator (EPR) is a valve used to keep the pressure up in the evaporator. It is located in the suction line, downstream of the evaporator. Because of the way this valve responds to pressure, Sporlan refers to it as an ORI (opens on rise of inlet) valve. The valve opens only when the evaporator pressure rises above the valve's setting (Figure 6-18).

FIGURE 6-18 **Evaporator pressure regulators.** *Compliments of Parker Hannifin–Sporlan Division.*

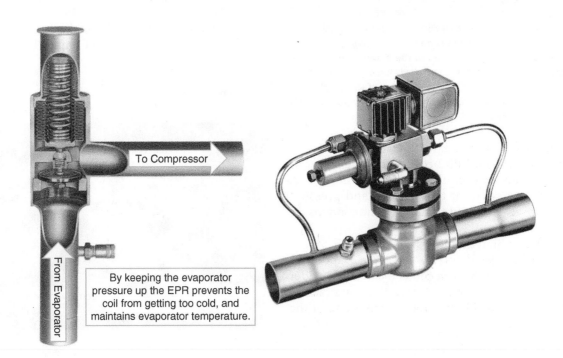

To Compressor

From Evaporator

By keeping the evaporator pressure up the EPR prevents the coil from getting too cold, and maintains evaporator temperature.

An EPR is an evaporator temperature controller. Its purpose is to keep the evaporator from becoming too cold rather than too warm. The following is an example of an EPR application where there are several refrigeration units on one compressor.

EXAMPLE: 6

A 36,000-Btuh R22 compressor has three 12,000-Btuh systems attached to it, operating at 25°F suction temperature and 49-psig pressure. Each system has a tstat and pump-down solenoid. If two of the units are satisfied, the third provides only a 12,000-Btuh load on a 36,000-Btuh compressor. The compressor acts like a big vacuum cleaner, sucking the refrigerant back faster than the one evaporator can feed it. The capacity of that evaporator would be so overpowered by the large-capacity compressor that the decrease in pressure would drop the evaporator's temperature lower than its designed 25°F. This could cause both coil and product freezing. However, an EPR would keep the evaporator pressure up to the necessary 49 psig and evaporator temperature at a minimum 25°F.

This system would probably need a hot gas bypass valve feeding into the main suction line. When only one evaporator is feeding vapor into the suction line, there would not be enough refrigerant to properly cool the compressor.

An alternate approach is to use an EPR in each system and to install a hot gas bypass valve as capacity control to keep the compressor running at all times. EPRs as tstats are discussed more fully in Chapter 10, "Supermarket Refrigeration."

The following is an example of two different temperature units on one compressor.

EXAMPLE: 7

A compressor is hooked up to a 35°F walk-in and a 55°F candy case. Whenever the compressor is running, it is trying to pull the suction pressure and temperature of the candy case down to the level of the walk-in. If it did, the candy case would be about as cold as the walk-in. An EPR in the outlet of the candy case keeps the pressure (and temperature) up in the evaporator so it will maintain a 55°F case.

In this example, the walk-in system probably does not need an EPR because it is the larger of the two systems and requires more run time. Also, the walk-in's evaporator temperature and pressure are more closely matched to those of the condensing unit (Figure 6-19).

To maintain high humidity, some cases, like those used in meat and deli displays, have large evaporators and low air movement. The coil is located just under the pans that display the product. If the evaporator is allowed to get too cold, the product can freeze or the case humidity can drop.

FIGURE 6-19 **Evaporator pressure regulator in high-temperature case.** *Courtesy of Refrigeration Training Services.*

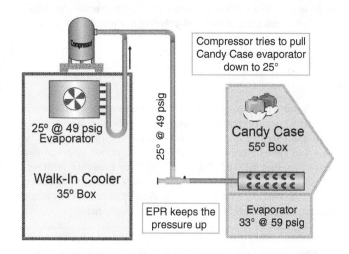

EXAMPLE: 8

A 36°F deli case requires an evaporator TD of 6°F to maintain high humidity inside the case. The coil temperature should stay at 30°F (6°F TD = 36°F – 30°F). Setting the EPR at 70 psig (R404A) would maintain the required 30°F evaporator temperature. The EPR would maintain both coil temperature and case humidity.

Some cases need to be shut down or isolated from other cases on the same system. If the system uses a liquid-line pump-down solenoid, the evaporator would become very cold during the long pull-down period. A pump-down solenoid in the suction line would solve this problem. Fortunately, there are EPRs that have a suction-line pump-down solenoid as part of the valve (see Figure 6-18, right side). Electronic EPRs like the one in Figure 6-20 can also be programmed to completely shut down the suction line when necessary. EPRs are used in parallel rack systems and will be covered in even more detail in Chapter 10.

FIGURE 6-20 **Electronic evaporator pressure regulator (EEPR).** *Photo by Dick Wirz.*

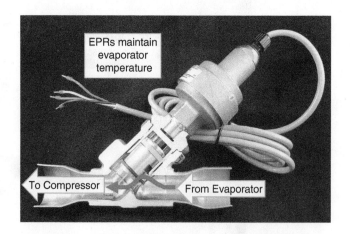

LOW-PRESSURE CONTROLS

An LP control opens the electrical circuit when the system pressure drops below the cut-out pressure (Figure 6-21). It also closes the circuit when the pressure rises to the cut-in setting. LP controls are primarily used for the following purposes:

1. To protect the compressor from damage due to loss of refrigerant
2. To turn the compressor off at the end of a pump-down cycle
3. As a tstat to regulate the temperature of a refrigerated box

The following are the primary concerns when setting an LP control:

1. Short-cycling of the compressor (cut-in and cut-out too close).
2. Compressor overheating before it shuts off (cut-in and cut-out too far apart).
3. Ambient conditions affecting the control's operation. For example, an LP control on an outdoor unit

is set to cut in at 50 psig or 26°F for an R22 system. When the ambient temperature drops to 0°F, the bellows on the control react to the lower ambient rather than the system pressure.

See Table 6-1 for the recommended LP settings for outdoor units.

NOTE: **For units located indoors, use the settings for 30°F minimum outdoor ambient.**

NOTE: **Many controls are labeled "high event" and "low event" instead of cut-in and cut-out.**

Certain types of refrigeration equipment utilize an LP control to maintain box temperature, in addition to being a safety in case of refrigerant loss. This method is used on some reach-in refrigerators and self-contained display cases. Always check with the manufacturer for the proper settings of the control. Design engineers use a specific combination of cut-in, cut-out, and differential to maintain the case temperature on their particular unit.

Having an LP control act as a temperature control is very similar to using a tstat that has its bulb attached to the suction line or in the evaporator coil. The LP control senses the pressure of the refrigerant, which is closely related to the evaporator temperature. This arrangement maintains very accurate case temperature and allows the evaporator to fully self-defrost before the compressor restarts.

Only TEV systems can use LP controls as temperature controls because their high- and low-side pressures do not equalize during the off cycle. When the LP control cut-out is reached, the compressor shuts off. Without the compressor running, the evaporator pressure pushes up on the valve diaphragm, closing the valve by pulling the valve needle tight into its seat. Because there is no equalizing of high- to low-side pressure through the valve, the only rise in suction-line pressure is from the refrigerant boiling as the box temperature rises. When the evaporator pressure climbs to the control's cut-in point, the compressor starts.

FIGURE 6-21 **Low-pressure controls.** *Photo by Dick Wirz.*

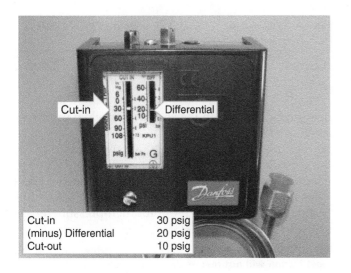

Cut-in	30 psig
(minus) Differential	20 psig
Cut-out	10 psig

TABLE 6-1 **Low-pressure control settings for outdoor condensing units**

Minimum Outdoor Ambient (°F)	R22 (+35°F Refrigerator) (psig)		R404A–R507 (+35°F Refrigerator) (psig)		R404A–R507 (–10°F Freezer) (psig)		R134a (+35°F Refrigerator) (psig)	
	Cut-in	Cut-out	Cut-in	Cut-out	Cut-in	Cut-out	Cut-in	Cut-out
30°F	50	30	60	40	20	5	25	10
10°F	30	10	40	20	20	5	12	1
0°F	20	1	25	5	20	5	8	1
–10°F	15	1	20	1	20	5	—	—
–20°F	10	1	15	1	15	1	—	—

FIGURE 6-22 **Illustration of how a low-pressure control can be used as a thermostat. Example is a reach-in refrigerator using R134a.** *Courtesy of Refrigeration Training Services.*

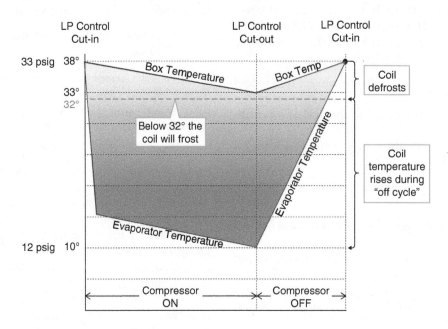

The equipment manufacturer should be able to provide the proper control setting for its equipment. However, if this information is not available, you may want to try the following general settings for reach-in refrigerators:

1. Set the cut-in pressure equivalent to the box temperature you want to maintain.
2. Set the cut-out pressure equivalent to 28°F below the box temperature.

EXAMPLE: 9

Assume a reach-in using R134a with a design box temperature of 38°F (Figure 6-22). The LP control cut-in would be 33 psig (38°F), and the cut-out would be 12 psig (10°F). The 28°F difference (38°F – 10°F) comes from the sum of the following:

- 5°F box temperature swing (38°F – 33°F)
- 20°F evaporator TD (for reach-ins)
- 3°F temperature equivalent of a 2-psig pressure drop through the evaporator

The sum of the above is the LP control's differential (5°F + 20°F + 3°F = 28°F) between cut-in and cut-out.

HIGH-PRESSURE CONTROLS

Dirty condensers, snow, fan problems, overcharge, and air in the system are some of the more common causes for system high pressure. To protect the compressor, the first line of defense is the high-pressure control that

TABLE 6-2 **High-pressure maximum control settings for outdoor condensing units**

Maximum Condensing Temperature	R22	R404A–R507	R134a
155°F	400 psig	475 psig	275 psig

opens a circuit to stop the compressor when pressures are above the control setting.

For most refrigeration units, anything over a maximum condensing temperature of 155°F would damage the compressor. Therefore, the maximum cut-out pressures should be those provided in Table 6-2. The cut-in is about 50 psig below cut-out for high-pressure refrigerants like R22 (Figure 6-23).

NOTE: **On manual-reset high-pressure controls, pushing the reset button will not restart the compressor until the system pressure has dropped to the control's cut-in pressure.**

OIL SEPARATORS

Oil separators are encountered on some large compressors over 10 hp and on parallel rack compressor systems. Oil separators are being introduced in this chapter in preparation for the systems to be discussed in Chapter 10 on supermarket refrigeration.

Assume a compressor and its receiver are located in the basement of a store. The compressor discharge line runs three stories up to the roof condenser. The liquid line leaving the condenser has to come back

FIGURE 6-23   **High-pressure control with manual reset.**
Photo by Dick Wirz.

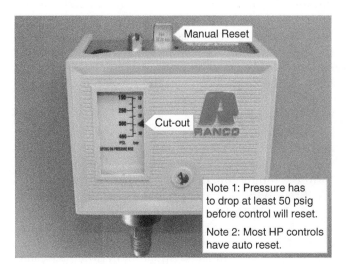

down to the receiver. The liquid then goes to some refrigerated cases on the other side of the store over 100 feet away. The refrigerant and the oil eventually return to the compressor in the suction line.

During the initial startup, oil may have to be added to the crankcase to make up for the oil both trapped and traveling through the long run of piping. As a result, this system would probably have too much oil sometimes and too little at other times.

Oil separators remove, or separate, the oil from the discharge gas almost immediately after it leaves the compressor. The high-pressure vapor is run through a canister of screens or baffles that sling the oil out of the discharge gas before it leaves the separator. A float inside the separator maintains the level of oil in the crankcase (Figure 6-24).

There is usually one problem area for oil separators: debris in the float valve. If the float cannot fully

close, the crankcase will overfill with oil. An oil filter assembly will prevent this problem. Therefore, it is recommended with each oil separator installation.

Oil separators are only 50 percent to 90 percent efficient, which means some oil gets into the system piping and must be returned to the compressor. For this reason, installing an oil separator is not a cure for poorly installed refrigeration piping. Oil separators are used in parallel rack systems and will be covered in even more detail in Chapter 10.

OIL SAFETY CONTROLS

Oil safety controls, also called oil failure controls, monitor the net oil pressure. These controls measure the difference between the suction pressure in the crankcase and the oil pump discharge pressure to determine net oil pressure (Figures 6-25 through 6-27).

FIGURE 6-25   **Oil pressure control mounted on a compressor.** *Courtesy of Emerson Education–Copeland.*

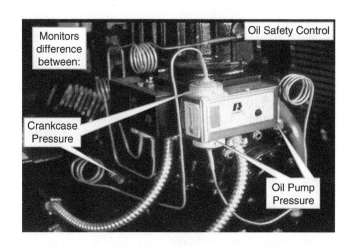

FIGURE 6-24   **Oil separator.** *Courtesy of Refrigeration Training Services.*

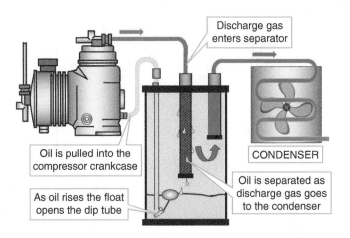

FIGURE 6-26   **Standard oil pressure control.** *Photo by Dick Wirz.*

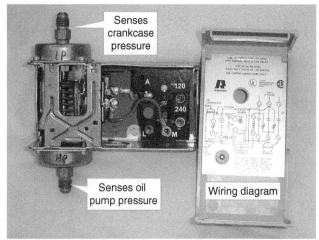

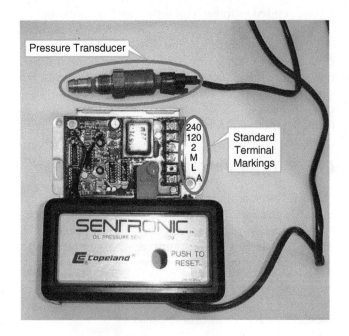

FIGURE 6-27 **Solid-state oil pressure control.** *Photo by Dick Wirz.*

The proper net oil pressure must be somewhere between 10 psig and 60 psig. Table 6-3 illustrates how both suction pressure and oil pump discharge pressure are used to determine the net oil pressure.

The last row in Table 6-3 is an example of low net oil pressure (10 psig). It is evident the oil pump is not moving enough oil. Low oil pressure is usually due to a low level of oil in the crankcase, too much foam in the oil, or a bad pump.

 REALITY CHECK 2

When a tripped oil safety control is reset and the compressor starts running, some technicians' first thought is the control is bad because the compressor seems to be operating correctly. Actually, the fact that the control tripped shows that it is just doing its job. It is up to the technician to figure out why it tripped.

Another incorrect assumption is to condemn the oil pump without actually checking the net oil pressure. Oil pumps are very rugged and they seldom fail.

TABLE 6-3 **Examples of suction pressures and net oil pressure**

Suction Pressure (*A*)	Pump Discharge Pressure (*B*)	Net Oil Pressure (*B – A*)
25	50	25
50	75	25
75	125	50
100	110	10

It is not unusual for the net oil pressure to drop below 10 psig during compressor startup. To prevent *nuisance trips*, the control has a built-in 120-second timer that delays cutting off the compressor. If the pressure has not reached at least 10 psig in that time period, the oil safety control shuts the compressor off, and the control must be reset manually. An optional set of contacts on the control can be wired to a remote alarm. The alarm will alert the customer if the compressor has gone out on oil failure.

The timer circuit (also known as the *timer* or *time delay*) is the key component (see Figure 6-28). As shown in Figure 6-28, it requires two hot legs to perform the timing function. One of the legs comes from L1, through the operating controls, through terminals L and M, and finally through terminal 2 to energize one side of the timing circuit. The common leg comes from L2, but it is on the load side of the contactor so that the contactor must be pulled in to provide power on that leg. The oil pressure opens and closes a switch in the common leg, which is a normally closed (NC) set of contacts. In Figure 6-28, the switch is open because the compressor is running and the oil pressure is high enough to keep the contacts open. If the oil pressure drops below 10 psig, the switch will close, and the timer is energized. If after 2 minutes the pressure is still low, contacts L and M open, which de-energizes the compressor contactor (see Figure 6-29).

Occasionally, an oil safety control will trip because of a problem other than the condition it was designed to prevent. The oil pressure control will trip if the compressor stops running while its contactor is energized. For instance, if the compressor goes out on an internal overload, the compressor will stop without de-energizing the compressor contactor. The pump, which is attached to the compressor crankshaft and develops pressure from the rotation of the crankshaft, will not

FIGURE 6-28 **Oil failure control (OFC): normal operation, compressor running.** *Courtesy of Refrigeration Training Services.*

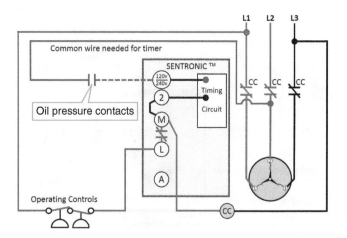

FIGURE 6-29   **Low oil pressure: OFC times out, contactor opens, compressor shuts off.** *Courtesy of Refrigeration Training Services.*

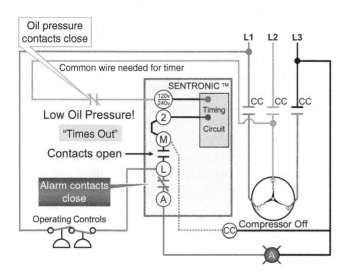

FIGURE 6-30   **Current-sensing relay.** *Photo by Dick Wirz.*

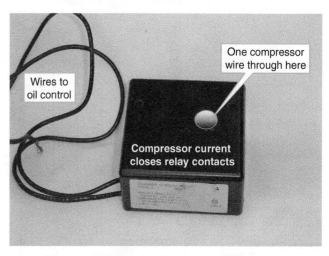

FIGURE 6-31   **Cutaway view of a current-sensing relay.** *Photo by Dick Wirz.*

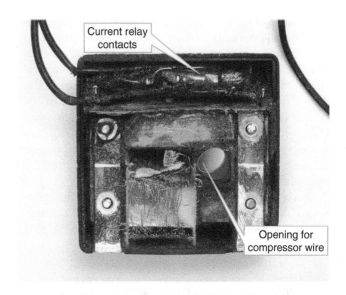

pump. However, the contactor remains energized so there is still power to both sides of the timer. Because there is no oil pressure and the control is energized, the control contacts will open in about 120 seconds, tripping the oil safety reset. Even if the condition corrects itself, the compressor cannot restart without the safety control being manually reset.

Assume the compressor overload problem was a brown out, or maybe due to too much pressure coming back to the compressor during an unusually hot pull-down, or some other temporary condition. By the time the technician arrives and resets the control, the compressor may start and run fine because the original condition no longer exists. However, the customer may be very upset because his equipment went down, and now there seems to be no reason for it. Worse, the technician cannot assure the customer it will not happen again.

A control called a *current-sensing relay* (CSR) can prevent this type of nuisance problem (see Figures 6-30 and 6-31). The CSR is a small black plastic box with a hole in its center and two wires coming out the sides.

The CSR is connected in series to one leg of the oil safety control. (See the oil failure control wiring diagram in Figure 6-32.) The normally open (NO) set of contacts inside the CSR is represented by the switch in the wire from L2 to the common side of the timer. One of the power wires feeding the compressor is fed through the hole in the middle of the box. In Figure 6-32, this is represented by what looks like a square block around the leg of L1 between the contactor and the compressor. As long as the compressor is running, the magnetic field around the wire energizes the CSR, closing the contacts.

FIGURE 6-32   **Add current-sensing relay (CSR): normal operation, compressor running.** *Courtesy of Refrigeration Training Services.*

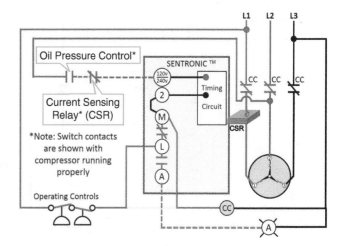

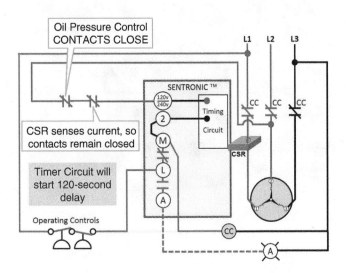

FIGURE 6-33 **Low oil pressure: OFC contacts close, CSR contacts remain closed, timer starts.** *Courtesy of Refrigeration Training Services.*

Figure 6-33 shows the initial stage of low oil pressure. The oil pressure contacts close because the net oil pressure has dropped below 10 psig. The switch contacts in the CSR are also closed because it is sensing the current of the compressor as it continues to run. If the oil pressure remains low, in about 120 seconds, the control will time out, opening the contacts between L and M, which cuts power to the contactor coil (CC). The compressor contacts will open, and the compressor will shut down. The oil failure control (OFC) must be manually reset to restart the compressor.

If the compressor goes out on overload, the compressor will stop, and the magnetic field around the wire will collapse. As shown in Figure 6-34, this

FIGURE 6-34 **Brown out, compressor off on internal overload: CSR does not allow OFC to trip on oil failure.** *Courtesy of Refrigeration Training Services.*

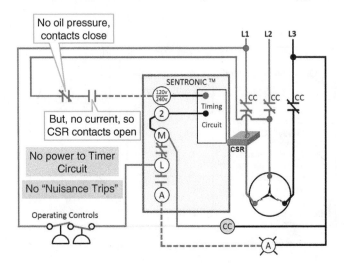

allows the CSR contacts to open the circuit to the timer, preventing the control from timing out and tripping the manual reset. As a result, the compressor will restart on its own when the overload problem is corrected. The customer will not have to wait for someone to reset the control, and the technician will not have a reason to suspect an oil problem.

Following are a few things to check after resetting a tripped oil safety control:

- Electrical diagnosis
  - Check voltage to the contactor.
  - Check voltage across the contacts of the energized contactor for signs of pitted contacts. (Zero volt is perfect, but any voltage indicates a problem.)
  - Check voltage at the compressor on startup.
  - Check the compressor amperage.

- Refrigeration system diagnosis
  - Monitor the net oil pressure and oil level through a complete cycle.
  - On a freezer, check the oil level just before the system goes into defrost.
  - Look for evidence of a flooded start or flood-back to the compressor.
  - Look for abnormally high or low suction pressures.
  - Check for short-cycling.

NOTE: Short-cycling of the compressor pumps the oil out into the system, without allowing enough compressor run time to return the oil to the crankcase. If the cut-in and cut-out of an LP control is set too close (small differential), this short-cycling can occur during pump-down.

RECEIVERS

A receiver is just a storage tank for refrigerant. All TEV systems must have some reserve capacity to make sure there is enough refrigerant to the TEV under all conditions.

The entire charge of the system should fill only 80 percent of the receiver. The receiver size depends on system requirements. Long refrigeration runs may require a greater receiver capacity, especially if the pump-down solenoid is located at the condensing unit. In that type of installation, all the refrigerant in the liquid line has to be pumped back to the receiver before the compressor shuts down. Also, units with head pressure–regulating (HPR) valves need more refrigerant to flood the condenser during cold weather than they need for refrigeration. During warm weather, the receiver is used to hold the excess refrigerant.

What is considered "short cycling"?

Compressor runs less than 2 minutes.

In an effort to prevent short-cycling during a pump-down sequence, some technicians set the LP cut-out just above 0 psig. Once the unit pumps down and shuts off, it will definitely stay off. However, compressor manufacturers warn against this practice because the duration of the pump-down will take so long that the compressor could overheat. Therefore, they recommend setting the LP control cut-out as high as possible. In addition, it is perfectly acceptable for the compressor to cycle on and off a few times before staying off.

For pump-down

Maximum of two additional cycles after initial pump-down.

The king valve, the service valve on the receiver outlet, is very useful for pumping down a system for service (Figure 6-35). To do so, front-seat the king valve with the compressor running. This shuts off the flow of refrigerant to the liquid line. Turn the compressor off when the suction pressure has dropped to 0 psig. All refrigerant should now be removed from the liquid line, the evaporator, and the suction line.

This procedure is useful for replacing filter driers, the evaporator, and the solenoid valve; for adjusting LP controls; or for repairing any part or leak from the receiver to the compressor. Of course, evacuate this part of the system before reopening the king valve.

Some larger receivers actually have something called a *queen valve* (Figure 6-36). It is on the receiver where the liquid line from the condenser enters the receiver. Shutting this valve off isolates the receiver from the high side. This prevents having to recover the entire system charge in the event it is necessary to open the high side of the system for repairs.

Safety Note: Never close off both valves if the receiver is filled with liquid. Upon a rise in the surrounding air temperature, the expansion of the liquid could cause the receiver to explode.

FIGURE 6-36 **Receiver with both king and queen valves.** *Photo by Jess Lukin.*

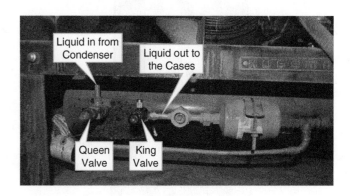

FIGURE 6-35 **Receiver and king valve.** *Courtesy of Refrigeration Training Services.*

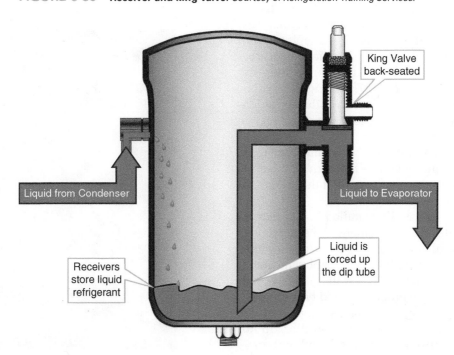

ACCUMULATORS

In low-temperature systems, floodback is very likely during a hot pull-down, especially after defrost. The accumulator is located in the suction line near the compressor, protecting it from liquid slugging (Figure 6-37).

FIGURE 6-37 **Suction-line accumulator.** *Courtesy of Refrigeration Training Services.*

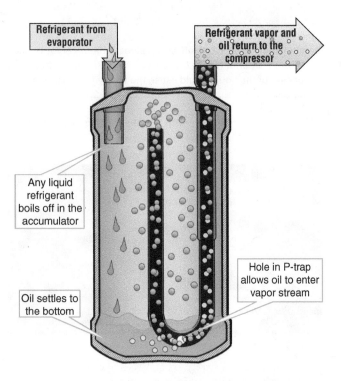

An accumulator gathers, or accumulates, liquid refrigerant from the suction line near the compressor. If liquid is present in the suction line, it will fall to the bottom of the accumulator. Ambient air warms the outside surface of the accumulator, boiling the liquid refrigerant into a vapor. The refrigerant vapor enters the U-tube opening at the top of the accumulator, which directs the vapor toward the suction service valve of the compressor.

Oil settles at the bottom of the accumulator tank. As it enters a hole in the side of the U-tube, the oil is lifted, with the refrigerant vapor, out of the accumulator and back to the compressor.

This accessory is usually installed in a system as part of a factory design requirement. If a system does not have an accumulator but flooding is occurring after defrost or at startup, check for system problems first. If an accumulator is required, discuss the situation with the manufacturer for the proper size and type to use.

Accumulators normally sweat because they are at cold suction-line temperatures. However, do not insulate them because heat from the ambient air is needed to vaporize any liquid in the tank. Accumulators are made of steel and will eventually rust from the continuous sweating. Severe rusting can cause pinhole leaks in the accumulator shell. Therefore, if the system has a refrigerant leak, the accumulator should be checked.

FILTER DRIERS

One of the first things a technician should check on an AC service call is the air filter. This is because technicians understand how the condition of the air filter affects system airflow and impacts overall system performance. Likewise, refrigerant filter driers perform just as important a function inside the refrigeration system, yet few service technicians check them. Admittedly, a filter drier in a properly installed and evacuated system should be good for many years (Figure 6-38). However, the filter drier should be checked as a standard part of every technician's service procedures and system maintenance.

A refrigeration filter drier has screens and fiber filters to trap particles traveling through the system. In addition, it has desiccants to trap contaminants such as acids, wax, and, most importantly, moisture. When moisture is combined with refrigerant, oil, and heat, it forms sludge and acids that cause system failures. Silica gel is the moisture-absorbing compound that makes up the majority of most filter drier desiccants.

FIGURE 6-38 **Filter drier.** *Compliments of Parker Hannifin–Sporlan Division.*

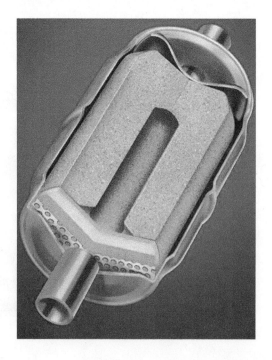

When a technician adds refrigerant or opens the system for repairs, there is a possibility of moisture entering the system. As the refrigerant circulates through the system, moisture is trapped in the filter drier's desiccant. At low liquid temperatures, the filter drier can hold more moisture than at higher temperatures. A cool drier that is near its water-holding capacity will give up some of its moisture when the liquid line warms up. That is why on the first warm day of summer a system can experience moisture-related problems, such as a stuck expansion valve from ice in the orifice.

Filter driers do not necessarily trap particles or moisture on the first pass, nor do they stay trapped when the filter drier has reached its capacity. A restriction caused by debris in the system is easy to determine.

If there is too great a pressure drop across the filter, it is due to debris blocking the fluid flow. Suction-line filter driers usually have pressure taps on both the inlet and the outlet so the pressure drop can be measured. As always, refer to the manufacturer's specifications, but the usual maximum for permanently installed suction filter driers is 2 psig for refrigerators and 1 psig for freezers. Pressure drop on the suction line is more critical than on the liquid line. Most equipment manufacturers recommend suction-line filter driers for cleanup after a compressor burnout. They also recommend replacing the filter drier when the pressure drop across it is excessive. The filter drier should be removed when the system is clean. In no case should a temporary suction filter drier be left in the system for more than three days. If suction filters are installed permanently, they must be monitored regularly for pressure drop.

In addition to cleanup after compressor burnouts, suction-line filter driers are very good for moisture removal. Some manufacturers claim that suction-line filter driers are actually more effective at removing system moisture than those in the liquid line. Moisture readily condenses in the cold suction line and is easily trapped by the suction filter drier.

Liquid-line driers must also be replaced when the pressure drop through them is too high. However, it is difficult to measure the pressure drop across liquid-line filter driers because they do not come with pressure taps. Measuring the temperature difference between filter drier inlet and outlet is almost as good as taking a pressure difference. If there is a temperature difference, there is a pressure drop. To get an accurate reading, technicians should use a good dual-temperature electronic thermometer with two thermistors.

 T . R . O . T

Filter driers must be replaced whenever the system is opened for repairs.

This rule works well for high-pressure refrigerants (R22 and R404A) at about 110°F condensing. A 3°F difference is approximately equal to a 10-psig pressure drop. The 3°F rule also works for lower-pressure refrigerants like R134a. As always, refer to the manufacturer's specifications for the exact figures.

Some technicians say they check filter drier restrictions by placing their hands on the inlet and outlet of the filter drier. In fact, tests have proven that most people cannot detect a temperature difference of less than 10°F between their two hands. That would translate to about a 30-psig difference on an R22 system.

Pressure drop across a filter indicates a buildup of system trash, but what about moisture? Don't depend on pressure drop to indicate moisture. There may be no pressure drop even when the filter drier is filled to its capacity with water. A moisture indicator–type sight glass will help. However, as stated earlier, the system may show dry when checked in the cool morning and then show wet when the moisture is released during the hot afternoon.

 R E A L I T Y   C H E C K   3

If the filter drier is hard to get to, most service technicians will be reluctant to check it, much less replace it.

 T . R . O . T

For filter drier replacement
Liquid-line filter driers

- Whenever the system is opened for repairs
- When sight glass indicates moisture in the system
- Whenever there is more than a 3°F temperature drop across the filter drier
- When in doubt, change it out

Permanent suction-line filter driers

- 2-psig drop on medium-temperature units
- 1-psig drop on low-temperature units

Installing the liquid-line filter drier just before the TEV is considered the best location. However, in some refrigerated cases filled with product, it can be very difficult to get to the filter drier. In such instances, install it at the condensing unit, where it will be easier to check and replace.

Flare-type filter driers are easy to replace. For that reason, they are preferred by service technicians who must maintain systems. However, flares must be made correctly. A technician is wise to purchase the best flaring tool possible. It will pay for itself many times over by preventing expensive refrigerant leaks.

Sweat-type filter drier maintenance requires proper procedures. When replacing a sweat-type filter drier, it is best to cut the old one out. Technicians must not attempt to unsweat it. Torch heat will release the moisture trapped in the filter drier back into the system.

When installing a new filter drier, remember that overheating the filter shell can damage internal parts in the filter. Technicians can avoid this problem by putting a damp rag around the filter body and making sure the flame is pointed away from the filter drier.

Replaceable core-type filter driers can be installed in the suction or liquid line (Figure 6-39). Replacement is easy; just pump down the system and unbolt the housing cover. There are different combinations of desiccant cartridges to choose from, based on what kind of contaminant removal is necessary—moisture, acid, wax, or just filtration of installation debris.

FIGURE 6-39 **Replaceable core filter drier.** *Compliments of Parker Hannifin–Sporlan Division.*

The numbers on filter driers usually show the amount of desiccant in cubic inches and the line size in eighths of an inch. For instance, in Figure 6-40, the Sporlan filter drier C-163-S has 16 cubic inches of desiccant. The line size is three-eighths of an inch, or 3/8 inch OD pipe size. Only sweat driers have the letter S at the end; flare driers have no letter.

Refer to the filter drier manufacturer's chart for proper sizing. It is better to be too large than too small. However, cap tube systems are usually limited to only 3-cubic-inch filter size. A larger filter drier in this type of system could act as a receiver and alter the critical charge of the unit. Note that the use of polyolester (POE) oil in refrigeration systems has increased moisture-related problems in refrigeration systems. Therefore, replacing filter driers and adding sight glasses with moisture indicators have become an important part of good refrigeration installation, service, and maintenance.

SIGHT GLASS

A sight glass with a moisture indicator is recommended on all TEV systems. The sight glass shows the flow of refrigerant at that point where it is installed in the liquid line. When a moisture indicator is part of the sight glass, additional information is available about whether the system is dry or has moisture in it (Figure 6-41).

FIGURE 6-41 **Sight glass with moisture indicator.** *Courtesy of Refrigeration Training Services.*

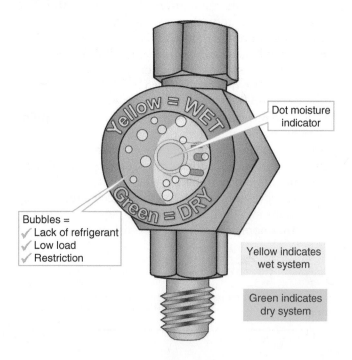

According to factory specifications, the sight glass can be installed either before or after the filter drier. Some technicians believe it is essential to install the sight glass after the filter drier because if the filter drier is restricted, it will show bubbles.

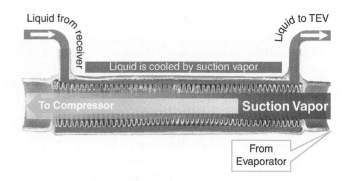

| EXAMPLE: 10 |

R22 at 226 psig has a condensing temperature of 110°F. If the restricted filter drier drops the liquid pressure by 10 psig to 216 psig, it will boil at 107°F. The 110°F liquid will begin flashing off, and bubbles will probably appear in the sight glass. However, if the liquid were subcooled only 5°F to 105°F, no bubbles would show in the sight glass even with a 10-psig pressure drop.

REALITY CHECK 4

A restricted filter drier will not show bubbles if there is enough subcooling.

REALITY CHECK 5

Moisture reaches equilibrium in the refrigeration system and will show up on the moisture indicator, no matter where it is placed.

If the sight glass is on the condensing unit and the evaporator is above the unit or far away, the factory suggests a second sight glass at the evaporator. As stated in Chapter 5, "Metering Devices," technicians must be able to verify that the TEV has a full column of liquid. Two sight glasses on a system will not increase liquid-line pressure drop but will increase the service technician's accuracy when diagnosing the system's operation.

Some technicians add refrigerant whenever bubbles appear in the sight glass. This is not good practice because it can result in an overcharged system. It is not unusual for the sight glass to show bubbles during startup and low load conditions. These are temporary situations and do not necessarily require the addition of refrigerant.

When a system is opened, the moisture indicator turns color to show a wet system. Even when a proper vacuum is pulled, the indicator will not show completely dry. This is normal, and it may take up to 12 hours of the system running for the moisture indicator to show dry. If it does not, the technician should replace the filter drier.

HEAT EXCHANGERS

Chapter 3, "Condensers," covered how pressure drop and heat added to liquid lines can cause the refrigerant to flash off before it gets to the TEV.

To prevent flashing, the liquid can be subcooled by a heat exchanger located in the suction line leaving the evaporator (Figure 6-42).

In addition, lowering the temperature of liquid refrigerant actually increases the capacity of a TEV. System designers sometimes take advantage of this added feature of heat exchangers to use a smaller TEV to feed the unit's evaporator.

In some cap tube systems, the manufacturer solders the capillary tube to a section of suction line. This arrangement acts as a heat exchanger that helps maintain consistent metering in the cap tube even during high-ambient conditions such as a hot kitchen.

VIBRATION ELIMINATOR

Semihermetic compressors actually rock when they start up. This is similar to the twisting motion of a car engine when starting or accelerating. A vibration eliminator, or vibration absorber, absorbs twisting and vibration in order to prevent damage and leaks in the refrigeration lines connected to the compressor (Figure 6-43).

FIGURE 6-43 **Vibration eliminator.** *Photo by Dick Wirz.*

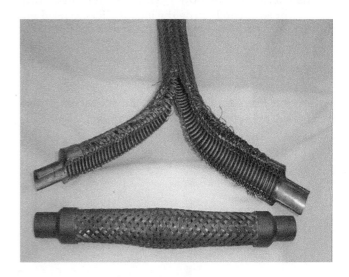

The vibration eliminator is a corrugated copper pipe that looks like the bellows of an accordion. For added strength, a mesh of woven copper wires covers the flexible tubing. To absorb the twisting motion on startup, the vibration eliminator must always be installed in-line (parallel) with the crankshaft. If it is at a right angle (perpendicular) to the crankshaft, the rocking of the compressor on startup will compress the absorber like the accordion bellows it resembles. Unfortunately, this will cause the thin flexible tubing to break.

When a technician checks a system for refrigerant leaks, it is a good idea for her to always include the vibration eliminator in the leak check.

SUMMARY

AC tstats only have a 2°F temperature swing because a wider difference would be noticeable to the people in the room. However, refrigeration tstats have at least a 3°F to 5°F temperature swing, which allows enough time for most medium-temperature evaporators to self-defrost. The majority of refrigeration tstats sense air temperature, but some sense evaporator temperature.

Compressor service valves are useful for checking pressure and the condition of its own reed valves and for isolating the compressor in order to perform repairs or maintenance. Remember, a technician should never front-seat a discharge service valve while the compressor is running.

Solenoid valves are often used to pump down a system before the compressor shuts off. Pump-downs prevent refrigerant migration during the off cycle. Therefore, solenoid valves should be installed on all remote refrigeration condensing units. A pump-down solenoid is not needed on systems using an EEV because an EEV completely stops the flow of refrigerant when powered down.

Hot gas bypass valves inject refrigerant into the low side of the system to prevent compressor overheating during low-load conditions. They can also be used to prevent coil freeze-ups by keeping evaporator pressures and temperatures up.

During hot pull-downs, CPRs automatically throttle the suction vapor flow from the evaporator until the compressor can handle the load. CPRs are used on low-temperature compressors.

EPRs keep the suction pressure up in the evaporator. As long as the suction pressure does not drop, neither will the evaporator temperature. EPRs maintain case temperatures by maintaining the case's evaporator temperature.

LP controls shut the compressor off when the suction pressure falls. This feature is useful during system pump-down or when there is a loss of refrigerant. LP controls can also be used as a temperature control for some reach-ins and refrigerated cases.

High-pressure controls shut off the compressor when the head pressure rises. The primary causes of high head pressure are high ambients, dirty condensers, and condenser motor problems.

Oil safety controls protect compressors with oil pumps from lack of lubrication. The primary reasons for low oil pressure are system-related problems such as floodback, short-cycling, coil freezing, and improper piping. Seldom is the oil pump or the oil failure control defective.

Oil separators return the majority of oil entrained in the discharge vapor back into the compressor crankcase. This accessory is primarily used on large systems with remote condensers and long piping runs.

Receivers are refrigerant storage tanks for systems with a TEV. The size depends on the amount of refrigerant charge in the system. The king valve is useful for pumping the system down before making repairs in the liquid and suction lines.

Accumulators protect low-temperature compressors from liquid floodback during hot pull-down after defrost.

SERVICE SCENARIOS

Following are some actual service scenarios. Most happened many years ago, and some facts may be a little fuzzy. However, they are important on-the-job experiences, and the lessons learned are worth sharing.

SERVICE SCENARIO #1

SMOKING SOLENOID COIL

By Dick Wirz

I was doing the startup and final leak check on a cheese display case in a new wine and cheese shop. The thermostat was mounted on the back of the case near the electrical junction box, but the liquid line solenoid was under the case. My leak detector signaled a leak on the threaded packing nut of the plunger housing on the top of the valve. To tighten it, I needed a wrench on the packing nut and another on the valve body. The solenoid's electric coil was in the way, and I needed to remove it from the valve to get my wrench on the nut. I would have to shut off the power, or at least turn up the thermostat, but that would mean crawling out from under the case. So I removed the small hold-down screw from the coil, pulled it off, and set it to the side while I tightened the nut.

That's when the coil started smoking, and I came out from under the case coughing and confused. The job electrician happened to be walking by and asked what had happened. I told him the coil just burned up when I took it off the solenoid. He said, "I bet you forgot to put a screw driver inside the coil while the power was on." I asked him why that was important. He explained that the solenoid coil is an electromagnet when power is applied. If there is nothing for it to magnetize, the electric field goes crazy, and the coil quickly burns up. I'm sure there is a more technical explanation than that, but it was good enough for me. In the future, I will not make that mistake again.

Lesson Learned: Make sure the solenoid coil is mounted on the valve when power is on. If power must be applied to a coil that is not on a valve, then insert a screwdriver, bolt, or some other metallic object for the coil to magnetize before turning on the power.

SERVICE SCENARIO #2

GO WITH THE FLOW

By Dick Wirz

Service Call: New walk-in cooler is too cold.

Equipment: An 8 × 15 walk-in cooler with 5 display doors and a condensing unit on top of the box.

Preliminary Checks: Box temperature is 25°F. Thermostat set for 35°F, its contacts are open, and the solenoid valve is de-energized. However, the condensing unit is still running and cooling the box.

How Is It Supposed to Work? This unit has a thermostat and liquid line pump-down solenoid next to the evaporator in the walk-in. They control the remote condensing unit on top of the box in response to temperature in the walk-in. When the box is down to 35°F, the tstat contacts open, the solenoid valve coil is de-energized, the plunger in the valve drops to block the flow of liquid refrigerant to the TEV, the suction pressure falls as the unit pumps down, the low pressure control opens when the cut-out pressure is reached, and the condensing unit shuts off.

What Was It Doing? When the box temperature reached 35°F, the thermostat satisfied and cut power to the pump-down solenoid. The plunger in the solenoid dropped and stopped refrigerant flow. Well, not all of the flow. Some of the liquid was getting through the valve, as evidenced by the fact that the suction pressure was above the cut-out of the low pressure control. The valve had some frost on it, which indicated liquid flashing off as it passed through the valve.

What Was the Preliminary Diagnosis and Repair? Something in the valve was not allowing the valve to close tight against the valve seat. We pumped down the unit by front-seating the king valve on the receiver and jumping out the low-pressure control. This allowed all the refrigerant in the liquid line and suction line to be pulled out by the time the pressure got down to 0 psig. We then turned off the electrical disconnect to stop the unit and front-seated the suction service valve to keep air and moisture out of the crankcase when we opened the system at the valve. With no refrigerant in the line, we could take the solenoid valve apart to check it out.

Final Diagnosis and Repair: When we took the valve apart, we did not notice any blockage in the valve. However, we did realize that the arrow on the side of the valve was pointing in the opposite direction of the liquid refrigerant flow. We had installed the valve backward!

We removed and reinstalled the valve correctly, pulled a 500-micron vacuum from the suction service valve, pressurized and checked for leaks, opened the king valve and suction service valve, removed the jumper wire from the low-pressure (LP) control, started the unit, and check out the operation. This time we made sure the tstat satisfied at 35°F and that the unit pumped down and shut off on the LP control.

Epilogue: We had finished installing a walk-in cooler in a liquor store inside Union Station in Washington, D.C. It was late, and we were getting ready to leave. We had to return the next day to finish up by installing the shelving behind the doors among other things, so we thought it would be a good idea to let the unit run overnight. By the time we were ready to leave, the temperature in the box was at 50°F and coming down. The last thing I did was to make sure the pump-down solenoid was working by turning the tstat up to 70°F then back down to 35°F, the box temperature the customer wanted. I heard the click of the solenoid plunger when the solenoid coil was energized as the thermostat setting was slowly lowered past 50°F on the way to the final set point of 35°F, so I assumed it was working properly.

When we came in the next day, the box temperature was down to 25°F, the condensing unit was running, and there was white frost from the solenoid valve to the TEV. Not a good sign.

The problem with installing the valve backward is when the tstat de-energizes, the solenoid coil and the plunger drop on top of the valve seat to stop the flow. The pressure of the liquid is supposed to be on top of the valve seat to help close the valve. However, with the valve reversed, the pressure of the liquid was under the seat and was able to push it up enough to let a small amount of refrigerant flow to the evaporator. It was enough refrigerant to prevent the unit from pumping down by raising the suction pressure above the low-pressure control cut-out setting, thereby keeping the condensing unit running and cooling the walk-in below the 35°F setting.

Lessons Learned: Check the arrow on the pump-down solenoid to make sure it is pointing in the direction of flow. In this case, it should have pointed toward the TEV. Also, don't depend on the "click" of the solenoid plunger to signify the solenoid will work correctly. Wait until the unit pumps down and shuts off on the low pressure control, to be sure.

SERVICE SCENARIO #3

HOT PULL-DOWN

By Dick Wirz

Please keep in mind that this service scenario was encountered over 30 years ago. However, the lessons learned were useful and helped me better understand the relationship between suction pressure and its effect on low-temperature compressor operation.

Problem: I cannot keep the new freezer compressor running. It keeps going out on high pressure.

Situation: We had just installed a new open-front ice cream case and a remote 7.5-hp low-temperature condensing unit in the mechanical room about 50 feet away. The temperature in the new store was about 85°F, and the temperature inside the mechanical room with the other equipment running was over 100°F. After evacuating the system, I charged liquid into the receiver until the pressures equalized. Then I switched the charging hose over to the suction service valve, started the unit, and opened the valve on my refrigerant cylinder. After a while, the unit shut off on what I assumed was the high-pressure control (HPC). I did not think it too unusual, since there was a high heat load on both the case and the condensing unit. When the pressure came down, the unit restarted. However, when it went out again, I got an idea. I cooled the condenser with water from a pump sprayer and when it cycled back on by itself, the compressor seemed to work well. The unit kept running as long as I sprayed the fins. Eventually, the sight glass cleared, I quit spraying, the head pressure was normal for the ambient, and the unit stayed on line. I figured that I just learned a valuable lesson. My assumption was that under load, all you had to do was cool down the condenser to keep it running.

A couple days later, I shared this with our lead technician George, who was much more experienced than I. He listened to what I had to say and then told me how he deals with that kind of situation. He said what I did may well have worked, but my good result may have been just a coincidence. He has found that a high load on a low temperature compressor will often cause the compressor to shut down on its compressor motor overload. When he runs into that situation, he front-seats the suction service valve almost closed, puts his ammeter on one of the compressor leads, and watches the amps as he is charging to make sure they do not rise above RLA. He said that I may have been overloading the compressor if I was charging liquid into the suction service valve. Well, he might be right because I was alternating between liquid and vapor in an attempt to charge the unit faster. This is a fairly common practice.

Epilogue: In the years since then, I have found his method works well. Even if the refrigerated case has a pressure-limiting expansion valve or a crankcase pressure regulator (CPR), you can overload it by liquid charging if you don't keep an eye on the amperage. That is because charging at the suction service valve is done after those pressure-reducing valves and the pressure of the liquid entering the crankcase may be more than the compressor can handle.

Another thing I learned is how to tell the difference between a single compressor condensing unit that goes off on a pressure control and one that goes off on motor overload. If the compressor goes off on high or low pressure, the condenser fan also shuts off. However, if the compressor motor overload opens, the compressor will stop, but the condenser fan continues running because the contactor is still energized. This, of course, is true only as long as the fan receives its power from the same contactor as the compressor.

SERVICE SCENARIO #4

ALL ACTIONS HAVE CONSEQUENCES

By Jess Lukin
Bypassing safety controls is never a good practice. Following is an example of what can happen.

Service Call: No AC, second floor of a historical building
The second-floor AC air handler was running but was blowing warm air. The remote condensing unit located outside on the same level was not running. There was proper line voltage to the compressor contactor, but the contactor was not pulled in. I checked for power at the contactor coil, but there was none. The next step was to find out what in the control circuit was preventing voltage to the coil. Checked the thermostat, dual pressure control, oil safety control, and time delay.

The dual pressure switch was open. It had an auto reset, which meant that if it was off on high head, it would eventually cool down and start running for a little while before going off again. So it must have been off on the low-pressure part of the control. I used a stubby low-side gauge to check the low side, the gauge read 0 psig—no refrigerant in the system.

Where Did All the Refrigerant Go? I picked up four bottles of nitrogen and some soap bubbles to find the leak. To lose the whole charge of refrigerant, it had to be a major leak. After adding a couple tanks of nitrogen, a leak was evident in the compressor area.

After spraying the entire compressor down with soap bubbles, leaks were all over the place. The discharge service valve gasket was blown out, the motor terminals stubs were leaking, the unloader was leaking, and both compressor valve plates were leaking. One leak was bad enough, but what could cause all these leaks? See Figures 6-44, 6-45, and 6-46.

What Caused the Refrigerant Loss? All the leaks had one thing in common—they could be caused by high head pressure. The cut-out on the high pressure control was set for 500 psig. That was insane! In this area, we usually set R22 units for a cut-out at 350 psig, which is equivalent to a condensing temperature of 143°F. Assuming this old unit had a standard condenser split of 30°F, the ambient would have to get to 113°F for

FIGURE 6-44 **Leaking valve plate gasket.** *Photo by Jess Lukin.*

FIGURE 6-45 **Motor terminals leaking under orange seals.** *Photo by Jess Lukin.*

FIGURE 6-46 **Unloader valve with blown gasket.** *Photo by Jess Lukin.*

the control to trip. I know some techs set these controls for 400 psig (154°F condensing). But 500 psig is like having no control at all. Why would someone do that?

Apparently, there was something causing high head pressure, but the former service tech did not figure it out or maybe did not want to deal with it. He just kept raising the cut-out setting to keep the compressor running.

An inspection of the condenser revealed one condenser fan motor out and an extremely dirty condenser. Those two problems would definitely cause high head pressures.

Correcting the Problem: The fan motor was replaced, and the condenser was thoroughly cleaned with a pressure washer. That should take care of what caused the high condensing pressure.

Repairing the Damage: Before recommending further repairs, the condition of the compressor motor needed to be investigated. Ohm readings between windings and from the windings to ground did not reveal shorts, open windings, or grounds. An order was placed for both valve plates, unloader and gasket, and discharge service valve gasket.

After making the necessary repairs, another leak test was done just to make sure there were no more leaks. Unfortunately, leaks were found in some of the same areas! How could this be?

My supervisor was called in to help figure out what was going on. After investigating several theories, it was determined that the high heat had warped the compressor body. Therefore, the only option was to replace the compressor.

After the compressor was changed out, the liquid line filter drier and oil filter were replaced. Next, the system was pressure tested with nitrogen. When no leaks were detected, I connected the vacuum pump and evacuated down to 500 microns. While the system was being evacuated, the three-phase contactor was re-placed. It is important to note that anytime a three-phase motor or compressor is replaced, the manufacturers insist the contactor be replaced at the same time. Old contactors with worn contacts can cause single phasing or voltage unbalance, which can damage a new compressor. The dual pressure control was replaced with one that had a manual reset on the high-pressure side. This will prevent the compressor from short cycling on high pressure in the future. The unit was charged and started. Volts and amps as well as loaded and unloaded operation were checked.

The wiring was in pretty bad shape (see Figure 6-47). An attempt was made to clean it up and to verify all controls were wired correctly. With the compressor running, the safety controls were tested. The low-pressure part of the control was checked by closing the suction service valve and pumping down the system to verify the cut-out pressure. The high-pressure control was checked by turning off the condenser fans and verifying the cut-out pressure. The oil failure control was checked by "dry running" (see the epilogue of this case for details on how to do this.) Not only was it wired wrong, but it was determined that the control was bad and had to be replaced. More on this in the epilogue of this case.

FIGURE 6-47 **Wiring is not too neat.** *Photo by Jess Lukin.*

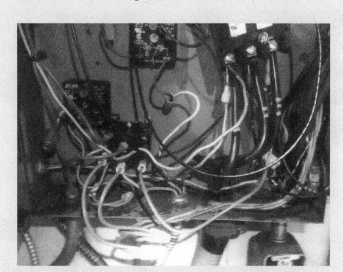

What Caused This Expensive Repair? The bottom line is that an improperly trained technician set the auto reset high-pressure control so high that he essentially jumped out a safety control in order to keep the unit running. That technician apparently never investigated what was causing the unit to go out on high pressure. In addition, a second technician replaced the oil failure control but never checked it out to see if it was working properly.

Epilogue: If a safety control is bypassed or set incorrectly, there will be consequences. Maybe not immediately, but someone is going to pay for it. If a safety control is tripping, the reason must be determined and corrected immediately. This is such an important and fundamental issue that some companies have a policy that jumping out a safety control is cause for immediate termination. Also, whenever a control or part is replaced, it should be checked to make sure it is operating as it should. During this final check, the reason for its failure may show up, or it may uncover another problem in the system that needs to be corrected, pre-venting a call-back.

Checking an Oil Failure Control by "Dry Running": Dry running is a method where you shut off the power, remove the compressor leads from the load side of the compressor contactor, then turn the power back on. Make sure the operating controls, such as the low-pressure switch, are closed and the contactor coil is energized.

Because there is power to the oil control, it thinks the compressor is running, but there is no oil pressure. The oil safety control times out in about 2 minutes; if longer than 3 minutes, the oil control is bad, or it is wired incorrectly. If the wiring checks out, then the control must be bad and needs to be replaced.

An additional check can be done with an infrared temperature gun while dry running. Remove the control cover. The heater resistor, which is typically white, is connected electrically between terminals L and M. When the compressor contactor is energized, aim the infrared gun at the resistor. Because there is no oil pressure, contacts 1 to 2 are closed, allowing voltage to the resistor, which will start getting hot. At around 250°F to 300°F, the heater will open up the contacts between L and M, cutting power to the contactor coil, which will disengage the contactor. If the temperature rises above 300°F and the normally closed contacts between L and M have not opened, then the oil safety control is bad, or the wiring is incorrect.

In the original service call, the oil failure control was checked with an infrared. The resistor started heating up (Figure 6-48) but went way beyond 300°F, up to 423°F (Figure 6-49); the contacts were still closed between L and M, and the contactor coil was still energized after 5 minutes. Because the wiring was incorrect, the control overheated, burning out the resistor and necessitating the replacement of the control.

FIGURE 6-48 **Oil control resistor heater is starting to heat up.** *Photo by Jess Lukin.*

FIGURE 6-49 **Resistor heated to 423°F, but switch had not opened between L and M.** *Photo by Jess Lukin.*

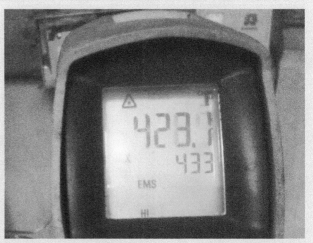

Measuring with your voltmeter between L and M should show source voltage (120 V or 240 V), which means the contacts are open after the resistor heats up enough to open the switch. Oil failure controls must be reset manually.

An ohm meter can be used to make preliminary checks of the oil failure control. With the compressor off and no power to the oil control, the normally closed contacts between 1 and 2 should be closed, and normally closed contacts between L and M should also be closed. If not, the control is bad.

Notes on Checking for Leaks: Bubbles are great, but oil residue is another good indication of a refrigerant leak. Notice on the compressor head (Figure 6-44) that the outline of refrigerant oil is present. To determine if this is a new refrigerant oil leak or an old leak that had been repaired, cleaning the area of refrigerant oil during a preventative maintenance can help. If this is your first time on the call and no maintenance contract exists, refrigerant oil that is lightly dusted with other debris, like dirt or dust, most likely indicates an old leak that may have been previously repaired. You can always verify your suspicions with an electronic leak detector or bubble solution.

REVIEW QUESTIONS

1. Commercial refrigeration thermostats that sense the box air temperature have a temperature swing of about how many degrees?

    a. 28
    b. 58
    c. 128
    d. 158

2. Why does a thermostat that senses coil temperature prevent frost buildup on a medium-temperature evaporator?

    a. It has a narrow temperature swing.
    b. Once the tstat is satisfied, any buildup of frost on the evaporator must be melted before the tstat restarts the compressor.
    c. The thermostat knows when to go into defrost.

3. What is the primary purpose of the cap on a service valve stem?

    a. To prevent refrigerant leaks
    b. To keep the valve stem from being hit accidentally
    c. To prevent dirt from damaging the packing when the valve is opened

4. What are two troubleshooting procedures that can be performed by front-seating the suction service valve?

    a. Check reed valves and high back pressure overloading
    b. Check superheat and subcooling
    c. Check for high head and suction pressures

5. What will happen if the compressor is started with the discharge service valve front-seated?

    a. The suction reed valves will break.
    b. The liquid line will get excessive pressure.
    c. The valve plate gasket will be damaged or even injure the technician.

6. With both service valves front-seated, what is the basic procedure for replacing a semihermetic compressor?

7. What controls the pump-down solenoid?

    a. Thermostat
    b. Liquid-line service valve
    c. Fan switch

8. Describe the pump-down sequence.

9. What will happen if the pump-down solenoid is installed backward?

    a. Nothing; it can be installed either way.
    b. The system will stay in a pump-down condition.
    c. The system will not pump down, and the compressor will not shut off.

10. What is the primary function of a hot gas bypass valve?

    a. Maintains a minimum suction pressure at the compressor
    b. Prevents compressor overload during hot pull-down
    c. Keeps pressures down in the evaporator

11. What is the primary function of a CPR valve?

    a. Maintains a minimum suction pressure at the compressor
    b. Prevents compressor overload during hot pull-down
    c. Keeps pressures up in the evaporator

12. What is the primary function of an EPR valve?

    a. Maintains a minimum suction pressure at the compressor
    b. Prevents compressor overload during hot pull-down
    c. Keeps pressures up in the evaporator

13. What does "ZP" stand for on a Sporlan TEV?

    a. A freezer valve that limits the pressure to the compressor
    b. A medium-temperature valve made for low-pressure freezing
    c. A multirange valve that keeps the evaporator pressures up

14. List the steps required to determine the maximum suction pressure a compressor can handle, and to also set a CPR valve.

15. Two cases are installed on one compressor. One is a 35° display case, and the other is a 50° display case. Which case would require an EPR?

    a. The higher-temperature case
    b. The lower-temperature case

16. What are three uses for a low-pressure control?

    a. Prevents damage due to loss of refrigerant and temperature control and shuts off compressor after pump-down
    b. Prevents damage due to loss of refrigerant and temperature control and keeps from overloading the compressor on pump-down

c. Prevents damage due to loss of temperature and pressure control and shuts off the compressor before pump-down

17. What are three concerns when setting a low-pressure control for safety or for pump-down?

    a. Short-cycling, compressor overheating, and low ambient preventing control from cutting in
    b. Short-cycling, compressor overheating, and high ambient preventing control from cutting out
    c. Safety, high head pressure, and low ambient preventing restart of compressor

18. If the minimum winter design temperature in your town is 0°F, what would the low-pressure cut-in and cut-out be for a walk-in refrigerator using R22 in an outdoor condensing unit?

    a. 50-psig cut-in and 30-psig cut-out
    b. 20-psig cut-in and 1-psig cut-out
    c. 20-psig cut-in and 5-psig cut-out

19. The same as question 18, but set the low-pressure control for a walk-in freezer using R404A in an outdoor condensing unit.

    a. 20-psig cut-in and 1-psig cut-out
    b. 25-psig cut-in and 5-psig cut-out
    c. 20-psig cut-in and 5-psig cut-out

20. The same as question 18, but set the low-pressure control for a walk-in refrigerator using R404A in an indoor condensing unit.

    a. 40-psig cut-in and 20-psig cut-out
    b. 30-psig cut-in and 20-psig cut-out
    c. 60-psig cut-in and 40-psig cut-out

21. Assume you replace a low-pressure control that acts as the temperature control on an R22 reach-in refrigerator that maintains 38°F. It is a Friday evening and the factory is closed, so you cannot get any information on settings. At what pressure cut-in and cut-out would you set the low-pressure control?

    a. 38-psig cut-in and 10-psig cut-out
    b. 66-psig cut-in and 33-psig cut-out
    c. 33-psig cut-in and 12-psig cut-out

22. What would be the high-pressure control cut-out for an outdoor R404A refrigeration unit?

    a. 155 psig
    b. 400 psig
    c. 475 psig

23. What is the function of an oil separator?

    a. Separates oil from the discharge gas
    b. Prevents floodback to the compressor
    c. Stops refrigerant migration to the oil during the off cycle

24. If a system has an oil separator, do you still need to slope and trap the suction lines for oil return? Why or why not?

    a. No, the oil is returned to the crankcase before it gets into the piping.
    b. Yes, some oil still gets into the piping.

25. What are the two pressures an oil safety control must monitor?

    a. Suction pressure and head pressure
    b. Crankcase pressure and oil pump discharge pressure
    c. Low side pressure and receiver pressure

26. Below what minimum net oil pressure will the oil safety control start its delay timer?

    a. 5 pounds
    b. 10 pounds
    c. 60 pounds

27. How long must the oil remain below its minimum pressure before the oil safety control trips?

    a. 10 seconds
    b. 60 seconds
    c. 120 seconds

28. Will the oil safety control trip if the compressor goes out on internal overload? Why or why not?

    a. Yes, it will trip because there is power to the control but no oil pressure.
    b. No, it won't trip because the overload cuts power to the oil safety control.

29. What can you do to prevent nuisance tripping of the oil failure control from electrical problems such as brown outs (low incoming voltage)?

    a. Install a compressor start relay
    b. Install a current-sensing relay
    c. Install a low-voltage oil safety control

30. Describe how a current-sensing relay works.

31. After manually resetting an oil safety control, what electrical and refrigeration checks should be made to determine the cause for the control to trip?

32. The entire refrigerant charge of a system should fill only what percentage of the receiver? Hint: It is the same percentage of maximum fill for a recovery cylinder.

    a. 80 percent
    b. 90 percent
    c. 100 percent

33. What is the service valve called on the outlet of the receiver?

    a. King valve
    b. Queen valve
    c. Solenoid valve

34. What is the service valve on the outlet of the receiver used for?

    a. To pump the system down for repairs and to check liquid pressure
    b. To check head pressure and subcooling
    c. To isolate the compressor before replacing it

35. What is the primary function of an accumulator?

    a. To accumulate liquid so the TEV has enough during startup
    b. To protect the compressor from floodback after a hot pull-down
    c. To keep the head pressure up during low-ambient conditions

36. If the accumulator is sweating, should it be insulated? Why or why not?

    a. No, because it needs heat from the ambient air to boil off the refrigerant.
    b. Yes, because it needs to keep the heat in the accumulator.

37. The primary desiccant (usually silica gel) in a filter drier is designed to remove what system contaminant?

    a. Acid
    b. Wax
    c. Sludge
    d. Moisture

38. Why do systems that operate fine all winter suddenly develop a moisture problem when the weather turns warm?

    a. Because the filter drier was saturated with moisture when it was cool but released some of the moisture when the filter drier warmed up
    b. Because the higher pressures force moisture out of the drier
    c. Because summer weather is more demanding than winter weather

39. Does a filter drier trap debris and contaminants the first time they try to pass through it?

    a. Yes
    b. No

40. Once contaminants are trapped in the filter drier, can they be released back into the system?

    a. Yes
    b. No

41. On a medium-temperature system, what is the maximum pressure drop across its permanent suction filter drier before it must be replaced?

    a. 1 psig
    b. 2 psig
    c. 3 psig
    d. 10 psig

42. If a suction filter is installed temporarily to clean up acid after a burnout, what is the maximum length of time before it must be removed?

    a. One day
    b. Two days
    c. Three days

43. Can suction filter driers be used to remove system moisture?

    a. Yes
    b. No
    c. Maybe

44. According to TROT, what is the maximum temperature drop across a liquid-line filter drier before it requires replacement?

    a. 18
    b. 38
    c. 68
    d. 108

45. How would you check to see if a filter drier has reached its moisture-removal capacity?

    a. Check for a pressure/temperature drop across the drier
    b. Check the sight glass moisture indicator
    c. Pump the system down and see if it will hold a vacuum

46. What is the best procedure for replacing a sweat drier, and why?

    a. Use a torch to sweat it out, because it is quicker
    b. Cut it out, because heat from a torch will release the moisture into the system

47. A Sporlan filter drier is a C052; what is its cubic inch desiccant capacity and pipe size, and is it flare or sweat connection?

    a. 2 cubic inches, 5/8-inch pipe, flare
    b. 5 cubic inches, 1/2-inch pipe, sweat
    c. 5 cubic inches, 1/4-inch pipe, flare

48. If you see a sight glass bubbling, should you add refrigerant immediately? Why or why not?

    a. Yes, it may overheat the compressor.
    b. No, you should wait until all the bubbles stop before adding refrigerant.

c. No, the system may have just started, or there may be a low load.

49. What is a heat exchanger used for?

a. It boils off liquid refrigerant to prevent flooding the compressor.
b. It subcools liquid to prevent flash gas before the TEV.
c. It keeps the suction pressure up to the compressor.

50. Should a vibration eliminator be parallel or perpendicular to the compressor crankshaft, and why?

a. It should be perpendicular because the bellows act like an accordion to absorb vibration when the compressor rocks as it starts.
b. It should be parallel because the corrugated tubing inside will crack if it compresses like a bellows, but will not if it twists as the compressor rocks.

REFRIGERATION SYSTEM TROUBLESHOOTING

7

CHAPTER OVERVIEW

This is the chapter most technicians look forward to. It is like having all the pieces of a puzzle finally fall into place. The previous chapters are designed to help develop a thorough understanding of the commercial refrigeration system. This chapter builds on the knowledge gained so far and now focuses on developing diagnostic skills.

Some important information from other chapters is restated in this chapter, but in a slightly different manner. Although it has increased the length of the chapter, repeating concepts in different words will hopefully make those concepts easier to understand. If you have already fully comprehended the ideas, consider this to be a reinforcement of what you have already learned.

The art of diagnostics involves putting understanding into practical use. It gives technicians a sense of accomplishment and pride. If troubleshooting were easy, anybody could do it. The truth is, troubleshooting is not easy, but by the end of this chapter, you should be able to properly diagnose most refrigeration system problems. Problem solving can be fun. And the better you become at troubleshooting, the more fun it is!

OBJECTIVES

After completing this chapter, you should have a good understanding of overall refrigeration system diagnosis and in particular the following system problems:

- Dirty or iced evaporator or low airflow over the evaporator
- Undercharge of refrigerant
- Inefficient compressor
- Restricted metering device
- Restriction in the liquid line after the receiver
- Restriction in the high side before the receiver
- Dirty condenser
- Air in the system
- Overcharge of refrigerant
- Oil-logged evaporator

REVIEW AND PREVIEW

Before getting into troubleshooting specific problems, it will be beneficial to briefly review some aspects of the refrigeration system and its four major components. In addition, this chapter investigates how conditions in the refrigeration system affect these components and how the components influence each other. Understanding how different parts of the system interact is essential to effectively diagnosing refrigeration system operation.

After learning to troubleshoot one type of system, it is fairly easy to apply most of the same diagnostic tools to any refrigeration system. First, establish at what temperatures and pressures the unit being serviced is supposed to be operating, when it is running correctly. In other words, what is it supposed to be doing?

After deciding which part of the system is not functioning properly, the next step is to use diagnostic skills to find out what is causing the problem. Following is a list of the minimum information needed for troubleshooting:

- Ambient temperature entering the condenser
- Condensing temperature
- Condenser split
- Condenser subcooling
- Refrigerated space temperatures entering the evaporator
- Evaporator temperature
- Evaporator TD
- Evaporator superheat

Gauges and thermometers give most of the necessary information, but to use that data properly, a technician must know the system's design conditions. Unless the technician knows what the condenser split and evaporator TD are supposed to be, he really would not know if the system is working properly or not.

An R22 system on a 95°F day has a head pressure of 278 psig, which would be a condensing temperature of 125°F. Is that what it should be?

Yes, if the unit has a 30°F condenser split (125°F − 95°F = 30°F) like a 10 SEER A/C unit or a medium-temperature walk-in box with a standard condensing unit. However, the answer is no if the unit is a 14 SEER A/C unit with a 20°F condenser split, a freezer with a 25°F condenser split, or a high-efficiency commercial refrigeration remote condenser with a 10°F split. As indicated by this example, the technician must have an idea of what the designed condenser split is supposed to be in order to troubleshoot the high side of the system.

For the low side of the system, the suction pressure is an indicator of the evaporator temperature, and a thermometer will show what the box temperature is. However, to know if the coil temperature is correct for the conditions, the technician must know the designed evaporator TD.

To keep this chapter focused on the basics, the following two applications are used for both review and in the examples later in the chapter:

- Medium-temperature walk-in at 95°F ambient (Figure 7-1)
  - 30°F condenser split
  - 125°F condensing temperature

FIGURE 7-1   **Walk-in refrigerator with TEV.** *Courtesy of Refrigeration Training Services.*

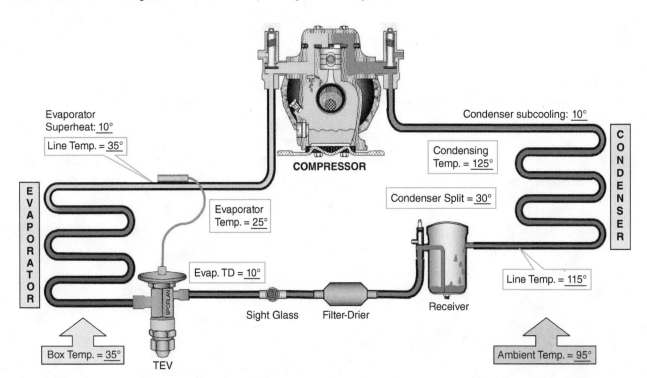

- 10°F condenser subcooling
- 35°F box temperature
- 10°F evaporator TD
- 25°F evaporator temperature
- 10°F evaporator superheat
- Thermal expansion valve (TEV) metering device

- Medium-temperature reach-in at 95°F ambient (Figure 7-2)
  - 30°F condenser split
  - 125°F condensing temperature
  - 10°F condenser subcooling
  - 38°F box temperature
  - 20°F evaporator TD
  - 18°F evaporator temperature
  - 10°F evaporator superheat
  - Capillary tube metering device

Notice that there are no refrigerant types or pressures listed in the two sample systems. That is because the type of refrigerant is not important, just the saturation temperatures are. For instance, the two systems illustrated would have the same saturation temperatures if they were using R134a, R404A, or even R410A. Only the pressures would be different. Of course, the compressor and metering device would have to be designed for the refrigerant used.

Evaporator

Refrigerant boiling in the evaporator absorbs heat from the refrigerated space. When all the liquid droplets of refrigerant have vaporized, the saturated vapor can pick up only sensible heat. By measuring the superheat, a technician can determine if the evaporator is operating effectively and if there is any damaging liquid returning to the compressor. Superheat is calculated by subtracting the evaporator temperature from the temperature of the suction line at the outlet of the evaporator.

This chapter considers 10°F superheat as typical for most of the sample refrigeration systems.

NOTE: **Normal superheat can be as high as 15°F in air conditioning units, as low as 6°F in freezers, and only 3°F in ice machines. No superheat (0°F) means floodback in all cases.**

Therefore, in examples where there is an excess of refrigerant entering the evaporator, enough to drop the superheat under 5°F, it is usually considered a flooding situation. If there is not enough refrigerant entering the evaporator so that the superheat is above 20°F, the assumption is the evaporator is definitely starving.

NOTE: **For the examples in this chapter, use the following superheat guidelines:**

- **flooding = superheat under 5°F**
- **starving = superheat over 20°F**

An 8,000-Btuh evaporator in a 35°F walk-in box under design conditions absorbs 8,000 Btuh and sends the heat to the condenser to be removed from the system. If warm product is put into the walk-in, more than 8,000 Btuh will be absorbed. The temperature of the refrigerant in the evaporator will rise, and the liquid will boil faster. The heat causes an increase in the movement of refrigerant molecules, which causes the evaporator pressure to increase.

On the other hand, if the evaporator is covered with frost, then no warm air will get through the evaporator

FIGURE 7-2 **Reach-in refrigerator with cap tube metering device.** *Courtesy of Refrigeration Training Services.*

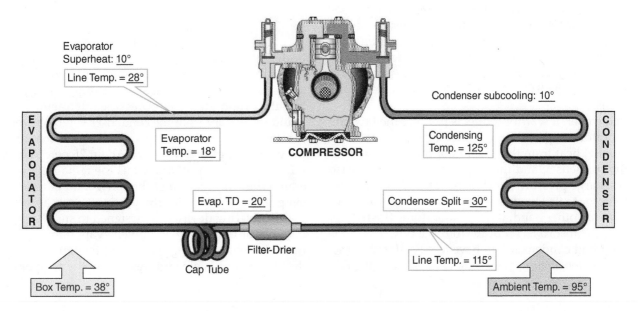

to warm the refrigerant. The frost acts like an insulator and prevents, or at least reduces, heat transfer between the warmer air in the box and the colder refrigerant in the evaporator. If no heat is added to the refrigerant, its temperature and pressure will fall.

Analyzing evaporator temperature can be confusing at times. For instance, the air leaving an evaporator that is frozen, dirty, or starved for refrigerant is certainly warmer than normal. This condition would show up as a low ΔT if we were measuring the difference between the air entering and leaving the evaporator.

However, in this chapter, we are diagnosing based on the temperature of the refrigerant inside the evaporator. The evaporator temperature is a function of the refrigerant pressure. The pressure is a function of the movement of molecules of refrigerant. The less heat absorbed, the slower the molecules move, the lower the refrigerant pressure.

If the evaporator is frozen, dirty, or starving for refrigerant, heat is not being absorbed into the evaporator. Less heat in the refrigerant means the molecules slow down, resulting in a lower pressure. A drop in pressure means a drop in evaporator temperature. Therefore, a lower-than-normal evaporator temperature would indicate that heat is not getting to the refrigerant, or there is not enough refrigerant in the evaporator.

Compressor

The compressor is a pump that increases the pressure and temperature of the entering suction vapor. The temperature of the leaving vapor must be high enough to condense when cooled by the ambient air entering the condenser. In the sample systems, the compressor raises the suction vapor temperature to 125°F so that the 95°F ambient air is cool enough to condense the hot vapor into a liquid.

Assume a 35°F walk-in cooler has a compressor rated for 8,000 Btuh at 25°F saturated vapor and 125°F condensing temperature. As long as the suction temperature or the condensing temperature does not rise above those maximum values, the compressor has no problem moving 8,000 Btuh from the evaporator to the condenser.

However, if warm product is put into the box, the evaporator temperature will increase above 25°F. The additional heat from the evaporator increases the heat in the condenser, and the compressor has to increase its discharge pressure to move the refrigerant from the evaporator to the condenser.

On the other hand, if the evaporator is frosted up, it will not absorb 8,000 Btuh from the box. The compressor and condenser will have less Btuh to process, so the condensing temperature and pressure will fall.

These two examples illustrate the fact that what happens in the evaporator has a very similar effect on the condenser; as the evaporator temperature rises, so does condensing temperatures, and vice versa.

Condenser

The condenser is sized to reject the evaporator latent heat and superheat as well as the compressor motor heat and heat of compression. At 70°F ambient conditions, a condenser designed with a 30°F split can reject its rated capacity of heat at a 100°F condensing temperature. If the ambient rises to 95°F, the condensing temperature must rise to 125°F to accomplish the same task.

Warm product in the box will cause a TEV system to feed more refrigerant into the evaporator. The increased refrigerant flow absorbs more heat, causing a rise in the evaporator pressure. The additional heat and quantity of refrigerant fills more of the condenser (taking up more condensing space) and results in higher condensing pressure and temperature.

If the evaporator is covered with frost, there will be less heat for the condenser to process; therefore, the condensing temperature and pressure will fall. The less heat absorbed by the evaporator, the less heat the condenser has to reject. Conversely, if there is a load on the evaporator (hot product or hot pull-down), the condensing temperature will rise. A dirty condenser will also raise the condensing temperature because the refrigerant will not be able to reject all its heat.

The condensing pressure and temperature are also affected by the amount of refrigerant in the condenser. This can be determined by measuring subcooling.

NOTE: **For the examples in this chapter, use the following subcooling guidelines:**

- **undercharge of refrigerant = subcooling under 5°F**
- **overcharge of refrigerant = subcooling over 20°F**

Metering Devices

Both TEVs and fixed metering devices (cap tubes) lower the temperature of condensed refrigerant and act as nozzles to spray droplets of liquid refrigerant into the evaporator. High-pressure, high-temperature liquid refrigerant from the condenser enters the metering device and is forced through an orifice that lowers the pressure of the refrigerant. About 25 percent of the refrigerant flashes off and cools the remaining refrigerant. The temperature of the refrigerant entering the evaporator is based on the pressure drop across the valve. Metering devices are designed to supply a given quantity of refrigerant at a given evaporator temperature. However, the temperature of the liquid entering the evaporator is also influenced by the temperature of the refrigerant entering the metering device. A later chapter explains how supermarket refrigeration uses this fact to increase evaporator efficiency through mechanical subcooling.

The pressure drop across one valve or cap tube is not going to produce the same evaporator temperature for every refrigerant. Therefore, metering devices are rated at a specific Btuh capacity, within a particular evaporator temperature range, for a particular refrigerant.

EXAMPLE: 2

A Sporlan model G 1/2 C expansion valve is designed for an R22 medium-temperature (25°F evaporator) system. Under most conditions, it will provide enough refrigerant to the evaporator to accomplish 0.5 ton (6,000 Btuh) of refrigeration effect.

HOW FIXED METERING DEVICES REACT TO SYSTEM CONDITIONS

Following is an example showing how fixed metering devices use pressure drop to create a specific evaporator temperature. Also, it shows how higher-pressure refrigerants require a greater pressure drop to produce the same evaporator temperature.

EXAMPLE: 3

A reach-in refrigerator has a 2,500-Btuh rating at a 20°F evaporator temperature and 120°F condensing. If the system operates with R134a, what would be the pressure drop across the cap tube, and what size cap tube would be used? What if it was an R22 system?

According to a P/T chart, R134a would have to drop 153 psig (171 psig at 120°F − 18 psig at 20°F). Assume that a cap tube chart showed that 9 feet of 0.049 ID cap tube would provide the necessary amount of refrigerant at the required pressure drop.

R22 would require a greater pressure drop of 217 psig (260 psig at 120°F − 43 psig at 20°F) to accomplish the same condensing to evaporator temperature drop. The sample cap tube chart that recommended 9 feet of 0.049 ID cap tube for R134a may recommend something like a 9-foot piece of 0.042 ID cap tube for R22. The smaller ID cap tube would provide the greater pressure drop needed for an R22 system.

The next example shows how increased condensing pressure and temperature directly affect the evaporator and box temperatures of a fixed metering device system.

EXAMPLE: 4

The previous R22 cap tube unit has a dirty condenser (or the ambient temperature increases) that raises the head pressure from 260 psig to 290 psig.

A fixed metering device is designed to have a fixed pressure drop; therefore, the higher the head pressure, the higher the evaporator pressure. In this example, assume an increase of 30 psig in head pressure will increase the evaporator pressure by 30 psig. If the evaporator temperature was 20°F at a suction pressure of 43 psig, then an increase of 30 psig to a suction pressure of 73 psig would increase the evaporator temperature to 43°F, which would raise the box temperature.

The evaporator superheat will drop because the refrigerant is not boiling away. To boil, the refrigerant temperature has to be lower than the box temperature. Another reason evaporator superheat drops is because high condensing pressures flood the evaporator with so much refrigerant that it has no chance to boil away.

After seeing how a cap tube system reacts to a load on the condenser, let us see what happens if the heat load is on the evaporator:

EXAMPLE: 5

What would happen if hot product was put in the box?

Since cap tubes allow only a fixed amount of refrigerant to enter the evaporator, the increased heat load quickly boils off the limited amount of refrigerant. Evaporator pressure stops rising when the refrigerant is completely vaporized, or totally saturated. If there is a limited amount of liquid refrigerant to absorb latent heat, then the evaporator pressure cannot rise very much due to an increased heat load.

For a similar reason, the condensing pressure does not rise that much. The quantity of refrigerant supplied to a warm evaporator is restricted to about the same amount supplied under a normal box load. Therefore, the evaporator can absorb only about the same amount of latent heat as usual and pass it on to the condenser.

Since the limited amount of refrigerant boils off quickly in the first part of the evaporator, it does pick up more sensible heat. This is verified by the higher suction-line temperature leaving the evaporator. In addition, the sensible heat picked up in the evaporator does not add enough load to the condenser to raise its pressure. To summarize, heat added to an evaporator with a cap tube raises the evaporator and condensing temperatures only a little. Superheat, however, increases a lot.

Now that you have seen what happens when there is a high heat load on the evaporator, let us see what happens when there is a low heat load on the evaporator.

EXAMPLE: 6

What would happen if the evaporator were frosted up?

If there is no air flowing through the evaporator fins, there is no load on the evaporator and little heat to boil the refrigerant. If the refrigerant is not boiling, the pressures and temperatures fall. Also, there is little or no superheat because the refrigerant has not boiled away.

If the evaporator is not picking up heat, the condenser has little heat to reject. As a result, the condensing temperature and pressure are low.

HOW A TEV SYSTEM REACTS TO SYSTEM CONDITIONS

Unlike cap tubes, TEVs have the ability to adequately feed the evaporator and maintain a constant superheat under most load conditions. The sensing bulb of the TEV is attached to the suction line at the evaporator outlet. If the bulb senses a rise in suction-line temperature (high superheat), the valve opens and feeds more refrigerant into the evaporator. When the suction line cools down (low superheat), the TEV cuts back on the amount of refrigerant it is feeding the evaporator.

The refrigerant inside a TEV sensing bulb boils within a certain temperature range. The superheat adjustment allows the valve to maintain a specific suction-line temperature. For instance, at standard conditions, a walk-in refrigerator evaporator at 25°F with 10°F superheat will have a suction-line temperature of 35°F where the sensing bulb is attached. If the evaporator temperature rises to 35°F, the temperature at the sensing bulb will be 45°F. The TEV will open wide enough to handle the additional 10°F of evaporator heat while still maintaining 10°F superheat at the evaporator outlet.

Following is an example of how a TEV system responds to an increased evaporator load:

EXAMPLE: 7

What happens when hot product is put into a walk-in refrigerator that has a TEV metered evaporator?

Initially, the refrigerant boils off quickly in the evaporator. The TEV bulb at the evaporator outlet senses the higher suction-line temperature. The valve opens and feeds more refrigerant in an attempt to lower the evaporator temperature. The more heat the refrigerant absorbs, the more violently it boils. The increased boiling rate of the refrigerant raises evaporator pressure and temperature.

Within limits, the more refrigerant the TEV feeds the evaporator, the more heat the evaporator can absorb from the product in the walk-in. All the heat absorbed in the evaporator is transferred directly to the condenser. The larger the quantity of heat added to the condenser, the greater the increase in condensing temperature and pressure.

As the temperature of the product falls, the temperature of the evaporator also drops. The TEV senses the colder suction line and cuts back on the amount of refrigerant it feeds the evaporator. When the box and product temperatures are down to design conditions, the TEV meters only enough refrigerant to the evaporator to maintain its designed suction-line temperature. The temperature of the suction line at the sensing bulb is the evaporator temperature plus the desired superheat. In summary, heat added to a TEV-metered evaporator raises both the evaporator temperature and the condensing temperature. The next example shows how the TEV system reacts to an increased load on its condenser.

EXAMPLE: 8

What happens if the condenser is dirty, or if the ambient temperature increases?

The condensing pressure and temperature increase, putting more pressure on the inlet of the valve. However, a TEV responds to its sensing bulb at the evaporator outlet, not the condensing pressure at the valve inlet. The evaporator TD and superheat remain about the same. However, the evaporator temperature will rise slightly.

The higher the condensing temperature, the more refrigerant is needed to flash off (adiabatic expansion) in order to maintain the designed evaporator temperature.

EXAMPLE: 9

Liquid refrigerant at 100°F enters a 25°F evaporator. Assume 25 percent of the refrigerant flashes off to cool down the rest of the refrigerant (75%) to the desired evaporator temperature. If the liquid temperature rises to 125°F, it may take 33 percent of the refrigerant to flash off, leaving less refrigerant (67%) to absorb heat from the space.

Moderately high condenser loads will raise head pressure but not affect evaporator temperature that much. However, the TEV does have its limits. If the condenser is very dirty or the ambient temperature is very high, the head pressure could force some higher-temperature refrigerant into the evaporator, raising evaporator temperature.

Now let us see how a TEV system responds to a low evaporator load.

EXAMPLE: 10

What happens if the evaporator is frosted?

Frost prevents the evaporator from absorbing much heat, so the evaporator temperature falls. The TEV bulb senses the lowered evaporator temperature and cuts back on the flow of refrigerant to maintain superheat.

If the evaporator is not picking up heat, then the condenser does not have to reject much heat. The condensing temperatures and pressures are low, and the subcooling remains normal.

SUMMARY OF HOW EXTERNAL TEMPERATURE CHANGES AFFECT THE REFRIGERATION SYSTEM

The following rules for temperatures apply to refrigerant that is saturated (in the process of boiling to a vapor or condensing to a liquid):

EXAMPLE: 11

Imagine a properly operating 35°F walk-in refrigerator with a TEV on an 80°F day. The walk-in is inside the building, and the condensing unit is outside.

From what we learned about evaporators, walk-ins have an evaporator TD of 10°F. Therefore, the evaporator temperature is 25°F (35°F − 10°F TD = 25°F). If the system is R22, the suction pressure is 49 psig, according to the P/T chart.

A standard refrigeration unit has a 30°F condenser split. With an 80°F ambient, the condensing temperature would be 110°F at a discharge pressure of 226 psig.

The compressor is increasing the R22 refrigerant pressure/temperature from 25°F at 49 psig to 110°F at 226 psig to maintain a 35°F box at an ambient of 80°F.

If refrigerant temperature increases, so does its pressure.

As the day gets warmer, the air temperature entering a condenser increases, and so does the head pressure. If warm product is put in a refrigerator, the box temperature increases, and so does the suction pressure.

If the refrigerant temperature decreases, so does its pressure.

When the sun goes down, the air temperature entering a condenser goes down, and so does the head pressure. As the temperature of the warm product inside the refrigerator goes down, so does the suction pressure.

With these two rules in mind, use the following example to help you visualize what influence external temperature changes have on a system.

The previous descriptions of system operation use fairly general terms to describe how temperature influences a system. The following examples are more specific in order to represent a "real-life" approach to the situation. Also, the results are given in a short but descriptive format.

Following are some common conditions that can affect a walk-in system, with descriptions of what happens and why:

- The 80°F outdoor ambient at the condenser rises to 100°F. What happens?
  - Head pressure rises to 130°F condensing at 297 psig. Why?
    - The technical answer is that the molecules of refrigerant move about faster as temperature rises. This increased movement increases the pressure.
    - The simpler explanation is that the refrigerant temperature and pressure rise with the ambient in direct proportion to the condenser split designed for that unit.

- Even if the head pressure increases, the suction pressure remains at 25°F and 49 psig. Why?
  - The expansion valve is not affected by increases, or decreases, in head pressure under normal conditions.
  - The ambient is neither increasing nor decreasing the load on the evaporator because it is not affecting what happens to the walk-in refrigerator inside a building.

- Someone leaves the walk-in door open, or they put warm product in the walk-in box. What happens?

  - The pressure and temperature in the evaporator rise. Why?
    - The increased product heat is added to the evaporator's refrigerant. This increases the motion of molecules in the refrigerant, which raises the pressure. (**EXAMPLE:** 35°F walk-in box rises to 50°F and raises evaporator pressure.)

  - The pressure and temperature in the condenser rise. Why?
    - This is a significant concept to understand. If a compressor receives suction vapor at an increased pressure and temperature, it simply passes on the heat to the condenser. The condenser is handling more heat than it is designed for at that condensing temperature; therefore, the pressures and temperatures start rising above normal. (**EXAMPLE:** The normal condensing temperature at 100°F ambient is 130°F. However, additional load from the evaporator may raise the condensing temperature to 145°F.)

It is important to understand how adding heat to the condenser may affect only part of the system (on a TEV system). Yet, adding heat to the evaporator affects both sides of the system. Understanding this relationship is a fundamental part of proper troubleshooting.

SUMMARY OF HOW PRESSURE CHANGES AFFECT THE REFRIGERATION SYSTEM

Without trying to oversimplify some complex laws of nature, the following basic rules may clarify some important pressure/temperature relationships:

> Pressure Rule #1: If refrigerant pressure is increased, so is its temperature.
> The compressor increases suction pressure to raise its temperature.
> Pressure Rule #2: If refrigerant pressure is decreased, so is its temperature.
> The metering device drops liquid pressure to lower its temperature.

Other than restricted airflow through a condenser and higher ambient temperatures, there are only two main causes for high head pressure:

- High suction pressure: This is a sign of high evaporator load.
- Refrigerant overcharge: Excess refrigerant takes up condenser space and increases the head pressure.

The opposite is also true:

- Loss of refrigerant means excess space in the condenser, so the condensing pressure falls.
- Lack of refrigerant to the evaporator results in less boiling action, which means less evaporator pressure.
- Lower suction pressure means lower discharge pressure leaving the compressor.

SUPERHEAT AND SUBCOOLING

Technicians use superheat and subcooling like a doctor uses body temperature. An increase or decrease from normal indicates something about what is going on inside the system.

When refrigerant has fully vaporized (boils off) before leaving the evaporator, any additional heat it picks up from air blowing across the evaporator does not raise the refrigerant's pressure, just the vapor's temperature. The temperature above its saturation temperature (from the P/T chart) is superheat. The pressure in the evaporator and in the suction line remains essentially the same. Only by taking the temperature of the suction line leaving the evaporator and comparing it to the saturation temperature on the P/T chart can we determine superheat. If the temperature of the suction line is the same or colder than the saturation temperature, then the refrigerant has not boiled off, and we still have liquid in the suction line. This is called flooding. However, if superheat is present, then all the refrigerant has boiled off in the evaporator, and there is no liquid in the suction line.

SUPERHEAT RULE: If the refrigerant is fully vaporized, any heat added will increase its temperature (superheat), but not its pressure.

If an evaporator with a fixed metering device is operating at 10°F of superheat, adding warm product increases the superheat (initially), with little increase in its pressure. In fact, the compressor will pull the pressures lower. This may seem strange, but a cap tube will allow only a fixed amount of refrigerant into an evaporator. If that amount is not enough to fill the compressor cylinders, the pressures will fall.

This is normal for a cap tube system during hot pull-down. The system just needs to run for a while until the box temperature comes down. Inexperienced technicians sometimes misdiagnose this situation and add refrigerant. It may help the system initially, but it causes problems after the technician leaves.

On the other hand, *low* superheat means there is too much refrigerant being fed to the evaporator. Some things to look for on a flooding TEV system are:

- TEV superheat adjustment too low
- TEV bulb not sensing the suction line
  - Thermal bulb not mounted properly
  - Thermal bulb may need to be wrapped with insulation
- TEV stuck open
- TEV oversized (usually it will hunt or vary between flooding and starving)

Some things to look for when a cap tube system is flooding are:

- High head pressure (overcharge, dirty condenser, high ambient)
- Incorrect cap tube size (too big)

High superheat on a TEV system indicates a lack of refrigerant going through the evaporator. Some things to look for on a starving TEV system are:

- Superheat adjustment too high
- TEV restricted
- Not enough refrigerant to the valve (low charge, clogged filter drier, liquid line restricted)

High superheat on a cap tube system could be caused by:

- Low charge
- Restricted filter drier or cap tube
- Incorrect cap tube size (too small)

SUBCOOLING RULE: **If the refrigerant is fully condensed, any heat removed will decrease its temperature (subcool), but not its pressure.**

When refrigerant has fully condensed from a hot discharge vapor to a warm liquid before leaving the condenser, any additional heat it rejects into the air blowing across the condenser coil does not lower the refrigerant's pressure, just the liquid's temperature. The temperature below its saturation temperature (from the P/T chart) is subcooling. The pressure in the condenser and in the liquid line remains essentially the same. Only by taking the temperature of the liquid line leaving the condenser and comparing it to the saturation temperature on the P/T chart can we determine subcooling. If the temperature of the liquid line is the same or higher than the saturation temperature, then the refrigerant has not fully condensed to a liquid, and we still have discharge vapor in the liquid line. This indicates there is not enough refrigerant in the system. However, if subcooling is present then all the refrigerant has condensed into a liquid.

Manufacturers design condensers for a specified condenser split and subcooling. A self-contained reach-in needs only about 5°F of subcooling, while most remote units for walk-ins have about 10°F. Condenser manufacturers use enough condenser tubing to accomplish complete condensing, plus the extra tubing necessary to give the desired amount of subcooling.

High subcooling usually occurs for only two reasons:

- Overcharge, which causes an excess of refrigerant in the condenser
- Noncondensables (air), which take up space in the condenser

Low subcooling occurs for only one reason:

- Undercharge because there is not enough refrigerant to condense and subcool

TROUBLESHOOTING TIP: **Each part of a refrigeration system can be affected by another part of the system. If a component is not working properly, do not replace it until first checking to see if one of the other system components is causing the problem.**

EXAMPLE: 12

If a TEV is not feeding refrigerant, it may not be the valve's fault. The system could be low on refrigerant, have an inefficient compressor, or have a stopped-up filter drier among other things.

DIAGNOSING NINE SYSTEM PROBLEMS

When a refrigeration system is completely down, it is fairly easy to diagnose the problem. For instance, if all refrigerant has leaked out of the unit, the compressor valves are completely shot, the condenser is completely covered with dirt, or the evaporator fan is not running, the reason for the unit's inability to refrigerate can be quickly determined. However, the examples given are intended to simulate more difficult situations. In these problems, assume the unit is running, but it is just not down to the temperature the customer needs. Troubleshooting these types of conditions requires a more thorough understanding of refrigeration systems.

The examples cover the following nine system problems technicians are most likely to encounter. Where applicable, an explanation is included of how the same problem may show different symptoms between a cap tube system and a TEV system.

- Undercharge of refrigerant
- Overcharge of refrigerant
- Dirty condenser, or low condenser airflow
- Noncondensable in the condenser
- Inefficient compressor
- Restricted metering device
- Restriction in liquid line after the receiver
- Restriction before the receiver
- Dirty or iced evaporator, or low evaporator airflow

The examples are limited to a reach-in refrigerator designed for 38°F and a walk-in refrigerator designed for 35°F. Both units are operating at 95°F ambient and using R22 refrigerant. Obviously, there are many different applications, from 55°F wine coolers to −40°F blast freezers. Also, the pressures would be different for other ambient temperatures and refrigerants. However, by using examples of only two types of systems with the same problems, it is hoped that the reader will be able to apply the same diagnostic procedures to most any other type of system and operating condition. With a good understanding of how specific refrigeration units respond to problems, the same general diagnostic principles can be applied to most any other system and condition.

First, the tech must know how the system operates when everything is running properly. Following is a list of the proper operating conditions of the two units that are used in the diagnostic exercises:

- Medium-temperature reach-in at 95°F ambient (Figure 7-3):
  - 30°F condenser split
  - 125°F condensing temperature
  - 10°F subcooling
  - 38°F box temperature
  - 20°F evaporator TD
  - 18°F evaporator temperature
  - 10°F superheat
  - Capillary tube metering device

- Medium-temperature walk-in at 95°F ambient (Figure 7-4):

  - 30°F condenser split
  - 125°F condensing temperature
  - 10°F subcooling

- 35°F box temperature
- 10°F evaporator TD
- 25°F evaporator temperature
- 10°F superheat
- TEV metering device

FIGURE 7-3 **Proper operation of an R22 reach-in refrigerator with cap tube.** *Courtesy of Refrigeration Training Services.*

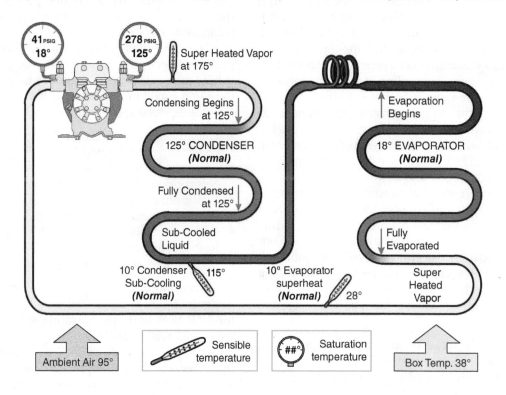

FIGURE 7-4 **Proper operation of an R22 walk-in refrigerator with TEV.** *Courtesy of Refrigeration Training Services.*

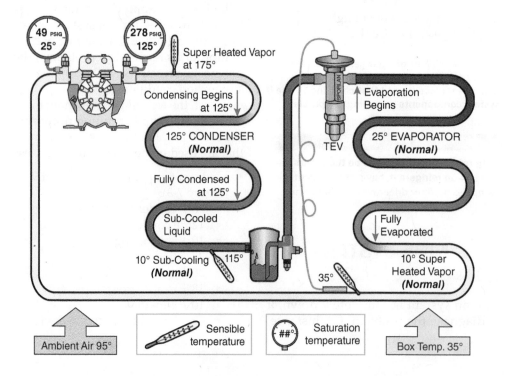

A unit's current operating temperatures must be compared to what they should be. To help with this, each figure has the original pressures and temperatures, but they are slightly shadowed by the current conditions. The current conditions must be analyzed in order to answer the following four questions:

1. Condensing temperature: Is it normal, high, or low?
2. Condenser subcooling: Is it normal, high, or low?
3. Evaporator temperature: Is it normal, high, or low?
4. Evaporator superheat: Is it normal, high, or low?

The ability to properly diagnose a refrigeration system is simply a matter of condensing all the operating information into these four key factors, then determining what can cause the four conditions to vary from normal operation.

UNDERCHARGE OF REFRIGERANT

How an Undercharge Affects a Reach-in with Cap Tube System

A unit completely out of refrigerant would be easy to diagnose. The following example of low charge on a cap tube system assumes it has some refrigerant in it, just not enough to keep the pressures up to normal. In a critically charged unit, anything below 10 percent of the rated charge will cause the system to malfunction. In a reach-in that only holds a pound of refrigerant, undercharging by 2 ounces would cause a problem. Low charge on a cap tube system is due more often to technicians putting their gauges on the unit than it is to leaks. This is easy to do, considering the average high-side hose on a gauge set holds about an ounce of liquid. If there is any doubt about the unit having the correct charge, the most efficient procedure is to recover the existing refrigerant and weigh in the proper amount of new refrigerant. Any further diagnosis will be much more accurate once you know the unit is properly charged.

Following are the questions that need to be answered for each of the troubleshooting examples. They are the same questions you must ask yourself when servicing equipment in the field:

- Evaporator:
  - Evaporator temperature high or low? Why?
  - Superheat high or low? Why?

- Condenser:
  - Condensing temperature high or low? Why?
  - Subcooling high or low? Why?

In Figure 7-5, the evaporator temperature has dropped from 18°F to 10°F, and the evaporator superheat is up to 30°F (40°F − 10°F). There is no condenser split because the condensing temperature is the same as the ambient air. There is no subcooling in the condenser; however, there may be a little in the liquid line.

FIGURE 7-5 **Cap tube system low on charge.** *Courtesy of Refrigeration Training Services.*

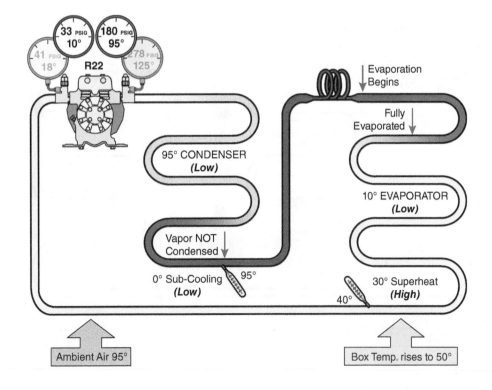

Following are symptoms of a cap tube system low on charge:

- Evaporator:
  - Low evaporator temperature. Lack of refrigerant lowers pressures and temperatures.
  - High superheat. Limited refrigerant boils off quickly in the evaporator.

- Condenser:
  - Low condenser temperature. Little heat is picked up by the starved evaporator, resulting in insufficient heat to raise the condensing temperature.
  - Low subcooling. There is little refrigerant to subcool.

NOTE: *Low* is the key word for units that are low on charge. Everything is low except the superheat, which is high. Low subcooling occurs only when the unit is low on charge.

How an Undercharge Affects a TEV System

The symptoms of a TEV system low on charge (Figure 7-6) are the same as a cap tube system. However, a TEV system often has a liquid-line sight glass that will be bubbling.

- Evaporator:
  - Low evaporator temperature. Lack of refrigerant lowers pressures and temperatures.

- High superheat. If there is not enough liquid to the TEV, the evaporator will starve, and the superheat will rise.

- Condenser:
  - Low condenser temperature. If there is little heat picked up by the starved evaporator, there is not enough heat to raise the condensing temperature.
  - Low subcooling. There is little refrigerant to subcool.

Just like on the cap tube system, the only time low subcooling occurs on a TEV system is when the unit is low on charge.

NOTE: **Low charge on a TEV system does not always mean there is a refrigerant leak. Assume the system has a remote condensing unit located in a cold ambient condition. If the unit has a head pressure regulating valve that backs up refrigerant in the condenser, there may just not be enough refrigerant in the system. If the unit was charged in warm weather, the technician may not have added enough extra refrigerant for cold ambient conditions. Chapter 9 covers charging for low-ambient conditions in great detail. Also, if the unit is using a fan cycle control, the fan off setting may not be adjusted low enough.**

OVERCHARGE OF REFRIGERANT

How an Overcharge Affects a Cap Tube System

In Figure 7-7, the condensing temperature has increased from 125°F to 145°F, and the subcooling has

FIGURE 7-6 **TEV system low on charge.** *Courtesy of Refrigeration Training Services.*

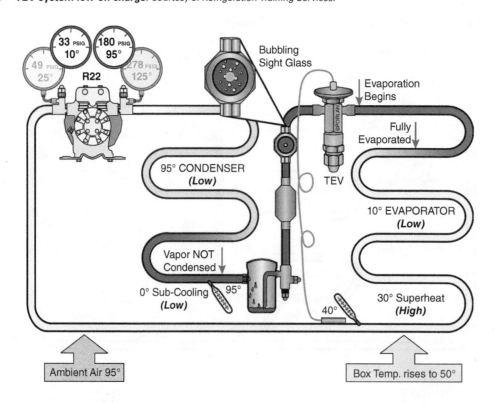

FIGURE 7-7   **Cap tube system overcharged.** *Courtesy of Refrigeration Training Services.*

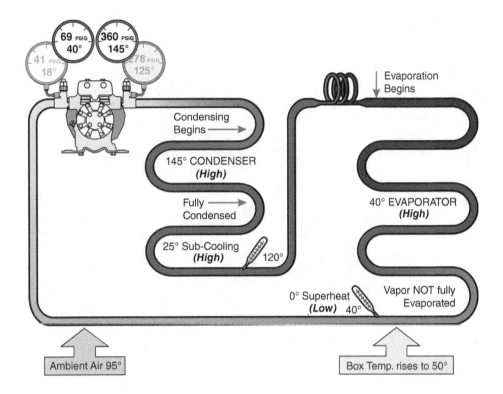

gone up to 25°F (145°F condensing − 120°F liquid line). The evaporator temperature has risen from 18°F to 40°F, and there is no superheat. Both of the evaporator conditions occur because the excess high-pressure liquid from the condenser is being forced through the evaporator in a greater quantity than the evaporator can boil away.

Following are symptoms of an overcharged cap tube system:

- Condenser:
  - High condensing temperature and pressure. Excess refrigerant is taking up space in the condenser.
  - High subcooling. There is more liquid to subcool.

- Evaporator:
  - Higher-temperature evaporator. The high pressure from the condenser is pushing more refrigerant through the cap tube.
  - Low superheat. The refrigerant being pushed through the cap tube is flooding the evaporator.

NOTE: **It does not take much to overcharge a cap tube system. Above 10 percent of the rated refrigerant charge can cause the above conditions. Check with the customer to see if another technician recently serviced the unit and added refrigerant. She may have mistaken the normally long pull-down period for lack of refrigerant. Or she may have put her gauges on the unit and assumed the low suction pressure during the hot pull-down was due to low charge.**

How an Overcharge Affects a TEV System

In Figure 7-8, the TEV system has similar conditions to the overcharged cap tube system. However, the TEV can hold back the pressure and meter refrigerant into the evaporator based on the temperature of the suction line leaving the evaporator.

Following are symptoms of too much refrigerant in a TEV system:

- Condenser:
  - High temperatures and pressures. The excess refrigerant is taking up space in the condenser, and the superheated discharge vapor does not have room to cool and condense.
  - High subcooling. The excess refrigerant is just piling up and being subcooled by the ambient air pulled through the condenser.

- Evaporator:
  - Slightly higher temperatures, depending on the amount of overcharge. The TEV is good at holding back even high condensing pressures.
  - Normal superheat. The TEV closes down if it senses a cold suction line.

NOTE: **Someone may have overcharged the system during the winter because the unit does not have a head pressure control, or the head pressure control does not work. Or maybe refrigerant was added during a low-load situation when some normal bubbling of the sight glass occurred, and the technician mistook it for being low on charge.**

FIGURE 7-8   **TEV system overcharged.** *Courtesy of Refrigeration Training Services.*

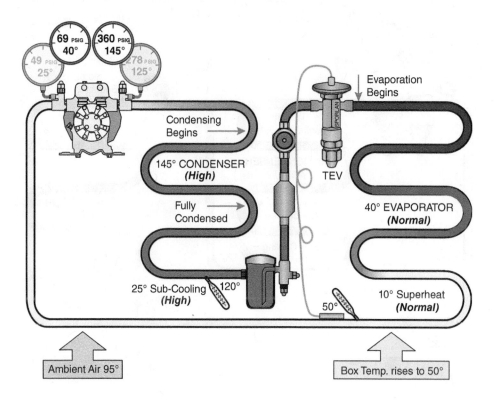

CONDENSER AIRFLOW PROBLEMS

Condenser Problems on a Cap Tube System

High ambient conditions, condenser discharge air recirculating back into the condenser air intake, discharge air from another unit entering the condenser, fan motor or blade problems, and dirt buildup on a condenser will all show similar symptoms. Since dirt buildup is by far the most common cause, all of the condenser problems just stated will be referred to as *dirty condenser* in this section.

In Figure 7-9, the reduced airflow through the condenser has increased the condensing temperature to 145°F. The subcooling has remained about normal because the amount of refrigerant has not changed and is allowed to cool the usual 10°F (145°F − 135°F) before leaving the condenser.

Following are symptoms of a dirty condenser on a cap tube system:

- Condenser:
  - Higher temperature. The ambient air cannot transfer the heat from the dirty condenser as efficiently as when the condenser is clean.
  - Normal subcooling. The amount of refrigerant has not changed.

- Evaporator:
  - Higher temperatures. High condensing pressure forces higher pressure refrigerant through evaporator.
  - Low to no superheat. High condensing pressure pushes refrigerant through the evaporator faster than it can boil away.

Condenser Problems on a TEV System

A condenser may look clean on the surface but could have the fins blocked with dirt deep inside. There are specific indicators that the heat inside a condenser is not being properly rejected to the ambient air.

In Figure 7-10, poor heat transfer raises the condensing temperature, but subcooling remains about average. As in the overcharge, the TEV once again tries to hold back higher condensing pressures and feed only enough refrigerant to the evaporator to maintain the proper superheat. However, the evaporator temperature may be forced to rise somewhat (and the superheat to fall) as the high condensing pressure affects the TEV more than it can handle.

Following are symptoms of a dirty condenser on a TEV system:

- Condenser:
  - Higher temperatures. The dirt on the condenser reduces proper heat transfer.
  - Subcooling is about normal because the amount of refrigerant is unchanged.

FIGURE 7-9   **Cap tube system with a dirty condenser.** *Courtesy of Refrigeration Training Services.*

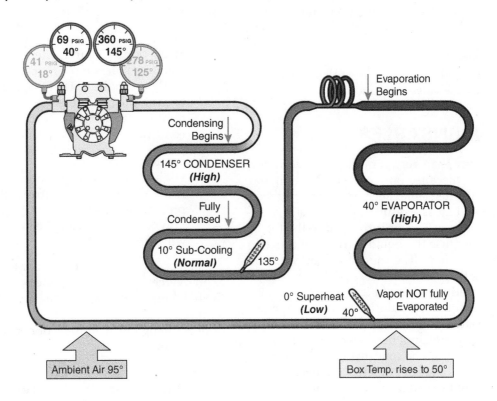

FIGURE 7-10   **TEV system with a dirty condenser.** *Courtesy of Refrigeration Training Services.*

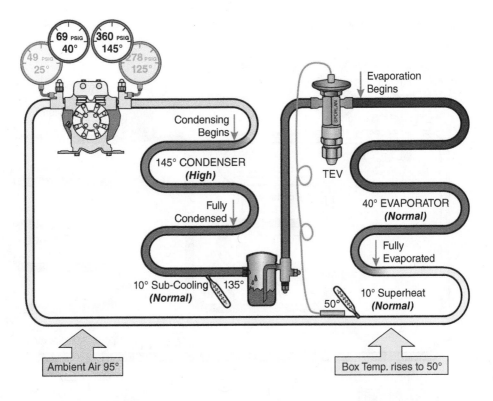

- Evaporator:
  - Slightly higher temperature. A little more refrigerant is being pushed through the valve, or the lower system efficiency is not able to cool the space properly.
  - Normal superheat until the pressure becomes more than the valve can handle.

NONCONDENSABLES

Any vapor, other than refrigerant, is considered a noncondensable because only refrigerants boil and condense. Noncondensables such as air and nitrogen may be in the system as a result of poor service or installation procedures. The air is trapped in the top of the condenser because the liquid refrigerant acts like a p-trap to keep it there. The vapor takes up valuable space, reducing the effective surface of the condenser. Therefore, the head pressure and temperature must increase in an attempt to reject the heat in the refrigerant from a smaller area of the condenser. The amount of subcooling is very high, not because of more refrigerant, but simply because the condensing temperature is so high, and there is a greater difference between it and the temperature of the liquid leaving the condenser.

Just like an overcharge or dirty condenser, the cap tube system with noncondensables will force more refrigerant through the evaporator, raising temperatures and lowering superheat.

Following are symptoms of noncondensables in a cap tube system (Figure 7-11):

- Condenser
  - High temperature. The air takes up space, resulting in less room for condensing.
  - High subcooling. There is a greater difference between the condensing temperature and the temperature at which the liquid leaves the condenser.

- Evaporator:
  - High temperature and pressure. The cap tube has a fixed pressure drop. High condensing pressure means high evaporator pressures.
  - Low to no superheat. The evaporator is flooded with refrigerant.

Figure 7-12 shows how similar air in the system is to an overcharge or dirty condenser, except that relatively higher pressures and higher subcooling can be found.

Following are symptoms of noncondensables in a TEV system:

- Condenser
  - High temperature. Air takes up space, leaving less room for condensing.
  - Very high subcooling. Greater difference between high condensing temperature and the leaving liquid temperature.

FIGURE 7-11 **Noncondensables in a cap tube system.** *Courtesy of Refrigeration Training Services.*

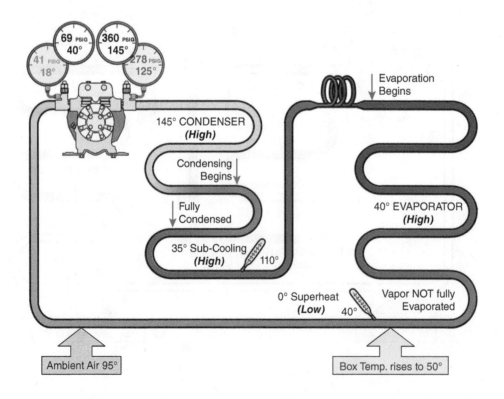

FIGURE 7-12 **Noncondensables in a TEV system.** *Courtesy of Refrigeration Training Services.*

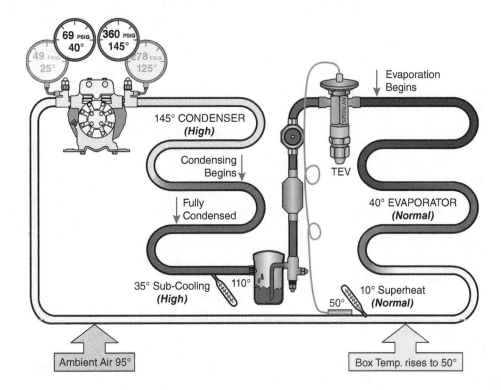

- Evaporator
  - Normal to slightly higher temperature. The system is less efficient and not able to maintain box temperature.
  - Normal to slightly lower superheat. The valve is doing its best to maintain the suction-line temperature. However, the increased pressure in the condenser due to noncondensables may be great enough to force too much refrigerant through the TEV, flooding the evaporator.

NOTE: **How can you tell if there is air in the system, rather than just an overcharge or hidden dirt in the condenser fins?**

Noncondensables cause higher subcooling than dirt or overcharge, but how much higher is difficult to say. A critically charged reach-in has so little refrigerant that it is easy to recover it and pull a good vacuum. This procedure will ensure that there is no air in the system. Also, weighing in new refrigerant verifies that the unit is correctly charged. If the technician is curious about how much refrigerant was originally in the unit, he can weigh the refrigerant that was removed.

On larger systems, diagnosing noncondensables can be accomplished by shutting off the compressor and observing the head pressure on the high-side gauge. On a normal system, the head pressure steadily falls as the condenser cools. However, air in the system prevents the pressure from dropping like a normal system. If the head pressure remains high for a few minutes after the compressor is turned off, there is

probably air in the system. It is recommended that this procedure be practiced on systems known to be free of noncondensables. This will help develop a feeling for just how fast the head pressure should drop in a proper system, after the compressor is turned off. Some technicians save diagnostic time by keeping the condenser fan running after they shut off the compressor. The condenser fan will make the head pressure fall even sooner if there is no air in the system.

A more exact method of checking for noncondensables is to shut the system off for about 15 minutes. This should give the refrigerant in the condenser time to drop to the ambient temperature. In order to speed up the process, temporarily wire the condenser fan(s) to run while the compressor is off. When the temperatures of the copper tubing entering and leaving the condenser are the same as the ambient temperature, check the head pressure. Use a P/T chart to find the condensing temperature based on head pressure. If the condensing temperature is the same as the ambient, there is no air in the system. However, if the two temperatures are different, then noncondensables need to be removed from the system.

INEFFICIENT COMPRESSOR

An inefficient compressor is one with valves or piston rings damaged badly enough to affect its pumping ability. The conditions will be the same for both cap tube and TEV systems.

Following are symptoms of an inefficient compressor on a TEV or cap tube system (Figure 7-13):

- Condenser:
  - Discharge pressure is low. Leaking valves or rings do not allow the pressure to build up during the compression stroke.
  - Normal subcooling. The refrigerant is not moving and remains in the condenser to be subcooled.

- Evaporator:
  - Suction pressure is high. The discharge gas is being pushed back into the suction side of the system.
  - High superheat. The evaporator is starving because the compressor is not pumping refrigerant through the metering device.

NOTE: **The primary indicator of an inefficient compressor is that it will be operating at a higher-than-normal suction pressure and at the same time have a lower-than-normal head pressure.**

A technician troubleshooting these symptoms should note that a compressor with this problem may not pump down into a vacuum. But if it does, it will not hold the vacuum for more than a few minutes before the low-side gauge pressure starts rising.

NOTE: **It is not the amount of vacuum the compressor will pull that is important. It is whether it will pull any vacuum at all and if it can maintain it. A rise in pressure of more than 3 psig per minute is generally considered excessive valve leakage.**

A good extra-low-temperature compressor will normally pull a very low vacuum, whereas a high-temperature refrigeration compressor may pull only a 5-inch Hg (column of mercury) vacuum. However, if the compressor will not pull a vacuum, it has valve or ring problems.

A word of caution: Even if the compressor will pull a vacuum and maintain it, this does not necessarily mean that the compressor is good. Even a compressor with a broken connecting rod may pass the vacuum test. Less drastic—but equally difficult to pinpoint—is the excessive bearing wear in the connecting rods or wrist pins. This condition will prevent the piston from rising as high as it should on the compression stroke. The increased clearance volume between the top of the piston and the valve plate will require excessive re-expansion before the suction valve will open. The result is a decrease in the compressor's volumetric efficiency.

NOTE: **Most air conditioning compressors are not designed to pull a vacuum. Carrier, for one, suggests pulling down to only about 5 psig before checking if the valves will hold.**

NOTE: **Scroll compressors should never be pumped into a vacuum. For one thing, they do not have valves to check. For another, pulling into a vacuum will damage the compressor.**

Compressor manufacturers recommend using their specification (spec) sheets to check a compressor's operation. The sheets provide a table or a graph that plots a compressor's capacity based on pressures

FIGURE 7-13   **Inefficient compressor on a TEV or cap tube system.** *Courtesy of Refrigeration Training Services.*

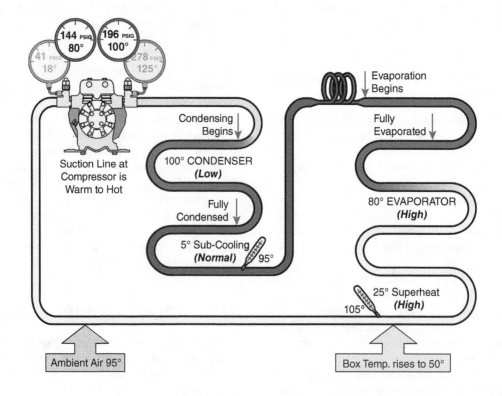

TABLE 7-1   Revised Copeland Compressor Performance Chart

RATING CONDITIONS 65°F Return Gas 0°F Subcooling 95°F Ambient Air Over		COMPRESSOR Model 9RJ1-0765-TFC COPELAMETIC® HCFC-22 208/230-3-60 Evaporating Temperature (°F)		
		15	20	25
130°F Condensing	Capacity (Btuh)	52,500	59,500	66,500
	Amperage (A)	27.8	29	30.2
120°F Condensing	Capacity (Btuh)	57,000	64,000	72,000
	Amperage (A)	27.1	28.1	29.2
110°F Condensing	Capacity (Btuh)	61,500	69,000	77,500
	Amperage (A)	26.3	27.2	28.2
100°F Condensing	Capacity (Btuh)	66,000	74,500	83,500
	Amperage (A)	25.5	26.3	27.1

Amperage ratings are valid within ±5 percent at 230V 3-phase.

and/or temperatures, voltage, and amperage under a wide range of conditions. Using the manufacturer's information will definitely verify whether the compressor in question is operating as it should be. If the compressor is not drawing the indicated current according to the spec sheets, then it definitely has a problem. Table 7-1 is an example of a Copeland compressor performance chart for a specific compressor.

RESTRICTED METERING DEVICE

Partially Restricted Capillary Tube

In Figure 7-14, the evaporator temperature drops and the superheat rises because there is not enough refrigerant feeding the evaporator. The condensing temperature is low because there is little heat being picked up in the evaporator for the condenser to reject. The subcooling is higher than normal because the reduced amount of refrigerant to the evaporator is backed up in the condenser. An excess amount of refrigerant in the condenser means higher subcooling.

Following are symptoms of a partially restricted cap tube:

- Evaporator:
  - Low temperature. Lack of refrigerant lowers pressure and saturation temperature.
  - High superheat. There is less refrigerant to boil away.
- Condenser:
  - Low condensing temperature. Little heat from the evaporator.
  - High subcooling. Refrigerant not used in the evaporator is stored in the condenser.

 R E A L I T Y   C H E C K   1

It is difficult to find a "partially" restricted cap tube because by the time the customer notices a problem, the small tubing will be completely blocked. A completely blocked cap tube is easy to diagnose. The evaporator will be in a vacuum, yet there will be enough head pressure and subcooling to indicate the condenser has refrigerant in it.

The most common restriction in a cap tube is a white powder. It comes from the beads of desiccant in the filter drier and builds up at the entrance of the cap tube. The usual solution to this problem is to cut off the first inch or two of the cap tube and replace the filter drier.

NOTE: **Some technicians believe a restricted cap tube should raise the head pressure.**

Those technicians believe that if the refrigerant is not in the evaporator, it must be backing up in the condenser and raising the head pressure. Yes, it is backed up in the condenser. No, it does not increase head pressure. The entire amount of refrigerant in a critically charged system is not enough to fill the condenser. In fact, the head pressure may even drop below normal because there is no heat from the evaporator being returned to the condenser.

The only way the head pressure will increase on a cap tube system is:

- High ambient temperatures
- Poor airflow through the condenser (dirt, grease, or fan problems)
- Overcharge of refrigerant

FIGURE 7-14   **Partially restricted cap tube.** *Courtesy of Refrigeration Training Services.*

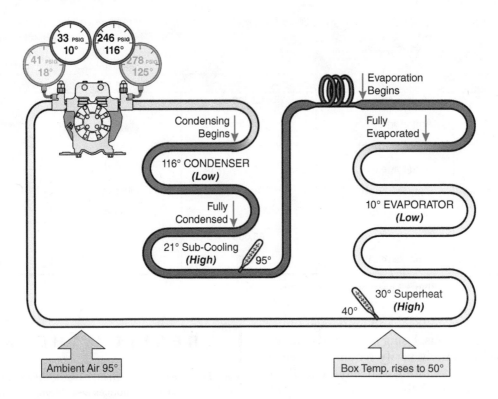

If a unit has a stopped-up cap tube with a high head, it is probably because a previous technician misdiagnosed the low suction and head pressures, and added refrigerant. That is why a good technician never just "adds refrigerant" on a critically charged unit. The technician should always remove the old refrigerant and weigh in new refrigerant per the rating plate.

Partially Restricted TEV

In Figure 7-15, the partially restricted TEV has similar symptoms to the restricted cap tube system. The main

FIGURE 7-15   **Partially restricted TEV.** *Courtesy of Refrigeration Training Services.*

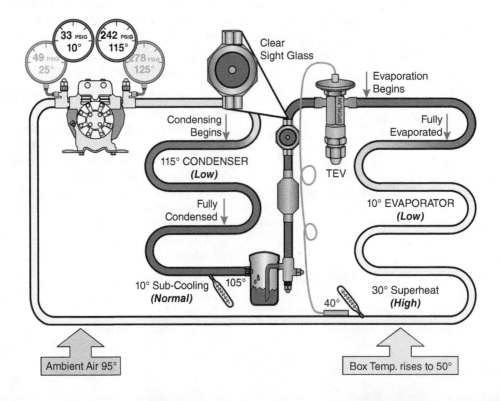

difference is that the subcooling would be normal in the TEV unit because the excess refrigerant is not trapped in the condenser but flows to the storage receiver. Also, a clear sight glass verifies there is liquid to the metering device.

Following are symptoms of a partially restricted TEV:

- Evaporator:
  - Lower temperature. Lack of refrigerant means lower pressure, which means lower saturated temperature.
  - High superheat. Limited refrigerant is boiled away quickly.

- Condenser:
  - Lower temperature. Little heat is picked up in the evaporator.
  - Normal subcooling. The refrigerant is being stored in the receiver, not in the condenser.

RESTRICTION IN THE LIQUID LINE AFTER THE RECEIVER

Only TEV systems have receivers. In Figure 7-16, the restriction is after the receiver, but before the metering device. There are two indicators of a restriction in the liquid line. First, the condenser has normal subcooling, but the liquid leaves the condenser at 105°F, only

to drop drastically to 85°F by the time it gets to the metering device. A drop in pressure creates a drop in temperature. Only a metering device or a restriction could make enough of a pressure drop to lower the temperature by 20°F. If the technician uses an electronic thermometer and takes the temperature between the receiver and the metering device, she will soon find the restriction.

The second indicator is bubbling in the sight glass, if it is located after the filter drier. If a kinked line is not evident, the most likely location for a restriction is in the filter drier. A temperature drop of more than 3°F between the inlet and outlet of the drier will verify the problem.

Following are symptoms of a restriction after the receiver:

- Evaporator:
  - Lower temperature. Lower pressure due to lack of refrigerant.
  - High superheat. Limited refrigerant boiling away quickly.

- Condenser:
  - Low condensing temperature. The evaporator is picking up very little heat.
  - Normal subcooling. The restriction backs up the liquid refrigerant into the receiver, not the condenser.
  - The liquid line would be relatively cool entering the TEV.

FIGURE 7-16 **Restriction after the receiver.** *Courtesy of Refrigeration Training Services.*

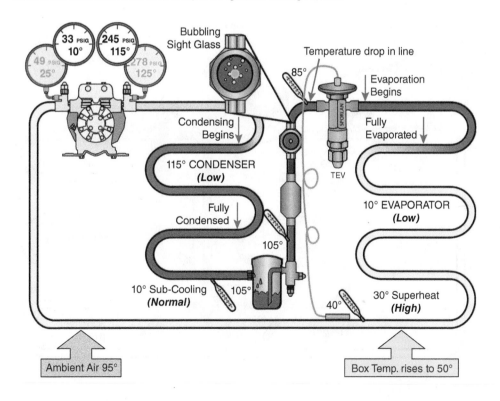

REALITY CHECK 2

Restrictions before the receiver are very rare. However, this chapter covers the diagnosis because symptoms occasionally point to a high-side restriction. More importantly, a technician should be able to quickly confirm, or eliminate, this type of restriction as a possible problem.

RESTRICTION IN THE HIGH SIDE BEFORE THE RECEIVER

Figure 7-17 shows similar symptoms to the restriction after the receiver, except the drop in liquid-line temperature is at or before the receiver.

Following are symptoms of a restriction before the receiver:

- Evaporator:
  - Low evaporator temperature and pressure due to lack of refrigerant getting to the evaporator.
  - High superheat because the limited refrigerant is boiled away quickly.

- Condenser:
  - High temperature and pressure. Excess refrigerant, normally stored in the receiver, is all packed into the condenser.
  - High subcooling due to more liquid in the condenser.

NOTE: The most likely place for this kind of restriction is where the liquid line from the condenser is connected to the receiver

inlet. This is especially true in smaller condensing units where the liquid line is only 1/4-inch to 3/8-inch copper. The only other place it might occur is damaged or flattened u-bends on the condenser. Once again, using an electronic thermometer to determine exactly where the temperature drop occurs will pinpoint the restriction.

DIRTY EVAPORATOR, ICED EVAPORATOR, OR LOW AIRFLOW

Evaporator Problems on a Cap Tube System

In Figure 7-18, the evaporator temperature is low because there is little heat transfer when the evaporator is covered with dirt, frost, or cellophane wrap, or when the fan motor is bad. If heat is not being absorbed in the evaporator, the condensing temperature is low because there is little heat to reject. The subcooling is normal since the refrigerant charge is correct and flowing freely to the cap tube.

Following are symptoms of a dirty or iced evaporator, or low airflow on a cap tube system:

- Evaporator:
  - Lower temperature and pressure. The refrigerant cannot pick up enough heat to boil the refrigerant very much.
  - Low or no superheat. The refrigerant has not vaporized.
  - Some superheat at the compressor suggests the refrigerant is vaporizing in the suction line, but not in the evaporator.

FIGURE 7-17   **Restriction before the receiver.** *Courtesy of Refrigeration Training Services.*

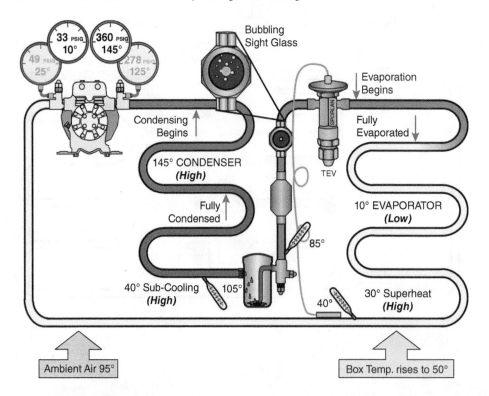

FIGURE 7-18 **Dirty or iced evaporator, or low airflow on a cap tube system.** *Courtesy of Refrigeration Training Services.*

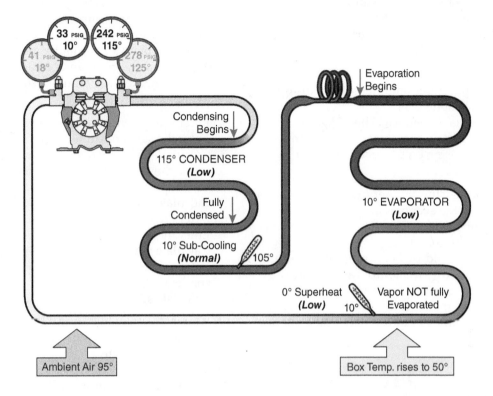

- Condenser:
  - Low temperature and pressure. The evaporator is not picking up evaporator heat for the condenser to reject.
  - Normal subcooling. The system has the correct amount of refrigerant, and it is flowing through the evaporator back to the condenser.

Evaporator Problems on a TEV System

In Figure 7-19, the evaporator temperature is slightly lower, yet not as much as the cap tube system. Once again, the TEV bulb is sensing the evaporator outlet temperature and adjusts the amount of refrigerant entering the evaporator. Given that the evaporator is

FIGURE 7-19 **Dirty or iced evaporator, or low airflow on a TEV system.** *Courtesy of Refrigeration Training Services.*

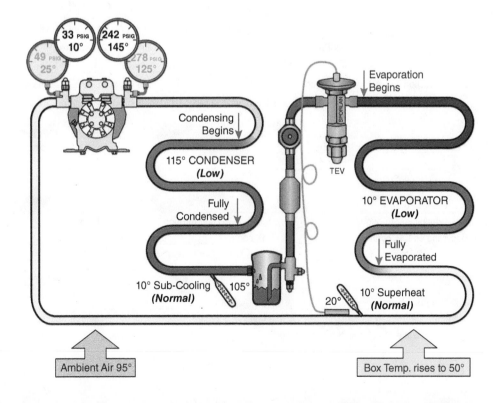

not picking up much heat, the condenser is cooler because it has less heat to reject. The subcooling is normal because the system is properly charged.

Following are symptoms of a dirty or iced evaporator, or low airflow on a TEV system:

- Evaporator:
  - Lower temperatures and pressures. This is due to reduced heat transfer.
  - Low to normal superheat. The TEV is trying to maintain a certain temperature based on what the bulb is sensing on the suction line.
- Condenser:
  - Lower condensing temperature. Little heat is picked up in the evaporator.
  - Normal subcooling. The system has the proper amount of refrigerant.

By now, you should be beginning to see a definite pattern to reasons for temperature and pressure changes throughout a system. Understanding these concepts well makes it much easier to troubleshoot a refrigeration unit and to arrive at a correct diagnosis.

GETTING THE RIGHT INFORMATION

To properly diagnose a problem, a technician must first collect the correct information. This includes the type of system, how it is supposed to operate, and how it is currently operating. To help with this information gathering, you can use the System Information sheet in Figure 7-20. The more important required data is listed first under Pressures & Temperatures.

Filling in blanks 1 through 5 makes it easy to calculate superheat, and blanks 6 through 10 are used for subcooling. Refer to Figure 7-20 while reading the following:

1. Enter the air temperature entering the evaporator.
2. Enter the suction temperature.
3. Calculate the evaporator TD (line 1 − line 2).
4. Enter the temperature of the suction line leaving the evaporator.
5. Calculate the superheat (line 4 − line 2).
6. Enter the air temperature entering the condenser.
7. Enter the condensing temperature.
8. Calculate the condenser split (line 7 − line 6).
9. Enter the temperature of the liquid line leaving the condenser.
10. Calculate subcooling (line 7 − line 9).

All the information on the sheet is necessary for proper diagnosing. However, after using it a few times, most technicians realize they are running most of the checklist mentally, rather than in writing. This is the desired outcome because it is much quicker to think through the steps than to write them down. It is still good practice to write down pressure, temperature, and electrical information on service tickets or work orders. It helps document what you find and will serve as a history of the equipment's operation.

FIGURE 7-20 **System Information sheet.** *Courtesy of Refrigeration Training Services.*

Pressures & Temperatures: (Makes calculations easy)

Design temp. of box ________ Air temp. entering evaporator (1) A   Air temp. entering the condenser (6) X

Refrigerant Type ________ Suction: Pressure ______ Temp. (2) B   Condensing: Pressure ______ Temp. (7) Y

Evaporator TD (3) A−B   Condenser split (8) Y−X

Temp. of suction line at TEV bulb (or coil outlet, if cap tube) (4) C   Temp. of liquid line leaving condenser (9) Z

Superheat (Suction line temp. − Suction temp.) (5) C−B   Subcooling (Cond. temp. − Liq. line temp.) (10) Y−Z

System Components & Accessories:

Cap Tube? ________ Exp. Valve? ________ Pumpdown Sol? ________ Evap. below compressor? ________

Remote? ________ Where? ________ HPR? ________ Fan Cycle? ________ P-trap? ________

Current Operation & Condition:

Evap. coil clean clear through? ________ Condenser clean clear through? ________ Drain clear? ________

Comp. cycling on L.P. control? ________ ,on H.P. control? ________ ,on Compr. O.L.? ________ Comp. noisy? ________

Sight glass full? ________ Door closed & gasket sealing? ________

Additional information relative to compressor operation:

Nameplate Amps (RLA) ________ Actual Amps ________

Nameplate Volts ________ Actual Voltage @ comp. ________ Actual Voltage on Start-up ________

Temp. of Suction line 3"– 6" from the compressor ________ Temp. of Discharge line 3"– 6" from the compressor ________

USING A DIAGNOSTIC CHART

Troubleshooting is the practice of mentally processing observations in a logical sequence in order to diagnose the problem. When available, a chart can help organize those observations into a pattern that assists the user in determining the solution. Figure 7-21 is a diagnostic chart created to help troubleshoot the nine system problems discussed in this chapter.

The four main symptoms are alterations in condensing temperature, subcooling, evaporator temperature, and superheat. To use the chart, just determine if the conditions at which the unit is operating are normal, higher than normal, or lower than normal. Each symptom has a statement about what is considered normal and how many degrees from normal is too high or too low.

The last category is the sight glass for TEV systems: Is it bubbling or clear?

To use the chart, just circle all the Xs in the row that applies to the observed condition. For instance, if the unit's condensing temperature is more than 10°F above normal, circle the Xs in the first row (HIGH) in columns 6 through 9. If the evaporator temperature is more than 10°F above normal, circle the Xs in the HIGH row in columns 4, 7, 8, and 9. An X means the condition applies to both TEV systems and fixed metering device systems or cap tube systems. The X_{FM} means the condition applies only to fixed metering

(FM) device or cap tube systems. FM is used because this chart can be used for air conditioning, and A/C techs are familiar with using the terms *fixed orifice* and *pistons* to describe the fixed metering devices. However, commercial refrigeration techs work with capillary tubes (or "cap tubes"), which are another type of fixed metering device. So, we have lumped all these devices under the term *fixed metering*, or FM. Likewise, X_{EV} applies only to systems using expansion valves.

After all the appropriate Xs are circled for each category, total the number of Xs in each column and enter the numbers in the blank boxes at the bottom of the chart. If the information has been collected and entered correctly, the column with the highest total of Xs is the problem.

Figure 7-22 is an example of how to use the diagnostic chart to find one of the problems discussed in this chapter.

Assume the temperatures and pressures of a problem TEV system are taken. The system exhibits the following conditions:

- **Low condensing temperature**
- **Low subcooling**
- **Low evaporator temperature**
- **High superheat**
- **Sight glass is bubbling**

FIGURE 7-21   **Diagnostic chart.** *Courtesy of Refrigeration Training Services.*

Determine which symptoms apply, then circle all the X's in the row for each symptom.
Total all the X's in each column. The column with the most X's is the problem

"X" symptoms apply to both Fixed Metering and Expansion Valve systems
"FM" symptoms apply only to Fixed Metering device systems
"EV" symptoms apply only to Expansion Valve systems

	Column Number ▸	1	2	3	4	5	6	7	8	9
Ambient + Cond. split / **Condensing Temperature** / Standard Units = Ambient + 30°	**HIGH** (10° higher than Normal)						X	X	X	X
	NORMAL									
	LOW (10° lower than Normal)	X	X	X	X	X				
Cond. temp. - Liquid line temp. / **Condenser Subcooling** / Normal Subcooling 5° to 20°	**HIGH** (Subcooling above 20°)		X FM				X		X	X
	NORMAL	X	X EV		X	X		X		
	LOW (Subcooling below 5°)			X						
Air Entering Evap. - TD / **Evaporator Temperature** / TD for A/C (35°), R/I (20°), W/I (10°)	**HIGH** (10° higher than Normal)				X			X FM	X FM	X FM
	NORMAL							X EV	X EV	X EV
	LOW (10° lower than Normal)	X	X	X		X	X			
Suction line at evap. - Evap. temp. / **Evaporator Superheat** / Normal Superheat 5° to 20°	**HIGH** (Superheat above 20°)		X	X	X	X	X			
	NORMAL	X EV						X EV	X EV	X EV
	LOW (Superheat below 5°)	X FM						X FM	X FM	X FM
Sight Glass	**FULL**	X	X		X	X		X	X	X
	BUBBLING			X		X	X			
Total the Circled X's in each Column =										
DIAGNOSIS (problem):		DIRTY or ICED EVAP	RESTR TEV or FM	LOW CHG	COMP VALVES	RESTR AFTER RECV	RESTR BEFOR RECV	DIRTY COND	AIR IN SYSTEM	OVER CHG.

FIGURE 7-22 **Using the diagnostic chart to find a problem.** *Courtesy of Refrigeration Training Services.*

Determine which symptoms apply, then circle all the X's in the row for each symptom.
Total all the X's in each column. The column with the most X's is the problem

"X" symptoms apply to both Fixed Metering and Expansion Valve systems
"FM" symptoms apply only to Fixed Metering device systems
"EV" symptoms apply only to Expansion Valve systems

		Diagnostic Chart								
	Column Number ▶	1	2	3	4	5	6	7	8	9
Ambient + Cond. split **Condensing Temp. LOW** Standard Units = Ambient + 30°	*HIGH* (10° higher than Normal)						X	X	X	X
	NORMAL									
	LOW (10° lower than Normal)	Ⓧ	Ⓧ	Ⓧ	Ⓧ	Ⓧ				
Cond. temp. - Liquid line temp. **Cond Subcooling LOW** Normal Subcooling 5° to 20°	*HIGH* (Subcooling above 20°)		X FM				X		X	X
	NORMAL	X	X EV		X	X		X		
	LOW (Subcooling below 5°)			Ⓧ						
Air Entering Evap. - TD **Evaporator Temp. LOW** TD for A/C (35°), R/I (20°), W/I (10°)	*HIGH* (10° higher than Normal)				X			X FM	X FM	X FM
	NORMAL							X EV	X EV	X EV
	LOW (10° lower than Normal)	Ⓧ	Ⓧ	Ⓧ		Ⓧ	Ⓧ			
Suction line at evap. - Evap. temp. **Evaporator Superheat HIGH** Normal Superheat 5° to 20°	*HIGH* (Superheat above 20°)		Ⓧ	Ⓧ	Ⓧ	Ⓧ	Ⓧ			
	NORMAL	X EV						X EV	X EV	X EV
	LOW (Superheat below 5°)	X FM						X FM	X FM	X FM
Sight Glass BUBBLING	FULL	X	X		X	X		X	X	X
	BUBBLING			Ⓧ		Ⓧ	Ⓧ			
Total the Circled X's in each Column =		2	3	5	2	4	3	0	0	0
DIAGNOSIS (problem):		DIRTY or ICED EVAP	RESTR TEV or FM	LOW CHG	COMP VALVES	RESTR AFTER RECV	RESTR BEFOR RECV	DIRTY COND	AIR IN SYSTEM	OVER CHG.

Look at Figure 7-22 to see how the appropriate Xs are circled for the following conditions:

- For Condensing Temperature in the row to the right of LOW, circle all five Xs.
- For Subcooling in the row to the right of LOW, there is only one X to circle.
- For Evaporator Temperature in the LOW row, there are five Xs to circle.
- For Evaporator Superheat in the HIGH row, there are also five Xs.
- For Sight Glass in the BUBBLING row, there are three Xs.

After totaling all the Xs in each column, the result is that column 3 has more Xs (five) than the other columns. The diagnosis is therefore low charge.

NOTE: **If this were a cap tube system (no sight glass), column 3 would still have the most Xs. Even without the sight glass information, the same low-charge symptoms apply for both TEVs and cap tubes.**

Using the diagnostic chart for troubleshooting refrigeration is similar to using a calculator in mathematics. It is not necessary for every problem, but it is good for difficult situations and to verify your calculations. The chart is just another tool; the more you

master troubleshooting techniques, the less you will use it. The diagnostic chart works very well if:

- The user knows what the proper operating conditions are supposed to be for the system being working on
- The user takes accurate readings of the system's current operations

The symptoms of HIGH *or* LOW on the diagnostic chart are deviations from the normal operation of the system under consideration. For example, most refrigeration condensers have a 30°F TD, but a freezer may have only a 25°F condenser split, and a large remote condenser may have only 10°F. The important thing to remember is that 10°F above the norm for the unit being serviced is too high, and 10°F below it is too low. Likewise, 10°F superheat is normal for a refrigerator. But a freezer is closer to 5°F, and many A/C units are close to 15°F. Whenever you are checking superheat, 5°F above or below the norm may indicate a problem. In all cases, a superheat below 5°F means there is a real danger of flooding the compressor, and superheat above 20°F means the evaporator is starving.

NOTE: **Superheat readings are accurate only if the space temperature is within 5°F of its design conditions.**

EVAPORATOR OIL LOGGING

Oil trapped in an evaporator is a unique problem that is often difficult to troubleshoot. It is usually diagnosed as a result of eliminating all other possibilities. If the conditions observed do not fit any of the nine problems discussed here, the problem may very well be an oil-logged evaporator.

The evaporator traps oil if the suction line is not piped for proper oil return. Also, low coil temperatures slow the flow of oil to the point that it collects in the evaporator. Defrosting freezer coils is as important for oil return as it is to melt the frost from the fins.

Following are some common symptoms of an oil-logged evaporator:

- The system has a history of "just does not seem to maintain temperatures."
- The symptoms do not fit any of the normal system problems.
- The system has TEV hunting or floodback, or the superheat cannot be adjusted.

The above symptoms remain even after the TEV is replaced. The reason is that the oil insulates the suction line. The TEV sensing bulb cannot detect true suction-line temperature. The insulating effect of the oil will also prevent you from getting an accurate temperature with your thermometer on the suction line. Therefore, your superheat calculation will not be accurate.

- Some u-bends near the evaporator outlet are not sweating or frosting like the others.
- There is a history of compressor changes.

The first compressor failed due to lack of oil or from damage caused by its oil trapped in the evaporator or restricting flow somewhere in the system. Subsequent compressor replacements do not solve the problem; they only provide more oil to increase the oil-logging problem.

- The oil level in the compressor's sight glass varies greatly during the cycle.

The oil is being lost in the system and then returns during startup or after defrost.

- The compressor is noisy or vibrating.

Lack of oil produces noise from bearing wear and piston slap. Too much oil can cause noise and vibration from the crankshaft hitting the oil in the crankcase.

- There is evidence of flooding or slugging.

Low to no superheat at the compressor inlet indicates floodback. Also, flooding damage and evidence of oil on the piston will be observed during a teardown of the failed compressor.

NOTE: Oil slugging damages more compressors than liquid slugging. It is difficult to diagnose when the system is operating. However, during teardown, look for excessive oil accumulation in the piston and valve areas.

- There is an absence of a pump-down solenoid on a remote commercial refrigeration condensing unit.

Pump-down helps to move the oil and refrigerant out of the evaporator and suction lines before the compressor shuts off.

- The freezer has only one defrost.

In very dry climates, a freezer coil may adequately remove frost with only one or two defrosts a day. However, if oil return is a problem, adding defrosts may be the answer.

- There is evidence of incorrect piping practices:
  - There are no p-traps when the compressor is above the evaporator.
  - Suction lines must be sloped in the direction of refrigerant flow.
  - The suction lines may be too large, especially suction risers. Suction lines must maintain enough velocity to move the oil with the refrigerant vapor.

Troubleshooting without Gauges

It is possible, and often desirable, to troubleshoot some units without gauges. Units with critical charges (those with a specific operating weight of refrigerant) should not have gauges attached to them unless absolutely necessary. Checking pressures opens the system to possible contamination and loss of the refrigerant charge. Critically charged units will not perform properly if they are undercharged (or overcharged) by more than 10 percent of the rated amount stamped on their nameplate. For instance, some commercial single-door reach-in refrigerators have a charge of less than 10 ounces. A loss of just an ounce of refrigerant (10% of the total charge) can cause high product temperatures and more energy costs due to longer run times.

A standard 3-foot gauge hose will hold approximately an ounce of liquid refrigerant. If the refrigerant is not returned to the system, the loss can be significant on small systems. Although the vapor capacity of a 1/4-inch gauge hose is almost negligible (0.02 ounce per foot), repeated use of gauges every time a unit is serviced will eventually cause problems.

Troubleshooting without gauges is the best practice, but the technician must know what the air and refrigeration line temperatures should be for the specific unit. This information is best obtained from the manufacturer's service department. Also, a calibrated set of electronic thermometers is required.

FIGURE 7-23 **Example of factory recommendations for diagnosing a reach-in.** *Illustration by Irene Wirz, RTS.*

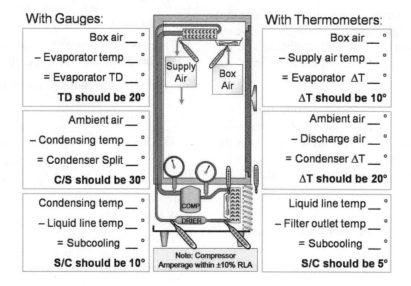

To illustrate how to use only thermometers to troubleshoot a refrigeration unit, assume the information shown in Figure 7-23 is provided by the manufacturer of a single-door reach-in. The first box on the left indicates that if gauges were used, there would be an evaporator TD of 20°F for a properly operating refrigerator or freezer. On the right side, the first box indicates that when using only thermometers, a properly operating unit should have a 10°F evaporator ΔT (the difference between the evaporator's entering and leaving air temperatures).

The second box down on the left shows that when checking with gauges, a properly operating unit will have about a 30°F condenser split. The second box on the right shows that by using thermometers, the ΔT across the condenser should be about 20°F.

NOTE: The temperature of the air leaving the condenser must be taken in the space between the condenser coil and the fan blade. It is almost impossible to get an accurate leaving air temperature at the outlet of the condenser fan because of the air turbulence caused by the fan blades. See Figure 7-24 for thermometer placements.

The bottom box on the left indicates a proper subcooling of 10°F if gauges are used. If thermometers are used according to the box on the right, there should be only a 5°F difference between the liquid line temperature leaving the condenser coil and the line temperature at the outlet of the filter drier.

In addition, compressor amperage less than 90 percent of the RLA (rated load amps) indicates the system is not functioning properly.

Assuming accurate temperatures are taken, the following diagnostics can help determine specific problems with the refrigeration systems. For each problem, all of its listed indicators will be occurring at the same time.

1. Low refrigerant charge:
   a. Low subcooling (less than 5°F)
   b. Low evaporator ΔT (less than 10°F)
   c. Low condenser ΔT (less than 20°F)

2. Stopped-up capillary tube:
   a. Slightly high subcooling (7°F to 10°F)
   b. Low evaporator ΔT (less than 10°F)
   c. Low condenser ΔT (less than 20°F)
   d. Suction line is only slightly cool

Note a restricted filter-drier will have a temperature drop across it of 3°F or more.

3. Bad compressor discharge valves:
   a. Little or no subcooling
   b. Low or no evaporator ΔT
   c. Low or no condenser ΔT
   d. Warm suction line at the compressor
   e. Hot compressor dome

A set of refrigeration gauges gives the most accurate evaporator and condensing temperatures. However, electronic thermometers can provide adequate information to diagnose critically charged units and those that do not have factory installed service valves. No matter what method is used to gather information, the technician must first know what to look for. Although rules of thumb are helpful, exact diagnostic information can come only from the equipment manufacturer.

FIGURE 7-24 **Thermometer locations for diagnosing without gauges.** *Illustration by Irene Wirz, RTS.*

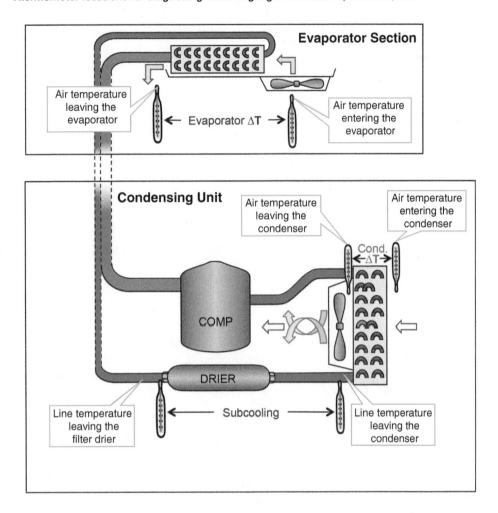

SUMMARY

A doctor diagnoses a patient's ailments based on what the patient says, as well as on the basis of the patient's medical history, body temperature, blood samples, X-rays, and other tests. If the diagnosis proves to be incorrect, the doctor tries something else, for an additional fee. As one doctor explained, "That's why they call it the medical 'practice,' because we keep practicing until we find out what cures the patient."

However, people with refrigeration problems will not pay someone to "practice" on their equipment. They expect service technicians to find and repair the problem correctly the first time. This is accomplished by using the proper instruments to gather all the information necessary to make a proper diagnosis. It is just a matter of comparing how a system is currently operating to how it should be operating.

The system problems discussed in this chapter are for those service calls where the customer says, "It's cooling, just not down to its normal

temperature." The situation may not be critical yet but is more likely inconvenient or annoying. Service calls of this nature are much more difficult to diagnose than when the box is hot and nothing seems to be working. For instance, a unit with a completely dirty condenser is easy to diagnose because it would be going out on the high head pressure control or compressor overload. Similarly, a system with no refrigerant in it would be off on low pressure or show no pressure in the system.

This section on troubleshooting draws on the information learned in the previous chapters to help you learn how to solve the tough problems that baffle most service techs. This chapter covered a lot of ground, but ideally it allowed you to apply what you have learned. With a good basis of the fundamentals, your troubleshooting skills will quickly increase with experience.

The diagnostic chart can be helpful on the more difficult system problems. The primary benefit of the chart is that it requires the user to enter all the necessary information, which is the most critical part of service analysis. A final diagnosis is made by simply choosing the problem that has the most symptoms. Remember, the chart will provide the correct diagnosis, but only if the information collected is accurate and is analyzed properly.

"We condemn what we do not fully understand; we suspect what we cannot verify." —Dick Wirz

SERVICE SCENARIOS

Following are some actual service scenarios. All the information may not have been provided, and the diagnostics may not have been "textbook." However, they are important on-the-job experiences, and the lessons learned are worth sharing.

SERVICE SCENARIO #1

WALK-IN FREEZER IS NOT COLD ENOUGH

By Dick Wirz

This service scenario took place over 35 years ago. Although I thought I was a good technician, I did not fully understand superheat or subcooling. Nor did very many other technicians in the trade at that time. Even if we understand it, we could not accurately measure it because we could not even buy electronic thermometers at that time. Yes, this was the stone age of refrigeration.

Service Request: Walk-in freezer not cold enough. Running at 0°F, should be −10°F.

The customer said the unit has been running fine for years. They noticed the higher temperature when they closed up last night but thought it might come down by morning. It did not.

Equipment: This was a 10 × 10 walk-in freezer with a water-cooled 2-hp semihermetic condensing unit using R502 refrigerant. The walk-in and condensing unit were inside the building in the same area, with the unit on the floor about 25 feet from the walk-in.

Current Operating Information:

Ambient air: 85°F

Inside box temperature: 0°F

Suction pressure: 40 psig (+10°F evaporator)

Head pressure: 246 psig (110°F condensing)

Incoming water temperature: 70°F

Leaving water temperature: 120°F

Sight glass was clear (full, no bubbles)

Compare Existing Operation to How It Should Be Operating: The evaporator should have a 10°F TD or 10°F below the box temperature. Which it did.

The water-cooled condenser should have a condensing temperature of about 105°F to 120°F depending on how the water regulating valve was set. This unit was running at about 110°F, which was okay.

Water-cooled units should have a 10°F TD (some call it *approach*) between the condensing temperature and the leaving water temperature. This condenser had a good condensing temperature of 110°F, leaving water at 120°F, and a 10°F TD, which was perfect.

NOTE: Higher than 10°F may indicate low water flow and the need to open the WRV to bring it down. Lower than that means either too much water flow or dirty condenser tubes.

Initial Diagnosis: All readings seemed okay. It just was not bringing the box temperature down to −10°F. Since it had been running fine for years, the equipment should be sized correctly for the application. Things changed suddenly, so that means something in the equipment messed up. All temperatures and pressures seemed good, and the compressor was running, but the symptoms pointed to the compressor not doing its job. So I decided to pump it down to a vacuum to check the compressor valves.

This was a low-temp compressor, and when I front-seated the suction service valve, the compressor did not seem to have too much trouble pumping down to 10-inch Hg vacuum. It held a vacuum for at least 5 minutes; therefore, I concluded that the compressor reed valves were good. I opened the suction service valve and restarted the unit. From a kneeling position, I put my hand on one of the compressor feet to push off as I tried to stand up. My weight compressed the spring, and I noticed for the first time how badly the compressor was vibrating. I was not sure what to make of it, but it made me suspicious. I went to my truck to see if I had a valve plate gasket kit for this unit. I wanted to pull the head and see what I could find.

Luckily, I had the correct gaskets. So, I pumped the unit down to 0 psig, turned it off, and pulled the head. Everything seemed in order. Reed valves looked good, pistons looked good, and the cylinder walls were not scored. I turned on the power to the unit and noticed only one piston moved. The other was stuck. Now I knew why the compressor was vibrating. Two moving pistons maintain a balanced cycle, but with only one moving, the unit vibrated noticeably.

I put the head back on with the new gaskets and left the unit operating at a 0°F box temperature to prevent any further loss of product. The office quoted the customer a new compressor, and we replaced it the same day. With the new compressor, the box temperature came down to −10°F. The oil level in the compressor sight glass was good. I checked it through a complete defrost, and all heaters and controls worked like they should. My assumption for the failure was that the compressor was just old. That was back in the day when we thought replacing the compressor was the end of our problems.

Had that happened today, I would have checked superheat for floodback and for starving. I would have also made sure there was no refrigerant migration that might lead to refrigerant in the crankcase that would cause a flooded start. I would also have torn the compressor down and analyzed the damage in the crankcase area to try to determine the cause of failure. We know now that compressors don't just die; we kill them.

Lesson Learned: The pump-down test has its limits. The compressor can be bad, even if it holds a vacuum. Also, heavy vibration in a compressor is not normal and means there is an internal compressor problem.

SERVICE SCENARIO #2

REACH-IN FREEZER WARM

By Mike Haynes

This service scenario comes from Mike Haynes, an A/C technician who was asked by his air conditioning customer to take a look at the self-contained reach-in freezer that was running warm.

Initial Inspection: The condenser was plugged with dirt, and the evaporator was moderately dirty, so he cleaned both. The unit still did not seem to get any colder. So, he put on his Fieldpiece SMAN3 digital gauge with temperature probes and provided the following information (with comments):

Refrigerant: R-404A

Metering device: cap tube

Ambient air temp: 84°F

Box temp: 71°F (high)

Suction pressure: 12 psig (evaporator temperature: −26°F) (low)

Suction line temperature: 80°F

Superheat: 106°F (high)

Discharge pressure: 195 psig (condensing temperature: 88°F) (low)

Liquid line temperature: 82°F

Subcooling: 6°F (normal)

Diagnosis: It acted like a restriction. Normal subcooling indicates there is refrigerant in the unit. High superheat indicates a starving evaporator. The refrigerant was not getting from the condenser through the evaporator. The only two things between those two points was a filter drier and a cap tube. One or both must be restricted. So, take care of both at the same time.

Repairs: Recover the refrigerant, unbraze the cap tube, cut off an inch or so and re-braze, replace filter drier and pull vacuum to sub 500 microns and hold for 30 minutes, weigh in the factory charge, and let it work.

After the work was completed, the customer was ecstatic. He told Mike that he was the third service tech that week to work on it. Mike was just glad to get the reach-in back online for the customer and for his business.

Epilogue: Mike was smart not to immediately put his gauges on the unit. He did a visual inspection and saw a dirty evaporator and condenser. He took care of the obvious problems first before going further to diagnose a refrigeration problem.

This would be a fairly simple diagnosis for an experienced commercial refrigeration technician. However, an A/C technician was able to diagnose it because he had a good understanding of the basic refrigeration cycle and what subcooling and superheat indicate.

Even if he was not totally comfortable with his diagnosis, Mike could have determined the operating conditions (below) and used the diagnostic chart in this book.

Operating Conditions:

ambient + condenser split = condensing temperature

What should the condensing temperature be?

ambient 84°F + 30°F (normal split) = 114°F

What was the actual condensing temperature?

ambient 84°F + 4°F (88°F − 84°F) = 88°F

Symptom: LOW condensing temperature (normal is ambient + 30°F ± 10°F)

At an ambient temp of 84°F, the condensing temp should be between 104°F and 124°F.

condensing temperature − liquid line temperature = subcooling

What should the subcooling be?

condensing temperature 88°F − 78°F = 10°F

What is the actual subcooling?

condensing temperature 88°F − 80°F = 8°F

Symptom: NORMAL subcooling (normal is 10°F with a max of 20°F and a min of 5°F)

box temperature − TD (box temp − evap temp) = evaporator temperature

What should the evaporator temp be?

box temperature 71°F − 20°F (TD for reach-in) = 51°F

What is the actual evaporator temp?

box temperature 71°F − 97°F [71°F − (226°F)] = −26°F

Symptom: LOW evaporator temperature (normal is box temp − 20°F ± 10°F)

At a box temp of 71°F, the evaporator temp should be about 51°F. Or, at least an evaporator temp between a low of 41°F and a high of 61°F.

suction line temperature − evaporator temperature = superheat

What should the superheat be?

suction line temp $-16°F$ − evaporator temp $-26°F = 10°F$

What is the actual superheat?

suction line temp $80°F$ − evaporator temp $-26°F = 106°F$

Symptom: HIGH superheat (normal is $10°F$, with a max $20°F$ and a min $5°F$)

Using the Diagnostic Chart

See Figure 7-25. Enter the findings by circling all the appropriate Xs to the right of the symptom. For this service call, there will be three total Xs in several columns, but four only in column 6 (Restriction after the Receiver). Although a cap tube system does not have a receiver, the chart is correct that there is a restriction.

There is a reason the diagnosis did not have four Xs in column 2 (Restricted TEV or FM). In Figure 7-25, there is a red square in column 2 around X_{FM} and X_{EV}. It shows normal subcooling for an EV (expansion valve system, TEV), but the subcooling would have to be high on an FM (fixed metering system, cap tube). A properly charged cap tube system with a restricted cap tube will empty the evaporator and back up all its refrigerant into the condenser, just like a pump-down. This should raise the subcooling, but it may—or may not—go above $20°F$ in order to classify it as high. The actual subcooling depends on the amount of charge in the unit. The primary reason high subcooling is a symptom in this chart is because even if the factory charge was not enough, from experience, we have found this type of problem usually has at least two service calls. The first technician misdiagnoses it as low on refrigerant, so he adds some, which raises the subcooling well over $20°F$. The next technician sees high subcooling, which makes it obvious that the unit has refrigerant. The high superheat leads her to the correct diagnosis, which is a restriction. In this service example, Mike was the third technician to look at the problem.

FIGURE 7-25 **Diagnostic chart for Service Scenario #2, "Reach-in Freezer Warm."** *Courtesy of Refrigeration Training Services.*

Determine which symptoms apply, then circle all the X's in the row for each symptom.
Total all the X's in each column. The column with the most X's is the problem

"X" symptoms apply to both Fixed Metering and Expansion Valve systems
"FM" symptoms apply only to Fixed Metering device systems
"EV" symptoms apply only to Expansion Valve systems

		1	2	3	4	5	6	7	8	9
Ambient + Cond. split	**HIGH** (10° higher than Normal)						X	X	X	X
Condensing Temp. LOW	**NORMAL**									
Standard Units = Ambient + 30°	**LOW** (10° lower than Normal)	X	X	X	X	X				
Cond. temp. - Liquid line temp.	**HIGH** (Subcooling above 20°)		X FM				X		X	X
Cond Subcooling NORMAL	**NORMAL**	X	X EV		X	X		X		
Normal Subcooling 5° to 20°	**LOW** (Subcooling below 5°)				X					
Air Entering Evap. - TD	**HIGH** (10° higher than Normal)				X			X FM	X FM	X FM
Evaporator Temp. LOW	**NORMAL**							X EV	X EV	X EV
TD for A/C (35°), R/I (20°), W/I (10°)	**LOW** (10° lower than Normal)	X	X	X		X	X			
Suction line at evap. - Evap. temp.	**HIGH** (Superheat above 20°)		X	X	X	X	X			
Evaporator Superheat HIGH	**NORMAL**	X EV						X EV	X EV	X EV
Normal Superheat 5° to 20°	**LOW** (Superheat below 5°)	X FM						X FM	X FM	X FM
Sight Glass	**FULL**	X	X		X	X		X	X	X
	BUBBLING		X			X	X			
Total the Circled X's in each Column =		3	3	3	3	4	2	1	0	0
DIAGNOSIS (problem):		DIRTY or ICED EVAP	RESTR TEV or FM	LOW CHG	COMP VALVES	RESTR AFTER RECV	RESTR BEFOR RECV	DIRTY COND	AIR IN SYSTEM	OVER CHG.

1. What is the first thing a technician needs to know about the system he is trying to troubleshoot?

   a. How the system is supposed to operate
   b. How much the service charge is supposed to be
   c. The system's model and serial numbers

2. What are eight pieces of information needed to diagnose a system problem?

3. If warm product is added to a box, how do the evaporator temperature and pressure respond?

   a. The temperature rises, and the pressure falls.
   b. The temperature and pressure rise.
   c. The temperature and pressure fall.

4. Why does raising the temperature of refrigerant raise its pressure?

   a. It makes the compressor pump harder.
   b. Higher temperature lowers the boiling point of the refrigerant.
   c. The added heat makes the refrigerant boil faster.

5. If the evaporator is covered in frost or dirt, or the fan motor stops, how do the evaporator temperature and pressure respond?

   a. The evaporator temperature and pressure fall.
   b. The evaporator temperature falls, but the pressure rises.
   c. The evaporator temperature and pressure rise.

6. If warm product is added to a box, how do the condensing temperature and pressure respond? Why?

   a. The condensing temperature and pressure rise because the evaporator heat causes the condensing pressure to rise more than normal to reject the additional heat.
   b. The condensing temperature and pressure remain the same because the condenser split remains the same.
   c. The condensing temperature rises, but the pressure remains the same because the condenser surface remains constant.

7. If the evaporator is covered in frost or dirt, or the fan motor stops, how do the condensing temperature and pressure respond? Why?

   a. The condensing temperature increases, but the pressure falls because the evaporator is cold.
   b. The condensing temperature and pressure both fall because there is little or no heat being picked up from the evaporator.

   c. The condensing temperature falls, but the pressure rises because pressure always rises even if the temperature does not.

8. What is the pressure drop across a metering device using R404A if the condensing temperature is 110°F and the evaporator temperature is 25°F?

   a. 178 psig
   b. 200 psig
   c. 210 psig

9. If the ambient temperature entering the condenser increases from 70°F to 90°F, will the condensing temperature increase? Why or why not?

   a. Yes. The condenser split is a constant value; therefore, an increase in ambient increases the condensing temperature.
   b. Yes, because the ambient is the only thing that determines condensing temperature.
   c. No, because the condenser split limits the increase in condensing temperature.

10. Refer to Question 9. What will be the effect on the evaporator temperature, evaporator pressure, and superheat if it is a cap tube system? Why?

11. Refer to Question 9. What will be the effect on the evaporator temperature, pressure, and superheat if it is a TEV system? Why?

For Service Problems 7-1 through 7-6, you will need to imagine you are working on a walk-in with a TEV. Standard conditions are 35°F box temperature for medium temperature, and –10°F for freezing, 10°F evaporator TD, 10°F superheat, 30°F condenser split for medium temperature, 25°F condenser split for a freezer, and 10°F subcooling for both. From the information given, you can fill in the numbered blanks from 1 to 6. From this data, you should be able to answer the questions in 7, 8, and 9. Make your initial diagnosis without using the troubleshooting chart in Figure 7-21. Then use the chart to verify your answer.

Here is a hint: based on the existing conditions, determine if the condensing temperature, evaporator temperature, superheat, and subcooling are considered normal, high, or low.

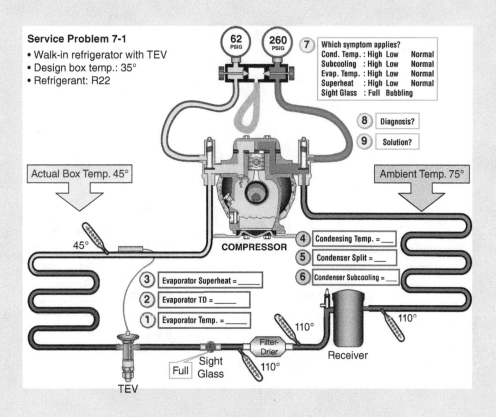

Service Problem 7-1
- Walk-in refrigerator with TEV
- Design box temp.: 35°
- Refrigerant: R22

62 PSIG   **260** PSIG

7 Which symptom applies?
Cond. Temp. : High Low Normal
Subcooling : High Low Normal
Evap. Temp. : High Low Normal
Superheat   : High Low Normal
Sight Glass : Full Bubbling

8 Diagnosis?

9 Solution?

Actual Box Temp. 45°

Ambient Temp. 75°

45°

4 Condensing Temp. = ___
5 Condenser Split = ___
6 Condenser Subcooling = ___

3 Evaporator Superheat = _____
2 Evaporator TD = _____
1 Evaporator Temp. = _____

COMPRESSOR

110°

110°

Full Sight Glass   110°

Filter-Drier

Receiver

TEV

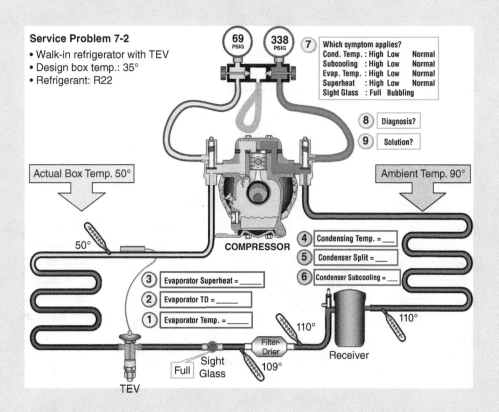

Service Problem 7-2
- Walk-in refrigerator with TEV
- Design box temp.: 35°
- Refrigerant: R22

69 PSIG   **338** PSIG

7 Which symptom applies?
Cond. Temp. : High Low Normal
Subcooling : High Low Normal
Evap. Temp. : High Low Normal
Superheat   : High Low Normal
Sight Glass : Full Bubbling

8 Diagnosis?

9 Solution?

Actual Box Temp. 50°

Ambient Temp. 90°

50°

4 Condensing Temp. = ___
5 Condenser Split = ___
6 Condenser Subcooling = ___

3 Evaporator Superheat = _____
2 Evaporator TD = _____
1 Evaporator Temp. = _____

COMPRESSOR

110°

110°

Full Sight Glass   109°

Filter-Drier

Receiver

TEV

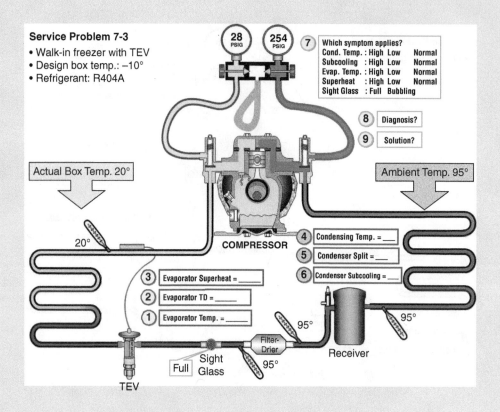

Service Problem 7-3
• Walk-in freezer with TEV
• Design box temp.: −10°
• Refrigerant: R404A

28 PSIG   254 PSIG

(7) Which symptom applies?
Cond. Temp. : High Low Normal
Subcooling : High Low Normal
Evap. Temp. : High Low Normal
Superheat : High Low Normal
Sight Glass : Full Bubbling

(8) Diagnosis?

(9) Solution?

Actual Box Temp. 20°

Ambient Temp. 95°

20°

COMPRESSOR

(4) Condensing Temp. = ___
(5) Condenser Split = ___
(6) Condenser Subcooling = ___

(3) Evaporator Superheat = _____
(2) Evaporator TD = _____
(1) Evaporator Temp. = _____

95°
95°

95°

Full   Sight Glass

Filter-Drier

Receiver

TEV

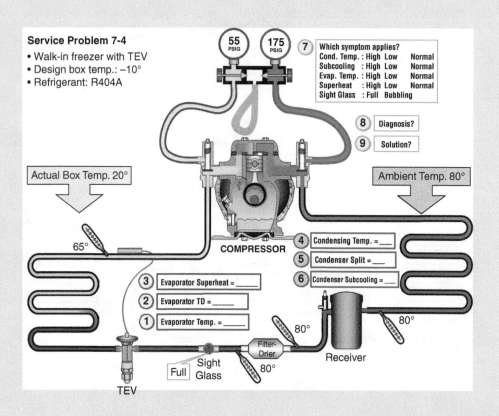

Service Problem 7-4
• Walk-in freezer with TEV
• Design box temp.: −10°
• Refrigerant: R404A

55 PSIG   175 PSIG

(7) Which symptom applies?
Cond. Temp. : High Low Normal
Subcooling : High Low Normal
Evap. Temp. : High Low Normal
Superheat : High Low Normal
Sight Glass : Full Bubbling

(8) Diagnosis?

(9) Solution?

Actual Box Temp. 20°

Ambient Temp. 80°

65°

COMPRESSOR

(4) Condensing Temp. = ___
(5) Condenser Split = ___
(6) Condenser Subcooling = ___

(3) Evaporator Superheat = _____
(2) Evaporator TD = _____
(1) Evaporator Temp. = _____

80°
80°

80°

Full   Sight Glass

Filter-Drier

Receiver

TEV

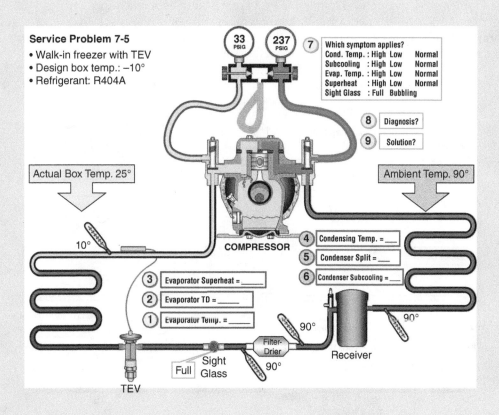

Service Problem 7-5
- Walk-in freezer with TEV
- Design box temp.: −10°
- Refrigerant: R404A

33 PSIG   237 PSIG

(7) Which symptom applies?
Cond. Temp. : High Low   Normal
Subcooling : High Low   Normal
Evap. Temp. : High Low   Normal
Superheat   : High Low   Normal
Sight Glass : Full Bubbling

(8) Diagnosis?

(9) Solution?

Actual Box Temp. 25°

Ambient Temp. 90°

10°

(4) Condensing Temp. = ___
(5) Condenser Split = ___
(6) Condenser Subcooling = ___

(3) Evaporator Superheat = _____
(2) Evaporator TD = _____
(1) Evaporator Temp. = _____

COMPRESSOR

90°
90°
90°

Filter-Drier

Receiver

Full   Sight Glass

TEV

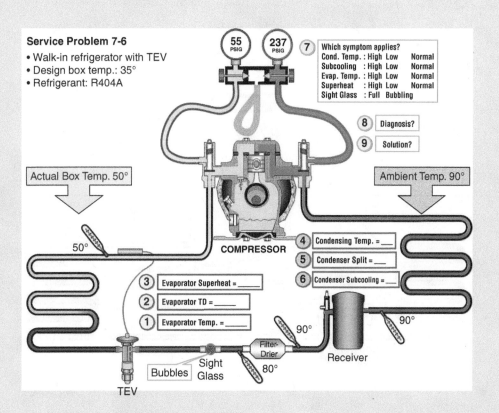

Service Problem 7-6
- Walk-in refrigerator with TEV
- Design box temp.: 35°
- Refrigerant: R404A

55 PSIG   237 PSIG

(7) Which symptom applies?
Cond. Temp. : High Low   Normal
Subcooling : High Low   Normal
Evap. Temp. : High Low   Normal
Superheat   : High Low   Normal
Sight Glass : Full Bubbling

(8) Diagnosis?

(9) Solution?

Actual Box Temp. 50°

Ambient Temp. 90°

50°

(4) Condensing Temp. = ___
(5) Condenser Split = ___
(6) Condenser Subcooling = ___

(3) Evaporator Superheat = _____
(2) Evaporator TD = _____
(1) Evaporator Temp. = _____

COMPRESSOR

90°
90°
80°

Filter-Drier

Receiver

Bubbles   Sight Glass

TEV

For Service Problems 7-7 through 7-12, imagine you are working on a reach-in with a cap tube system. Most reach-in refrigerators are designed for a 38°F box temperature, and the reach-in freezer will be about 0°F. The evaporator TD for these examples will be 20°F, superheat will be 10°F, condenser split will be 30°F for refrigerators and 25°F for freezers, and 10°F subcooling for both.

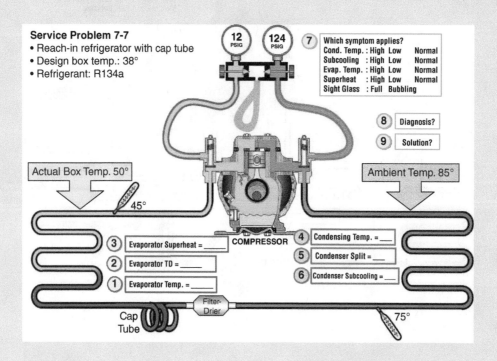

Service Problem 7-7
• Reach-in refrigerator with cap tube
• Design box temp.: 38°
• Refrigerant: R134a

12 PSIG   124 PSIG

7 Which symptom applies?
Cond. Temp. : High Low   Normal
Subcooling : High Low   Normal
Evap. Temp. : High Low   Normal
Superheat   : High Low   Normal
Sight Glass : Full Bubbling

8 Diagnosis?
9 Solution?

Actual Box Temp. 50°   Ambient Temp. 85°
45°

3 Evaporator Superheat = _____   COMPRESSOR   4 Condensing Temp. = ___
2 Evaporator TD = _____   5 Condenser Split = ___
1 Evaporator Temp. = _____   6 Condenser Subcooling = ___

Cap Tube   Filter-Drier   75°

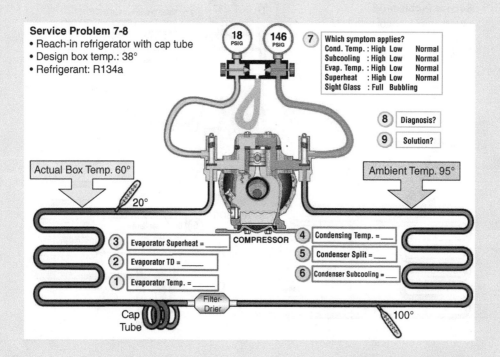

Service Problem 7-8
• Reach-in refrigerator with cap tube
• Design box temp.: 38°
• Refrigerant: R134a

18 PSIG   146 PSIG

7 Which symptom applies?
Cond. Temp. : High Low   Normal
Subcooling : High Low   Normal
Evap. Temp. : High Low   Normal
Superheat   : High Low   Normal
Sight Glass : Full Bubbling

8 Diagnosis?
9 Solution?

Actual Box Temp. 60°   Ambient Temp. 95°
20°

3 Evaporator Superheat = _____   COMPRESSOR   4 Condensing Temp. = ___
2 Evaporator TD = _____   5 Condenser Split = ___
1 Evaporator Temp. = _____   6 Condenser Subcooling = ___

Cap Tube   Filter-Drier   100°

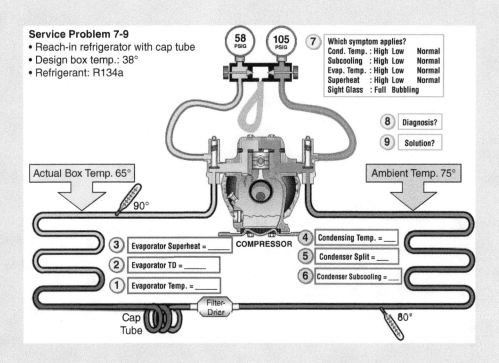

Service Problem 7-9
• Reach-in refrigerator with cap tube
• Design box temp.: 38°
• Refrigerant: R134a

58 PSIG   105 PSIG

(7) **Which symptom applies?**
Cond. Temp. : High Low Normal
Subcooling   : High Low Normal
Evap. Temp. : High Low Normal
Superheat   : High Low Normal
Sight Glass : Full Bubbling

(8) Diagnosis?
(9) Solution?

Actual Box Temp. 65°   90°   Ambient Temp. 75°

COMPRESSOR

(3) Evaporator Superheat = _____
(2) Evaporator TD = _____
(1) Evaporator Temp. = _____

(4) Condensing Temp. = ___
(5) Condenser Split = ___
(6) Condenser Subcooling = ___

Filter-Drier

Cap Tube   80°

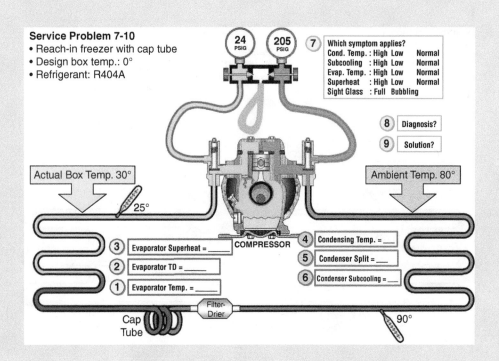

Service Problem 7-10
• Reach-in freezer with cap tube
• Design box temp.: 0°
• Refrigerant: R404A

24 PSIG   205 PSIG

(7) **Which symptom applies?**
Cond. Temp. : High Low Normal
Subcooling   : High Low Normal
Evap. Temp. : High Low Normal
Superheat   : High Low Normal
Sight Glass : Full Bubbling

(8) Diagnosis?
(9) Solution?

Actual Box Temp. 30°   25°   Ambient Temp. 80°

COMPRESSOR

(3) Evaporator Superheat = _____
(2) Evaporator TD = _____
(1) Evaporator Temp. = _____

(4) Condensing Temp. = ___
(5) Condenser Split = ___
(6) Condenser Subcooling = ___

Filter-Drier

Cap Tube   90°

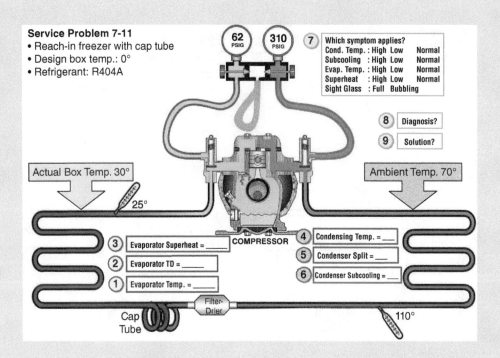

Service Problem 7-11
- Reach-in freezer with cap tube
- Design box temp.: 0°
- Refrigerant: R404A

62 PSIG   310 PSIG

⑦ **Which symptom applies?**
Cond. Temp. : High Low   Normal
Subcooling : High Low   Normal
Evap. Temp. : High Low   Normal
Superheat   : High Low   Normal
Sight Glass : Full Bubbling

⑧ Diagnosis?
⑨ Solution?

Actual Box Temp. 30°   25°

Ambient Temp. 70°

③ Evaporator Superheat = _____   COMPRESSOR
② Evaporator TD = _____
① Evaporator Temp. = _____

④ Condensing Temp. = ___
⑤ Condenser Split = ___
⑥ Condenser Subcooling = ___

Cap Tube   Filter-Drier   110°

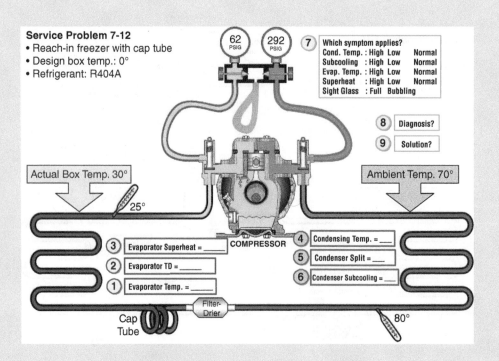

Service Problem 7-12
- Reach-in freezer with cap tube
- Design box temp.: 0°
- Refrigerant: R404A

62 PSIG   292 PSIG

⑦ **Which symptom applies?**
Cond. Temp. : High Low   Normal
Subcooling : High Low   Normal
Evap. Temp. : High Low   Normal
Superheat   : High Low   Normal
Sight Glass : Full Bubbling

⑧ Diagnosis?
⑨ Solution?

Actual Box Temp. 30°   25°

Ambient Temp. 70°

③ Evaporator Superheat = _____   COMPRESSOR
② Evaporator TD = _____
① Evaporator Temp. = _____

④ Condensing Temp. = ___
⑤ Condenser Split = ___
⑥ Condenser Subcooling = ___

Cap Tube   Filter-Drier   80°

COMPRESSOR MOTOR CONTROLS

8

CHAPTER OVERVIEW

A large percentage of compressor problems are electrical. Therefore, a good understanding of what controls a compressor motor is essential for troubleshooting.

You learned in Chapter 7, "Refrigeration System Troubleshooting," that in order to diagnose a component in a refrigeration unit, a technician must understand how other parts of the system affect that component. The same is true for motors. By following the path of electricity, a technician can determine exactly what control is preventing the motor from running, or if the motor itself is bad.

OBJECTIVES

After completing this chapter, you should be able to understand and explain the following:
- Three-phase motors
- Contactors and starters
- Single-phase motors
- Start relays and capacitors
- Motor overloads
- Troubleshooting motors

THREE-PHASE MOTORS

There are quite a few advantages to using three-phase (3Ø) compressors rather than single-phase (1Ø) compressors:

- Three-phase motors cost less to operate than 1Ø motors.
- A 3Ø motor draws less current; therefore, the motor circuit requires smaller wire.
- The starting torque of a 3Ø motor is more than twice that of a 1Ø motor.
- A 3Ø motor does not require start components. The three windings are only 120 degrees out of phase from each other, whereas the start and run windings of 1Ø motors are 180 degrees apart.

With all the strengths of 3Ø motors, their primary weaknesses are phase loss and voltage imbalance. Phase loss means one of the phases (one of the three power circuits) is interrupted (opened). With only two of the windings energized, the motor will draw high amperage and overload the motor, possibly burning the motor windings.

A voltage unbalance, or imbalance, is a difference between the voltages applied to the three power circuits. The three windings of a 3Ø motor have the same resistance, or ohm reading. Therefore, if different voltages are applied to the windings, the electrical unbalance can cause the motor windings to overheat. In fact, a voltage unbalance of more than 2 percent can damage a 3Ø motor.

Voltage unbalance is determined by the following formula:

$$\text{voltage unbalance } (V_u) = \text{voltage deviation } (V_d)$$
$$\div \text{ voltage average } (V_a),$$

where voltage deviation (V_d) is the greatest difference from the average.

EXAMPLE: 1

L1 → L2 = 230 volts, L2 → L3 = 240 volts, and L1 → L3 = 245 volts

voltage average (V_a) = (230 + 240 + 245) 4 ÷ 3 = 238 volts

voltage deviation (V_d): 238 − 230 = 8 volts;
240 − 238 = 2 volts; 245 − 238 = 7 volts

The greatest deviation from the average is 8 volts.

$$V_u = V_d \div V_a = 8 \div 238 = 0.034$$

Multiply by 100 to convert to percent: 0.034 × 100 = 3.4 percent voltage unbalance (V_u).

Motor overloads open all three circuits inside the motor. However, sometimes motor damage occurs long before the internal overloads can react. Motor starters and phase monitors are recommended for more precise 3Ø motor protection.

According to the Carlyle Compressor Service Guide, a voltage unbalance will cause a current unbalance, but a current unbalance does not necessarily mean that a voltage unbalance exists. For instance, a loose terminal connection or pitted contacts on L1 would cause a higher resistance on L1 than on the other two legs. Because current follows the path of least resistance, there would be higher current on L2 and L3 than on L1. Higher current causes heat buildup in the motor windings.

The maximum acceptable current unbalance is 10 percent. Current unbalance is calculated the same way as voltage unbalance.

Obviously, accuracy in measuring current is critical. However, standard ammeters can no longer be trusted to give accurate readings because they do not take into account the effects microprocessors have on the electrical systems from which they are powered. Some sort of computer system is used in just about every building in which air conditioning (A/C) and refrigeration equipment is located. Because the electronic impulses are intertwined with the alternating current, measurements with a standard ammeter can be in error by as much as 40 percent. True Root Mean Square (True RMS) meters have been developed to allow technicians to read the correct amperage on systems affected by computer electronics. Of course, True RMS meters can also be used on standard systems.

CONTACTORS

A contactor is a large relay that opens and closes the circuit to a motor. An electromagnetic coil inside the contactor pulls the contacts together, and power is sent to the compressor. The thermostat (tstat), high-pressure controls, and low-pressure controls are wired in series with the contactor coil.

To stop a 3Ø motor, all three power wires must be opened at the same time. Single-phase compressors need only one of the two power legs opened to stop the motor.

Following are factors to consider when choosing a contactor:

1. Poles.
   A. Three-phase motors require three poles.
   B. Single-phase motors require one or two poles.

2. The contactor's amperage rating must be equal to or greater than the compressor's rated load amps (RLA) displayed on the compressor nameplate *plus any other fans or accessories also operated through the contactor*. Most contactors have two ratings. Use the inductive load amp rating for compressors (see Figure 8-1).

FIGURE 8-1 **Amperage ratings listed on a contactor.** *Photo by Dick Wirz.*

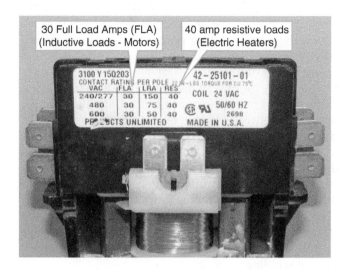

A. Inductive load amps: high inrush current (motors).
B. Resistive load amps: no inrush current (resistance heaters).

3. Line voltage rating of load circuit.
4. Coil voltage rating.

Figure 8-2 shows a contactor that is pulled in (energized) and supplying power to a 1Ø motor. If the contacts are good, there will be a voltage reading of 0 volts between L1 and T1 and between L2 and T2. The contacts are burned or pitted if there is any voltage reading across them. The contactor, or contacts, should be replaced.

NOTE: Whenever a compressor is replaced, the contactor must also be replaced. This is especially important on 3Ø units.

There are two types of contactors; NEMA (National Electrical Manufacturers Association) rated general purpose contactors and definite purpose contactors. NEMA contactors are listed by sizes that generally relate to motor horsepower (hp) and amperage. They are built for the most severe industrial usage and are designed for a minimum life of 2 million cycles. Because they must be adaptable for many different applications, NEMA-rated contactors must have a high safety factor and as a result are both large and expensive.

Definite purpose contactors are designed specifically for refrigeration and air conditioning applications and rated in amperes (amps). They must still be designed to meet harsh conditions such as rapid cycling, extended overloads, and low voltage. However, a life of 250,000 cycles is adequate for their specific purpose; therefore, the contactor is smaller and less expensive. Refer to Table 8-1 for an example of the different ratings of NEMA and definite purpose contactors.

FIGURE 8-2 **Contactor cutaway view and wire diagram.** *Courtesy of Refrigeration Training Services.*

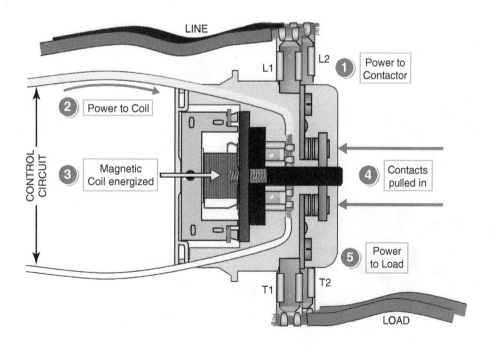

TABLE 8-1 NEMA (general purpose) and definite purpose contactor current ratings. *Courtesy of Refrigeration Training Services.*

NEMA SIZE	NEMA AMP RATING	DEFINITE PURPOSE AVAILABLE AMP RATINGS
1	27	25
7.5 hp @ 230V		30
10 hp @ 460V		40
2	45	50
15 hp @ 230V		60
25 hp @ 460V		75
3	90	90
30 hp @ 230V		120
50 hp @ 460V		150

MOTOR STARTERS

A motor starter is basically a contactor with overloads (Figure 8-3). The overloads in the starter are selected according to both the voltage and the RLA of the compressor. These protectors respond to a change in heat caused by voltage drop and amperage increases. If the starter senses an overload condition in any one of the motor circuits, it will cause the contactor to open and interrupt the power to all the circuits. The starter will usually interrupt power before the motor's internal overload opens. Motor starters are designed to be manually reset after they trip on overload. The assumption is that the person resetting the control will then monitor the motor operation to determine what caused the motor protection to open.

SINGLE-PHASE MOTORS

The windings of a 1Ø motor are 180 degrees apart. As a result, when the *stator* (stationary winding) is energized, the *rotor* (rotating permanent magnet) cannot move because there is an equal but opposite attraction on both sides of the rotor.

For the rotor to start turning, it needs to have some of the stator's magnetic power exerted at a slight angle, rather than directly opposite the rotor's positive and negative poles. To do this, a separate start winding is used. By diverting some of the incoming voltage into the high-resistance start winding, a phase shift occurs, creating a magnetic pull on the rotor from a slightly different angle. This forces the rotor to start turning. Within a few seconds, the motor is up to speed, and the start winding is then de-energized. The rotor continues spinning within a magnetic field created through only the run windings.

The 1Ø motors used in compressors are called split-phase motors because they are split into two windings, the start winding and the run winding.

The motor in Figure 8-4 shows the smaller start windings, one at the top and one at the bottom. The two larger bundles of wires are the run windings, one on the right side and one on the left side of the motor. Because the motor has two sets of run windings, it is called a two-pole motor. If the motor nameplate is missing, the technician can determine the approximate speed of the motor by counting the number of poles:

synchronous motor speed = (120 × frequency) ÷ poles,

where 120 is a constant and the frequency is 60 cycles (120 × 60 = 7,200).

FIGURE 8-3 Motor starter and wire diagram. *Courtesy of Refrigeration Training Services.*

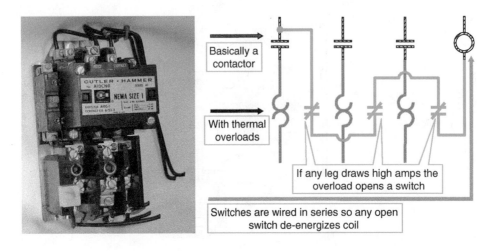

Basically a contactor

With thermal overloads

If any leg draws high amps the overload opens a switch

Switches are wired in series so any open switch de-energizes coil

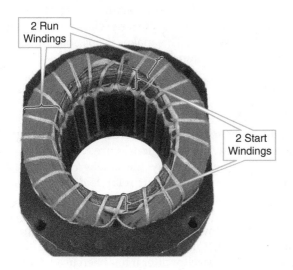

FIGURE 8-4 **Single-phase motor with start and run windings.** *Courtesy of Copeland Corporation.*

A two-pole motor runs at 3,600 revolutions per minute (rpm) (7,200 ÷ 2). The actual rpm is closer to 3,450 after slippage is taken into account. Hermetic compressors use two-pole motors.

A four-pole motor runs slower at only 1,800 rpm (7,200 ÷ 4), with an actual speed closer to 1,725 rpm. Semihermetic compressors use four-pole motors.

There are two basic methods used to start 1Ø compressors:

1. A start relay energizes the start winding then de-energizes power from the start winding when the motor is running.
2. A run capacitor is permanently connected between the run and start terminals. It creates a slight phase shift in the start winding both for starting and for running efficiency. This method is used only on permanent split capacitor (PSC) motors of compressors and fans.

START RELAYS AND CAPACITORS

Compressor motors draw locked rotor amperage (LRA) each time they try to start. As the rotor picks up speed, the amperage falls. When the motor is nearly at full speed, the start relay opens, shutting off power to the start winding. The motor continues running, powered only through its run windings. The current draw is down to the motor's RLA (see Figure 8-5).

Figure 8-6 illustrates how a start relay "borrows" power from L1 and directs it to the start terminal. For added torque (force to start the rotor turning), a start capacitor can be added in series with the relay. The start relay allows full power to the start winding on

start-up. When the compressor motor is up to speed, the relay drops out, or de-energizes, the start winding.

There are three types of start relays:

1. Current relay
2. Potential relay
3. Positive temperature coefficient relay (PTCR)

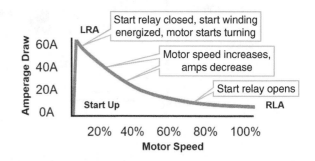

FIGURE 8-5 **Motor current draw during startup.** *Courtesy of Refrigeration Training Services.*

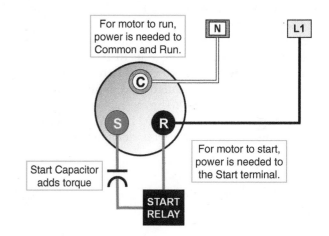

FIGURE 8-6 **Simple start relay wiring diagram.** *Courtesy of Refrigeration Training Services.*

Current Relays

For fractional horsepower (below 1 hp) compressors, the current relay works very well. The wire connections are designated by

- *L* for line power
- *M* for main winding (run winding)
- *S* for start winding

Figure 8-7 can help clarify the following description of motor starting with a current relay. On startup, the amp draw is near LRA before the rotor starts to turn. The high current in the power leg going from M, through R, to the run winding, energizes the relay

FIGURE 8-7 **Current relay operation with a start cap.**
Adapted from Copeland Corporation by RTS.

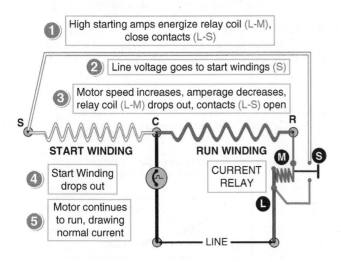

However, on thermal expansion valve (TEV) systems, the pressures do not equalize, and the compressor has quite a load to start against. A start capacitor, added in series with the current relay, provides an additional phase shift. This makes it possible for the compressor to start against high pressures.

Start Capacitors

A start capacitor (start cap) increases starting torque by increasing the phase angle (Figure 8-9). It is wired in series with the start relay and is taken out of the electrical circuit when the start relay is de-energized. All capacitors sold by Copeland for its compressors come with a bleed resistor across the capacitor terminals to prevent arcing when the relay contacts open. The resistor "bleeds off," or eliminates, any remaining electrical charge in the capacitor after it has been used to energize the start winding. Not only does the resistor extend the life of the relay, but it also protects the capacitor.

NOTE: Remove the bleed resistor when checking a start cap. If the resistor is left in the line, the capacitor checker will incorrectly read a much higher value than the rated capacitance.

A capacitor collects a large amount of electrical energy when voltage is applied. This energy must be discharged before the capacitor can be checked. Some technicians discharge capacitors by placing a screwdriver across the terminals. This short circuit can damage the capacitor and possibly injure the technician. The proper way to discharge a capacitor is to use a 20,000-ohm, 5-watt resistor across the connectors for about 5 seconds.

The microfarad (MFD) reading on a start capacitor is a range; for example, 145–174 MFD. When using

coil between terminals L and M. This coil pulls in the normally open contacts L and S. Line voltage is sent through the closed contacts to the start winding (S), which starts the rotor turning.

When the motor is up to speed, the amperage in the power leg to R starts to fall. The decrease in current in the relay coil between L and M reopens the circuit to the start winding (L to S). The motor continues operating on only its run winding at its RLA.

A current relay by itself is sufficient to start a refrigeration compressor installed on a capillary tube (cap tube) system (Figure 8-8). The cap tube allows pressures in the system to equalize during the off cycle. Therefore, the compressor does not start under a high load.

FIGURE 8-8 **Current relay cutaway.** *Courtesy of Refrigeration Training Services.*

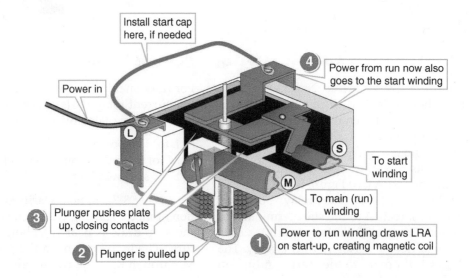

FIGURE 8-9   **Start cap with resistor.** *Photo by Dick Wirz.*

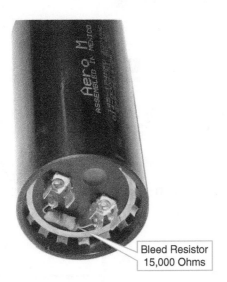

Bleed Resistor
15,000 Ohms

a capacitor checker, the middle value is ideal, but the capacitor should still operate properly at approximately 10 percent over or under that value. For instance, the previous capacitor's mid-range is about 160 MFD. Ten percent (16 MFD) higher would be 176 MFD, and 10 percent lower would be 144 MFD (see Figure 8-10).

The VAC (Volt Amp Capacity) of a replacement capacitor must be the same, or greater, than the original. A capacitor's VAC rating is determined by the motor manufacturer and is greater than the motor's rated operating voltage because it is subjected to the back electromotive force (back EMF) generated in the start winding. In Figure 8-11, there are four capacitors with the same MFD rating but with different VAC.

FIGURE 8-11   **Examples of 145–174 MFD capacitors with different VAC.** *Photo by Dick Wirz.*

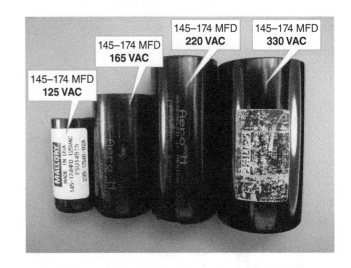

145–174 MFD
125 VAC

145–174 MFD
165 VAC

145–174 MFD
220 VAC

145–174 MFD
330 VAC

Potential Relays

Where a current relay has normally open (NO) contacts and senses starting current, a potential relay has normally closed (NC) contacts and senses voltage (Figure 8-12). In fact, the voltage it senses is the *back EMF* generated in the start windings due to the closeness of the

FIGURE 8-10   **Checking a start cap.** *Photo by Dick Wirz.*

Measured capacitance

Capacitor rating

FIGURE 8-12   **Potential relay (removed from case).** *Courtesy of Copeland Corporation.*

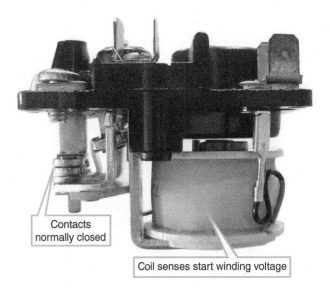

Contacts normally closed

Coil senses start winding voltage

spinning rotor to the stator (see Figure 8-13). As the back EMF increases during startup, it energizes the relay coil (between 5 and 2) and opens the contacts between terminals 1 and 2 to drop out the start winding and start capacitor. The continuous back EMF keeps the contacts open as long as the compressor is running.

NOTE: **Just about all potential relays use the same numbering system. Wires from the compressor terminals are connected to the numbered terminals on the potential relay in the following order: common to 5, start to 2, and run to 1. A good rhyme used to remember these terminals is "5, 2, 1, can she run?"**

Potential relays have several voltage ratings:

1. Pickup voltage—the voltage that opens contacts 1 and 2
2. Dropout voltage—the minimum voltage required to keep the contacts open
3. Continuous voltage—the maximum continuous voltage the coil can handle at terminals 5 and 2

FIGURE 8-14 **Wiring diagram with run cap added.** *Adapted from Copeland Corporation by RTS.*

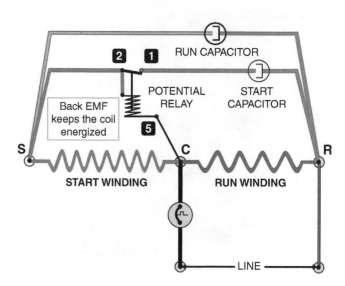

FIGURE 8-13 **Wiring diagram with potential relay and start cap.** *Adapted from Copeland Corporation by RTS.*

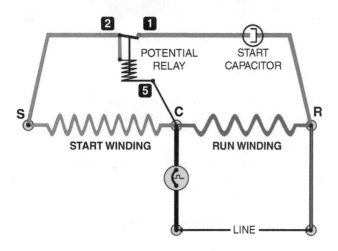

Run Capacitor

A run capacitor (run cap) assists a motor's operation by providing a phase shift. Unlike the start relay or the start cap, the run cap is designed to remain in the circuit the entire time the compressor is running. It provides a small phase shift by slightly energizing the start winding. This helps the compressor run efficiently.

In Figure 8-14, the run cap is wired so that it will stay in the circuit all the time, even after the potential relay has opened.

One indicator of a bad run capacitor is that although the compressor will start normally, it will run at slightly higher amperage. This can be checked by measuring the running amps of a unit first with, then without, the benefit of its run capacitor.

For a motor to operate properly, a run cap, like a start cap, must also have a capacitance of at least 10 percent of its rating. This can be determined easily with a capacitor checker. Figure 8-15 is not only an illustration of how to check a run cap, but this particular manufacturer recommends a maximum deviation of 6 percent

FIGURE 8-15 **Checking a run cap.** *Photo by Dick Wirz.*

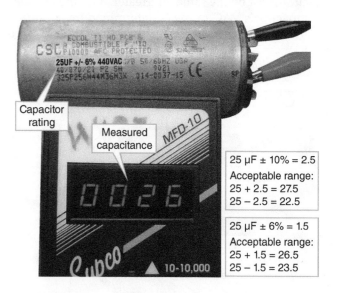

rather than the normal 10 percent. The text boxes on the figure illustrate the difference between the two factors.

In an emergency, if a technician does not have the exact MFD (μF) replacement run cap, he can use a combination of capacitors (see Figures 8-16 and 8-17).

NOTE: **Make sure the VAC rating of the capacitors is equal to or greater than the capacitor that is being replaced. This goes for both run and start caps.**

FIGURE 8-16 **Two capacitors wired in parallel for greater capacity.** *Courtesy of Refrigeration Training Services.*

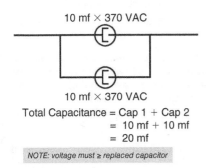

10 mf × 370 VAC

10 mf × 370 VAC

Total Capacitance = Cap 1 + Cap 2
= 10 mf + 10 mf
= 20 mf

NOTE: voltage must ≥ replaced capacitor

FIGURE 8-17 **Two capacitors wired in series for lower capacity.** *Courtesy of Refrigeration Training Services.*

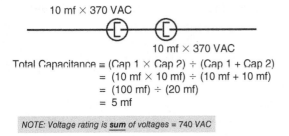

10 mf × 370 VAC

10 mf × 370 VAC

Total Capacitance = (Cap 1 × Cap 2) ÷ (Cap 1 + Cap 2)
= (10 mf × 10 mf) ÷ (10 mf + 10 mf)
= (100 mf) ÷ (20 mf)
= 5 mf

NOTE: Voltage rating is **sum** of voltages = 740 VAC

PTCR

The PTC relay is a solid state relay that increases its resistance to current flow when it heats up (Figure 8-18). When the compressor motor starts, there is full power through the PTCR to the start winding. Within a few seconds, the current draw of the compressor heats up the ceramic disc inside the relay. Heat transforms the conducting properties of the disc to an insulator. The material stops the flow of power to the start winding as

FIGURE 8-18 **Cutaway of a PTCR.** *Courtesy of Refrigeration Training Services.*

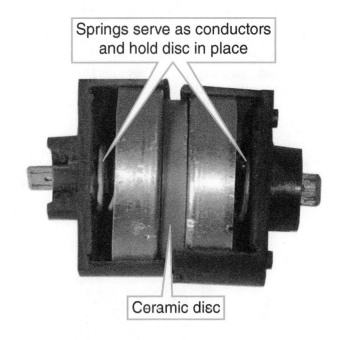

Springs serve as conductors and hold disc in place

Ceramic disc

effectively as the opening of a set of contacts, but without using moving mechanical parts.

The PTCR is usually used in conjunction with a run capacitor (Figure 8-19). When the circuit through the relay is opened, the voltage is redirected through the run cap. This provides a path for some current to travel to the start winding, providing the slight phase shift necessary to increase the compressor's running efficiency.

This arrangement has been common for years in residential A/C hermetic compressors with PSC motors. However, it is now being used with great success in Manitowoc commercial ice machines.

FIGURE 8-19 **PTCR and run cap.** *Courtesy of Refrigeration Training Services.*

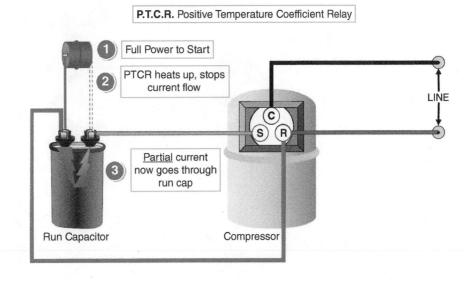

P.T.C.R. Positive Temperature Coefficient Relay

1 Full Power to Start

2 PTCR heats up, stops current flow

3 Partial current now goes through run cap

LINE

Run Capacitor

Compressor

TYPES OF MOTOR OVERLOADS

To help prevent compressor motor damage, the manufacturers provide motor overload protection. These overloads open in response to excessive heat, amperage, or a combination of the two. Some overloads are located inside the compressor. They are called internal or inherent motor overloads and are buried in the motor windings so that they can accurately detect motor heat (Figure 8-20).

FIGURE 8-20 **Internal or inherent motor overload.** *Courtesy of Refrigeration Training Services.*

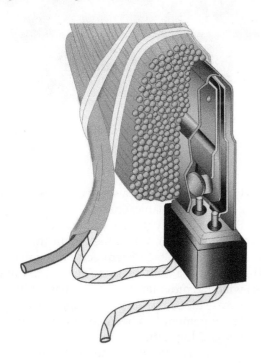

Usually, an external overload is located in the compressor terminal box (Figure 8-21). This type of overload is a thermo disc that senses both the heat of the compressor body and the circuit heat generated by the amperage draw.

NOTE: **The compressor terminal box cover must remain closed so that the overload properly senses the compressor temperature. The cover also prevents dirt and oils from entering the terminal box, where they could present a fire hazard. In addition, it prevents the technician from accidentally being electrocuted by exposed wires.**

Magnetic overloads are used on some of the larger compressors. They are small black boxes, about the size of a deck of cards, attached to each of the power legs of the compressor. These overloads are usually located in the condensing unit's control box, not in the electrical terminal box on the body of the compressor. A magnetic overload is a very sensitive electronic device that accurately responds to compressor amperage, but is not influenced by the temperature of the air around it.

This has been a very brief explanation of the types of overloads. A more complete explanation would be part of an electrical course. However, the most important point is that an overload disconnects power to the compressor motor whenever it senses a problem that could damage the motor.

Checking Overloads

Overloads must have time to cool before they will reset. This may take as little as a few minutes for an external overload to as much as several hours for an internal overload. If the overload is a bimetal type that is cycling the compressor on and off, the fact that it is

FIGURE 8-21 **External compressor overloads.** *Photo by Dick Wirz.*

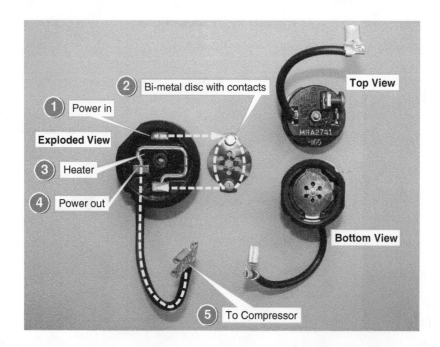

resetting itself is proof that the overload is still good. These protection devices are very tough, and rarely do they fail. When replacing a relay and capacitor, it is not necessary to replace an external overload, unless it is obviously bad.

NOTE: Never jump out an overload.

Bypassing the overload takes away the only emergency shutoff and electrical protection that the motor has. On many compressors, the compressor terminals are the emergency pressure relief in case the overload does not cut off the compressor. If the terminals blow out because the overload was bypassed, the compressor will be ruined. Even more serious, the technician could be injured or killed.

TROUBLESHOOTING MOTORS

The hardest part of troubleshooting is to visualize what is going on inside the motor windings. Figure 8-22 is a diagram of a 1Ø motor with examples of the resistances of the start and run windings.

FIGURE 8-22 **Resistances of 1Ø motor windings.** *Courtesy of Refrigeration Training Services.*

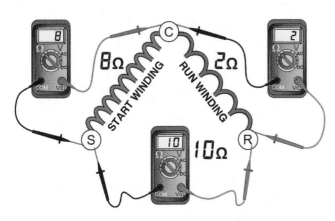

Three-phase motors have three run windings, each with the same resistance. However, 1Ø motors have only one run winding plus a high-resistance start winding. Following is the method used to check the windings on a 1Ø motor:

- Run to common (R–C) is the lowest ohm reading.
- Start to common (S–C) is about three to six times higher than the run winding.
- Run to start (R–S) is the sum of both windings, R–C and S–C.

$$R - S = [R - C] + [S - C]$$

Verifying the readings helps determine which terminals are run, start, and common. In addition, deviations from normal ohm readings indicate problems such as open windings or shorted windings, or if the internal overload is open.

There are two basic conditions when a compressor is not running:

1. No sound. It is not even trying to start.
2. It sounds like it is trying to start, but the rotor does not turn.

No Sound

If there is no sound, check for correct power to the compressor. If there is no power, check the circuit breaker and fuses. Tripped circuit breakers and blown fuses usually indicate a short circuit. Therefore, check the motor windings for shorts or grounds, as illustrated in Figures 8-23 and 8-24. Shorted windings will show up on an ohm meter as 0 ohms (or close to 0).

FIGURE 8-23 **Windings shorted together.** *Courtesy of Refrigeration Training Services.*

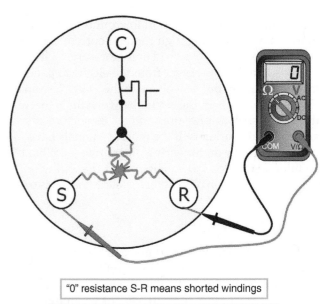

"0" resistance S-R means shorted windings

C-S and C-R will be higher than rating in manual

To check for a grounded compressor, take an ohm reading from each terminal to ground (usually the suction line). There should be an open circuit, or infinite resistance with a meter reading of OL (infinite ohms). However, some contamination in the crankcase oil may provide a small resistance reading. Therefore, most compressor manufacturers state that the minimum allowable resistance to ground is 1,000 ohms per operating volt. An ohm reading less than 240,000 ohms for a 240-volt motor or 120,000 ohms for a 120-volt motor

FIGURE 8-24 **Windings shorted to ground.** *Courtesy of Refrigeration Training Services.*

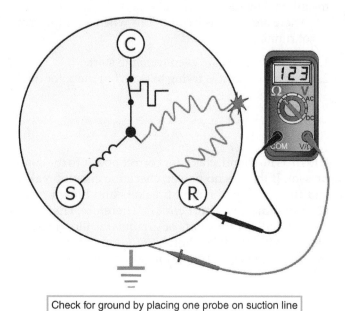

Check for ground by placing one probe on suction line

would indicate a possible short to ground situation. The motor may be drawing high running current and occasionally tripping the overload or motor starter if it is 3Ø.

Assume there is power from the electrical panel on the motor circuit but no voltage to the motor terminals. Check all controls in the circuit between the panel and the motor, such as pressure controls, contactors, relays, and external overloads. If the motor terminals have full power, check the motor wiring diagram to see if there is an internal overload. If so, disconnect the wiring to the overload and use an ohmmeter to determine if it is open.

Motor Tries to Start

A motor that is trying to start makes a humming sound. The hum stops when the overload opens the motor circuit. Make sure the voltage to the motor is correct. A voltage drop of more than 10 percent when the motor tries to start indicates a power problem, not a motor problem.

NOTE: **Be careful when taking start-up voltage at the compressor terminals.**

Checking voltage with meter probes at the terminals can be dangerous, especially on small compressors where the space is limited. One slip of the probe may accidentally create a short and blow the terminals out of the compressor. Short circuits of this nature can be lethal to the technician.

NOTE: **It may be safer to use jumper wires with insulated alligator clips between the meter probes and the compressor terminal posts.**

Figure 8-25 illustrates an open internal overload when the ohmmeter is showing OL from R to C and also from S to C. Single-phase compressors have their overloads (internal and external) in the common circuit.

Figure 8-26 illustrates an open internal overload on a 3Ø compressor when the ohmmeter is reading OL between all three terminals. Three-phase compressor overloads must open all three legs to the windings. If one of the windings were still energized, it would overheat and burn out.

It often takes hours for an internal overload to reset. Allow enough time for the motor to cool down,

FIGURE 8-25 **Open internal overload on a 1Ø compressor.** *Courtesy of Refrigeration Training Services.*

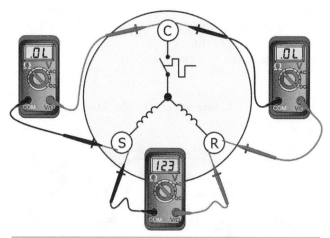

If both R & S are open to Common, the Internal Overload is open

FIGURE 8-26 **Open internal overload on a 3Ø compressor.** *Courtesy of Refrigeration Training Services.*

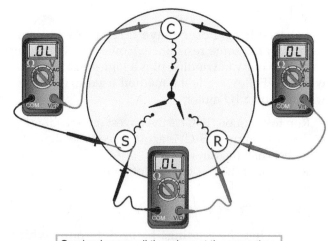

Overload opens all three legs at the same time.

and close the overload. Too many compressors are erroneously replaced just because the technician thought the compressor had three bad windings.

NOTE: **An OL reading in only one winding is a sign of an open winding, indicating that the compressor would have to be replaced.**

When the overload resets, verify that the windings are not open, shorted, or grounded. If the ohm tests are good, then the only two reasons the compressor will not start are improper starting voltage, or the compressor is mechanically locked up.

If a 3Ø compressor seems to be locked up, switching any two power wires will reverse the motor's rotation and may result in the motor starting. If not, check the voltage to the motor when it tries to start. If the starting voltage is good but the compressor still doesn't run and goes out on overload again, this will verify that the compressor is locked up. Replace the compressor.

NOTE: **Do not reverse leads on a 3Ø scroll or screw compressor. They are rotation sensitive and can be damaged if run backward.**

NOTE: **Compressors don't just die, they're murdered!**

There is almost always an external cause for a compressor failure. Unless the reason for the failure is determined, the replacement compressor is bound to fail also.

Single-Phase Compressor Start Components

Up to this point, 1Ø and 3Ø compressor checking has been very similar. However, a 1Ø compressor has start components that require an additional diagnostic procedure before it can be condemned for not running. After verifying proper voltage, good windings, and that the overload is closed, the technician must still "listen" for what the compressor is trying to "tell" him.

1. There is no sound. The motor does not even try to start.

    This is the same result as a no power problem. However, this time, the technician must verify that there is power to the start components and then through them to the motor terminals.

    - On a potential relay (Figure 8-13), the contacts between 1 and 2 should be NC. If not, the relay is bad.
    - On a current relay (Figure 8-7), the contacts between M and S should be open, and then they should close when the motor tries to start. If not, the relay is bad.

Diagnosing compressor start relays can be performed with an ohmmeter or by jumping them out, or they can be replaced with a start kit. These options are discussed later in the chapter.

2. It sounds like the motor is trying to start, but the rotor does not turn and trips the overload.

    If the motor is trying to start, then power is getting through the start relay but still not turning the rotor. First, check to make sure it has the proper voltage. Next, if there is a start capacitor in the circuit, check it out with a capacitor tester. If the correct power is going through the relay and the capacitor checks out good, then the compressor is locked up. Replace the compressor.

3. The compressor runs for a few seconds and then trips the overload.

    Because the compressor starts and runs, power is being supplied to the start terminal. The start cap (if used) is probably good, as it has enough capacity to get the motor started. It sounds like the relay contacts are not opening the start circuit when the motor is up to speed. Therefore, the start relay is bad.

Checking Start Relays

The only two functions of a start relay are to (1) provide power to the start winding and (2) open that circuit when the motor is up to speed. A start relay can be checked with an ohmmeter. Following are some guidelines for ohming out each type of relay:

1. Current relays (see Figure 8-7): L–S switch contacts should be open (OL). L–M is the relay coil that should ohm out with some measurable value. If the ohms are 0, the coil is shorted; if infinite (OL), the coil is open.
2. Potential relays (see Figure 8-13): The switch contacts, terminals 1 and 2, should be closed (0 ohms). The relay coil, terminals 5 to 2, should be checked the same as a current relay.
3. PTCRs (see Figure 8-18): The reading should be 15 ohms to 45 ohms, depending on the manufacturer. Open and shorted readings are the same as other relays.

NOTE: **Check a PTCR when it has cooled to ambient temperature. This may take up to 5 minutes. The ceramic disc inside a solid state relay opens the circuit at about 80,000 ohms, when its temperature reaches 200°F to 300°F. A properly operating PTCR keeps the starting circuit open by staying hot until the compressor cycles off. The relay must cool down before the compressor can try to restart.**

Many technicians are very comfortable verifying a defective start relay by using a jumper wire in place of the suspected start relay. A jumper wire connects the run (R) terminal to the start (S) terminal. If there is a capacitor, it is wired in series with the jumper wire. The technician applies power, counts to three, and then disconnects one of the jumper wire leads. This procedure is very effective for experienced technicians, but it can be dangerous if not performed properly.

The recommended method for jumping out suspected start relays is to use an electronic potential relay like the one shown in Figure 8-27, or the combination current relay and capacitor replacement kit in Figure 8-28. If using the proper start kit does not bring satisfactory results, the compressor needs to be replaced.

FIGURE 8-27   **Electronic potential relay start kits.**
Photo by Dick Wirz.

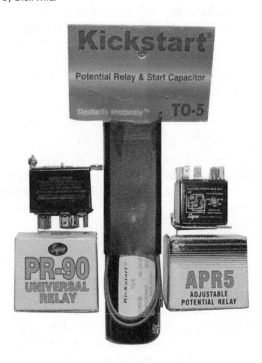

The original equipment manufacturers (OEMs) of compressors do not recommend leaving these electronic start components on their units. The factories provide specific electromechanical relays to respond to the particular starting current and voltages of their different compressors. Replacement electronic relays open the circuit more as a function of time rather than current and are therefore not as accurate as the OEM relays. The compressor manufacturers probably do not mind the use of electronic relays in an emergency or to verify if the compressor is bad. However, the factories

FIGURE 8-28   **Electronic current relay start kit.**
Photo by Dick Wirz.

recommend using OEM start components for maximum compressor life.

Responsible technicians communicate all this information to their customers and let them make the decision. Does the customer want the technician to return later with the correct parts (at an additional charge), or would he rather have the technician leave the existing replacement kits on the compressor?

NOTE: Whenever a start relay is replaced, the capacitor should also be replaced, and vice versa. When one of these components fails, it has likely put a strain on the other component.

There is one more situation technicians may experience with a 1Ø compressor. It may start and run but then draw higher amperage than normal.

This is a fairly unusual problem and is especially difficult to diagnose when the motor seems to run fine for a relatively long cycle then trips the overload. Mechanically, the motor bearings may be seizing. Electrically, it could be the run cap, if it has one. Once a motor has started, the run cap is in the circuit only to help the compressor run more efficiently. If the run cap is bad or weak, the compressor will draw higher RLA than normal and may go out on overload.

NOTE: A run cap does not have to be open or shorted to cause a problem. It could just be weak, putting out less than its rated capacitance. Check it with a capacitor tester to be sure.

SUMMARY

Three-phase motors are powerful, efficient, and fairly simple electrically. However, they can be damaged by loss of a power leg or a phase imbalance of more than 2 percent. Phase monitors, magnetic motor starters, and motor overloads provide adequate protection.

Correct voltage and phase at the compressor terminals is the first requirement for proper motor operation. Without it, the compressor will not start. If a 3Ø compressor has proper voltage but still does not run, the compressor most likely needs to be replaced.

Single-phase motors need a start winding to shift the phase of the incoming current to help start the rotor turning. Start relays are used to provide momentary power from the run winding to the start winding.

When the motor is up to speed, the relay opens the circuit, and the motor continues running on the run winding.

Single-phase compressors on systems with TEV metering devices often require start caps to provide additional torque during startup. Run caps are used on some 1Ø motors to help them run more efficiently.

Start windings, start relays, run caps, and start caps help the 1Ø compressor start and run properly. However, these extra parts add to the difficulty in properly diagnosing a 1Ø compressor's electrical problems. Capacitor testers and electronic relay kits help make troubleshooting 1Ø compressors easier and safer.

SERVICE SCENARIOS

Following are some actual service scenarios. All the information may not have been provided, and the diagnostics may not have been "textbook." However, they are important on the job experiences, and the lessons learned are worth sharing.

SERVICE SCENARIO #1

CONDENSER MOTOR PROBLEM

By Dick Wirz

Service Call: No A/C. Indoor fan blowing hot air, and the outdoor unit is not running.

Equipment: One 5-ton Carrier split system for a bagel shop in a strip mall. Gas furnace located in the back room, with the condensing unit on the ground outside the rear of the store.

Preliminary Checks: Thermostat set for cooling, indoor fan running, but the outdoor unit was off.

While removing the cover to the control box, the compressor came on, but the condenser fan did not. The unit soon cycled off, apparently on the auto reset high pressure control.

Using a multimeter, it was determined that the motor had the proper 220 volts. Checking the fan's run capacitor with the meter's capacitance function verified that it was good.

Diagnosis: Condenser motor was bad.

Repair Procedure: No quote to the customer was necessary because the equipment was covered under our company's full-coverage maintenance contract for the refrigeration, ice machine, and A/C. The manager was informed that a new motor was needed, and it would be repaired within two hours. A new motor and capacitor were picked up from the Carrier distributor, which was not far from the job. Always use a new capacitor when replacing a motor.

The motor was replaced and the unit started. Voltage and amp draw of the new motor was checked just to make sure it was operating correctly. The Carrier unit nameplate and the motor FLA both called for 1.5 amps. The motor was drawing 1.8 amps, which seemed strange. Condenser motors often draw less than their stated FLA but should never draw more.

The fan's run capacitor was checked and found to be good. The condenser was inspected for dirt or blockage of some kind, but it was clean clear through. The distributor was called and asked to check that they had provided the correct motor. The person at the parts counter verified that the 1/4-hp motor was correct for that model unit.

Our office was contacted and asked to check the service history of this customer. It turned out that two condenser motors had been replaced within the last two years. Condenser motors should not have to be replaced each year.

Another call was made to the Carrier distributor, and this time, the service manager got involved. He checked through his Carrier factory service bulletins. He finally found one stating that a fan blade change had affected the operation of some units, specifically the one we were working on. At some point, the assembly line ran out of fan blades from their original source, so they used fan blades from their alternate source. It was later discovered that the second blade had a greater pitch than the original one. The good news is that the blade moved more air. The bad news is that it was too much for the 1/4-hp condenser motor, making it draw high amps. This led to premature motor failures.

The distributor did not have the correct replacement condenser fan blade. So they provided a 1/3-HP fan motor rated at 2.0 amps. After installing the bigger motor and starting the unit, the motor was drawing just below its rated FLA. This motor should last a long time.

A complete check was made of both the electrical and the refrigeration of this A/C system to make sure there would not be any call-backs. The customer was upset enough as it was. Also, everything was documented in the service records for future reference, and a special note was made to bring up this scenario in the next service meeting.

Epilogue: The technician did something all technicians should do, but few actually do it. He checked to make sure that the part he replaced was running correctly. In this case, it was the few extra minutes it took to put the amp meter on the motor leads and check the current draw against the nameplate FLA.

Although the amperage was only about a quarter amp over the rating, he knew from experience that something was not right. Had the two previous techs done as he did, maybe this problem would have been caught sooner.

By tracking service call histories, he verified multiple replacements that further indicated a problem with something other than just the motor. Finally, he took the time to contact the distributor and asked them to check their service bulletins relating to that unit with that particular serial number.

SERVICE SCENARIO #2

POWER AND POTENTIAL RELAYS

By Dick Wirz

Service Call: Walk-in freezer warm.

Symptoms: Compressor tries to start but goes out on motor overload.

Initial Diagnosis: Potential relay and/or start capacitor is bad.

Repairs: Replaced relay and start capacitor. The compressor started and ran. Walk-in came down to temperature.

Call Back: The next day about the same time, the same problem. Walk-in warm.

Symptoms: The same symptoms—compressor tries to start but goes out on motor overload.

Final Diagnosis: Checked incoming voltage. It was only 185 volts on a 208-volt to 230-volt single-phase compressor. The services of an electrician were needed to restore the proper voltage to the unit.

Solution: The customer had his electrician correct the power problem, and the unit operated just fine.

Epilogue: Nearly 40 years ago, I was a technician that had a lot to learn. For one thing, I tended to skip important steps. Especially on this particular hot afternoon when I was rushing to complete my list of calls. That day my focus was misdirected to believing that a compressor starting problem indicated bad start components. I forgot the primary rule of electrical troubleshooting: "Check incoming voltage first!" Had I done that, I would have solved my customer's problem on the first call and saved myself and my company an expensive call-back.

The walk-in was on Captain White's Seafood barge docked at the Fisherman's Wharf on the Potomac River in Washington, D.C., on a hot July day when temperatures were near 100°F. At that time, the voltage in Washington, D.C., was 208, on a good day. But during that hot afternoon, the power was probably under 200 volts. To add to the problem, the barge had 100 feet of flexible cord running between the main power at the dock and the electrical panel on the barge. Due to improper wire size, loose electrical connections, or both, the voltage was not high enough to prevent compressor starting problems.

We have always been told that motors should operate within plus or minus (±) 10 percent of their rated voltage. On a 208-volt to 230-volt motor, that would mean a minimum of 187 volts (208 volts × 90%). That is true for Copeland's three-phase compressors but not their single-phase compressors. Copeland's minimum allowable voltage on their 208-volt to 230-volt single-phase compressor is 5 percent below 208 volts, or 198 volts. I had only about 185 volts at the compressor contactor, so there was no way that compressor was going to work properly. I'm surprised changing the start components even helped it to start on the first call.

Lesson Learned: I relearned a valuable lesson that day to always check the voltage first when a compressor is having a hard time starting. By sharing this experience, I hope to keep other technicians from making the same mistake.

REVIEW QUESTIONS

1. What are the two main weaknesses of 3Ø motors?

   a. Phase loss and voltage unbalance
   b. High amperage and low voltage
   c. Wire size and circuit breaker size

2. What formula is used to measure the percentage of voltage unbalance?

   a. $V_u = V_a \div V_d$
   b. $V_u = V_d \div V_a$
   c. $V_u = V_d \div (V_a + V_d)$

3. What are four things a technician needs to know when choosing a contactor?

4. The amperage load rating of a compressor contactor is listed under which rating, the inductive rating or resistive load rating?

5. What indicates the contacts are bad on a contactor?

   a. The contactor chatters or hums.
   b. There is a voltage reading across the contacts of the contactor when pulled in.

   c. There is no voltage reading (0 volts) across the contacts of the contactor when pulled in.

6. The contactor must be replaced whenever a 3Ø compressor is replaced.

   a. True
   b. False

7. How is a motor starter different from a contactor?

   a. A motor starter has overloads.
   b. A motor starter breaks and makes all three circuits.
   c. A motor starter shuts off the circuit breaker then resets it.

8. Why does a 1Ø motor need a phase shift to start the motor turning?

   a. The windings are 120 degrees out of phase and therefore are difficult to realign.
   b. Single-phase motors do not have equal resistance; therefore, they need something to equalize the phase shift.

c. The windings are 180 degrees out of phase; therefore, on startup, the stator and rotor have equal, but opposite, attraction.

9. How does a 1Ø motor accomplish its phase shift?

a. Voltage is delayed in order to energize the run winding sooner.
b. A run winding is used to change the resistance when the motor starts.
c. The start winding has more resistance, which causes a phase shift, creating a slightly different angle to the magnetic pull of the stator.

10. What is the difference between the windings in a 3Ø motor and a 1Ø motor?

a. A 1Ø motor has both start and run windings.
b. A 3Ø motor has only start windings.
c. A 1Ø motor has only run windings.

11. What are considered the poles of a motor?

a. The rotor and the stator
b. The bundles of wires for the run windings, or for the start windings
c. The negative and positive fields of magnetic reactors

12. What is the nominal rpm of a two-pole motor? A four-pole motor? A six-pole motor?

a. 3,000; 1,600; and 1,200 rpm
b. 3,600; 1,800; and 1,200 rpm
c. 7,200; 3,200; and 1,600 rpm

13. During startup, as the compressor motor speed increases, the amperage does what?

a. Increases
b. Decreases
c. Stays about the same

14. A motor draws LRA only when the motor is mechanically locked up.

a. True
b. False

15. What are the two basic functions of a start relay?

a. The start relay provides full power to the start winding on startup and then de-energizes the start windings when the motor is up to speed.
b. The start relay provides full power to the start windings on startup and then provides a phase shift to keep the motor turning.
c. The start relay provides a phase shift in one direction to start the motor then reverses the phase shift to keep the motor running.

16. What are three types of start relays?

17. What size compressors use current relays?

a. Compressors over 1 hp
b. Compressors up to 3 hp
c. Fractional hp compressors

18. On a current relay, what do the terminal designations L, M, and S stand for?

a. Line, main winding, and start winding
b. Load, middle winding, and short winding
c. Line, minor winding, and start winding

19. A start cap is added to a start relay for compressors on what type of system—cap tube or TEV? Why?

a. A cap tube system because it needs pressure to push through the tube
b. A TEV system because the system does not equalize during the off cycle
c. A TEV system because the valve opens on startup and pressurizes the system

20. How is a start capacitor wired in the electrical circuit?

a. In series with the run winding
b. In parallel with the run winding
c. In series with the start relay

21. Why do some manufacturers add a resistor to a start cap?

a. It makes the capacitor cheaper to manufacture.
b. It bleeds off excess electrical charge when the capacitor is not in the circuit.
c. It adds excess electrical charge to the capacitor in order to provide more of a phase shift during startup.

22. Why is the resistor removed before using a meter to check the capacitor?

a. It will cause the meter to give an incorrect reading.
b. It will harm the electrical meter.
c. It could overheat and cause a fire.

23. Are the contacts (L–S) of a current relay normally open (NO) or normally closed (NC)?

a. NO
b. NC

24. Are the contacts (1–2) of a potential relay NO or NC?

a. NO
b. NC

25. What is a good rhyme to remember the wiring terminals of a potential relay?

26. What three voltage ratings are on potential relays? What do they stand for?

27. How is the operation of a run cap different from that of a start cap?

    a. A start cap has a higher VAC.
    b. A start cap remains in the circuit longer, but a run cap has a lower MFD rating.
    c. The run cap remains in the circuit, but the start cap drops out after the compressor is up to speed.

28. Assume a technician has an emergency situation and needs a 15 µF × 370 VAC run capacitor. He has two of each of the following capacitors: 5 µF, 7.5 µF, and 10 µF, all rated at 370 VAC. What combinations could he use, and how would he wire them, in series or parallel?

    a. Two 7.5 µF capacitors wired in series, and a 5 µF wired in series
    b. A 10 µF and a 5 µF capacitor wired in series, and a 7.5 µF wired in parallel
    c. Two 7.5 µF capacitors wired in parallel, or a 10 µF and a 5 µF wired in parallel

29. When replacing a capacitor, what is the rule about capacitor VAC ratings?

    a. The new capacitor's VAC rating must be equal to or greater than the capacitor it is replacing.
    b. The new capacitor's VAC rating must always be greater than the capacitor it is replacing.
    c. If the MFD rating is the same as the original capacitor, the VAC can be lower or higher than the original.

30. What is a PTCR? How does it open the circuit to the start winding?

31. Motor overloads open a circuit in response to what?

    a. Voltage only, if it is an external overload
    b. Amperage only, if it is an internal overload
    c. Heat, amperage, or a combination of the two, depending on the overload

32. Why does the cover on a compressor terminal box need to be left on?

    a. So the overload will properly sense motor temperature and open during an overload condition
    b. To prevent hazards by fire if dirt and oil get into the control box

    c. To prevent accidental contact with live wires by the technician
    d. All of the above

33. How is a magnetic overload different from a thermo-disc-type external overload?

    a. Magnetic overloads are not as accurate as thermal-type overloads.
    b. Magnetic overloads accurately respond only to amperage and are not affected by ambient air temperature.
    c. Thermal overloads must be used on open motors; magnetic overloads must be used on compressors.

34. Why is it dangerous to jump out a compressor motor overload?

    a. Without the overload, there is no motor protection. The motor could be damaged and the technician injured.
    b. It could trip the circuit breaker.
    c. The technician might short the overload while trying to disconnect it.

35. On a 3Ø compressor motor, if the ohm reading from T1 to T2 is 2 ohms, what would be the ohm reading between T2 and T3, and between T3 and T1?

    a. T2–T3 = 6, T3–T1 = 8
    b. T2–T3 = 8, T3–T1 = 6
    c. T2–T3 = 2, T3–T1 = 2

36. On a 3Ø compressor, if the ohm readings for T1–T2, T2–T3, and T3–T1 are all infinite resistance, what is the problem?

    a. The compressor motor is shorted to ground.
    b. All the compressor windings have shorted open.
    c. The motor overload is open.

37. On a 3Ø compressor, if the ohm readings for T1–T2 and T2–T3 are each 3 ohms, and T3–T1 shows infinite resistance (OL or infinite ohms), what is the problem?

    a. The T3–T1 winding is shorted to ground.
    b. The T3–T1 winding is open.
    c. The motor overload is open.

38. If the ohm reading from T2 to ground (on the suction line) is infinite (OL), is there a problem? Why or why not?

    a. Yes, there should be measurable resistance to ground.
    b. No, there should not be any measurable resistance to ground.
    c. It depends; the motor would have to be running to check it.

39. On a 1Ø motor, if the ohm reading for R–S is 6 ohms and the ohm reading for R–C is 1.5 ohms, what should be the ohm reading for S–C?

    a. 4.5 ohms
    b. 7.5 ohms
    c. 3 ohms

40. Checking startup voltage at the compressor terminals can be dangerous. Why?

    a. Because voltage must be checked with the power turned off
    b. Because a slip of the meter probe could cause a short circuit
    c. Because it is difficult to tell which is the start terminal and which is common

41. If a 3Ø compressor is not even trying to start, what is the first thing to check?

    a. If the motor has full voltage to the compressor terminals
    b. If the internal overload is open
    c. If the motor is mechanically seized

42. If a 3Ø compressor is getting full voltage, trying to start, but going out on an internal overload, what is the likely problem?

    a. The compressor is grounded.
    b. The compressor windings are open.
    c. The compressor is mechanically seized, or locked up.

43. If a 3Ø compressor seems to be mechanically locked up, what is the next step?

    a. Quote the customer for a new compressor.
    b. Reverse the motor rotation to see if the compressor will run.
    c. Rebuild the compressor.

44. A 1Ø motor makes no sound and does not try to start. What is the first thing to check?

    a. If the motor has full voltage to the compressor terminals
    b. If the internal overload is open
    c. If the motor is mechanically seized

45. What may be the problem with a 1Ø motor that makes a humming sound, tries to start, but goes out on overload?

    a. Low voltage to the compressor
    b. Bad start cap
    c. Locked up compressor
    d. All of the above

46. What may be the problem with a 1Ø motor that starts and runs for a few seconds but goes out on overload?

    a. Start relay contacts stuck closed
    b. Weak overload
    c. Bad start cap

47. What are the two functions of a start relay?

48. How is a current relay checked with an ohmmeter?

    a. L–S contacts should be open, and the relay coil (L–M) should have measurable resistance.
    b. L–S contacts should be closed, and the relay coil (L–M) should have infinite resistance.
    c. L–S contacts should be open, and the relay coil (L–M) should have no resistance.

49. How is a potential relay checked with an ohmmeter?

    a. The switch contacts (1–2) should be closed, and the relay coil (5–2) should have measurable resistance.
    b. The switch contacts (1–2) should be open, and the relay coil (5–2) should have infinite resistance.
    c. The switch contacts (1–2) should be closed, and the relay coil (5–2) should have no resistance.

50. If a compressor start relay needs to be replaced, should the start capacitor be replaced as well? Why or why not?

    a. Yes, because a new capacitor always comes with a replacement relay.
    b. No, because you should replace only what you know is bad.
    c. Yes, because if the relay fails, it has probably put a strain on the capacitor as well.

RETROFITTING, RECOVERY, EVACUATION, AND CHARGING

9

CHAPTER OVERVIEW

Eventually, all units with R12 and R502 will either be replaced or be retrofitted with a hydrogen, chlorine, fluorine, and carbon (HCFC) or hydrogen, fluorine, and carbon (HFC) refrigerant. This chapter should help in understanding what options are available when it comes to changing refrigerants in reach-ins and walk-ins. Retrofitting often requires changing the compressor oil. This chapter includes a detailed description of how to make this procedure quick and easy when performed on semihermetic compressors. With hermetic compressors, the process is difficult, and the cost in time and materials may outweigh its worth.

Refrigerant retrofitting also requires refrigerant recovery, system evacuation, and refrigerant charging. All these tasks are covered in detail in this chapter. In addition, this chapter offers an excellent opportunity to share some tips from experienced technicians on how to make these procedures a little faster and simpler.

OBJECTIVES

After completing this chapter, you should be able to

- Understand retrofitting CFC refrigeration systems
- Explain refrigerant recovery procedures
- Explain system evacuation
- Understand startup and system charging

RETROFITTING REFRIGERATION SYSTEMS

For years, R12 and R502 have been used in most domestic and commercial refrigeration systems. Starting in 1995, these chlorofluorocarbon (CFC) refrigerants were no longer produced because they damaged the ozone layer and contributed to global warming. (CFC is the acronym for the elements that make up those refrigerants: chlorine, fluorine, and carbon.)

Hydrogen, chlorine, fluorine, and carbon (HCFC) refrigerants have less chlorine than CFC refrigerants; in addition, hydrogen is in the compound. Hydrogen makes the molecules break up in the atmosphere, and this releases the chlorine before it can react with the ozone in the stratosphere. Although less environmentally harmful than CFCs, HCFCs are still scheduled for total phaseout by 2030. HCFC-22 (R22) is a special case. This refrigerant will no longer be used in new equipment as of 2010 and is scheduled for total phaseout by 2020.

Hydrogen, fluorine, and carbon (HFC)–based refrigerants are the long-term replacements for CFCs and HCFCs. Although HFC refrigerants do not harm the ozone layer, they do have a small global warming potential.

Refrigerant blends are combinations of refrigerants mixed together to give the properties and efficiencies that closely resemble the refrigerants they are designed to replace. HCFC-based blends are short-term replacements for CFC and other HCFC refrigerants.

Single-compound refrigerants such as HCFC-22 and HFC-134a, as well as azeotropic blends such as R502, have only one boiling point and/or condensing point for each given pressure. Zeotropic blends are mixtures of refrigerants whose individual components boil off at different pressures, or fractionate. Therefore, to ensure they maintain their original composition, these blended refrigerants must be charged in a liquid state. In addition, a technician calculating superheat and subcooling on systems with zeotropic mixtures may have to take into account the temperature glide of these blended refrigerants.

Retrofitting Basics

Refrigerant retrofitting is the term used to describe the replacing of one type of refrigerant in a system with a different type of refrigerant. Currently, most systems that are retrofitted have their original CFC refrigerant replaced with either an HCFC or an HFC refrigerant. This process would be very easy, if it were not for two factors:

1. Unequal oil miscibility
2. Unequal efficiencies, or weights, of replacement refrigerants

Oil Miscibility

Mineral oil is very miscible (mixes well) with CFC refrigerants. However, according to the refrigerant manufacturers, HCFC and HFC refrigerants do not mix well with mineral oil. As a result, HCFCs and HFCs will not move the mineral oil through the system, which would cause lubrication problems for the compressor. Therefore, refrigerant manufacturers have the following two recommendations:

1. When retrofitting a CFC system with an HCFC refrigerant, replace at least 50 percent of the existing mineral oil with alkylbenzene oil.
2. When retrofitting a CFC system with an HFC refrigerant, replace at least 95 percent of the existing mineral oil with polyolester oil.

Refrigerant Efficiencies

In most cases, the replacement refrigerants are more efficient by weight than the original CFC refrigerants. This sounds like an advantage, but it is very difficult to determine how much replacement refrigerant to weigh into a critically charged system originally designed for R12. Also, the ability of HCFCs to absorb more heat can overload the compressor during maximum operating conditions.

Most refrigerant manufacturers recommend using less than the stated amount of replacement refrigerant when retrofitting critically charged systems. On the other hand, thermal expansion valve (TEV) systems must be fully charged to ensure complete liquid to the metering device. Following is an example of how the increased refrigeration effect of a replacement refrigerant can overload a compressor if the TEV is oversized.

EXAMPLE: 1

Consider an 8,000-Btuh R12 system with a 1-ton (12,000-Btuh) rated TEV. When retrofitted with R401A, the compressor would go out on overload when the ambient temperature at the condenser was above 90°F. After the 1-ton TEV was replaced with a smaller 1/2-ton TEV, the compressor could handle the load, and the walk-in operated as well as it did before the retrofit.

NOTE: This actual occurrence can be used to justify the claim that some replacement refrigerants are more efficient than R12. Also, if the unit has an oversized TEV, the refrigerant in the evaporator may absorb more heat than the compressor can handle during high-load conditions. Although this condition is something technicians should be aware of, it definitely does not mean every retrofit requires a smaller TEV.

Oil Replacement

When retrofitting a CFC unit with an HCFC refrigerant, at least half of the original mineral oil must be replaced with alkylbenzene oil. Usually, one change of all the original mineral oil in the crankcase with alkyl-benzene is enough to accomplish this requirement.

If the replacement refrigerant is an HFC, the factory recommends that at least 95 percent of the original mineral oil be replaced with polyolester oil. This usually takes a minimum of three oil changes. After each change of all the oil in the crankcase, the system should be operated for at least several hours to make sure the oil is circulated throughout the equipment. A refractometer is needed to verify the required minimum of polyolester in the oil mixture.

The requirement of 95 percent purity of polyolester and the difficulty and expense of three oil changes discourage most technicians and their customers from retrofitting to HFC refrigerants. Therefore, any further references to retrofitting will be about only HCFCs.

Replacing oil in a semihermetic compressor is fairly easy with the help of a small copper tube and some putty to seal around the tubing. Refer to Figure 9-1 while reading the following description of the process:

1. With the compressor running, front-seat the suction service valve.
2. Shut off the compressor when the crankcase pressure drops to zero.
3. Remove the oil plug on the side of the compressor.
4. Insert a piece of 1/4-inch or 3/8-inch tubing into the hole so that one end is touching the bottom of the crankcase.
5. Make a temporary seal around the tubing with some putty.
6. Open the suction valve slightly to pressurize the crankcase.

NOTE: It only takes about 2 psig of force on the top of the oil to push it up the tube and out of the crankcase.

7. Measure the amount of oil removed from the crankcase.
8. Put the same amount of fresh replacement oil into the crankcase.
9. Reinstall the oil plug and evacuate the crankcase.
10. Open the suction valve, start the compressor, and check the oil sight glass.

NOTE: There are oil removal kits available containing plastic tubing and the fittings necessary to make this procedure even easier.

Changing oil in a hermetic compressor is much more difficult. The most effective way is to recover the refrigerant, cut the compressor out of the system and turn it upside down to drain the oil from the suction line, put in the replacement oil, and then reinstall the compressor.

Retrofit Refrigerants

There are quite a few replacement refrigerants on the market, and there will probably be more in the near future. However, the following retrofit discussion is limited to some of the replacement refrigerants that have been out long enough for most technicians to have had at least some experience with them.

The primary problem with replacement refrigerants is that those units retrofitted in mild ambients seem to have more problems than usual when operating in high ambients or heavy box loading.

FIGURE 9-1 **Changing oil in a semihermetic compressor.** *Courtesy of Refrigeration Training Services.*

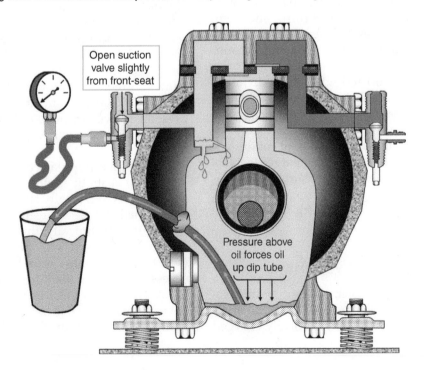

REALITY CHECK 1

It may be more cost-effective to replace a hermetic compressor than to attempt changing its oil. Both tasks take about the same amount of time, and the new compressor comes with the proper oil. Besides, an older hermetic compressor may not re-start after going through the rough handling of an oil-changing procedure

REALITY CHECK 2

Many service companies leave the mineral oil in the hermetic compressor when retrofitting small self-contained reach-ins. The evaporator and condensing unit are so close together (close-coupled) that, apparently, the mineral oil moves through the system without any problem.

The compressors go out on overload on some units, and others experience extreme refrigerant floodback.

For the purpose of demonstration in this text-book, three R12 reach-in units were shop-tested by the author for more than two weeks in an attempt to recreate the problems with replacement refrigerants and to find a solution. The test units consisted of a single-door refrigerator with a cap tube system, a sin-gle-door freezer with a cap tube system, and a three-door refrigerator with a TEV system. All temperatures and pressures at the factory-recommended charge of R12 were first recorded at their design conditions. The next condition was the simulation of warm prod-uct being placed in the boxes. The third, and final, condition was the compressor operation under the high-ambient temperatures the units would encoun-ter in a hot kitchen.

Each unit was retrofitted with three different HCFCs. After each retrofit, the units were tested under the three different operating conditions just mentioned. A full charge of the replacement HCFCs seemed to al-ways cause problems under high-ambient conditions. Therefore, the amounts of the replacement refrigerants were varied to determine what percentage brought the operating conditions closest to the original R12. The three HCFC refrigerants chosen for the tests were R401A, R409A, and R414B.

Retrofitting Cap Tube Systems

Manufacturers of most replacement refrigerants rec-ommend using a percentage (usually %) of the origi-nal R12 charge when retrofitting critically charged

systems. The factory recommendations were followed. However, it was soon apparent that the percentage actually varied, based on the refrigerant and whether the test unit was a medium- or a low-temperature box. Each replacement refrigerant would use one percent-age for refrigerators and a different percentage for freezers. Trying to decide which refrigerant was best, and easiest to use, was a challenge.

Eventually, some interesting similarities became evident. The findings were verified by a senior applica-tions engineer of a large compressor manufacturer. As a result of these tests, some definite procedures have been developed for retrofitting R12 cap tube units with R401A, R409A, or R414B.

NOTE: To use the charging procedures given in the TROT, the condenser must be absolutely clean.

EXAMPLE: 2

Assume a unit's rated charge is 20 ounces of R12. Its head pres-sure would be 136 psig at a condensing temperature of 110°F (refer to the R12 column on the P/T chart in the appendix).

To retrofit this unit with an HCFC at an ambient of 80°F would require also charging to a head pressure of 136 psig. If the HCFC chosen is R401A, it probably will only take 14 ounces (70%) to reach 136 psig. If R414B is used, most likely 18 ounces (90%) will be needed to reach 136 psig. Either refrigerant will work very well; they just do the same job with different amounts.

The amount of subcooling is also an indicator of the amount of refrigerant in the system. If there is more than 20°F subcooling, the system is probably a little overcharged. However, do not remove any refrig-erant before checking the evaporator superheat under high-ambient conditions.

To simulate high-ambient conditions, block the condenser and raise the condensing temperature to about 125°F, which is equivalent to a 95°F ambient for a medium-temperature reach-in. If there are still a few degrees of superheat, that means the system is not overcharged, and it is not flooding the com-pressor. However, if there is no superheat, then a little refrigerant may need to be removed to prevent floodback.

R408A is a popular choice for retrofitting R502 systems; it is about as close to a drop-in replacement as possible. On critically charged systems, R408A can be weighed in at 100 percent of the nameplate, and it works fine. Technically, this refrigerant is an HCFC and should be used with alkylbenzene oil. However, most technicians have found that an oil change is not necessary, even on walk-ins with remote condensing units. The reason for this may be that R502 systems are usually freezers; therefore, the defrost cycle helps return the oil to the compressor.

T . R . O . T

Charging a retrofitted cap tube system

1. Charge the system to the equivalent R12 head pressures.

NOTE: Use a 30°F condenser split for refrigerators and a 25°F split for freezers.

EXAMPLE

For a reach-in refrigerator, if the ambient is 80°F, charge to 110°F at 136 psig.

1. Subcooling should be 5°F to 10°F.
2. When the box is at design temperature (38°F for refrigerator and 0°F for freezer), the superheat should be about 5°F to 10°F.

NOTE: It is difficult to mount a temperature probe inside the box at the evaporator outlet. Therefore, take the suction-line temperature where it exits the wall of the box.

1. Block the condenser enough to get the condensing temperature up to 125°F.
2. Superheat will drop to about 1°F or 2°F based on the suction-line temperature about 6 inches from the compressor. If there is no superheat, remove a little refrigerant until there is some superheat.

The most important point of this retrofitting procedure for cap tube systems is to charge whatever retrofit refrigerant is being used as if it were R12—not by weight, but by head pressure. When charging the HCFC refrigerant to the proper head pressure for R12, the weight of the retrofit refrigerant used will be about 70 percent to 90 percent of the critical charge of R12 on the unit's data plate.

Retrofitting TEV Systems

A retrofitted R12 TEV system is charged the same as any other TEV system that is not critically charged. In other words, charge to a 30°F condenser split (or 25°F on a freezer) and look for about 10°F subcooling (use the P/T chart in the Appendix for the replacement refrigerant). The strange thing about blended refrigerants is that the sight glass may bubble slightly even when the system is fully charged. To check if the bubbling is due to a low charge, measure the amount of subcooling at the sight glass. As long as there is subcooling, there should be a full stream of liquid entering the sight glass. However, blended HCFCs and blended HFCs are made of mixtures of refrigerants that boil off at different temperatures and pressures. Therefore, only one of the refrigerants in the blend may be boiling off in the sight glass, which would not necessarily mean all the refrigerant is flashing off.

Another reason for bubbles is that the inside diameter of the sight glass may be larger than the pipe leading to it. This wide area in the pipe gives the refrigerant a chance to momentarily flash off and then return to liquid as it continues into the liquid line. In this case, bubbling from blended refrigerants does not necessarily mean the TEV will be affected. Therefore, adding refrigerant just to clear the sight glass could result in overcharging the unit and high head pressure problems.

REFRIGERANT RECOVERY PROCEDURES

This section begins with the basics of refrigerant recovery and is followed by a description of the push–pull method. This technique speeds the recovery process on systems with a large amount of liquid refrigerant. The final portion of this section describes how, in some instances, just a cold-evacuated recovery tank can be used to remove liquid refrigerant, without the use of a recovery machine (Figure 9-2).

Recovery is always done with the refrigeration unit's compressor turned off. Following is the hookup procedure for the recovery unit and tank:

1. Connect the suction and discharge hoses from the charging manifold to the service valves on the refrigeration unit.
2. Connect a charging hose from the center tap on the manifold to the inlet of the recovery machine.
3. Connect a charging hose from the outlet of the recovery machine to the vapor valve on the recovery cylinder.
4. Open both the inlet and the outlet valves on the recovery unit.
5. Open the vapor valve on the recovery tank and start the recovery machine.
6. Open the suction service valve on both the refrigeration unit and the gauge manifold.
7. After about two minutes, open the discharge service valve on the refrigeration unit and on the gauge manifold.
8. All of the refrigerant has been recovered when the low-side gauge shows the system is at 0 psig, or lower.

NOTE: Watch the weight of the recovery tank as the recovered refrigerant fills it. Stop the process when the tank is 80 percent full, by weight. Replace the tank with an empty tank, and finish the recovery process.

A recovery machine is basically just a small compressor with a condenser. The recovery unit compressor sucks vapor out of the unit to be retrofitted. The vapor is compressed and discharged into the recovery machine condenser where the vapor changes to liquid. The liquid is pushed out of the recovery machine and into a recovery cylinder.

Recovery unit compressors, just like refrigeration unit compressors, can be damaged by trying to compress liquid. Therefore, most recovery units must have

FIGURE 9-2 **Standard vapor recovery of refrigerant.** *Courtesy of Refrigeration Training Services.*

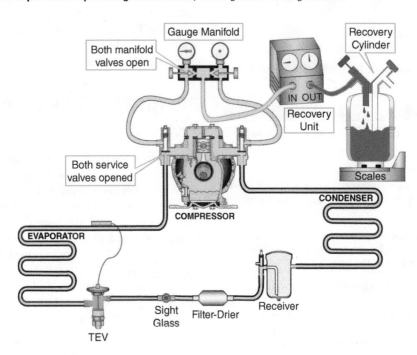

only vapor entering their inlet ports. Even though the literature with the recovery unit may claim that it can recover liquid, in actuality, it must vaporize the liquid before it reaches the recovery unit's compressor. To accomplish this, most recovery units have an orifice at the compressor inlet. If liquid does enter the recovery machine, it will be vaporized by the orifice, or metering device, before it enters the compressor.

The process of recovering vapor, or vaporizing liquid, is slower than moving liquid directly from the refrigeration unit to the recovery cylinder. Following are some ways to get most of the liquid out of a refrigeration system, without the recovery unit having to vaporize it first. Note that an exception to the rule is the Appion G5Twin recovery machine, which pumps vapor but can also pump liquid refrigerant without vaporizing it first.

Push–Pull Liquid Recovery

On systems with receivers, the recovery time can be shortened by first pushing the liquid out of the receiver and into a recovery cylinder. The remaining refrigerant is then removed by vapor recovery (see Figure 9-3).

After the gauges are installed on a unit, the following steps will hook up the recovery unit and tank for the push–pull procedure:

1. Connect a charging hose from the king valve on the receiver to the liquid valve on the recovery tank.
2. Connect a charging hose from the vapor valve on the recovery tank to the inlet of the recovery machine.

3. Connect the charging hose from the center of the gauge manifold to the outlet of the recovery machine.
4. Open (mid-seat) the discharge valve on the compressor and open the high-side valve on the gauge manifold.
5. Open both the inlet and the outlet valves on the recovery unit and the vapor and liquid valves on the recovery tank.
6. Start the recovery unit and then open the king valve fully to the front-seat position.

The recovery machine lowers the vapor pressure in the recovery tank while increasing the pressure in the refrigeration system. The liquid is forced out of the receiver's king valve and into the recovery cylinder. After liquid stops flowing out of the receiver, switch the hoses back to the standard positions and mid-seat the king valve. Continue recovery by the vapor method until all the remaining refrigerant has been pulled from the refrigeration unit.

There is a recovery machine on the market that has a very high liquid-pumping capability. The Appion G5 Twin will recover vapor at a rate similar to other machines; however, its liquid recovery rate is many times faster (Figure 9-4). The Appion G5 Twin, unlike other recovery machines, can pump liquid directly because it runs at a slower rotational speed, has heavier connecting rods, and utilizes spring-loaded poppet valves that are not damaged when pumping liquid. A rapid recovery is a benefit to the technician and to the customer.

FIGURE 9-3   **Liquid recovery by the push–pull method.** *Courtesy of Refrigeration Training Services.*

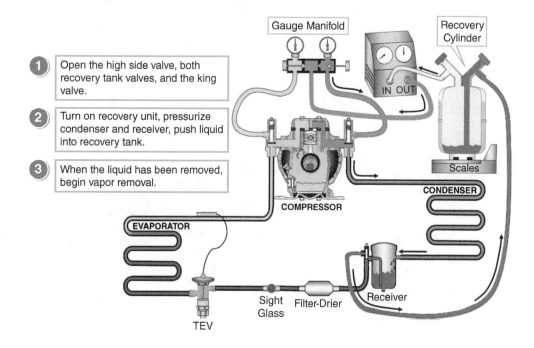

1. Open the high side valve, both recovery tank valves, and the king valve.

2. Turn on recovery unit, pressurize condenser and receiver, push liquid into recovery tank.

3. When the liquid has been removed, begin vapor removal.

FIGURE 9-4   **Appion recovery machine.** *Courtesy of Appion, Inc.*

Liquid Recovery with a Tank in a Vacuum

An evacuated recovery cylinder will pull a surprising amount of liquid out of the receiver (see Figure 9-5). On small systems, this method may be enough to remove all the liquid from the system. Simply use a vacuum pump to evacuate an empty recovery tank. Hook up a charging hose from the king valve on the receiver to the liquid valve on the recovery tank. Place the recovery tank on a scale and open the valves at both ends of the charging hose. Refrigerant always travels from a place of higher pressure to one of lower pressure. When the weight of the scale stops rising, it has recovered all the liquid it can. A recovery machine will quickly remove the remaining vapor.

If there is ice available, nearly all the liquid from even large systems can be recovered without the use of a recovery machine. Place the recovery tank in a container of ice water. As long as the liquid in the refrigeration system is above freezing, it will naturally migrate to the colder space in the recovery cylinder. In addition, as the warmer liquid from the unit's receiver is cooled in the cold recovery cylinder, its pressure falls as well. Therefore, the flow of refrigerant is assisted by the fact that refrigerant always moves from a higher-pressure location to an area of lower pressure.

Connect one end of a long-charging hose to the king valve on the receiver. The other end is attached to a recovery cylinder that is sitting in a container of ice water. When both valves are opened, the liquid from the receiver will flow into the colder tank.

Recovery Equipment Maintenance

Install a small (3-cubic-inch) drier on the inlet of the recovery machine. Most filter-drier manufacturers make them available with a 1/4-inch female flare outlet and a 1/4-inch male flare inlet, which works just fine (Sporlan C-032-F). Investing a few dollars in a

FIGURE 9-5 **Liquid recovery into an evacuated tank.** *Courtesy of Refrigeration Training Services.*

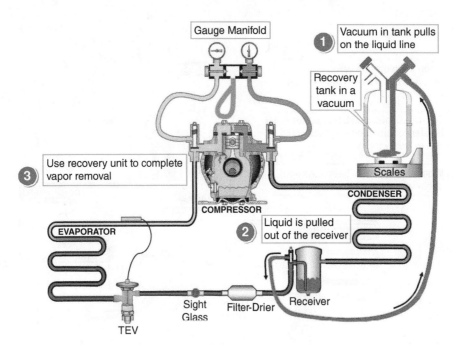

filter drier will protect an expensive recovery unit from acids, sludge, and debris that may be recovered with the refrigerants. Replace the filter drier after each burn-out recovery, or after every 10 normal recoveries.

Recovery Cylinders

The gray-and-yellow recovery cylinders used in refrigeration are familiar to most technicians, but there are a few interesting points that should be made. Two valves are on the top of a recovery cylinder: one for vapor and one for liquid. However, there is no standardization for the color of the valves. One manufacturer will use the red valve for liquid, and another may have a red vapor valve. Make sure you know if the valve you are connected to is for vapor or for liquid.

The most popular sizes of recovery cylinders carried on service trucks are the nominal 50-pound and 25-pound tanks. Each cylinder has a date stamped on it, which means it must be re-inspected if it is turned into a supply house after that date. The *TW* stamped on the tank handle represents "tare weight," or the weight of the tank when empty. If the weight is slightly above the TW, do not just assume that it must have some refrigerant left in it. It could be some accumulated oil from previous recovery operations. Also, tag your tanks with the refrigerant recovered so there is no chance of mixing different refrigerants.

SAFETY NOTE: Fill the recovery cylinder to only 80 percent of its maximum capacity.

The rule of 80 percent maximum fill is a safety factor. If the tank were filled to more than 80 percent of its capacity, and if its temperature were allowed to rise to 125°F, the expansion of the refrigerant could cause the tank to rupture. The capacity of a tank is rated by how much water it can hold, and this amount is stamped on the handle, preceded by the letters *WC* for "water capacity." The amount of refrigerant a tank can hold depends on the type of refrigerant and its temperature. If we wanted to be precise, we would need to calculate the specific weight of the refrigerant used, compare it with the WC, and subtract 20 percent. However, it is easier and safer to limit the amount of recovered refrigerant to 80 percent of its nominal capacity. For instance, if you have a 25-pound tank, do not put any more than 20 pounds (25 × 0.80 = 20) of refrigerant in it. To verify that a tank does not have any more than the maximum amount, just add 20 pounds to the tare weight and weigh it (20 pounds + *TW* = maximum total weight). See Figure 9-6 for an example of the information stamped on a recovery cylinder handle.

SYSTEM EVACUATION

Air in the system consists of oxygen, nitrogen, and any moisture that entered the system with the vapor. The individual pressures of the oxygen and nitrogen are added to the system's refrigerant pressure and will increase as their temperature rises. This pressure rise depends on the amount of air trapped in the system,

FIGURE 9-6 **Recovery cylinders with *WC, TW*, and date codes on the handle.** *Photo by Dick Wirz.*

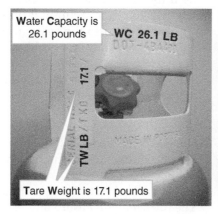
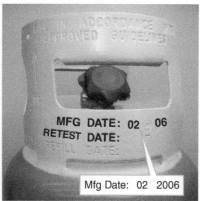

but it can easily add 50 pounds or more to the head pressure. Oxygen in the system combines readily with other elements in the system to cause rust, corrosion, and burning. When moisture is combined with oxygen, the two elements quickly attack refrigerant and oil, causing acid formation and sludge, which is evidenced by copper plating and burned windings.

Many recovery machines can pull a slight vacuum on the system from which they are recovering refrigerant. However, refrigeration systems must be pulled down into a micron-level vacuum to ensure that moisture and noncondensables are removed. Most refrigeration equipment manufacturers require the system to

hold a 500-micron-level vacuum before charging the system (Figure 9-7).

NOTE: **Make sure the evacuation equipment setup is free of leaks before trying to pull a vacuum on the refrigeration system.**

Hook up the gauges to the refrigeration unit, the vacuum pump, and the micron gauge, but do not open the refrigeration unit service valves. Start the vacuum pump and pull a 250-micron to 500-micron vacuum and then turn off the pump. A constant micron level indicates that there are no leaks in the evacuation equipment connections to the system to be evacuated.

FIGURE 9-7 **Single evacuation.** *Courtesy of Refrigeration Training Services.*

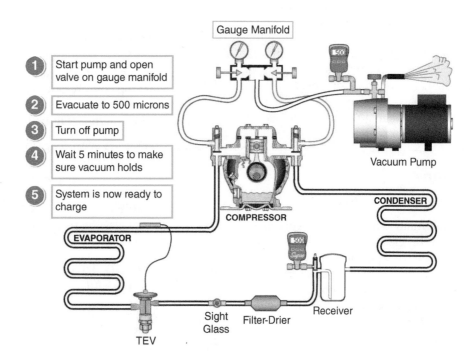

R E A L I T Y   C H E C K   3

Sometimes the micron gauge rises to 750 or 1,000 microns after the vacuum pump is turned off. This could be a very small amount of remaining moisture or refrigerant boiling off. As long as the micron level stops at one of those levels, the system vacuum is good. However, if the vacuum level keeps rising, there may be a leak.

Finally, open the refrigeration unit service valves and restart the vacuum pump. Any subsequent vacuum leaks will most certainly be in the refrigeration unit, not in the connections to it.

Most manufacturers recommend the vacuum gauge be on the system, rather than on the vacuum pump or gauge hose connections. With the system pumped down and the service valves closed, the system gauge is an accurate indicator of the level of evacuation. In Figure 9-7, the only available system access is the king valve on the receiver. With just one micron gauge at that location, it is a little more difficult to determine if a rise in vacuum after turning off the pump is in the manifold, hoses, or system. Of course, if the rising vacuum level stops after the service valves are front-seated, then you can be pretty confident the manifold gauge or connections were the source of the leak.

The ultimate setup is to have a second vacuum gauge on the system in addition to the one on the gauge manifold or vacuum pump. Some new vacuum gauges are so sensitive they will pick up minute losses of vacuum through old worn rubber gauge hoses, which can easily fool the technician into thinking the system is leaking. However, with a second vacuum gauge just on the system, the technician can pull a vacuum then front-seat both service valves to isolate the system from the manifold gauges and vacuum pump. By comparing the read-out of the two vacuum gauges, the technician will have a better idea of where the leak is, if there is one. Figures 9-7 through 9-10 show two micron gauges, which are the most accurate method of verifying the quality of the evacuation.

Triple Evacuation

All of the following statements are true:

- If the vacuum is maintained at 500 to 1,000 microns, the system is dry and is clear of noncondensables.
- Most single evacuations are performed as a preventative measure just in case there is a small amount of moisture or air in the system.
- A single evacuation to 500 microns is usually faster than a triple evacuation.

However, there are times when a triple evacuation is necessary. For instance, when a water-cooled condenser has burst and water has entered the system. Or if the system has developed a leak on the low side and has run in a vacuum, it definitely has pulled air and moisture into the system. And finally, contaminants will certainly be a problem if the system has been open to the atmosphere for some time.

The evacuation equipment is hooked up in the same manner for a triple evacuation as it is for a single evacuation, except that provision has to be made for introducing dry nitrogen to the system. The manufacturer of Copeland compressors (Emerson Climate Technologies, Inc.) recommends evacuating twice to 1,500 microns and the final time to 500 microns. The first two vacuums should be broken to 2 psig each time with dry nitrogen. The nitrogen will absorb any residual moisture, which makes the next evacuation more efficient. For example, if the first evacuation removed 98 percent of the contaminants, the next evacuation would remove 98 percent of the remaining 2 percent. After the third evacuation, the remaining contaminant percentage would be 2 percent × 2 percent × 2 percent, or .0008 percent. The residual contaminants are reduced to such a low level that they are no longer a danger to the system.

If a system has a high-moisture problem, pulling a good vacuum will remove most of it. However, moisture must be in a vapor state before it can be pulled out by a vacuum pump. Droplets of water remain under the refrigerant oil if they are not heated or stirred up somehow to give the water a chance to break free from the oil. The triple evacuation process takes care of this concern by running the compressor for a few seconds between the first and second evacuation.

The following is a summary of the three steps of a triple evacuation:

Step 1 – Evacuate the system to 1,500 microns (Figure 9-8).

Step 2 – Break the vacuum by adding nitrogen to the system until the pressures are about 2 psig. Turn off the nitrogen, start the compressor, and let it run for about 5 seconds. Shut the compressor off and pull a second vacuum to 1,500 microns (Figure 9-9).

Step 3 – Break the vacuum again with nitrogen. Turn off the nitrogen and pull a third vacuum to 500 microns (Figure 9-10).

The system should now be free of moisture and noncondensables. As long as the vacuum holds 1,000 microns or less the system is ready to be charged.

Vacuum Pump and Micron Gauge Operation and Maintenance Tips

Vacuum pumps remove only vapor; they cannot remove acid, sludge, debris, or liquid. Filter driers in the refrigeration system are designed to remove such non-vapor contaminants.

FIGURE 9-8 **Triple evacuation step 1.**
Courtesy of Refrigeration Training Services.

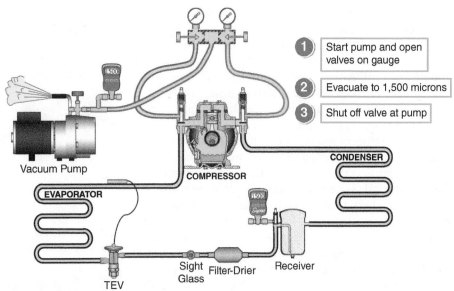

1 Start pump and open valves on gauge

2 Evacuate to 1,500 microns

3 Shut off valve at pump

Vacuum Pump

CONDENSER

COMPRESSOR

EVAPORATOR

Sight Glass Filter-Drier Receiver

TEV

FIGURE 9-9 **Triple evacuation step 2.**
Courtesy of Refrigeration Training Services.

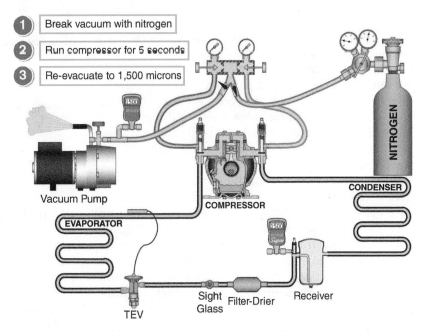

1 Break vacuum with nitrogen

2 Run compressor for 5 seconds

3 Re-evacuate to 1,500 microns

Vacuum Pump

COMPRESSOR

NITROGEN

CONDENSER

EVAPORATOR

Sight Glass Filter-Drier Receiver

TEV

FIGURE 9-10 **Triple evacuation step 3.**
Courtesy of Refrigeration Training Services.

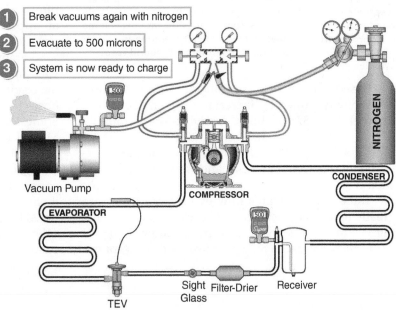

1 Break vacuums again with nitrogen

2 Evacuate to 500 microns

3 System is now ready to charge

Vacuum Pump

COMPRESSOR

NITROGEN

CONDENSER

EVAPORATOR

Sight Glass Filter-Drier Receiver

TEV

If the vacuum pump has a gas ballast knob, open it just before starting the pump. This will open the first stage of a two-stage pump and pull the moisture-laden primary air out of the system without contaminating the pump's oil. After a few minutes of running, close the ballast. This allows the pump to utilize its second stage to pull a deep vacuum. Leave the ballast closed while the pump is turned off.

Manufacturers recommend changing the vacuum pump before each evacuation. A vacuum pump can pull a vacuum only as deep as the vapor pressure of the sealing oil. When oil becomes saturated or contaminated, the vapor pressure rises, and the evacuation process can slow or halt. Do not wait until you can see a change in the oil sight glass. Moisture causes the oil to turn opaque or milky white, but it may be hidden below the level of the sight glass. The oil in a vacuum pump can absorb moisture just sitting in the truck or in the shop. Therefore, whenever the pump seems slow in pulling a vacuum, change the oil and try again. This can be done even in the middle of an evacuation process by isolating the pump from the system by shutting off the proper valves. The Appion™ TEZ8 vacuum pump has an external oil reservoir; therefore, changing the oil takes only a few seconds when you simply change the oil cartridge. One last thing: Do not leave oil in a vacuum pump that is badly contaminated with water or acid. These contaminants can damage the pump.

One factor that is not fully appreciated by most service technicians is how critical the performance of a vacuum pump is to the size of the connecting hoses and fittings. Although standard 1/4-inch ID hoses are fine for small systems, larger systems must use 3/8- to 1/2-inch hoses. Tables 9-1 and 9-2 are adaptations of a pump manufacturer's catalog information on pumping speed of rotary vacuum pumps.

NOTE: The internal volume in Tables 9-1 and 9-2, i.e., 5 cubic feet, represents a relatively large system. In comparison, only one-tenth of that size, or 0.50 cubic feet, is the internal volume of approximately 100 feet of 1 1/8-inch refrigeration tubing, or 920 feet of 3/8-inch tubing, or a nominal 25-pound recovery cylinder.

Based on the information in Table 9-1, more efficiency is gained by increasing the line size from 1/4 inch ID to 3/8 inch on the 1-cubic-feet-per-minute (CFM) pump than is gained by putting a 5-CFM pump on the original 1/4-inch connection. For the greatest benefit, both the line size and the gauge port size should be increased.

Table 9-2 illustrates how evacuation times are greatly reduced by using larger hoses. For instance, using 1/2-inch hoses on a 1-CFM pump allows it to pull a 500-micron vacuum almost as fast as a 5-CFM pump

TABLE 9-1 **Net pumping speed in CFM for a vacuum pump at 1,000 microns** *Adapted from Refrigeration Manual by Emerson Climate Technologies, Inc.*

VACUUM PUMP SIZE (CFM)	PUMPING CAPACITY WITH 6-INCH HOSE (CFM)		
	1/4-INCH ID HOSE	3/8-INCH ID HOSE	1/2-INCH ID HOSE
1	0.23	0.60	0.87
2	0.26	0.83	1.50
5	0.29	1.11	2.95

TABLE 9-2 **Estimated time for vacuum pull-down to 500 microns for 5 cubic feet internal volume** *Adapted from Refrigeration Manual by Emerson Climate Technologies, Inc.*

VACUUM PUMP SIZE (CFM)	PULL-DOWN TIME TO 500 MICRONS WITH 6-INCH HOSE (MINUTES)		
	1/4-INCH ID HOSE	3/8-INCH ID HOSE	1/2-INCH ID HOSE
1	78	51	45
2	56	29	23
5	43	16	10

System volume is 5 cubic feet.

using standard 1/4-inch hoses. Similarly, increasing the hose size from 1/4 inch to just 3/8 inch on a 5-CFM pump allows the evacuation time to be reduced by almost two-thirds.

Using larger hoses during evacuation could be the answer for those technicians who do not want to take the time to pull a good vacuum.

Calculations for determining pull-down time are very complicated, have many possible variables, and will differ based on the design and quality of any given pump. Therefore, the information in Table 9-2 is at best an approximation and should be used only for purposes of illustration.

Larger-gauge hoses allow the pump to pull a faster vacuum. Even though the large hoses are connected to the original small-gauge ports, the increased size of the hoses reduces the total restriction during evacuation. Some technicians use braided stainless steel hoses. Makers of the metal hoses claim they have low-pressure resistance and are less porous than regular rubberized gauge hoses. It is a good idea to have a four-valve gauge set just for evacuation. The extra hose makes it easier to introduce nitrogen during triple evacuations, for systems that require an extra tap to bypass a solenoid valve, or for charging after the evacuation is complete.

Micron gauges are very sensitive to oil and high pressure. Most digital micron gauges have an inlet and outlet port so they can be installed in the line between the refrigeration unit and the vacuum pump. If oil vapor is pulled out of the refrigeration unit and through the micron gauge, it will clog the electronic equipment in the gauge and cause the gauge to malfunction. However, a little alcohol poured into the gauge's ports should clean out the oil and return the gauge to its accurate operation.

Instead of having the vacuum pull through the micron gauge, use just one of the gauge ports and have it teed into the line. Also, a positive pressure over 150 psig can damage the electronics in the gauge. Make sure the gauge is isolated from the system during charging and while the system is in operation.

START-UP AND SYSTEM CHARGING

The standard method of charging CFCs has been to start the compressor and charge refrigerant vapor through the suction service valve. However, blended refrigerants will fractionate, or boil off into individual refrigerants, if released from the tank as a vapor. Therefore, all blended HCFC and blended HFC refrigerants must be charged in a liquid form through the suction valve (see Figure 9-11).

To prevent liquid damage to the compressor, monitor the compressor amperage during the charging process. If the compressor starts drawing more than 10 percent above nameplate, there may be too much liquid entering the compressor. Also, if the suction valve on the compressor starts to frost, slow down or stop charging until the frost disappears.

On larger systems, the charging process will happen more quickly if liquid is fed into the high side of the system. This can be done only if the compressor is off and the system is in a vacuum. The vacuum will initially pull a large amount of liquid into the unit from the refrigerant tank, through the high side of the charging valve, and into the compressor discharge service valve (or the receiver's king valve). When the pressures equalize (low and high side are the same pressure), shut off the high side of the charging manifold. Open the low-side valve on the charging manifold, start the compressor, and complete the charging through the suction service valve (Figure 9-12).

NOTE: Do not use this method on small systems that hold less than 5 pounds of refrigerant. The vacuum will pull too much refrigerant into the system, resulting in an overcharge. Instead, charge small systems only through the suction service valve while the compressor is running.

FIGURE 9-11 **Liquid charging through the low side, compressor running.** *Courtesy of Refrigeration Training Services.*

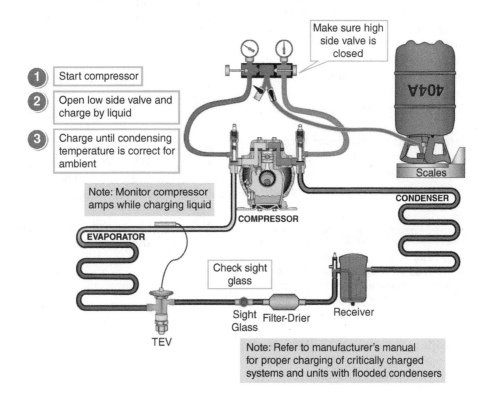

FIGURE 9-12   **Liquid charging through the high side, compressor off.** *Courtesy of Refrigeration Training Services.*

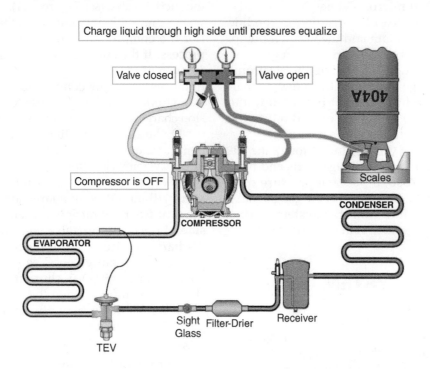

How to Know the Charge Is Correct

Critically charged systems have a weighed-in charge determined by the factory. If the nameplate on the unit calls for 20 ounces of R404a or 50 ounces of R134a, that means the charge is very important. Therefore, use an electronic scale to weigh the exact amount of refrigerant into the unit. As long as all the system components are working right, the unit will maintain the correct pressures and temperatures after charging. However, as little as a 10 percent overcharge or undercharge will prevent the system from operating properly.

On all other systems, charge refrigerant through the suction valve and watch the discharge pressure gauge. Stop charging when the condensing temperature is equal to the ambient temperature plus the condenser split for that unit.

EXAMPLE: 3

A standard walk-in refrigerator is being charged with R404A at an ambient temperature of 85°F. The condenser split is 30°F, so the condensing temperature should be 115°F (85°F + 30°F). At 115°F condensing, the pressure of R404A should be about 292 psig.

To verify the proper charge, check the subcooling. Anything lower than 5°F indicates a low charge, and more than 20°F subcooling points to an overcharge. If the unit has a sight glass, it should be clear without bubbles, except for the occasional minor bubbling of blended refrigerants described earlier in this chapter.

Figures 9-13 through 9-15 give a visual representation of what happens in the condenser as refrigerant is added to a unit during startup. Figure 9-13 illustrates the early stages of charging. The condenser basically has mostly vapor in it, which is evidenced by low condensing temperatures, low-to-no subcooling, and bubbles in the sight glass.

Figure 9-14 shows how liquid condenses out of the vapor as more refrigerant is added. As the amount of refrigerant in the system increases, enough liquid will be formed at the bottom of the condenser to fill up the liquid line leaving the condenser. When the liquid line is full, the bubbles of refrigerant vapor will clear from the sight glass.

A clear sight glass at the condenser does not necessarily mean the system is fully charged, especially if the unit has a long liquid line run to the evaporator. The system in Figure 9-14 may have a clear sight glass and 30°F condenser split, but the subcooling could be anywhere between 5°F and 20°F. If the evaporator is 10 feet from the condenser, then very little subcooling is needed, but a long run or vertical rise may require more subcooling to overcome the pressure drop.

FIGURE 9-13 **An example of how a low refrigerant charge affects the condenser.** *Courtesy of Refrigeration Training Services.*

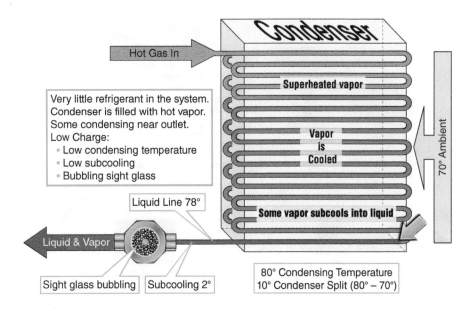

FIGURE 9-14 **An example of a condenser with the proper charge.** *Courtesy of Refrigeration Training Services.*

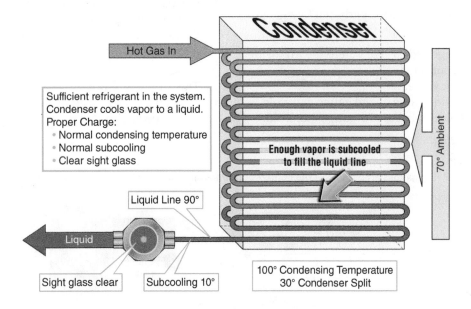

Overcharging can also be a problem, as Figure 9-15 shows. Too much refrigerant means more liquid is condensed, taking up space in the condenser. The reduced space will require the compressor to increase the discharge pressure in order to push vapor into the condenser. Increased condensing temperatures and high subcooling are indicators of an overcharge.

When sufficient refrigerant is added to the system, the condensing temperature is normal for the application, and the sight glass is clear of bubbles. But do not stop there. Check the subcooling to verify the amount of refrigerant in the condenser. Also, check evaporator superheat to make sure that there is enough refrigerant getting to the evaporator. Check the superheat when the system is down to its design space temperature. Allow the unit to cycle off and back on by its thermostat. If the unit has a defrost clock, make sure it is working properly and is set for the correct time of day.

FIGURE 9-15 **An example of a condenser with too much refrigerant (overcharge).** *Courtesy of Refrigeration Training Services.*

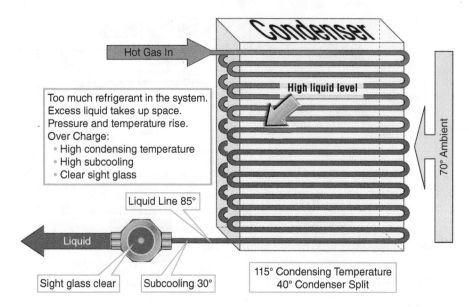

Removing Gauges from the Unit

The following are the steps to properly remove gauges from a system while the compressor is running:

1. Back-seat the high-side valve where the high-side gauge hose is connected.
2. Shut off the tank valve.
3. Open the high-side valve on the gauge manifold.
4. Open the low-side valve on the gauge manifold.

5. Open the suction service valve. This will purge the high-pressure refrigerant in the hoses over to the low side of the system.
6. Back-seat the suction service valve and remove the gauges.

This procedure not only saves refrigerant but prevents a sloppy discharge of hot gas and oil from the high-side hose when it is disconnected (Figure 9-16).

FIGURE 9-16 **Removing gauges properly with the compressor running.** *Courtesy of Refrigeration Training Services.*

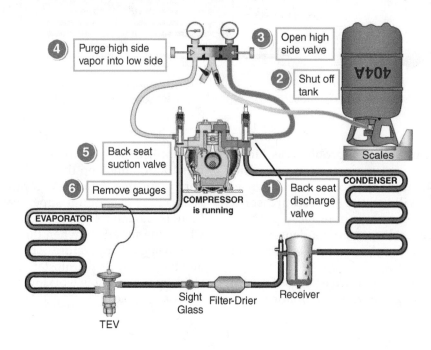

Charging for Low-Ambient Conditions

On a standard condenser without low-ambient controls, charging refrigerant only until the sight glass clears means the unit has just enough refrigerant to operate at the current ambient, or warmer. If the ambient temperature falls, so will the head pressure, which may cause flashing in the liquid line. For instance, clearing the sight glass at 100°F does not mean there will be enough refrigerant in a walk-in freezer condenser when the ambient drops to 70°F. However, technicians usually add more refrigerant "just to make sure," which sometimes increases the subcooling enough to ensure proper operation at lower temperatures. Moreover, when the ambient drops to about 60°F, most units start to have problems.

A unit that does not have any low-ambient controls will have low head pressure during low-ambient conditions. If the head pressure drops below a minimum (usually a pressure equivalent to 90°F condensing temperature), the liquid pressure entering a standard TEV will not be enough for the metering device to operate properly. Low inlet pressure at the TEV can result in valve hunting, starving, or flooding.

A condensing unit that has a head pressure regulating (HPR) valve will maintain its rated minimum condensing pressure in low-ambient conditions. The valve, located at the outlet of the condenser, will prevent refrigerant from leaving the condenser until it is at the HPR valve's minimum rating. Backing up liquid in the condenser decreases the condensing area available, which increases the head pressure. This is what happens in a standard unit if it is overcharged with refrigerant. The HPR valve will maintain head pressure at very low ambients, as long as the unit is properly charged. Proper charge means there has to be enough refrigerant to fill the condenser to a certain percentage of its capacity in order to keep the head pressure up. Although no condenser will need to be filled to its capacity, it is not uncommon for condensers to be 75 percent to 85 percent filled with liquid in order to maintain the valves' minimum head pressure at outdoor ambients below 0°F.

Most technicians charge units by clearing the liquid-line sight glass. The question remains, Once the sight class is cleared, how much more refrigerant will the system need in order for the valve to maintain the minimum head pressure at low ambients?

The answer depends mainly on all of the following three factors:

1. The size of the condenser
2. The ambient temperature when it is being charged
3. The lowest ambient in which the unit will be expected to operate

The best way to determine the correct charge is to contact the manufacturer of the equipment for their recommendations. For instance, Russell (part of the HTP Group of Witt, Kramer, Coldzone, and Carrier Commercial Refrigeration) has both a charging tag on their units and charts for each of their condensing units. If the condensing temperature is 105°F or greater, just clear the sight glass and then weigh in the additional amount as per their chart or tag amount for that unit. If charging in low-ambient conditions, just block the condenser and charge until the condensing temperature rises to 105°F. Then weigh in the additional amount as per the chart or tag for that unit.

If factory information is not available, the next best thing to do is to go to the Sporlan (Parker) Web site and download Bulletin 90-30-1 (http://sporlan .jandrewschoen.com/90-30-1.pdf). The four-page document gives explicit recommendations for charging condensing units with and without compressor cylinder unloading, with evaporator temperatures from +50°F to −35°F, that have 13 of the most common refrigerants, with condenser tubing sizes from 3/8 to 1 3/8 inch, and operating in outdoor ambients from +80°F to −40°F.

Technicians have their own favorite *rules of thumb* for charging units for low-ambient conditions. Some claim to have good results when charging in ambients of 70°F or above by clearing the sight glass and adding an additional 25 percent more refrigerant. Others claim that charging a system with about 4.5 pounds of refrigerant per horsepower works well. Still other technicians check the factory specifications for receiver size and weigh in an amount equal to 90 percent of the receiver capacity. This would certainly be the maximum charge one would want to put in the unit, no matter what the ambient conditions.

A word of caution should be made at this point. First, a system must never be charged to more than 90 percent of the receiver capacity unless recommended by the equipment manufacturer. Second, where the pump-down solenoid valve is located may be critical. The total charge of a system includes the refrigerant in the liquid line. Where that liquid is stored depends on where the pump-down solenoid is installed. If the valve is at the end of the liquid line near the TEV, then during pump-down, the liquid line becomes additional storage for the receiver. However, if the valve is located at the remote condensing unit near the receiver outlet, then all the liquid in the line must be pumped out and added to the receiver. If the liquid line is long, this additional refrigerant may just be enough to overfill the receiver and cause the system to go out on high head pressure during pump-down.

There is one more option for charging a flooded condenser system. It is a simplified version of the

Sporlan bulletin and should be fairly accurate for systems up to 7.5 HP with 3/8- and 1/2-inch condenser tubing.

1. Determine the total length of the straight condenser tubing.
2. Count the number of U-bends and multiply by 0.25 feet.
3. Add the two figures for the total equivalent length of condenser tubing.

EXAMPLE: 4

90 feet of tubing + 40 U-bends = 90 + (40 × 0.25) = 90 + 10 = 100 feet.

4. Multiply the length of tubing by the amount of refrigerant the tubing can hold. (0.05 pounds is the weight per foot of liquid refrigerant for 3/8-inch tubing, and 0.10 pounds per foot for 1/2-inch tubing.)

Refer to Figure 9-17 to calculate the tubing length and refrigerant capacity of a condenser with 100 feet of 1/2-inch tubing.

EXAMPLE: 5

100 feet × 0.10 pounds per feet = 10 pounds of refrigerant.

Assume that we are charging a medium-temperature unit at an outdoor temperature of 70°F. If we only clear the sight glass and there is less than 5°F of subcooling, the unit has just enough charge to fill the liquid line and operate at 70°F ambient or above.

If there is a total of 100 feet of 1/2-inch tubing in the condenser, it would take 10 pounds of refrigerant

after clearing the sight glass to fill the condenser with 100 percent liquid. Of course, we would never want to completely fill the condenser with liquid because the unit would go out on high head pressure safety.

However, as the ambient drops, the HPR valve will back up the condenser to raise the head pressure to its pressure rating (usually 180 psig for R22). If we know the lowest ambient the condenser will experience, then we can use the data in Table 9-3 to determine what percentage of condenser fill will be needed to maintain the head pressure rating of the HPR valve.

Table 9-3 summarizes data for low, medium, and high (air conditioner) condensers with HPR valves, or *hold-back* valves. The figures in each column are the percentages of condenser flooding required of an HPR valve at the ambient temperatures listed in the left column.

Previously, we cleared the sight glass when charging a medium-temperature walk-in at an ambient of 70°F. After calculating the total equivalent length of 3/8-inch condenser tubing, we determined that the condenser would take another 5 pounds to completely fill the condenser to 100 percent of its capacity. If the winter-design temperature for the unit's location is 0°F, then according to Table 9-3, we need to add enough refrigerant to fill it to 80 percent. Therefore, we would add 4 pounds (5 pounds × 80 percent = 4 pounds).

What if we are charging the medium-temperature walk-in refrigerator when it is 40°F outside? How much more would we add after clearing the sight glass to maintain a proper head pressure when the ambient drops to 0°F?

If we clear the sight glass at a 40°F ambient, it means the HPR valve has already backed up enough

FIGURE 9-17 **Calculating tubing length for condenser fill capacity.** *Photo by Dick Wirz.*

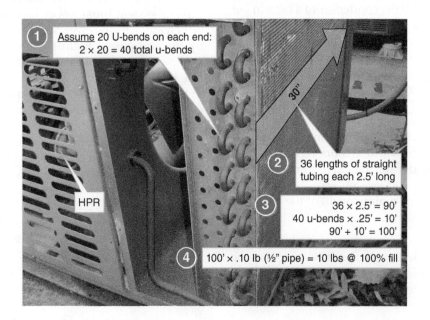

TABLE 9-3 **Percentage of condenser fill required in ambients from −10°F to 180°F**

AMBIENT (°F)	LOW −25°F EVAPORATOR (PERCENT FILL)	MEDIUM +20°F EVAPORATOR (PERCENT FILL)	HIGH (AC) +40°F EVAPORATOR (PERCENT FILL)
80	15	0	0
70	50	0	0
60	65	35	20
50	75	50	40
40	80	60	50
30	80	65	60
20	85	70	65
10	85	75	75
0	90	80	80
−10	90	85	85

refrigerant to fill 60 percent of the condenser. (Refer to Table 9-4 for this example of a medium-temperature unit at 40°F ambient.) Therefore, we need to add only another 20 percent for the condenser capacity to reach the required 80 percent fill at a 0°F ambient. This means that only 1 pound of additional refrigerant is needed (5 pounds × 20 percent = 1 pound).

The following are a few more examples:

1. A medium-temperature (+20°F evaporator) unit is charged to a clear sight glass at 80°F ambient, and the lowest winter ambient for the area is +10°F (Table 9-5).

   - The condenser has 190 feet of 3/8-inch condenser tubing and 40 U-bends.
   - The equivalent length of pipe is 190 feet + (40 U-bends × 0.25 feet per U-bend) = 200 feet
   - 200 feet × 0.05 pounds per foot (for 3/8-inch tubing) = 10 pounds of additional refrigerant to fill the condenser to 100 percent of its capacity.

- At 80°F ambient when charged to a clear sight glass, the condenser is considered 0 percent full.
- At +10°F ambient, the condenser needs to be 75 percent full of liquid.
- 75 percent × 10 pounds = 7.5 pounds more refrigerant is needed for low-ambient operation.

2. A low-temperature (−25°F evaporator) unit is charged to a clear sight glass at 70°F ambient, and the lowest winter ambient for the area is 0°F (Table 9-6).

   - The condenser has 90 feet of 1/2-inch condenser tubing and 40 U-bends.
   - The equivalent length of pipe is 90 feet + (40 U-bends × 0.25 feet per U-bend) = 100 feet.
   - 100 feet × 0.10 pounds per foot (for 1/2-inch tubing) = 10 pounds of additional refrigerant to fill the condenser to 100 percent of its capacity.

TABLE 9-4 **Example of charging a medium-temperature unit at 40°F ambient with enough refrigerant to operate at 0°F ambient**

AMBIENT (°F)	LOW −25°F EVAPORATOR (PERCENT FILL)	MEDIUM +20°F EVAPORATOR (PERCENT FILL)	HIGH (AC) +40°F EVAPORATOR (PERCENT FILL)
80	15	0	0
70	50	0	0
60	65	35	20
50	75	50	40
40*	80	60*	50
30	80	65	60
20	85	70	65
10	85	75	75
0†	90	80†	80
−10	90	85	85

Clear sight glass (60% full)

†*Add 20 percent more (80% − 60%)*

TABLE 9-5   Example of charging a medium-temperature unit at 80°F ambient with enough refrigerant to operate at +10°F ambient

AMBIENT (°F)	LOW −25°F EVAPORATOR (PERCENT FILL)	MEDIUM +20°F EVAPORATOR (PERCENT FILL)	HIGH (AC) +40°F EVAPORATOR (PERCENT FILL)
80*	15	0*	0
70	50	0	0
60	65	35	20
50	75	50	40
40	80	60	50
30	80	65	60
20	85	70	65
10†	85	75†	75
0	90	80	80
−10	90	85	85

*Clear sight glass (0% full)

†Add 75 percent more (75% − 0%)

- At 70°F ambient when charged to a clear sight glass, the condenser is already 50 percent full (it has taken 5 pounds of refrigerant to clear the sight glass).
- At 0°F ambient, the condenser needs to be 90 percent full (9 pounds of refrigerant).

- 90 percent − 50 percent = 40 percent more condenser fill is needed for 0°F ambient operation.
- 40 percent × 10 pounds = 4 pounds of refrigerant needs to be added to the unit.

TABLE 9-6   Example of charging a low-temperature unit at 70°F ambient with enough refrigerant to operate at 0°F ambient

AMBIENT (°F)	LOW −25°F EVAPORATOR (PERCENT FILL)	MEDIUM +20°F EVAPORATOR (PERCENT FILL)	HIGH (AC) +40°F EVAPORATOR (PERCENT FILL)
80	15	0	0
70*	50*	0	0
60	65	35	20
50	75	50	40
40	80	60	50
30	80	65	60
20	85	70	65
10	85	75	75
0†	90†	80	80
−10	90	85	85

*Clear sight glass (50% full)

†Add 40 percent more (90% − 50%)

SUMMARY

This chapter covered the basics of recovery and discussed the push–pull, tank-in-a-vacuum, and cold-tank methods of liquid refrigerant recovery. Evacuating a system to micron levels is important for removing all the moisture and noncondensables. Triple evacuation procedures are used on highly contaminated systems.

Critically charged systems should be charged by weight. However, replacing R12 with HCFCs has forced technicians to go back to charging by the condensing temperatures when retrofitting. In addition, the use of blended refrigerants requires technicians to carefully charge liquid through the suction valve. To speed the overall charging process on large units, technicians should add liquid into the high side if the compressor is off and the system is in a vacuum. The final charging must be done through the low side when the compressor is running.

Typically, a system is fully charged when the condensing temperature is equal to the ambient plus the designed condenser split for that unit. A full charge is confirmed based on subcooling measurements.

Making sure a system is free of moisture and non-condensables is an important part of installation and service procedures. Pulling a good vacuum on a system is also important. Using larger-gauge hoses and changing the vacuum pump oil regularly can help speed up the process.

Charging a unit that has an HPR valve can present some challenges. The proper method of charging a unit is to follow the procedures recommended by the equipment manufacturer. This chapter discusses some rules of thumb used by technicians. A simplified version of the factory procedure is presented. It consists of determining condenser capacity and then applying that information to a chart giving the percentage of condenser fill required at different ambient temperatures.

REVIEW QUESTIONS

1. What is refrigerant retrofitting?

2. Why is oil miscibility a concern in refrigerant retrofitting?

    a. If the refrigerant oil is not miscible with the refrigerant in the system, it may cause compressor lubrication problems.
    b. If the refrigerant oil is not miscible with the refrigerant in the system, it may cause a clog in the metering device.
    c. If the refrigerant oil is not miscible with the refrigerant in the system, it may cause head pressure problems.

3. What system problems arise when the replacement refrigerant is more efficient than the original refrigerant?

    a. The replacement refrigerant can damage the compressor because of lubrication problems.
    b. The replacement refrigerant can remove more heat from the space than the compressor can handle.
    c. The replacement refrigerant can cause the suction pressure to fall dangerously low.

4. How many oil changes of alkylbenzene will result in the required 50 percent mixture?

    a. Usually one oil change
    b. Usually two oil changes
    c. Usually three oil changes

5. How many oil changes for polyolester oil to reach about a 95 percent mixture?

    a. Usually one oil change
    b. Usually two oil changes
    c. Usually three oil changes

6. Describe the 10-step process of removing oil from a semihermetic compressor by inserting a tube into the crankcase.

7. How is the oil changed in a hermetic compressor?

8. Why do many service companies choose not to replace the mineral oil when retrofitting self-contained reach-ins?

    a. Because the system is close-coupled, the original mineral oil seems to move through the system adequately and return to the compressor
    b. Because replacing the oil on reach-ins can actually damage the compressor
    c. Because for the cost of an oil change, the customer could buy a new reach-in

9. Assume you are retrofitting a reach-in refrigerator that has 30 ounces of R12 refrigerant. The replacement refrigerant is R401A, and the ambient temperature entering the condenser is 75°F. What is the approximate head pressure when the unit is properly charged?

    a. 155 psig
    b. 145 psig
    c. 125 psig

10. Assume you are retrofitting a reach-in freezer that has a critical charge of 30 ounces of R502. How much R408A would you weigh into the unit? What would you do about the mineral oil in the hermetic compressor?

    a. Add 24 ounces of R408A and leave the mineral oil in the compressor
    b. Add 24 ounces of R408A and replace the mineral oil with alkylbenzene
    c. Add 30 ounces of R408A and leave the mineral oil in the compressor

11. Describe how to charge a retrofitted TEV system that is not critically charged.

12. Is it always necessary to clear all the bubbles from the sight glass of a retrofitted TEV system? Why or why not?

13. If liquid refrigerant will damage a recovery unit compressor, how can some manufacturers rate their recovery units on the amount of liquid they can recover?

    a. The manufacturer is not being truthful.
    b. Recovery unit compressors are strong and are able to pump liquid.
    c. The recovery units have an orifice that will vaporize liquid refrigerant before it enters the compressor.

14. Give a brief description of the push–pull procedure for recovering liquid from a system.

15. How does an evacuated tank remove liquid from a system?

    a. The evacuated cylinder pressurizes the system and forces the liquid out of the cylinder.
    b. The evacuated cylinder acts like a condenser and changes vapor refrigerant to a liquid refrigerant.
    c. The high-pressure liquid in the receiver is sucked into the evacuated cylinder.

16. Why does an empty tank in an ice bath pull liquid from a warmer refrigerant container or a warmer refrigerant receiver?

    a. By the law of thermodynamics, refrigerant is always attracted to ice.
    b. By the basic laws of nature, a warm refrigerant at a higher pressure will seek a colder location at a lower pressure.
    c. By the law of the infinite, cold seeks warmth.

17. What can protect a recovery machine from contaminants pulled out of refrigeration systems during the recovery process?

    a. Installing an accumulator on the outlet of the recovery machine
    b. Installing a receiver in the recovery machine
    c. Installing a filter drier on the inlet of the recovery machine

18. What level of vacuum do most manufacturers recommend to remove air and moisture from a system?

    a. 500 microns
    b. 1,000 microns
    c. 29 inches of Hg

19. Why is it a good idea to leak check the vacuum pump and gauge connections before evacuating a system?

    a. To make sure any vacuum leak is in the system, and not in the gauge or vacuum pump connections to the system

    b. To make sure all air is out of the vacuum pump and hoses before trying to pull a vacuum on the system
    c. To make sure the vacuum pump is working properly before attempting to pull a vacuum on the whole system

20. Briefly describe the triple evacuation procedure.

21. Is a vacuum pump supposed to be used to remove acid, sludge, and debris?

    a. Yes, vacuum pumps will suck everything out of the system.
    b. No, only refrigerant vapor and moisture. Filter driers should be used to remove the other contaminants.
    c. Yes, in order to get to a micron level, the pump has to remove the contaminants first.

22. On a two-stage vacuum pump, why is the gas ballast opened when the pump is started and closed a few minutes later?

    a. The first stage will pull out the moisture-laden primary air before it can contaminate the pump's oil. Closing it utilizes the second stage.
    b. The second stage will pull out the moisture-laden primary air before it can contaminate the pump's oil. Closing it utilizes the first stage.
    c. Opening the gas ballast makes it easier to start the pump. After the pump is started, closing the gas ballast utilizes both stages of the pump.

23. How often should the oil be changed in a vacuum pump?

    a. Only after evacuating a system with a lot of water, after a compressor burnout, or when the oil turns dark or white
    b. After evacuating a system with a lot of water, after a compressor burnout, or after five normal evacuations
    c. After every evacuation

24. If a micron gauge is not working properly, what could be the problem, and can it be corrected?

    a. Oil in the micron gauge can cause it to malfunction. Clean it out by flushing the gauge inlet with alcohol.
    b. If a micron gauge is not working because of oil contamination, it must be blown out by liquid refrigerant.
    c. Micron gauges are very delicate and break easily, so replace them when they malfunction.

25. What is the maximum positive pressure most micron gauges can tolerate before they get damaged?

    a. 50 psig
    b. 150 psig
    c. 250 psig

26. How do you prevent damage to the compressor when charging blended refrigerant in a liquid state through the suction service valve?

    a. Keep compressor amperage within 10 percent of RLA; also, slow down the charging if the suction service valve starts to frost
    b. Keep compressor amperage at its RLA and stop if the discharge service valve starts to frost
    c. Charge by liquid until the compressor starts knocking and then back it off a little

27. Describe the procedure for charging liquid through the high side when the compressor is off and in a vacuum.

28. If a reach-in freezer nameplate says "24 oz. R404A," how much R404A should be put into it, and how do you know you have charged it correctly?

    a. The unit will be charged correctly if the system is evacuated and 24 ounces of R404A is weighed into it.

    b. The unit is charged properly when the condensing pressure of R12 is reached.
    c. The unit is charged correctly if 80 percent of the nameplate weight, or 19.2 ounces of R404A, is weighted into the system.

29. Assume an R404A walk-in freezer with a TEV system and 90°F ambient air entering the condenser. What pressure(s) indicate it is properly charged? What temperatures would verify the system was fully charged?

    a. 146 psig head pressure and 10°F subcooling
    b. 292 psig head pressure and 10°F subcooling
    c. 312 psig head pressure, 16 psig suction pressure, and 5°F subcooling

30. What is the proper procedure for removing gauges from a unit while it is operating?

    a. Close the discharge service valve and bleed the high-side hoses back into the suction service valve
    b. Open the discharge service valve and bleed the low-side hoses back into the refrigerant tank
    c. Close the suction service valve and bleed the high-side hoses back into the discharge line

10

SUPERMARKET REFRIGERATION

OBJECTIVES

After completing this chapter, you should be able to explain and describe the following subjects that are important features in supermarket refrigeration:

- Product visibility and customer access
- Multiple refrigerated cases and temperatures
- Parallel rack systems
- Oil return
- Controls and controllers
- Energy efficiency
- Coil defrosting
- Mechanical subcooling
- Heat reclaim and reheat
- Display case airflow
- Installation, service, and maintenance
- Refrigerant leak detection
- New Technology

CHAPTER OVERVIEW

Although the basic refrigeration principles are the same, supermarket equipment differs in some aspects from the commercial refrigeration we have been discussing in the previous chapters.

When shopping for food, most customers are very price conscious. Consequently, supermarkets must keep their prices competitive. The combination of low pricing and high operating costs results in an average net profit per store of only 1 to 2 percent. In order to prosper, supermarkets must sell a large volume of products while keeping their operating costs to a minimum.

This chapter focuses on how properly engineered and installed supermarket equipment can accomplish the requirements of both the store and its customers.

PRODUCT VISIBILITY AND CUSTOMER ACCESS

The display of food in supermarkets is almost a science in itself. Many food purchases are made on impulse. Therefore, it is important the product is presented well and is easily accessible by the customer. Open front and glass door refrigerated cases make it easy for shoppers to pick out what appeals to them (Figure 10-1).

FIGURE 10-1 **Open-front refrigerated case at the Hussmann Training Center.** *Courtesy of Hussmann Training Center.*

MULTIPLE REFRIGERATED CASES AND TEMPERATURES

There is a wide variety of perishable foods in a supermarket. Each food product has an optimal temperature at which it must be refrigerated to keep it fresh for as long as possible. Following is a list of some familiar products and their recommended case temperatures:

- 36°F to 40°F Produce
- 34°F to 36°F Dairy
- 32°F to 34°F Deli
- 28°F to 32°F Meat
- −10°F to −5°F Frozen juice
- −20°F to −15°F Frozen food and ice cream

PARALLEL RACK SYSTEMS

A parallel rack shares the same refrigerant and oil with multiple compressors piped in parallel to a common discharge header (Figures 10-2 and 10-3). Oil is separated out and returned to the compressors while the discharge vapor continues to a single condenser. Liquid from the condenser flows back to a single receiver at the rack to feed a common liquid header. Multiple circuits from the header provide liquid to multiple evaporators in each circuit. Suction vapor from the evaporators is returned to a suction header at the rack which feeds the compressors, continuing the cycle (Figure 10-4). In comparison to the number of single systems that would be required in a supermarket, rack systems are very cost effective. Overall, less horsepower, piping, and refrigerant are needed. In addition, fewer accessories are needed, there's better capacity control, and operating costs are lower due to higher efficiencies.

OIL RETURN SYSTEMS FOR RACK REFRIGERATION

Oil is essential as a lubricant in refrigeration compressors. Every bearing surface has a thin film of oil separating the two metal parts. Visualize the horizontal crankshaft of a semihermetic reciprocating compressor (see also Figure 10-5). The entire weight of the rotor, crankshaft, pistons, and connecting rods are supported by two main bearing surfaces. Actually,

FIGURE 10-2 **Parallel rack system at the Hussmann Training Center.** *Photo by Dick Wirz.*

FIGURE 10-3   **Hussmann rack made for display purposes.** *Courtesy of Hussmann Training Center.*

FIGURE 10-4   **Illustration of a basic parallel rack system.** *Drawing by Refrigeration Training Services.*

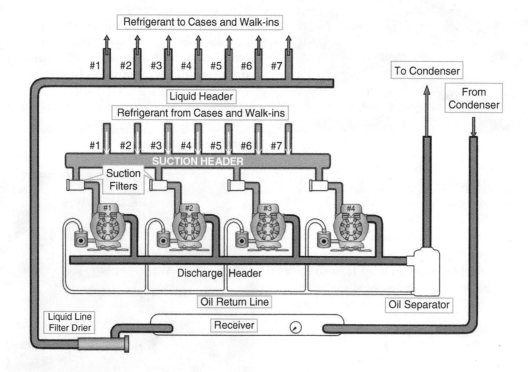

they are supported by a very thin film of oil at those points. If it were not for the oil separating them, the metal crankshaft would contact the metal bearings, the connecting rods to the crankshaft, the connecting rods to the pistons, the pistons to the cylinders, and so on. When rotating at 1,750 rpm or more, two adjoining metal surfaces without oil between them would generate friction and very quickly gouge and scrape the

FIGURE 10-5 **Cutaway diagram of oil path in semihermetic reciprocating Copeland suction-cooled compressor.** *Courtesy of Emerson and enhanced by Refrigeration Training Services.*

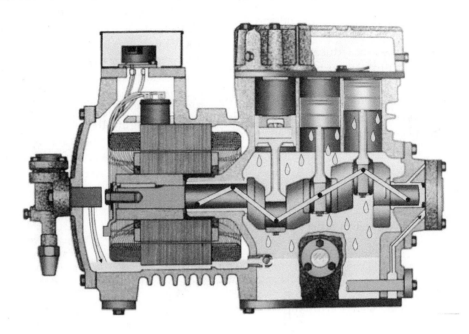

smooth surfaces, changing the tolerances or distance between them. If it continued, the heat of friction generated would become so great that, for example, the crankshaft and bearing could become fused or welded together.

Not only does the oil support, but it also separates moving metal parts. The thin film of oil in the compressor cylinder prevents the piston from scraping the walls of the cylinder. In addition, it also prevents the high pressure generated on the compression stroke from blowing down the sides of the piston and into the low-pressure area of the crankcase. By the way, this is known as blow-by. On large compressors, the piston rings help prevent this, but in most compressors under 2 hp, there are no piston rings. Just a thin film of oil between the piston and cylinder is enough to separate the high- and low-pressure sides of the system.

Keeping oil circulating in the compressor is not difficult when there is only one compressor in the system and the piping run is not very long. However, in a supermarket application, the average total length of copper piping is about 4 miles. It is easy to see that the compressor would run out of oil long before the original oil could return. Therefore, an oil separator is located near the compressor to remove most of the oil from the discharge vapor and return it to the compressor (see Figure 10-6).

Although the oil separator traps most of the oil before it enters the piping system, several other

FIGURE 10-6 **Oil separator mounted on the end of the compressor rack.** *Photo by Dick Wirz.*

components are necessary to ensure the oil is distributed properly to all the compressors in a parallel rack system. Each semihermetic compressor has an oil pump and oil safety control, but the *oil return system* as described in this section will consist of the following:

- **Oil separator and reservoir**—to capture oil and store before it enters the piping system
- **Pressure reducing valves**—to reduce high oil pressure in the reservoir to a lower pressure close to that of the crankcase
- **Oil level controls**—to prevent the compressor oil level from getting too low
- **Oil filters**—to help keep the oil system clean
- **Shut off valves**—to facilitate oil system service and maintenance

There are two types of oil return systems. The first is the *conventional* or *low pressure* oil system, which has an oil separator and a separate oil reservoir (see Figure 10-7). It uses an oil differential check valve (OCV) to lower the pressure in the reservoir from high discharge vapor pressure in the separator down to just above the crankcase pressure. Usually, this is about 20 psig above that in the crankcase, but optional OCVs are 5, 10, and 30 psig and are engineered for specific systems (see Figure 10-8).

The second type is the *high pressure* oil return system, which has a combination oil separator reservoir (see Figure 10-9). Reducing the pressure of the oil to the

FIGURE 10-7 Low-pressure oil return system. *Courtesy of Refrigeration Training Services.*

FIGURE 10-8 Oil pressure differential check valve for 20 psi above crankcase pressure. *Compliments of Parker Hannifin-Sporlan Division.*

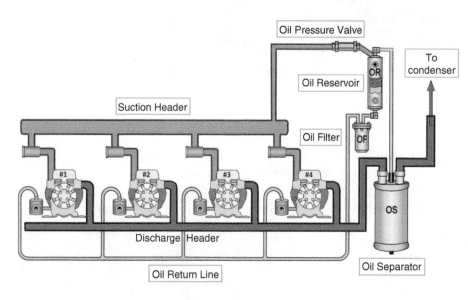

FIGURE 10-9 High-pressure oil return system. *Courtesy of Refrigeration Training Services.*

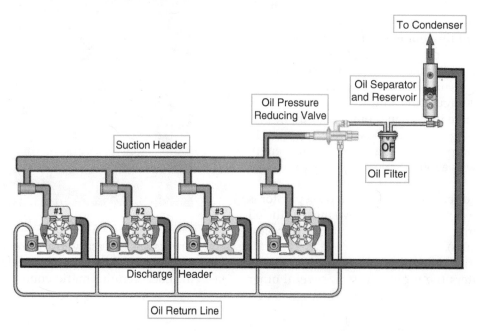

oil level control is accomplished by a Y-valve, which has three ports with pipes attached. Maybe this is where the "Y" part of the name came from. Sporlan's oil pressure differential valve is the Y1236-C (see Figure 10-10). This valve is adjustable between 10 and 25 psig with a factory setting of 17 psig above the suction pressure. One port is the high-pressure oil inlet from the separator/reservoir. One is the low-pressure oil outlet to the oil level controls. The last port senses the pressure in the suction header, which is necessary for the valve to raise and lower outlet oil pressure consistent with the differential pressure setting on the valve.

FIGURE 10-10 **Adjustable oil pressure reducing "Y" valve, Sporlan Y1236-C.** *Drawing by Refrigeration Training Services.*

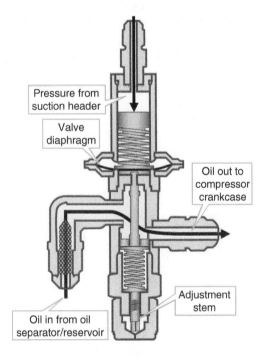

FIGURE 10-11 **Mechanical oil level control.** *Photo by Dick Wirz.*

FIGURE 10-12 **Electronic oil level control.** *Courtesy of Emerson.*

Oil level controls are either mechanical or electronic. Mechanical controls have a float ball that travels up and down with the level of oil in the crankcase (see Figure 10-11). As the level drops, the float mechanism allows more oil into the crankcase. Interestingly, the float maintains a minimum level of oil but does not prevent a high level of oil in the crankcase. There is a sight glass in the side that allows a visual inspection of the oil level. Electronic controls are similar, but instead of a float ball, they utilize an electronic sensor to determine the level and to energize a solenoid coil to open the inlet port to allow more oil into the crankcase (see Figure 10-12).

Oil filters are important to keep debris generated by installation and service out of the components that affect the oil level system. POE (polyolester) oil acts like a detergent or solvent to scour the walls of the piping and any other surface it contacts. If the piping is

not purged with nitrogen during brazing, there will be a black coating on the inside of the pipes called oxidation. POE oil strips this from the pipe and suspends it in the oil as it circulates through the system. Eventually, these particles will clog some orifice, to the detriment of the oil return system. Oil filters look very much like refrigeration filter-driers except that oil filters are much more efficient (see Figures 10-13 and 10-14). Refrigeration filters can capture particles down to about 30 microns; however, oil filters are 99.9 percent efficient down to about 3 micron particles. A micron is one-millionth of a meter. The diameter of a human hair is about 100 microns, and a red blood cell is about 5 microns. A standard coffee filter is designed to filter down

FIGURE 10-13 **Sporlan OF-303-T oil filter with external pressure tap or "T" and cutaway of OF-303-BP with internal bypass.**
Compliments of Parker Hannifin-Sporlan Division

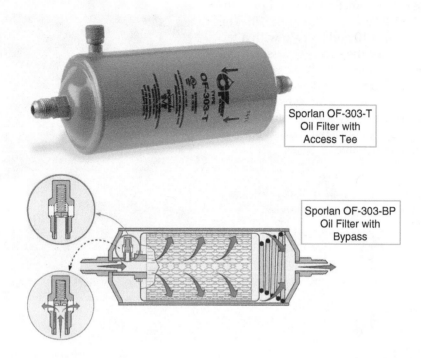

Sporlan OF-303-T
Oil Filter with
Access Tee

Sporlan OF-303-BP
Oil Filter with
Bypass

FIGURE 10-14 **Replaceable cartridge-type oil filter.** *Compliments of Parker Hannifin-Sporlan Division.*

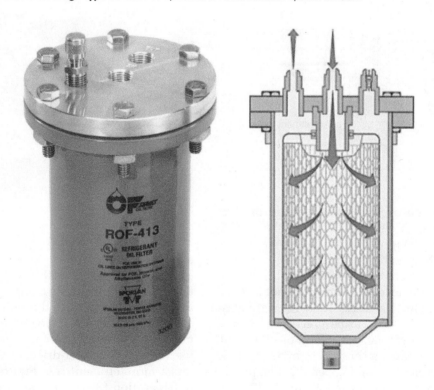

to 20 microns, and a MERV 13 pleated air filter is at least 90 percent efficient in filtering down to 3 microns.

Oil line shut-off valves are strategically placed throughout the oil return system. This makes it easier and safer for the technician to service the oil system. Unfortunately we do not have a clear picture to present in this text, but there is one behind the oil level control in Figure 10-11.

So what can go wrong with a parallel rack oil return system? The answer is . . . a lot. They are probably the most complicated and misunderstood part of rack refrigeration systems. And everything that could happen will adversely affect the compressor, from not enough oil to too much oil. Before condemning components such as the pressure differential valve, oil separator, or oil level control, keep your diagnostics simple. The primary cause of oil return problems is trash in the system. We like to think that every installation is done properly by professionals who do all the right things. However, that is not always the case. Also, even on a good installation, an overheated compressor can deposit burned carbon particles into both the refrigerant and the oil. Some, but not necessarily all, the particles will be caught by the refrigeration filters because refrigeration filters filter down to only 20 microns. The oil filters, which filter down to 3 microns, are the last line of defense against oil particles and should be inspected and/or replaced anytime there is an oil return issue. Although filters trap most particles, some still get through and can eventually clog the 100-micron mesh screens on the inlet ports of the valves and on the oil-level controls. They are similar to the inlet screens on expansion valves.

Blow-by is another problem for oil return. It can occur on suction-cooled semihermetic compressors that have a separate motor section and crankcase section. On suction-cooled compressors, the refrigerant and oil enter the suction service valve. The suction vapor cools the motor before entering the cylinders. The oil settles in the bottom of the motor compartment and flows through a check valve into the crankcase where the crankshaft and pistons are located. You can see the ball-type check valve near the sight glass in the drawing back a few pages in Figure 10-5. A slightly different style is shown in Figure 10-15. If the aluminum piston has worn down, then the space between the piston and steel cylinder walls is too great and cannot be properly sealed by the film of oil and the piston compression rings. Therefore, during the compression stroke, some of the high-pressure vapor is forced down into the crankcase, increasing its pressure. If the pressure in the crankcase is more than 3 psig higher than the pressure in the motor compartment, the oil check valve that allows oil to flow from the motor compartment will be pushed closed, and oil will not drain to the crankcase. If the compressor continues to run, it may use up all the oil in the crankcase, and the compressor will shut off on the oil safety switch.

Technically the problem is lack of oil, but diagnosing the cause is not as easy as one might think. After the compressor stops the pressure between the motor compartment and the crankcase equalizes. When it does, the oil buildup in the motor compartment will drain into the crankcase. When the service technician

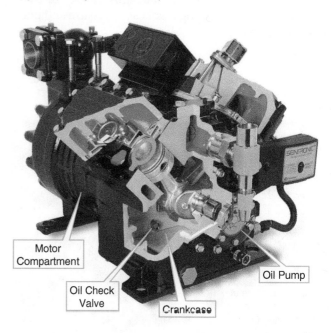

Motor Compartment

Oil Check Valve

Crankcase

Oil Pump

arrives, he will see there is oil in the sight glass, and when the compressor is started, the pump pressure will show normal. Only if he checks for blow-by will he know what caused the problem.

Blow-by can also cause the compressor motor overload to trip. Although not apparent as an oil issue, the buildup of oil in the motor compartment could cause the rotor to drag, which would raise both motor amperage and heat. These two factors are being monitored by the compressor overload and could cause the overload to trip. Once again, when the motor shuts down, the pressures equalize, and the excess oil in the motor compartment drains into the crankcase. Again, the technician may not discover the problem unless she checks for blow-by.

Most compressors have a little blow-by caused by normal wear. Copeland's oil check valve is designed to stay open as long as there is no more than 3 psig blow-by. To check for blow-by, one must access the pressure in the crankcase and compare it to the suction pressure entering the compressor at the suction service valve. If the pressure in the crankcase is more than 3 psig higher than the suction pressure, then more than normal blow-by is occurring. Most compressors have ports or plugs in the crankcase and on the low side of the compressor head. Fittings can be installed at these locations to be used for the low-pressure control or the oil failure control crankcase pressure access. Replacing the existing fitting with a tee fitting will allow the control to be reconnected, and there will be an access port for checking the crankcase pressure. It is recommended to have a Schrader fitting in the access side of the tee fitting to make it easier for future service technicians to check crankcase pressure. See Figure 10-16 for two ideal locations for checking crankcase pressure and one for suction pressure.

FIGURE 10-16 **Location of suction pressure port and two possible crankcase pressure access ports.** *Photo by Jess Lukin.*

Suction pressure at suction service valve

Crankcase Pressure

PARALLEL RACK SYSTEM CONTROLS

Only accurate control settings can maintain proper operation of large rack systems, and most use computerized programmable systems to accomplish this. There are many brands, but most often, techs use generic terms such as an EMS (energy management system) or CPS (computer-programmed system) to describe them. Extremely stable case temperatures are achieved by a combination of electronic and mechanical controls. The refrigeration system pressures in the suction header rise and fall based on the heat load in the cases. In response to these pressures, the compressors cut in and out and are staged by the EMS as necessary to maintain efficient overall operation.

In the following example, a store has enough medium-temperature refrigeration to require a rack of three 20-horsepower (hp) compressors with a smaller 10-hp as the "lead" or primary compressor. There are several walk-ins, a section of produce cases, some dairy cases, deli cases, and several meat cases. When talking about the refrigeration "circuit" or branch, the group of cases on that circuit have the same temperature and are referred to as a lineup. For instance, there could be a 56-foot lineup of 28°F meat cases. The lineup consists of three 12-foot cases, two 8-foot cases, and a 4-foot case.

Under full load, with every case and walk-in calling for refrigeration, all four compressors must run at

the same time to handle the load. The call for refrigeration for the compressors is based on the staging or stepping percentage in the EMS controls program. Figure 10-17 is an example of a step table from an EMS controls program.

FIGURE 10-17 **Example of a compressor step table.** *Drawing by Irene Wirz, Refrigeration Training Services.*

Comp Number	COMP 1	COMP 2	COMP 3	COMP 4	COMP 5	COMP 6	COMP 7	COMP 8	
H.P. ▶	10 h.p.	20 h.p.	20 h.p.	20 h.p.					
Share ▶	14.2%	28.5%	28.5%	28.5%					Load
Step 1	OFF	OFF	OFF	OFF					0%
Step 2	ON	OFF	OFF	OFF					14.2%
Step 3	OFF	ON	OFF	OFF					28.5%
Step 4	ON	ON	OFF	OFF					42.7%
Step 5	OFF	ON	ON	OFF					57.0%
Step 6	ON	ON	ON	OFF					71.2%
Step 7	OFF	ON	ON	ON					85.5%
Step 8	ON	ON	ON	ON					99.7%

In this example, the suction group consists of four compressors. If they are not all the same size, the smallest compressor is usually considered the lead compressor and is staged to take care of low or light loads. In this example, compressor 1 is 10 hp and can handle up to 14.2 percent of the load. The three other compressors are all 20 hp, and each is capable of handling 28.5 percent of the load, which is shown in the row marked "SHARE" for their portion of the total load. Each step in the left column (1 through 8) shows what combination of compressors operating will handle a specific percentage of the "LOAD" listed in the far right column. The load range is from step 1 at 0 percent when all compressors are off, to step 8 at essentially 100 percent when all compressors are running.

When the refrigeration load requirements are 14.2 percent, step 2 is initiated, and compressor 1 is energized. When the refrigeration load requirements increase to 71.2 percent, step 6 is in play, so compressors 1, 2, and 3 will come on. Note that when the load is at 85.5 percent, compressor 1 will turn off, and compressors 2, 3, and 4 will be running. The step table is based on an algorithm programmed into the controller. As the rack system settles in over time, the controller can be reprogrammed, should the need arise.

As the temperature in the cases starts coming down, so does the pressure on the rack system. Each compressor is equipped with a mechanical pressure control that cycles off the compressor when it reaches a certain pressure. These controls are primarily used as safety controls. In addition, there are pressure transducers in the suction header that feed information to the EMS controls. This provides redundancy and alarm capabilities. In this case, *redundancy* means having controls that duplicate the function of another; if one fails, the other will provide backup so that the entire system does not go down. For example, pressure transducers provide information to the controller in order for the rack to operate according to its programming. However, it will also have mechanical pressure controls to cycle the compressor off should something cause the suction pressure to drop drastically. In addition, it will set off an alarm through the EMS system to notify the manager that there may be a problem. The mechanical control is usually an automatic reset so that it will not shut the entire system down, but it may short cycle until someone arrives to check it out.

The beauty of the EMS system is that it has many more capabilities than the older-style mechanical control systems. However, it can be difficult to explain the algorithms necessary to control a multiple compressor rack system. Instead, the following examples will use the old-style mechanical low-pressure controls to show how the compressors in a rack system are cycled on and off based on system pressure, which indicates the load on the system.

Each low-pressure control is set to shut off one of the compressors when the system pressure drops to the setting on the control. The exact compressor cut-out pressures are determined by the factory according to the specific system requirements. However, the following pressure settings for a four-compressor rack (Figure 10-18) give an example of what the designers have in mind. Please note that this is only an example and is not intended to represent actual settings of rack low-pressure controls.

FIGURE 10-18 **Example of a four-compressor rack.** *Photo by Jess Lukin.*

System pressure is 70 psig, all compressors (A, B, C, and D) are running. System pressure drops to 65 psig; compressor A stops, whereas B, C, and D run.

- System pressure drops to 60 psig; compressor B stops, whereas C and D continue.
- System pressure drops to 55 psig; compressor C stops, whereas D continues.

Compressor D is set to cut out at 50 psig but often runs continuously because normal heat loads cause temperatures and pressures to keep the pressure above D's cut-out.

- System pressure rises above 55 psig; compressor C starts, runs with D.
- System pressure rises above 60 psig; compressor B starts, runs with C and D.

The compressors continue to cycle on and off as system pressures change.

Case temperatures are usually regulated by installing an evaporator pressure regulator (EPR) for each product section or circuit. An example of setting an EPR follows.

Assume a 36°F dairy case operates at a 26°F evaporator temperature. A 26°F evaporator would have a suction pressure at the evaporator of 64 psig for an R404 system.

In Chapter 6, it was stated that an EPR valve keeps the suction pressure up to the desired evaporator pressure. Therefore, if the EPR is set to maintain 64 psig, the evaporator will stay at 26°F. Assuming the evaporator is designed with a 10°F TD, this case would maintain a 36°F case and product temperature.

Also in Chapter 6, the EPRs were installed in the evaporator outlets of several individual cases served by a single compressor. The cases were often of different temperatures, with the EPRs in the higher-temperature cases only. In supermarkets, however, there are several cases in a row at the same temperature. Therefore, the most logical location for the EPR is where the suction line for that circuit taps into the suction header at the compressor rack. Even on long piping runs, the EPR can be adjusted to maintain the proper evaporator pressure in the cases.

Most new supermarket installations are using electronic EPR (EEPR) valves for more accurate temperature control. On single systems, the standard means of temperature control has been a thermostat and pump-down solenoid. On parallel rack systems with multiple circuits and multiple cases on each circuit, the standard since the 1970s has been the mechanical EPR or Sporlan's SORIT, which stands for a solenoid-operated (S) open on rise inlet (ORI) pressure valve with a pressure access tee (T). However, in the past 10 years, the EEPR has proven to be the most efficient means of controlling case temperatures. In fact, Sporlan's CDS valve, which is used as an EEPR, has a kit available to replace the guts of the mechanical SORIT with the CDS while leaving the originally installed brazed base of the SORIT in place on the rack system. CDS basically stands for close (C) on decrease (D) of temperature using a stepper (S) motor (see Figure 10-19).

On a pump-down system, there is about a ±3°F temperature swing from the intended set point. That means that if the case discharge air temperature is supposed to average 35°F, then the air temperature may rise 3°F to 38°F and then pull down to 32°F before the thermostat cycles off. When above the set point of 35°F, the product is getting warmer than it should. When the temperature drops below 35°F, the product does not benefit, and the cost of bringing the temperature below the set point is wasted energy. The fluctuating temperatures could adversely affect the condition of the product.

A mechanical EPR is more efficient. It senses the suction pressure, and there is only about a ±1°F swing in the discharge air temperature. However, after defrost, the discharge air temperature may not be down to the desired 35°F. The high load after defrost opens the EPR fully, and the evaporator pressure quickly drops. When the pressure is down to the EPR's set point, it throttles the suction pressure to maintain a constant evaporator pressure, regardless of whether the evaporator discharge air temperature is down to 35°F or not. Very often, the EPR prematurely throttles down before the discharge air temperature reaches 35°F since the EPR is maintaining a constant evaporator pressure and only indirectly maintains the discharge air temperature. Therefore, it takes longer to bring the discharge

air down to the temperature the product needs in order to provide the optimum shelf life.

The EEPR is even more efficient (see Figure 10-20). It responds directly to evaporator discharge air temperature and not to evaporator pressure. Discharge air sensors (thermistors) send a signal to the controller that in turn powers the valve's stepper motor open and closed to maintain discharge air temperatures within ±0.5°F. In addition, after defrost, the EEPR is fully opened to bring the discharge air temperature down to 35°F much faster than the mechanical EPR. Bringing down the temperature quickly after defrost and maintaining a very constant discharge air temperature increases product shelf life and reduces shrinkage of products such as fresh meat and produce.

Figure 10-21 is a graphic representation of the discharge air temperatures of three different cases, each with one of the previous three methods of temperature control. The thermostat with pump-down solenoid has the widest temperature swing. The mechanical EPR is much better but does not drop to the desired temperature on the initial pull-down after defrost. However, the EEPR pulls down quickly and maintains the most consistent discharge air temperature with the smallest temperature swing.

In addition, on shutdown, they completely close the suction line, often referred to as a suction stop. This feature is especially important during defrost when the evaporator temperature rises. By sealing off the suction line between the evaporators and the suction header,

FIGURE 10-19 **Mechanical Sporlan SORIT and electronic valve retrofit.** *Photo (a) by Jess Lukin and (b) compliments of Parker Hannifin-Sporlan Division.*

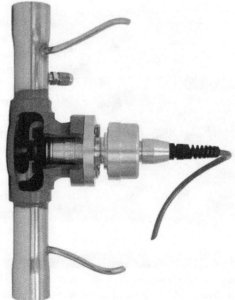

FIGURE 10-20   **Sporlan CDS valve electronic evaporator pressure regulator (EEPR).** *Photo by Dick Wirz.*

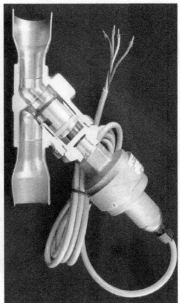

FIGURE 10-21   **Comparing three types of temperature control.** *Drawing by Irene Wirz, Refrigeration Training Services.*

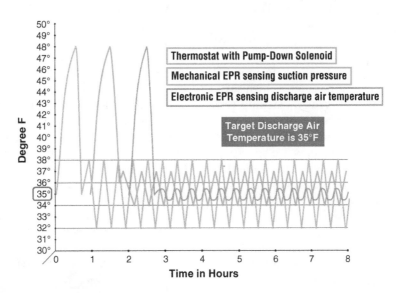

it keeps evaporator temperature and pressure up for better defrost, while keeping that warm suction vapor from entering the header, which could overload the compressors.

RACK SYSTEM CONTROLLERS

Computer-based systems originated in large commercial buildings and factories as building management systems (BMS) or building automation systems (BAS).

They are used to monitor and control the building's mechanical and electrical equipment such as ventilation, lighting, power, fire systems, and security to name a few. A BMS or BAS consists of software and hardware.

The next logical step was to include the building's HVAC system. Supermarkets had already been using control cabinets to house relays, contactors, and defrost time clocks, as seen in Figure 10-22. It was natural for them to embrace the new computer technology that had the potential to monitor and control power

FIGURE 10-22   **Old refrigeration rack with contactors, relays, and defrost timers.** *Photo by Dick Wirz.*

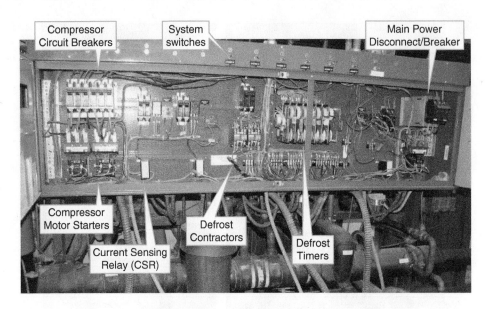

consumption and mechanical operation, as well as to improve both the quality and safety of their refrigerated products. The term *energy management system (EMS)* seemed to describe this function. Therefore, except in the case of specific equipment, we will refer to all of these types of controlled systems as the EMS or *controllers.*

Microprocessor-based controllers are designed to provide multiple controls in one package, which saves space and money. They also provide a tracking program to monitor temperatures, pressures, and alarms, which is especially important for rack refrigeration systems. Since these refrigeration systems are so expensive, monitoring their health is a wise investment to safeguard the systems. Doing so also provides a health safety standard by ensuring refrigerated product temperatures are maintained to prevent premature food spoilage.

Inputs and outputs of control devices are either *analog* or *digital.* A digital input or output is either "on" or "off." For example, a push button switch is either pressed or not, and an LED is either emitting light or not emitting light. Digital input devices are binary, which means on or off, open or closed. Examples are safety switches such as high- or low-temperature safety limits and pressure switches. Digital output devices include relays, contactors, heaters, and solenoid valves.

Analog devices sense continuous parameters such as temperature or pressure. They give information as a continuous range of values, not just an on or off indicator. For example, an analog input device such as a temperature sensor usually warms up slowly and cools down slowly, and the change in temperature is constantly increasing or decreasing. This input to a

controller may result in analog output signals that can vary the speed of an evaporator fan or the valve position of a stepper motor on an electronic expansion valve (EEV).

Control systems can be very simple or very complex. An example of simple is a stand-alone microprocessor-based control found on many self-contained refrigeration systems today (see Figure 10-23). In the past, a reach-in freezer usually had a thermostat, a defrost time clock, and maybe a pressure control—each with its own sensors and specific function. The current controllers integrate the thermostat and time clock by utilizing only a couple analog temperature sensors (thermistors) to provide information to the logic board,

FIGURE 10-23   **Traulsen reach-in freezer controller.** *Courtesy of ITW Refrigeration–Traulsen & Kairak.*

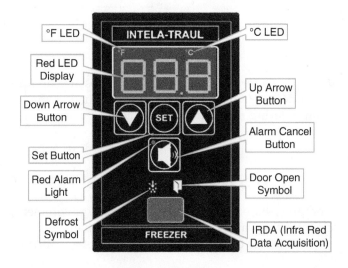

which makes all the decisions the thermostat and defrost clock used to have to make. The electronic control can monitor the box temperature, provide a means to adjust it from the control keypad, initiate and terminate defrost, and perform any other function the programmer decided was necessary. In addition, the new controls have digital displays to help managers monitor system operation, including visual and audible alarms to alert them to problems. The new controllers offer much more in a smaller package than what was available with the older controls in the past.

The Traulsen reach-in freezer controller in Figure 10-23 has "set points" such as temperature and defrost times that can be programmed by the technician. Even a simple controller like this requires the technician to consult the manufacturer's user manual for specific operation information. However, with a little reading, it soon becomes apparent that the new controllers have so much more to offer the customer and the technician in features. And they usually have several diagnostic programs to help the service technician. See the "troubleshooting tree" or diagnostic chart in Figure 10-24 for an explanation of what to check if that particular alarm is displayed. The high-temperature alarm diagnosis looks fairly simple because it directs the technician in a logical sequence toward solving the problem. Even more complex diagnostic charts are easier than they look at first because most of the steps illustrated are eliminated as the questions are answered in each segment.

Manufacturers usually have a technical support phone number to call if the technician needs help. Knowing these important phone numbers can aid in troubleshooting and eventually lead the technician into more effective troubleshooting techniques.

Controllers require analog inputs in the form of sensors or probes, as shown in Figure 10-25. As previously described, an analog input is a variable such as pressure or temperature that changes its values in a linear progression (in a straight line) from one logical sequence to the next. For example, a temperature sensor moves from one degree to the next as it heats up or cools down. Temperature analog probes are either an NTC (negative temperature coefficient) or a PTC (positive temperature coefficient) and have only two wires. Temperature changes the resistance of the probe in a very predictable sequence or "linear" manner.

An NTC probe has a negative temperature coefficient, which means that as the probe's temperature goes up (positive direction), its resistance in ohms goes down (negative). Manufactures supply charts such as those in Figure 10-26 so the technician, while on the job, can verify the sensor's condition and accuracy by comparing the sensor's actual resistance at a given temperature to what is on the chart. For example, at 32°F, the resistance of a Sporlan 2K probe will be 5.1 ohms, a

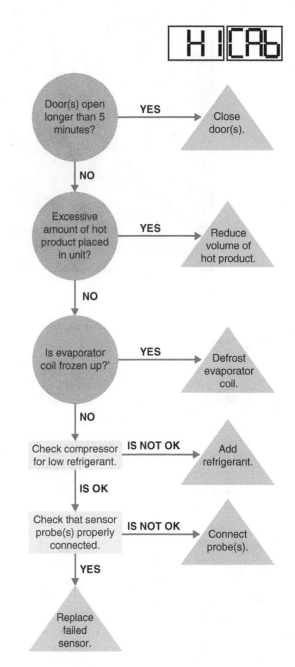

FIGURE 10-24 **Traulsen high-temperature troubleshooting tree.** *Courtesy of ITW Refrigeration–Traulsen & Kairak.*

3K probe would be 8.8 ohms, and a 98.6K probe would be 327.5 ohms. To get the probe to 32°F, the technician would fill a cup with ice then add enough water to fill the cup, stir the mixture, insert the probe in the ice bath, and check the resistance with an ohmmeter. Although the accuracy will be affected somewhat by the amount of minerals in the water, the result should be fairly close to the values in the table.

NTC probes are not interchangeable with PTC probes. NTCs respond to an increase in temperature with a decrease in resistance. A PTC is the complete

FIGURE 10-25 **Temperature sensors or probes.** *Compliments of Parker Hannifin-Sporlan Division (left), Photo by Dick Wirz (right).*

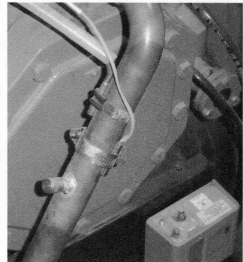

FIGURE 10-26 **Sensor temperature/resistance.** *Compliments of Parker Hannifin-Sporlan Division.*

Sporlan Temperature Sensor Temperature Vs. Resistance			
Temp °F (°C)	Resistance (kohms)		
	Resistance Rating		
	2K	3K	98.6K
–40 (–40)	32.3	100	NA
–20 (–29.9)	18.5	49.5	NA
0 (–17.8)	11.1	25.6	NA
20 (–6.7)	6.7	13.9	NA
32 (0)	5.1	9.8	327.5
40 (4.4)	4.3	7.8	261.1
60 (15.6)	2.8	4.6	152.1
80 (26.7)	1.9	2.8	91.6
100 (37.8)	1.3	1.7	56.9
120 (48.9)	0.91	1.1	36.4
140 (60)	0.66	0.35	23.8
160 (71.1)	0.48	0.25	16
212 (100)	NA	NA	6.3
300 (148.9)	NA	NA	1.7

opposite, so an increase in temperature equals an increase in resistance. Some controllers have the ability to work with either PTC or NTC sensors; therefore, the options available to the program or logic board increase, and all that is required is a change in the parameter set points.

An analog input pressure transducer typically uses three wires—one for ground, another for the input signal, and another for the separate output signal. Figure 10-27 is an example of a pressure transducer with a drawing of the three connections. Figure 10-28 is a copy of the table for voltage outputs for testing accuracy and application of the transducer.

Some systems have more inputs and outputs. These systems can also be stand-alone or connected to a monitoring system. The monitoring system can be in-house and/or remoted to a third party. One example is a Lennox Prodigy controller for RTUs (roof top units) (see Figure 10-29). This controller provides complete control for the RTU system. It controls compressor operation, fan operation, and staging, if available on the unit. One advantage is the USB port that provides access to the report functions. The servicing technician can view reports, save them on a flash drive, or print them out to be included with the service reports.

The Lennox Prodigy controllers, like most controllers, have a self-test feature integrated into them. This is very useful to the service technician, as it enables her to make sure that the controller is operating properly. In addition, control systems in HVAC and in commercial refrigeration utilize *redundancy* as a failsafe. For example, the units will have standard mechanical safety pressure controls in addition to the pressure transducers monitored by the controller. Even this backup device could fail, so the controller itself is programmed with additional safety features in place that sense system problems. For instance, should the mechanical pressure control be out of calibration, misadjusted, or fail, the controller will still sense an alarm condition.

FIGURE 10-27   **Pressure transducer.** *Compliments of Parker Hannifin-Sporlan Division (left and center). Photo by Dick Wirz (right).*

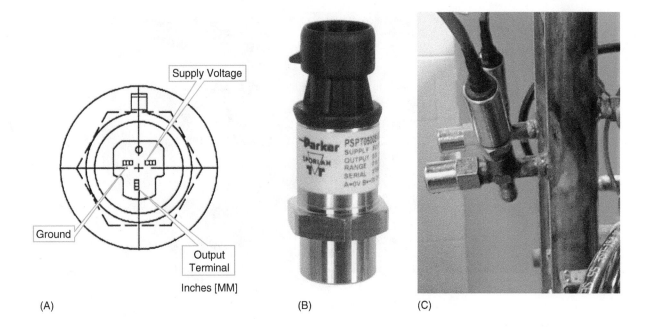

(A)   (B)   (C)

FIGURE 10-28   **Transducer pressure/voltage chart.**
Compliments of Parker Hannifin-Sporlan Division.

Pressure Transducer			
	DC Voltage (White & Green Leads)		
Pressure psig	952572 0–150 psig Green Label	952574 0–300 psig Silver Label	952576 0–500 psig Yellow Label
0	0.5	0.5	0.5
25	1.2	0.8	0.7
50	1.8	1.2	0.9
75	2.5	1.5	1.1
100	3.2	1.8	1.3
125	3.8	2.2	1.5
150	4.5	2.5	1.7
175	–	2.8	1.9
200	–	3.2	2.1
225	–	3.5	2.3
250	–	3.8	2.5
275	–	4.2	2.7
300	–	4.5	2.9
325	–	–	3.1
350	–	–	3.3
375	–	–	3.5
400	–	–	3.7
425	–	–	3.9
450	–	–	4.1
475	–	–	4.3
500	–	–	4.5

More sophisticated controllers are used on rack systems. These complex microprocessor-based controls may require additional boards for analog inputs (AI), analog outputs (AO), digital inputs (DI), digital outputs (DO), and relay outputs (RO) in order to be effective. These boards can be combined or separated depending on what the customer selects. They can be monitored as they control HVAC, refrigeration, store lighting, and just about everything inside and outside the building.

Most controllers will have a display screen and keyboard or HMI (human machine interface) for the technician to use for troubleshooting purposes or just gathering information. For example, Figure 10-30 is the E2 control system offered by Emerson, and Figure 10-31 is an example of how the information is displayed on its screen. It is capable of controlling multiple devices through additional separate AI/DI and AO/DO boards, as well as RO boards. Figure 10-32 is a view of some of those boards and other components behind the first door that has the HMI. Figure 10-33 is a picture of what is behind the second door. As you can see, there is a lot going on in there.

Relay outputs are on (1) or off (0) binary signals given to control components like compressors, defrost heaters, and condenser fans. A relay is an electrical switch under the control of another electrical circuit. Relay outputs are designed to handle more of an amp draw through the relay in comparison to digital outputs. That is an important distinction. Other configurations are available, but Figure 10-34 is a good example of a basic relay output board. This board has

FIGURE 10-29   **Lennox Prodigy controller for roof top unit.** *Courtesy of Lennox.*

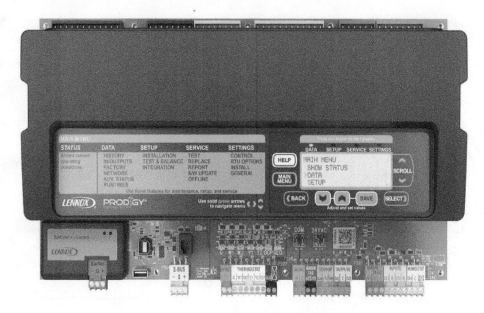

FIGURE 10-30   **Emerson controller on a rack system.** *Courtesy of Emerson.*

manual three-way toggle switches that can force or manipulate each one of the components listed. The toggle switch can be in a manual position of On or Off, or the Auto position, which allows the controls program to control the component operation. The availability of the manual positions of On or Off is an advantage when troubleshooting or servicing the equipment. Unless, of course, the technicians forgets to put it back into the Auto position. Technicians must take care when using the manual mode because sometimes placing the switch in the On positon will completely bypass the safeties controls. This can really cause some problems to the system and dangers to the technician. See Figure 10-35 for an example of a control panel with external toggle switches and lights.

FIGURE 10-31   **Display screen on an Emerson E2.** *Photo by Dick Wirz.*

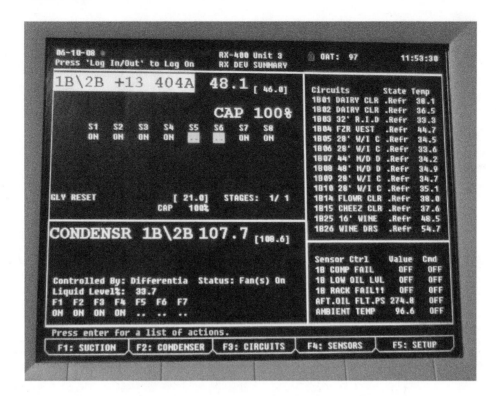

FIGURE 10-32   **Control boards behind HMI on the first door.** *Photo by Dick Wirz.*

FIGURE 10-33 **Inside the second door of the control panel.** *Photo by Dick Wirz.*

FIGURE 10-34 **Example of a relay output board.** *Drawing by Irene Wirz, Refrigeration Training Services.*

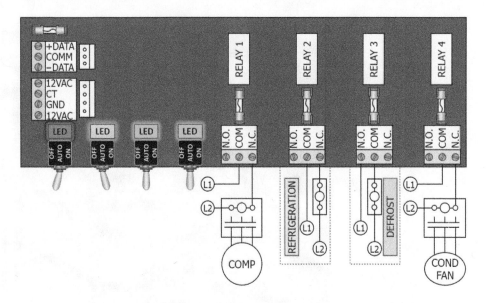

Notice in Figure 10-34 that the refrigeration relay is connected to a normally closed (NC) contact. This provides a failsafe so that if the board loses communication, it will fail to Refrigeration On because the contacts are already closed, a signal is not needed to close the switch.

It is important to review the types of relay outputs available. Typically, we have three types: NO, NC, and changeover.

NO or *normally open* connects the circuit when the relay is energized. It's a *Form A* contact or "make" contact because the circuit is electrically made when activated.

FIGURE 10-35　**Control panel toggle switches with lights.**
Photo by Jess Lukin.

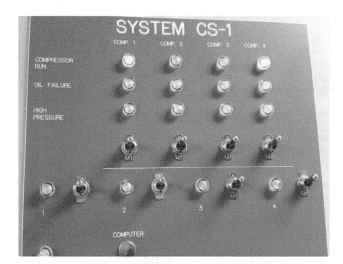

FIGURE 10-36　**Picture of a Micro Thermo compressor controller.** *Courtesy of Micro Thermo and Parker Sporlan.*

It is ideally used for applications that require a high current power source from a remote device like defrost.

NC or *normally closed* disconnects the circuit when the relay is energized. It's a *Form B* contact or "break" contact because power applied breaks the electrical line. It is ideally used for applications that require the circuit to remain closed until the relay is energized (e.g., with compressors or other load devices).

Changeover contacts control two circuits that consist of one NC and one NO with a common terminal. This type isn't foreign to technicians in the field. Most know the changeover contacts utilized in defrost time clocks. The concept is exactly the same in this instance and is called a *Form C* contact. This can be used for refrigeration load devices or defrost control. It also prevents either from being on at the same time.

Controller electrical diagrams can be shown either energized or de-energized. This fact is important because our example would be understood completely opposite of what we see with it de-energized.

In complex controller systems, each load component will have its own board. For example, Figure 10-36 a picture of a compressor controller. The diagram in Figure 10-37 shows the electrical connections and component designations.

As more functions are required, some boards have both analog and digital inputs. Figure 10-38 is an example of an analog input (AI) and digital input (DI) board combined, or an AI/DI board. In the diagram are two examples of analog inputs. One is a two-wire temperature sensor. The other is a three-wire pressure transducer with a "drain connection." To keep down interference from electromagnetic "noise," the cable

covering contains a foil shield. Wound inside the cable and in contact with the shield is a bare copper conductor known as a drain connection. The single wire is connected to the common terminal along with the common wire.

In this diagram the digital input is a *dry contact*. A dry contact could be a secondary contact of a relay circuit, which does not make or break the primary voltage. As an example, this dry contact may be an auxiliary switch on the side of the compressor contactor. When the contactor closes to start the compressor, the auxiliary switch closes. The power to or through the contactor does not go through the switch. The purpose of the switch is to close (or open). This action signals some device in the board or through the board that the contactor has opened (or closed). It is important to understand that no power is to be applied to this terminal strip; otherwise, it could damage the board.

Some controllers combine almost every function on one board. For example, Figure 10-39 shows a case controller. It combines the relay outputs on the left with analog and digital inputs as well as outputs on the right. There is a pressure transducer plus temperature sensors for superheat, defrost termination, and alarm. Even antisweat heaters are controlled from this board. An interesting feature not covered previously is the ability of this board to control the actions of the stepper motor in the electronic expansion valve (EEV).

FIGURE 10-37 **Micro Thermo wiring diagram of compressor controller.** *Courtesy of Micro Thermo and Parker Sporlan.*

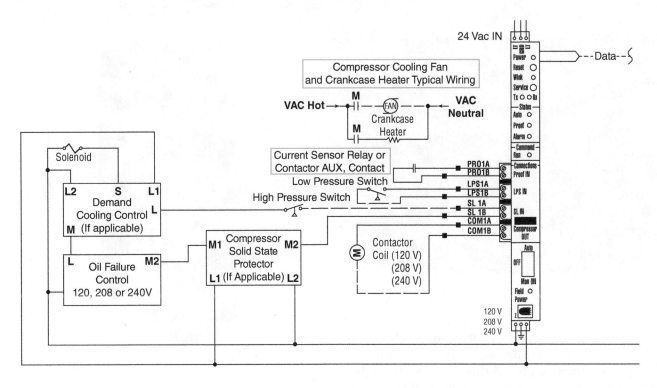

FIGURE 10-38 **A 16-channel analog/digital input board showing three types of inputs.** *Courtesy of Refrigeration Training Services.*

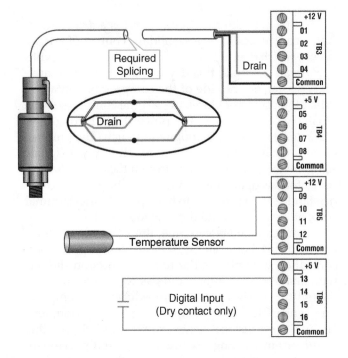

The four connections on the far right terminal board in Figure 10-39 are reserved for the EEV.

There are a few different network setups. *Daisy chain* networking is where all controllers are linked together in a series and is commonly used on the local area network (LAN). The example also has I/O network connection with corresponding I/O Net Tx and Rx LEDs for easy troubleshooting. These LEDs blink when communicating. It's best to refer back to the actual manual, but the majority of the time, red or no LED is loss of communication or another issue. "Heartbeat green," or a slow continuous blinking of a green LED, is always a good sign and confirms communication from the board in and out. A "lost board" can occur when the address is changed or communication is lost. This is typically seen during a power outage or spike on the initial restart-up.

Figure 10-40 is a drawing that includes a compressor board, suction pressure controller board, condenser controller, and circuit board. All these boards need to communicate with the main controls and monitoring program to work correctly. Each board can be located at the component or central location. Typically, the compressor controls and suction-pressure controller

FIGURE 10-39   **Picture of a case controller with labels.** *Courtesy of Micro Thermo and Parker Sporlan.*

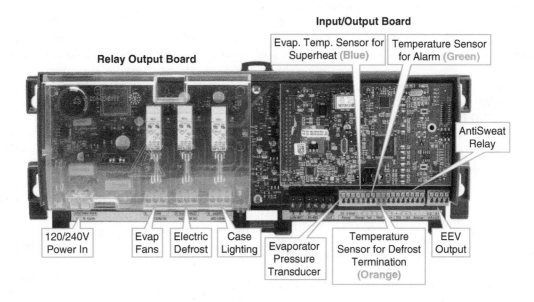

FIGURE 10-40   **Diagram of several controller boards working together.** *Courtesy of Micro Thermo and Parker Sporlan.*

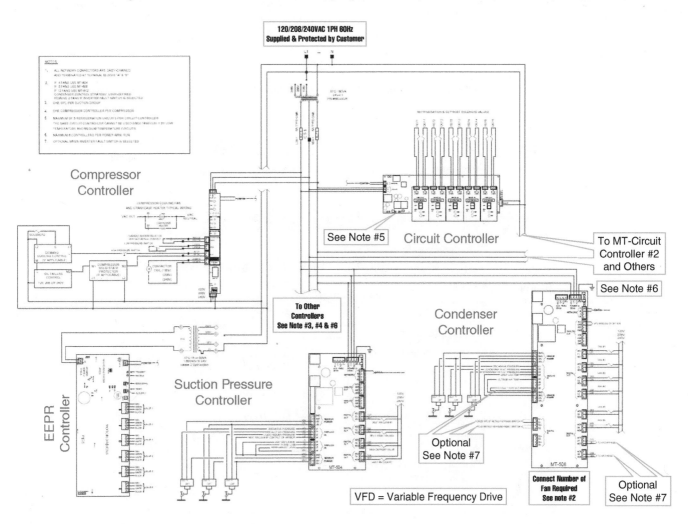

boards will be located at the electrical panel on the rack house. The condenser controller will be located inside the electrical panel at the condenser. The circuit boards may be located in a back room or on top of the circuit cases, depending on wherever the engineer designed them to be.

In the manager's office, there is a computer connected to the refrigeration system controller. The display screen or HMI (human machine interface) has a PID drawing of the store layout in customer-friendly view (see Figure 10-41). *PID* stands for piping and instrumentation diagram/drawing and is defined by the Institute of Instrumentation and Control as follows:

1. A diagram that shows the interconnection of process equipment and the instrumentation used to control the process. In the process industry, a standard set of symbols is used to prepare drawings of processes. The instrument symbols used in these drawings are generally based on International Society of Automation (ISA) Standard S51.
2. The primary schematic drawing used for laying out a process control installation.

HEAD PRESSURE CONTROL

Head pressure control on some rack systems is a combination of fan cycling and flooded condensers with head pressure regulation (HPR) valves. Smaller remote single systems use a single nonadjustable HPR or a headmaster valve (LAC-4) to maintain head pressure. The large systems in supermarket refrigeration utilize an adjustable pressure regulating valve (ORI/ORD) to keep the head pressure up and a differential pressure valve to make sure the receiver pressure is also maintained (see Figure 10-42).

In newer systems, the split condenser concept is used in conjunction with the HPR valves (see Figure 10-43). The condenser is split, or divided, into two separate sections (Figure 10-44). Both sections are utilized during the summer and at high-load conditions. During cold weather or low-load conditions, a valve automatically closes off the flow of discharge vapor to one section of the condenser. The advantage of the split condenser is that the total charge of the system is considerably reduced because the HPR valve only has to back up refrigerant in the active half of the condenser to keep the head pressure up.

FIGURE 10-41 **Example of a computer screen in a manager's office showing system alarm status.** *Courtesy of Micro Thermo and Parker Sporlan.*

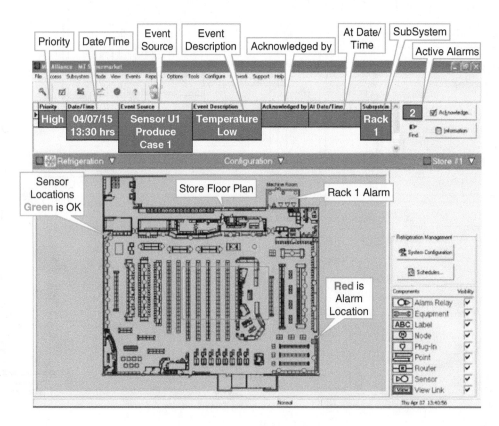

FIGURE 10-42 **HPR for smaller systems (top) and HPR with differential valve for supermarket equipment (bottom).** *Adaptation of Sporlan diagrams by Refrigeration Training Services.*

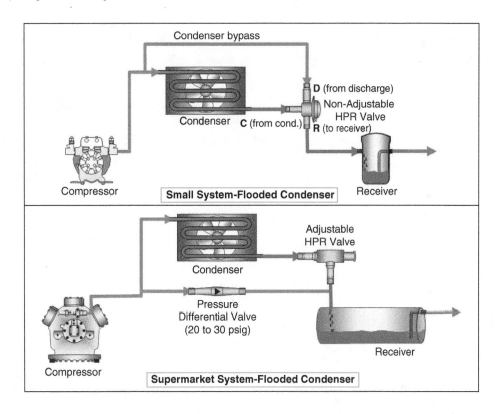

FIGURE 10-43 **Split condenser piping.** *Photo by Dick Wirz.*

ENERGY EFFICIENCY

With profit margins near 1 percent, supermarkets are very concerned about keeping their equipment operating costs as low as possible. The more efficient their equipment, the less it will cost to operate. Electronic controls and controllers provide accurate operation of refrigeration equipment, which increases efficiency. If the refrigeration preserves foods better, and longer, there is less waste and reduced expense. Following are some other methods of energy savings in supermarket equipment:

1. Multiple compressors use less energy and have better capacity control than a single large compressor.
2. Electronic thermal expansion valves (TEVs) and EPRs increase evaporator efficiency by lowering superheats to 3°F or less.
3. Cases defrosted with hot gas use less energy than those defrosted with electric heaters (see Figures 10-45 and 10-46).
4. Demand defrost, which involves defrosting only when needed, is more efficient than defrosting at regularly scheduled times.
5. Glass doors with a special film to prevent fogging rather than electric heaters.
6. Nonmetal door frames that do not require electric antisweat heaters.
7. Mechanical liquid subcooling (see Figures 10-47, 10-48, and 10-49).
8. Heat reclaim and reheat (see Figures 10-50 and 10-51).

FIGURE 10-44 **Split condenser diagram by Sporlan.** *Compliments of Parker Hannifin-Sporlan Division.*

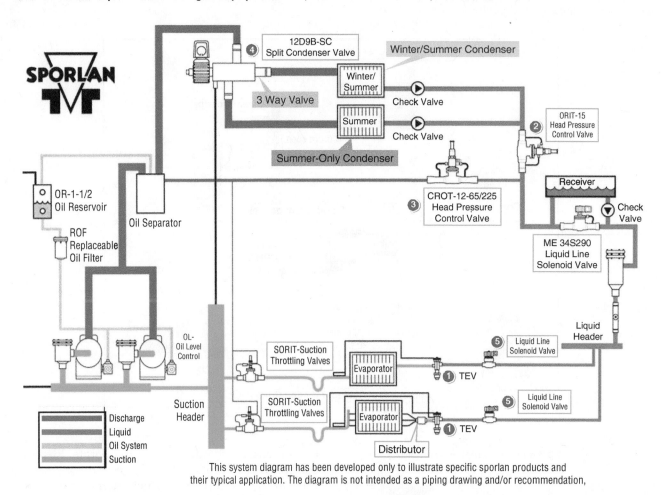

This system diagram has been developed only to illustrate specific sporlan products and their typical application. The diagram is not intended as a piping drawing and/or recommendation,

COIL DEFROSTING

Historically, the two most common methods of defrosting commercial refrigeration have been air defrost ("off-cycle") for medium-temperature coils and electric heaters for defrosting low-temperature coils. Unfortunately, both of these types of defrosting are rather slow. In supermarket refrigeration, it is important to defrost the evaporator as quickly as possible in order to prevent the product from rising in temperature any more than absolutely necessary.

There are two methods of internal defrost that are much quicker and more energy efficient than the old standards of air and electric defrost. One is to use the hot discharge gas from the compressor to defrost the evaporator. Another is to use the warm vapor off the top of the receiver. Although the hot discharge gas is warmer, the vapor from the receiver actually contains more heat per pound of vapor. In both methods, the advantage is that the defrosting takes place from

the inside of the evaporator near the source of the frost, rather than from outside the coil where heat must penetrate into the area that has frost. The internal method of defrosting is therefore much quicker and more efficient.

When defrost is required, a solenoid in the liquid line stops the refrigerant flow into the evaporator. At the same time, a valve at the defrost header opens and allows high-pressure refrigerant (hot gas or warm vapor) to be forced into the evaporator, warming it enough to melt the accumulated frost. One technique is to have the defrost vapor enter from the suction outlet of the evaporator and exit from a tee between the TEV and the distributor (see Figure 10-45). After the hot gas exits the coil and bypasses the TEV and liquid line solenoid, it enters the liquid line and quickly condenses to a liquid. Figure 10-46 is a photo of the check valve and the tee between the TEV and distributor used in this type of hot gas defrost. Another technique is similar but opposite in the direction of flow. The vapor enters

FIGURE 10-45 **A Sporlan diagram of several types of valves used to accomplish a reverse type of hot gas defrost that enters the evaporator outlet.** *Compliments of Parker Hannifin-Sporlan Division.*

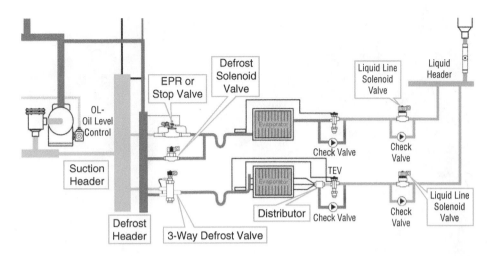

FIGURE 10-46 **The distributor tee and check valve assembly designed to allow hot gas defrost vapor to leave the evaporator and enter the liquid line.** *Photo by Dick Wirz.*

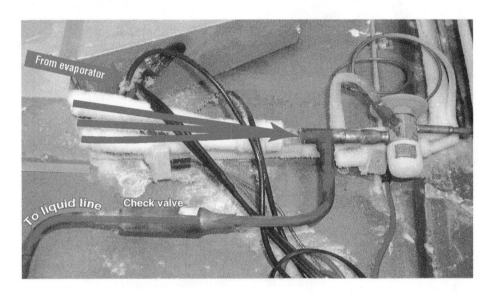

the evaporator inlet at the distributor tee and exits at the suction outlet of the evaporator. Both methods work well and which one is used is just a matter of the application engineer's preference based on the system design.

MECHANICAL SUBCOOLING

Most single-compressor systems are designed to have between 5°F and 15°F of condenser subcooling. This safety cushion of subcooling helps to prevent the flashing of liquid before it reaches the TEV. However, if

liquid refrigerant is subcooled when it enters the TEV, it will also increase the efficiency of the unit.

NOTE: **For every 1°F of subcooling at the TEV, there is 0.5 percent increase in evaporator capacity.**

In supermarket refrigeration, mechanical subcooling is used to provide up to 45°F of subcooling, which increases the system capacity more than 20 percent.

The plate subcooler in Figure 10-47 is located after the liquid receiver and just before the liquid enters the header where the case liquid lines begin. It is a special TEV-metered evaporator that cools liquid refrigerant from approximately 100°F to 55°F (see Figure 10-48). Some have

FIGURE 10-47   **Plate subcooler used by Hussmann.** *Photo by Dick Wirz.*

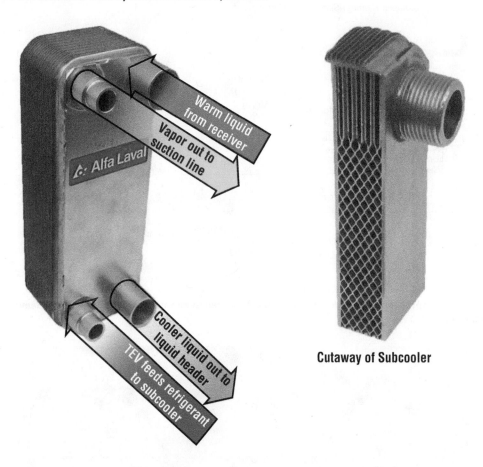

Cutaway of Subcooler

FIGURE 10-48   **Mechanical subcooling using one mechanical TEV.** *Courtesy of Refrigeration Training Services.*

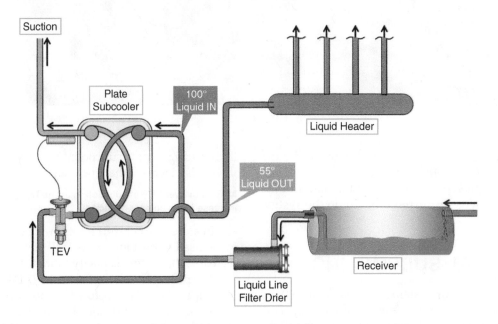

two TEVs and two circuits so they can be staged. During high-ambient conditions, both circuits are required. However, when the ambient drops, only one circuit is needed.

Figure 10-49 shows an electronic subcooling controller and a single EEV, which is even better at controlling subcooling than those with two mechanical TEVs.

FIGURE 10-49 **Mechanical subcooling using EEV and controller.** *Compliments of Parker Hannifin-Sporlan Division.*

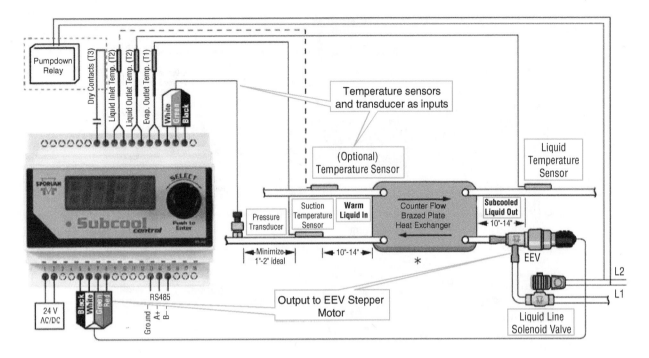

HEAT RECLAIM AND REHEAT

Additional energy savings can be realized if the hot gas from the compressors is used to help heat the building or to preheat water going to the domestic hot water heater. This is essentially a free energy supply (see Figure 10-50).

Instead of rejecting all the system and compressor heat to the outdoor air, a three-way valve diverts the hot gas to a heating unit. As the hot gas travels through a heat exchanger in the heating unit, it gives up the superheat from the heat of compression and motor heat. The de-superheated vapor then enters the condenser where the hot gas is condensed into liquid (see Figure 10-51).

Compressor heat can also be used to control the humidity in the building. The open cases in supermarkets will not operate correctly if the store humidity is above 60 percent. In humid geographical areas, this can be a real problem. Fortunately, the compressor hot gas can be diverted to reheat coils in the supply ducts of the A/C system. Slightly raising the supply air temperature lowers the relative humidity of the air discharged into the store. Relative humidity is a percentage of the total amount of moisture air can hold at a given temperature. If the amount of moisture in a pound of air remains constant, raising the air temperature decreases its relative humidity.

In addition, the A/C runs a little longer, which allows the evaporator to remove more moisture from

FIGURE 10-50 **Heat reclaim for hot water at the Hussmann Training Center.** *Photo by Dick Wirz.*

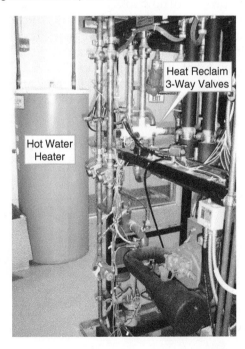

the return air. Lower building humidity not only allows the cases to perform better but also offers a more comfortable store environment for both employees and customers.

FIGURE 10-51   **Heat reclaim drawing by Sporlan.** *Compliments of Parker Hannifin-Sporlan Division.*

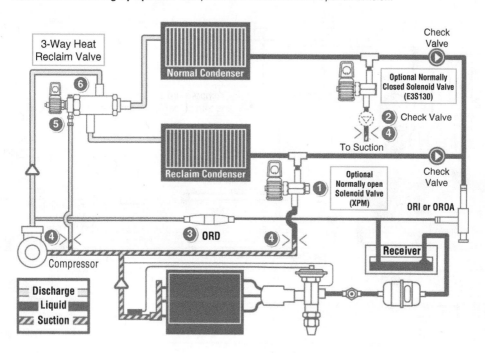

DISPLAY CASE AIRFLOW

Air conditioning technicians already know the importance of proper airflow in A/C and heating systems. However, few technicians realize how critical airflow is to supermarket open display cases (Figure 10-52). Curtains of air are forced out of an air grill, called a "honeycomb," at the top of the case (Figure 10-53). The air is directed down the front of the case and sucked into the return air vents at the bottom edge of the case

FIGURE 10-52   **Open display case air curtains.** *Courtesy of Refrigeration Training Services.*

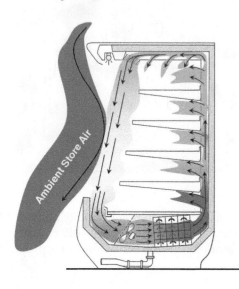

FIGURE 10-53   **Discharge air "honeycombs."** *Photo by Dick Wirz.*

opening. The airstream not only refrigerates the case but also provides an air curtain that prevents the warmer ambient air in front of the open display area from entering the case. Open vertical freezers require several air curtains. However, most stores currently use the more efficient glass door freezer cases. Horizontal open freezers, or "coffin cases," are still used because cold air falls; therefore, the refrigerated air stays in the case.

If the air curtain is disrupted in an open display case, it will lose its ability to refrigerate. Some common causes for airflow problems are dirty honeycombs, fan problems, or improper stacking of product that blocks the air curtain.

A different type of airflow is used in some glass enclosed display cases. Gravity coils, like the one in Figure 10-54, are very effective where high case humidity is essential to prevent the product from drying out. When air passes over uncovered food products, it removes moisture. In addition, moisture is removed from air as a result of the relatively high temperature difference (TD) of forced air fan coils. However, gravity coils rely on natural convection. The gentle circulation of air has very little drying effect on the product. In addition, the design of the coil is such that the temperature of the air passing through it is not lowered enough to remove very much of its moisture. To accomplish this, the evaporator is mounted at the highest point inside the case. As the air in the lower portion of the case is heated by the product, it rises to the top of the interior of the case and comes in contact with the cold evaporator coil. As the heat is removed by the coil, the cool air gently falls to the bottom of the case, where it refrigerates the product.

INSTALLATION, SERVICE, AND MAINTENANCE

Installation of large supermarkets can take a year to complete. It is difficult and very time consuming to set literally tons of cases, compressors, and condensers (Figure 10-55). In addition, installing large walk-ins, row after row of steel shelving, and literally miles of refrigeration piping and wiring is no easy task (Figure 10-56). Refrigerant and drain piping rough-in begins before the walls of the building are up. Tunnels under the floor, sometimes 6 feet high by 6 feet wide, are nearly filled with refrigeration and drain piping. After the equipment is put into operation, it can take weeks to check and balance the system. Supermarket installation technicians are considered specialists in their field because it is exacting work and requires good planning, excellent job skills, and a strong body.

Servicing supermarket equipment can be challenging, to say the least. Imagine how difficult it could be to find a refrigerant leak in a system that may have thousands of feet of piping in tunnels, above the ceiling, on the roof, and inside scores of cases. The average supermarket has approximately 4 miles of refrigeration piping. The service technician's job is not an easy one, but it will be less difficult if the installing technicians were

FIGURE 10-54 **Gravity coil in a deli case.** *Photo by Dick Wirz.*

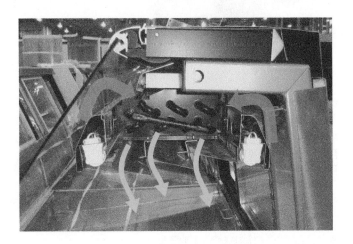

FIGURE 10-55 **Setting cases.** *Courtesy of Jerry Meyer, Hussmann Training Center.*

FIGURE 10-56   **Rack system piping.** *Courtesy of Jerry Meyer, Hussmann Training Center.*

highly skilled, took pride in their work, and checked the system thoroughly for leaks before start-up. A refrigerant leak in a supermarket system can easily cost the owner thousands of dollars.

Supermarket service technicians must also be thoroughly familiar with how the entire system works and completely understand the controls and controllers. The technician must be able to find the problem quickly and repair it properly.

To prevent the need for expensive emergency service, a comprehensive maintenance program should be part of every supermarket's budget. During the initial start-up of the systems, a record should be made of all equipment temperatures, pressures, voltages, and amperages. During regularly scheduled inspections, the maintenance technician compares the current operating conditions with the original start-up information. Any deviations from the original readings can help to identify possible problems that can be corrected long before they become an emergency.

REFRIGERANT LEAK DETECTION

Even with good installation and maintenance practices, there are refrigerant leaks in almost every system. The national average for loss of refrigerant in a supermarket is about 20 percent per year. The average store could easily lose 500 pounds of refrigerant annually, which is not only expensive but is damaging to the environment. Electronic leak detectors are being used in many stores to set off an alarm when a leak is sensed. The detector's remote sensors indicate where the leak is occurring. This type of early warning system can prevent excessive refrigerant and product loss, and it reduces the chances of equipment damage resulting from low-refrigerant charge. Figure 10-57 shows one of the electronic leak detectors that are being used.

NEW TECHNOLOGY

Electronic valves have been used in supermarket refrigeration for some time (see Figure 10-58). However, there have been significant advances in valve design and the devices that control them. A good example of new technology is the Superheat Management System (SMS) used by Hillphoenix, a manufacturer of refrigerated cases and refrigeration systems. Each case has an EEV, pressure transducer, temperature thermistor, and control module (see Figure 10-59). The SMS automatically adjusts the superheat in response to changing ambient conditions, case load, and variations in system parameters. If the superheat target values need to be changed, the control module inside the refrigerated case is reprogrammed from a laptop computer or handheld device by simply pointing an IR (infrared) device at the controller. The possibility of being able to adjust refrigerated case operation by simply pressing computer keys is very appealing to any technician who has had the difficult task of adjusting a standard expansion valve in the bottom of a refrigerated case.

FIGURE 10-57   **Electronic remote leak detector.** *Photo by Dick Wirz.*

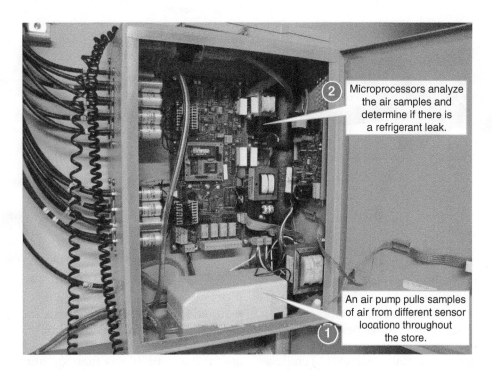

② Microprocessors analyze the air samples and determine if there is a refrigerant leak.

① An air pump pulls samples of air from different sensor locations throughout the store.

FIGURE 10-58   **Refrigerated case evaporator with electric valve and controls.** *Photo by Dick Wirz.*

Pressure Transducer

Thermistor

EEV

FIGURE 10-59   **The SMS by Hillphoenix.** *Photo by Dick Wirz.*

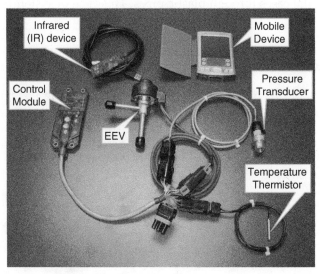

Infrared (IR) device

Mobile Device

Control Module

Pressure Transducer

EEV

Temperature Thermistor

Variable Frequency Drive (VFD) technology is being used extensively to control the speed of compressor, condenser, and pump motors for both capacity control and energy efficiency. However, motor technology is not just for the large horsepower equipment. The use of ECM (electrically commutated motor) evaporator fans in refrigerated cases is another technological advance. These fan motors use less

energy and can be programmed to vary their speed in order to regulate case-discharge air temperature (see Figure 10-60).

Proper lighting in refrigerated cases is very important for displaying merchandise in the best possible manner. However, the energy used and the heat generated by lighting has always been a concern to case manufacturers and their customers. New lighting technology has

FIGURE 10-60   **ECM evaporator fans.** *Photo by Dick Wirz.*

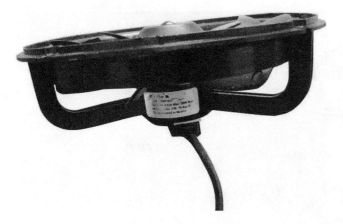

FIGURE 10-61   **LED case lighting (left) and LED bulb (right).**
Photos by Dick Wirz.

addressed both those issues. Light emitting diode (LED) lighting provides good lighting at lower operational costs and produces very little heat (See Figure 10-61).

The old standard of using chilled water for large A/C applications is now being used for commercial refrigeration purposes. Hillphoenix has developed these secondary refrigerant systems for supermarket applications. The primary refrigerant (e.g., R404A) chills a secondary refrigerant of propylene glycol. The primary refrigeration rack system is confined to the mechanical equipment room, where it chills the glycol solution. The chilled water is pumped from the mechanical room to the cases through plastic pipes. This type of system

uses only a fraction of the amount of refrigerant that is required for conventional systems and greatly reduces the cost of both installation labor and materials. Not only is there less chance of refrigerant leaks, but there are far fewer problems than are normally associated with direct expansion (DX) evaporators and expansion valves. Figure 10-62 illustrates the differences between a conventional refrigeration system and the Second Nature® system designed by Hillphoenix. Figure 10-63 is a picture of an actual chiller portion of a secondary

FIGURE 10-62   **Second Nature® system by Hill PHOENIX (left) and a conventional DX system (right).** *Refrigeration Training Services adaptation of Hill PHOENIX drawing.*

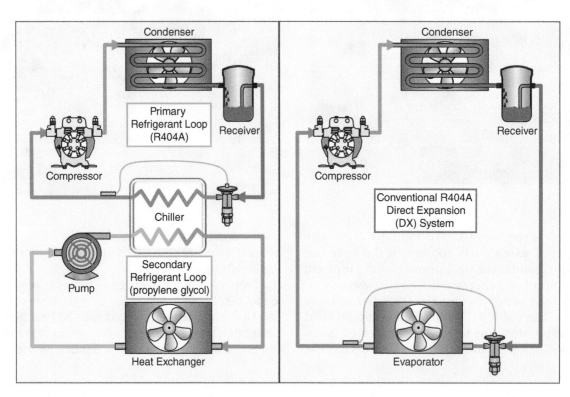

refrigerant system with circulating pumps. Figure 10-64 shows how low some chilled-water systems need to operate in supermarket applications. Figure 10-65 is a picture of a couple of evaporators set up for chilled water and waiting to be installed at the factory in a refrigerated case. Figure 10-66 shows a deli case with chilled water pans used to refrigerate the product.

Defrosting of evaporators is accomplished efficiently by circulating a warm glycol solution through the coils. The defrost solution is heated by the discharge vapor of the primary refrigeration loop.

Another new technology is the use of CO_2 as the secondary refrigerant. Figure 10-67 is a drawing of a Second Nature® CO_2 system with R404A as the primary refrigerant. The liquid CO_2, rather than HFC refrigerants,

FIGURE 10-63 Chiller and circulating pump used in a secondary refrigerant system. *Photo by Dick Wirz.*

FIGURE 10-64 Glycol temperatures for supermarket refrigeration. *Photo by Dick Wirz.*

FIGURE 10-65 Coils piped for chilled water. *Photo by Dick Wirz.*

FIGURE 10-66 Deli case with chilled water pans. *Photo by Dick Wirz.*

is circulated through the store to remove heat from cases and walk-ins. The CO_2 vaporizes some but not all of its liquid as it absorbs heat in the coil. The mixture of liquid and vapor is returned to the separator. From there, the vapor is pulled into the condenser-evaporator where the primary refrigerant, an HFC such as R404A, is used to condense the CO_2 vapor back into a liquid before it is returned to the separator.

CO_2 is also being used as the primary refrigerant in some low-temperature applications. In this arrangement, a conventional R404A system has two functions. It chills a glycol solution, which refrigerates medium-temperature cases. It also acts as the upper cascade portion of the low-temperature system by cooling the CO_2 condensers. Figure 10-68 is a drawing of a Second Nature® system using CO_2 as a primary refrigerant in the low temperature portion of a cascade system.

FIGURE 10-67 **Second Nature® system with CO$_2$ as the secondary refrigerant.** *Refrigeration Training Services adaptation of Hill PHOENIX drawing.*

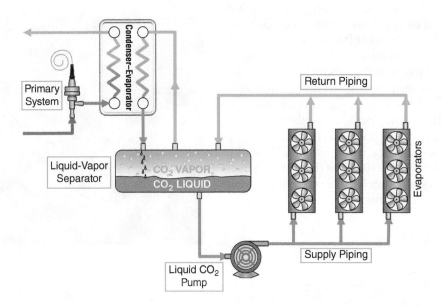

FIGURE 10-68 **Second Nature® system with CO$_2$ as the primary refrigerant in a cascade system.** *Refrigeration Training Services adaptation of Hill PHOENIX drawing.*

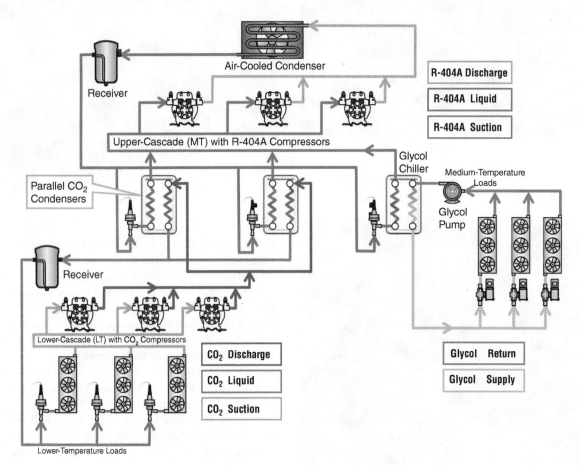

CO₂ AS A REFRIGERANT

CO$_2$ has been used in refrigeration for a very long time. It was introduced as a refrigerant in 1866 by T. S. C. Lowe and promoted as a safer refrigerant than ammonia and sulfur dioxide. Its use peaked in the 1920s and 1930s. In the 1930s, DuPont and General Motors jointly produced the synthetic refrigerant Freon-12. It eventually replaced CO$_2$, ammonia, and sulfur dioxide because it was safer and operated at a much lower pressure than CO$_2$, and the required refrigeration components were much less expensive to fabricate than those of the high-pressure CO$_2$.

In 1996, the Montreal Protocol banned CFCs and started phasing out HCFCs because they damaged Earth's ozone layer. Until something better could be discovered, HFCs such as R404A, R407A, R507A, and R134a were produced as acceptable "interim" refrigerants. Although they have a zero ODP (ozone depletion potential), they have a high GWP (global warming potential). For example, R404A (3922 GWP), R407A (2107 GWP), R507A (3985 GWP), and R134a (1430 GWP). With awareness of the threats to our environment caused by global warming, new refrigerants with low GWP are being sought.

Listed as refrigerant R744, CO$_2$ has a zero ODP and a very low rating of 1 GWP and has generated renewed interest in its use as a refrigerant. It has no regulatory liability as do HFCs, and there is no need to account for the amount used or lost. Further, it does not (in fact, cannot) be reclaimed by service technicians in the field. Although its high pressures require more expensive components, the fact that it is much cheaper than current refrigerants helps to offset some of the initial expense. Between 2004 and January 2014, Coca-Cola installed a million vending machines with CO$_2$ refrigeration equipment. Within a few years, they hope to have this equipment in all of their nearly 3 million machines. Not only are they environmentally friendly, but the CO$_2$ systems are up to 37 percent more efficient than vending machines that use HFC equipment. Although small, the vending machine refrigeration equipment uses basically the same transcritical CO$_2$ refrigeration process as much larger commercial refrigeration system using R744. In addition, a great number of hot water heaters in Japan are CO$_2$ heat pump systems.

Supermarket refrigeration systems in the United States contain thousands of pounds of refrigerant and have an annual average leak rate of 23.5 percent, according to a 2008 statistic from the EPA (Environmental Protection Agency). Therefore, CO$_2$ is a very viable alternative to synthetic refrigerants for these applications and is currently being utilized in many new supermarkets installations.

Before going any further in describing the CO$_2$ refrigeration process, the following terms need to be defined:

- *Critical point*—the temperature above which a refrigerant cannot be condensed to a liquid no matter how much pressure is applied
- *Triple point*—occurs when the refrigerant exists in all three states (vapor, liquid, and solid)
- *Subcritical* refrigeration process—takes place below the refrigerant's critical point
- *Supercritical* fluid—state of a substance above its critical point
- *Transcritical* refrigeration process—when the heat rejection takes place above the critical point while heat absorption is taking place below the critical point, or the subcritical area

Figure 10-69 is a phase diagram of CO$_2$ that may be a helpful visual reference as some of these terms

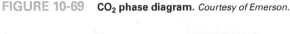

FIGURE 10-69   **CO₂ phase diagram.** *Courtesy of Emerson.*

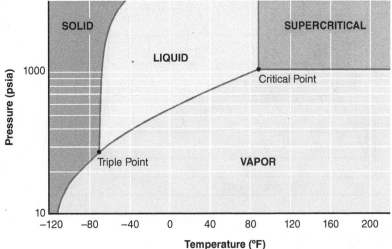

are further explained. Figure 10-70 is the PH (pressure enthalpy) diagram for CO_2. Enthalpy is the change in the amount of heat absorbed (Btu per pound) at specific pressures and temperatures. Therefore, the *H* in *PH* actually stands for "heat." It might be more appropriate to call it a pressure/heat diagram.

Critical point is nothing new, as all refrigerants have critical points. R404A is 162°F at 521 psig. R744 CO_2 is 87.8°F (round to 88°F) at 1055 psig. Even water has a critical point of 705°F at 3,185 psig.

Triple point is another term that did not come up when learning about HFC refrigerants, but they have them. R404A is –100°F at a vacuum of 165,487 microns. The triple point of R744 CO_2 is –69.9°F at 60.4 psig. Water (yes, it is a refrigerant—R718) has a triple point of 32.02°F in a vacuum at 4,588 microns.

Subcritical just means that the whole refrigeration process of compression, condensing, expansion, and evaporation takes place at temperatures and pressures below that refrigerant's critical point. This is the operating envelope in which all HFC refrigerants function.

Supercritical is the range of temperatures and pressures above the critical point. This is a realm in which HFC refrigerants do not function at all; therefore, it was never discussed in classes or conversations involving HFC refrigerants. However, under certain applications, R744 CO_2 systems reject heat in the supercritical stage.

Transcritical just means the heat rejection process is in the supercritical region, and the heat absorption is occurring in the subcritical area. For our purposes, transcritical is when R744 uses a DX (direct expansion) system with compressors capable of handling very high pressures. Under high ambient conditions, part of the system is capable of operating in the supercritical region, and in cooler ambient, the same system will operate entirely in the subcritical region. To reject the heat in the supercritical range, a "gas cooler" is used to lower the discharge vapor temperature rather than condensing the vapor into a liquid. For example, what Coca-Cola is using in their vending machines is an R744 transcritical CO_2 system.

Now that the terms have been defined, it is time to go a little deeper into subcritical and transcritical refrigeration to explain the different ways in which CO_2 is used as a refrigerant in commercial refrigeration.

Subcritical operates within a range that is similar to the HFC refrigerants with which we are familiar. In fact, all the HFC refrigerants used in commercial refrigeration and air conditioning operate in their subcritical envelope. Subcritical just means the refrigerants are operating below their critical point. We don't think of a refrigerant's critical point because we seldom operate even close to it. See Figure 10-71 for the PH (pressure enthalpy) chart of a medium-temperature

FIGURE 10-70 **CO_2 PH diagram.** *Courtesy of Emerson.*

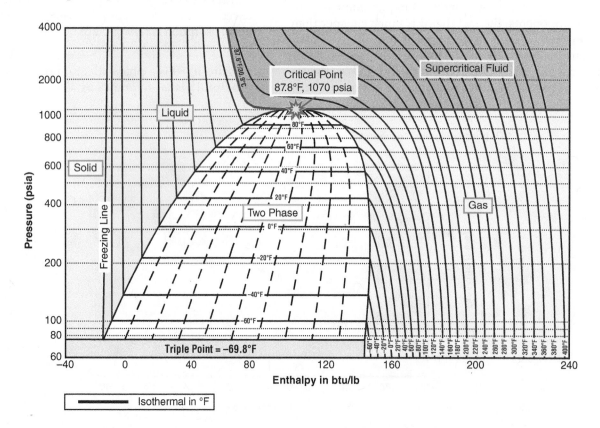

FIGURE 10-71   **Pressure enthalpy (PH) diagram of R404A medium-temp application.** *Courtesy of Refrigeration Training Services.*

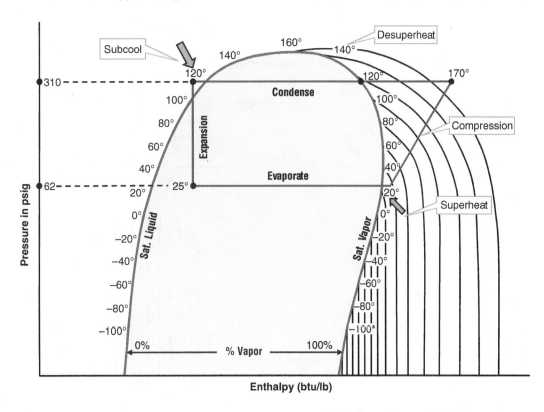

R404A system. The critical point is at 162°F (161.6°F) and 521 (520.5) psig. (To prevent confusion when reading numbers with decimal points, most pressures and temperatures in this section have been rounded off to the nearest whole number.) As the chart shows, the normal condensing temperature is about 120°F at 310 psig, which is far below the critical point. When R404A is used in a medium-temperature application, the evaporator temperature is 25°F, and the suction pressure is 62 psig. For a low-temperature application, the condensing temperature could remain the same; just the evaporator temperature would drop to –25°F at 12 psig.

Figure 10-72 is the PH diagram for R744 or CO_2 in a medium-temperature application. The evaporating temperature of 25°F is the same as it is for R404A; however, the condensing temperature may be only 68°F at 817 psig. Two things become very obvious—the R744 condensing temperature is very low, but the pressures are almost three times higher than R404A. The condensing temperature of R744 is low because it has a very low critical point of about 88°F at a pressure of 1,055 psig. Like all DX (direct expansion) systems, a solid column of liquid is required to feed the evaporator's expansion valve. A CO_2 transcritical system uses a high-pressure expansion valve (HPEV) located at the inlet of the flash tank (receiver) to drop the refrigerant pressure below its critical point to ensure that liquid falls out of the mixture inside the tank. The liquid is

then drawn from the bottom of the flash tank to feed the evaporators.

The first example of a CO_2 subcritical system is a simple secondary refrigerant system. See Figure 10-73 for a medium-temperature application. It is identical to the secondary refrigerant glycol system described earlier in this chapter, except that the pump is circulating liquid CO_2. The primary refrigeration system incorporates an HFC refrigerant such as R404A that cools a heat exchanger/condenser, which in turn cools the CO_2 vapor from the evaporators and condenses it into a liquid. See Figure 10-74 for a simplified PH diagram for the operating range or envelope of CO_2 used as a medium-temperature secondary refrigerant. This secondary refrigerant CO_2 system can also be used in a low-temperature application by utilizing a low-temperature primary HFC system. A secondary refrigerant system using R744 is more efficient than one using glycol. Some of the CO_2 will boil off in the evaporator as it absorbs latent heat and therefore removes heat faster than circulating glycol, which remains in a liquid state.

The second type of subcritical CO_2 system is a DX (direct expansion) system using equipment that functions the same as familiar HFC refrigerant systems:

1. Compression
2. Condensing, with subcooling
3. Expansion
4. Evaporation, with superheat

FIGURE 10-72 **Pressure enthalpy (PH) diagram of R744A CO₂ medium temp.** *Courtesy of Refrigeration Training Services.*

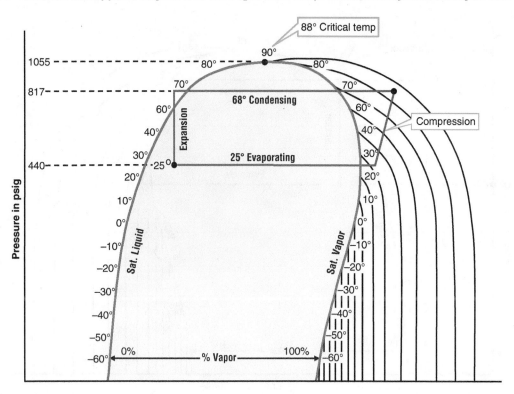

FIGURE 10-73 **CO₂ Secondary fluid.** *Courtesy of Hillphoenix.*

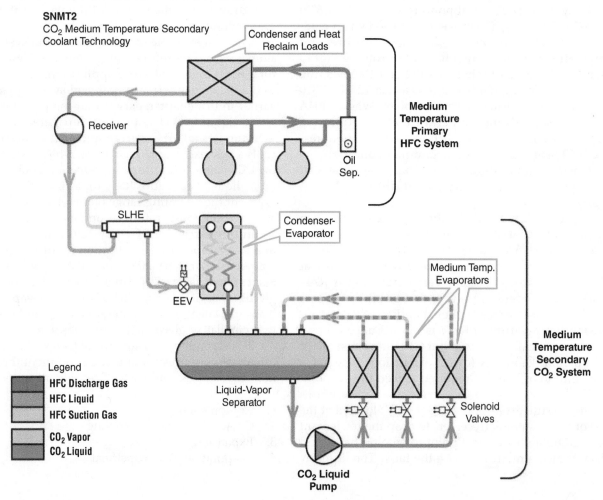

FIGURE 10-74 **Simplified PH diagram of subcritical medium-temp CO$_2$.** *Courtesy of Refrigeration Training Services.*

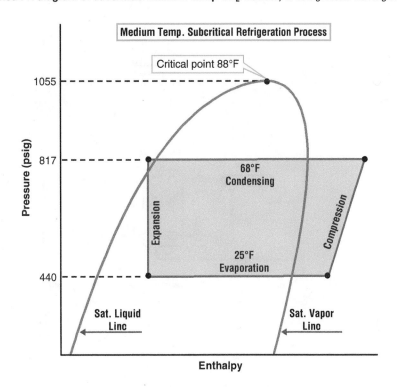

Figure 10-75 is a diagram of a CO$_2$ DX hybrid cascade system where CO$_2$ is used for a low-temperature application. The primary or upper HFC refrigeration system keeps the heat exchanger/condenser cold enough that the CO$_2$ will condense below its critical point. The R744 CO$_2$ compressor acts just like an HFC compressor to increase the refrigerant vapor pressure so that it will condense into a liquid in the heat exchanger, which functions as the condenser. The temperatures on the diagram are for illustration purposes only and should not be taken as design temperatures. Figure 10-76 is a simplified PH diagram of the low-temperature CO$_2$ portion of the cascade system. The diagram is useful for a better understanding of the operating envelope of the R744 CO$_2$ in the low-temperature system.

The term *transcritical* refers to the transitions that occur when CO$_2$ refrigeration systems shift from subcritical to supercritical operation. Many R744 transcritical systems operate above the critical point some or all of the time. This is not a problem; it just works differently. CO$_2$ systems operate subcritically when the condensing temperature is below 88°F, and the same system can operate transcritically when the gas cooler exit temperature rises above 88°F. Figure 10-77 is a diagram of a cascade system where both the primary and secondary systems use R744. The primary system using the gas cooler is the transcritical system, and the lower system is the subcritical. Figure 10-78 is a simplified PH

diagram of the transcritical process showing how the heat rejection is done in the supercritical region and the heat absorption in the medium-temperature evaporators is in the subcritical region. In Figure 10-79 is a transcritical booster system with controllers shown along with the valves they are controlling. In Figure 10-80 is a very descriptive PH diagram of a booster system.

It is important to understand that a transcritical system can operate as a subcritical system if the ambient temperatures are low enough or there is supplemental cooling for the gas cooler. For instance, if the daytime ambient is 80°F, the unit will operate as a transcritical system. However, if the evening temperatures drop to 60°F, it will operate as a subcritical system.

FIVE HAZARDS OF R744 CO$_2$

1. **Asphyxiation**
   R744 is odorless, heavier than air, and already in the air we breathe at about 390 ppm (parts per million). However, like HFC refrigerants, it displaces oxygen, which can cause asphyxiation (suffocation due to a lack of oxygen). As described in the ASHRAE (American Society of Heating, Refrigeration, and Air-Conditioning Engineers) manual, the practical limit of a refrigerant is the maximum charge of a system that can be installed in a room volume without additional safety precautions.

FIGURE 10-75   **Cascade system with CO₂ direct expansion for low-temp system.** *Courtesy of Hillphoenix.*

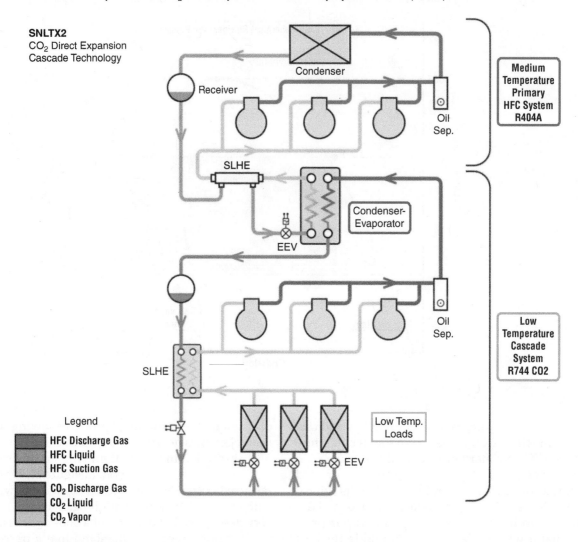

FIGURE 10-76   **Simplified PH diagram of low-temperature DX portion of cascade system.** *Courtesy of Refrigeration Training Services.*

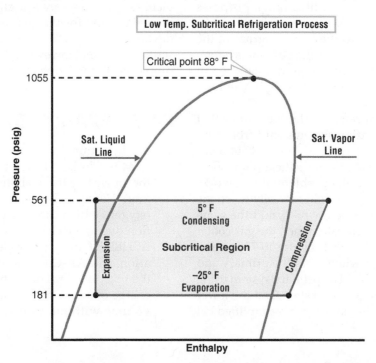

FIGURE 10-77   **Transcritical cascade low-temp refrigeration system.** *Courtesy of Hillphoenix.*

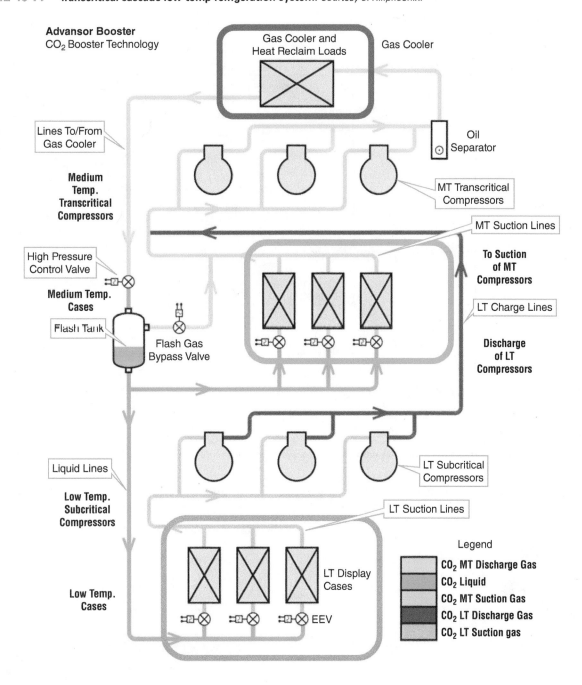

The charge and volume are in units of kg/m³. The practical limits of R404A are 120,000 ppm, but for R744, it is only 56,000 ppm. That concentration of R744 could be produced by releasing 6 pounds of CO₂ into an enclosed mechanical room 10 feet by 10 feet by 10 feet high. The TLV (threshold limit value) is 5,000 ppm and is the low-alarm setting for most CO₂ sensing systems. The TLV of a chemical substance is the level a worker can be exposed for a maximum of 8 hours without adverse health effects. A second or high alarm may be activated between 9,000 and 15,000 ppm. See Figure 10-81 for a summary of the effects of CO₂ in the air at various concentrations.

2.   **High Pressures**
R744 operates at a much higher pressure than conventional HFC systems. Therefore, system components, piping, and servicing equipment and tools must be rated at correspondingly high pressures. In Figure 10-82, the obvious indicators of an R744 compressor are the PRV (pressure relief valve) and many more bolts than you will see on an HFC

FIGURE 10-78   **Simplified PH diagram of transcritical process for cascade low-temp CO₂.** *Courtesy of Refrigeration Training Services.*

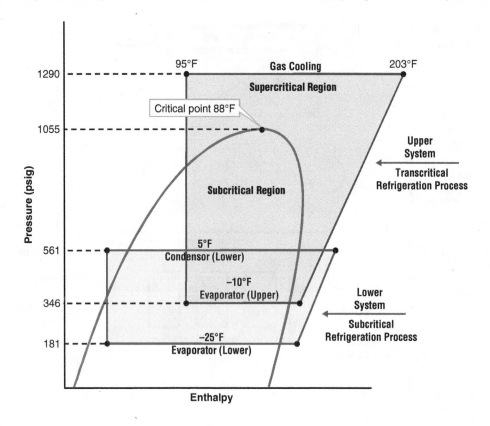

FIGURE 10-79   **Emerson transcritical booster with controllers system.** *Courtesy of Emerson.*

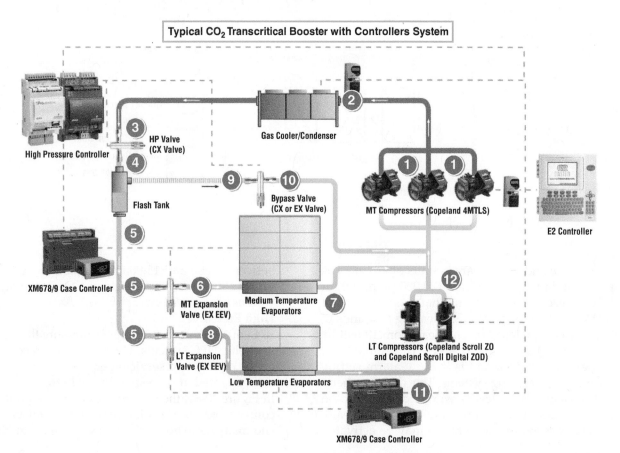

FIGURE 10-80 **PH Diagram of Emerson transcritical booster system.** *Courtesy of Emerson.*

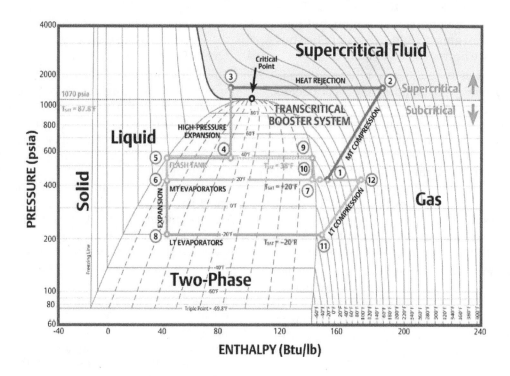

FIGURE 10-81 **Effects of R744 CO₂ at various concentrations of air.** *Courtesy of Emerson.*

CO₂ in PPM (Parts per Million)	Effects per Level of Concentration
390	None-Normal concentration in the atmosphere
5,000	TLV (Threshold Limit Value) Long-term exposure limit (8 hours)
15,000	Short-term exposure limit (10 minutes)
30,000	Discomfort, breathing difficult, headache, dizziness, etc.
100,000	Loss of consciousness, death
300,000	Immediate death

FIGURE 10-82 **Copeland transcritical CO₂ compressor and rack.** *Courtesy of Emerson and Rack photo by Andre Patenaude.*

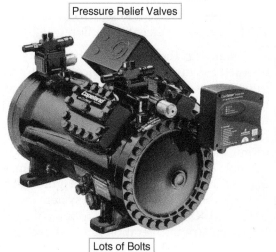

Pressure Relief Valves

Lots of Bolts

compressor. See Figure 10-83 for a listing of pressures associated with an R744 CO_2 system. Pressures are high during normal operation, but they can become damaging should the system be shut down for service or from a power failure.

The first line of defense during a power outage is an auxiliary refrigeration unit that is connected to the main power and switches over to the emergency electrical generator during a power outage. This unit has a cooling coil in the "flash tank" (receiver) of the CO_2 system receiver. By keeping the CO_2 cool, it keeps the system pressures low enough during a shutdown to prevent pressure relief. However, should the pressures get too high, there are pressure relief valves strategically placed throughout the system. It is important to note that all relief valves must be vented to the outside.

Rather than an auxiliary refrigeration unit, smaller systems may use a "fade-out vessel," which is a tank where excess pressure is discharged and captured instead of venting. However, for large systems, the space required and expense of the pressure tank can be difficult to justify.

3. **Trapped Liquid**

The coefficient for expansion for R744 is much higher than for other refrigerants. Explosive damage can occur if liquid CO_2 is trapped between closed valves. Assume a valve shuts off and traps CO_2 in the liquid line at 14°F. The pressure in that line is 638 psig. However, if the temperature of the liquid rises 36°F (from 14°F to 50°F), the pressure will go up to about 3,480 psig. As a rule of thumb, trapped liquid CO_2 will increase in pressure by approximately 80 psig for every 1°F of temperature rise. It is obvious that provisions must be made for

FIGURE 10-83 **R744 CO_2 conditions and associated pressures.** *Courtesy of Emerson.*

Condition or Component	Pressure (psig)
Standstill (system off) at 50°F	638
Standstill (system off) at 86°F	1,031
Low temperature evaporator (−25°F)	180
Medium temperature evaporator (15°F)	375
Cascade Condenser (at 25°F condensing)	440
Cascade high side high pressure cut-out (36°F)	522
Cascade high side pressure relief valve (43°F)	580
Transcritical high side operating pressure	1,305
Transcritical high side high pressure cut-out	1,566–1,827
Transcritical high side pressure relief valve	1,740–2,030

pressure relief anytime there is the potential for trapping liquid R744. This applies not only to system design but also to tools and equipment used in the servicing of these systems.

4. **Dry Ice**

Solid CO_2 or dry ice is formed when R744 pressure and temperature drop below its triple point of 61 psig and −69°F. This can happen during the processes of venting vapor through a pressure relief valve, venting vapor during service, or charging a system that is below 61 psig. Because the formation of dry ice will block lines and valves, it is important to make sure that the correct pressure relief valves are used. Also, when venting during service, make sure to vent as a liquid and monitor system pressures so they do not drop below 61 psig. Also, to prevent asphyxiation, make sure that R744 is always vented to the atmosphere outside the building.

5. **Freeze Burns**

Because the surface temperature of dry ice is −109°F, it is important to avoid touching it with bare skin. Always use appropriate gloves when working with CO_2. It should be obvious, but the wearing of safety glasses is also recommended.

ADDITIONAL NOTES ON CO_2 SYSTEMS

CO_2 used as R744 should be at least Grade 4 or "Coleman" grade CO_2 that is 99.99 percent pure and has less than 10 ppm moisture in it. Following is a ranking of both acceptable (YES) and nonacceptable (NO) grades of CO_2:

- NO—99.5 percent Industrial Grade
- NO—99.8 percent Bone Dry
- NO—99.9 percent Anaerobic
- YES—99.99 percent Grade 4, Coleman, or Instrument Grade
- YES—99.999 percent Research Grade
- YES—99.9999 percent Ultra Pure

Even with the best CO_2, filter-driers are recommended because normal service can introduce moisture to the system. Liquid CO_2 can hold up to 20 times more moisture than CO_2 vapor. After liquid changes to vapor in the expansion device, free water can be released. Not only is there a chance of regular water ice forming in the device, but water and CO_2 vapor form carbonic acid, which is very corrosive. Damage to piping and components is likely (see Figure 10-84).

Be aware that when a CO_2 system is shut down, the liquid R744 will migrate to the coldest sections. If an auxiliary refrigeration unit has been installed, it

FIGURE 10-84   **R744 Systems must be dry.** *Courtesy of Emerson.*

Emerson sight glass moisture indicator color chart

R744 CO₂ Sensitivity (PPM)				
Liquid Temperature	14° F	32° F	41° F	68° F
Very Dry	8	11	13	20
Dry/Caution	14	19	22	34
Caution/Wet	29	39	46	72
Wet	46	63	75	116

(A)

Carbonic acid corrosion on steel pipe

(B)

(C)

will come on and cool down the receiver. At the time of system shutdown, the evaporators are cold so they may fill with liquid and cause flooding when the system is started. In the event of a power outage, electronic stepper valves will remain in the position they were at that time. If open, that means the liquid will be able to enter and fill the evaporator. When the system power is restored and the compressors start, there is a very good chance of slugging. Since evaporators are not rated for high pressure, they must be protected by relief valves in case any trapped liquid becomes too warm.

Like HFC systems, CO₂ systems should also be evacuated to 500 microns. Before charging with liquid R744, it is important to vapor charge the unit to approximately 145 psig. After the pressure has equalized throughout the system, then liquid charging can begin. The reason for this is that charging liquid into a system in a vacuum or at very low pressure will quickly reach the triple point around 60 psig and turn to dry ice. Although dry ice does not expand like water ice, it will still plug up lines and controls. This does not happen when charging with vapor past the 60 psig triple point.

When the vapor pressure is up to 145 psig and liquid charging is in progress, take it slow because the pressure in the CO₂ tank (725 psig at 59°F) is greater than normal system pressure and could set off the pressure relief valves. This is especially true in low-temperature cascade systems.

If a CO₂ system component needs to be opened, it must first be pumped down to atmospheric pressure. For example, a technician has to replace the cartridge in a liquid line filter-drier. After removing the bolts and end plate, the tech discovers the filter cartridge and housing are covered with a snow-like frost. There was still some liquid CO₂ in the canister and when the plate was removed, it allowed the pressure to drop, and dry ice was formed. Although dry ice does not expand like water ice, it still has to melt or "sublimate," which is basically vaporizing without going through a liquid stage (see Figure 10-85).

Piping in transcritical CO₂ systems is much different than in HFC systems. R744 has a much higher volumetric efficiency than HFCs. What this means is that a pound of R744 can hold much more heat than a pound

FIGURE 10-85   **Dry ice formation and water condensation from failure to remove all liquid CO2 before opening.** *Courtesy of Emerson.*

Dry Ice Forms

Moisture Forms

of HFC refrigerant. This is especially true in vapor form. For example, assume an R404A refrigeration system requires a 5/16-inch liquid line and a 1 1/8-inch suction line. For the same capacity, an R744A system would require only a 1/4-inch liquid line and just a 1/2-inch suction line.

To withstand the high pressures encountered in transcritical systems, the wall thickness of the pipe must be increased. No longer is ACR or L copper sufficient. There is a K65 copper tubing available that has a MAWP (Maximum Allowable Working Pressure) of 1,740 psig. It cannot be bent or swaged, so it is joined using proprietary brazed fittings. Steel has a much higher pressure rating than copper. However, regular steel requires arc welding (stick welding), and steel is subject to corrosion. Stainless steel is being used on many systems. It resists corrosion but requires TIG (tungsten inert gas) or MIG (metal inert gas) welding (see Figure 10-86).

European countries started using CO_2 systems in supermarkets about 2005 and have thousands of successful installations in operation. In the United States, we are just beginning to take advantage of its capabilities. There is a learning curve that we must work through, and many things we use and know now will probably change. However, if you are going to work on

FIGURE 10-86 **Stainless steel piping for R744 systems.**
Photo by Jess Lukin.

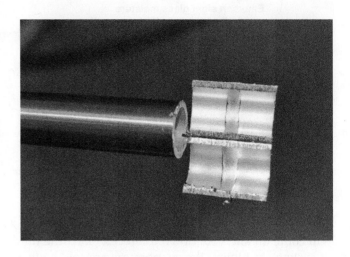

CO_2 systems, you must take advantage of all the training you can and keep up with the changes and updates by reading the papers and articles that are becoming available on both the Internet and through our HVACR (Heating, Ventilation, Air Conditioning, and Refrigeration) trade organizations.

SUMMARY

Supermarket refrigeration is designed to keep food products fresher longer. This not only makes the merchandise more appealing to store customers but also lowers product costs by reducing food spoilage and waste. In addition, the equipment must operate efficiently in order to keep utility costs to a minimum. In order to accomplish these goals, all the refrigeration principles discussed in this book are incorporated into the most efficient equipment design, installation, and service possible.

Supermarket refrigeration is challenging and specialized. This chapter cannot do justice to the training and expertise required for this type of work. However, food store equipment installation, service, or maintenance can be a very satisfying and lucrative line of work.

New technology is transforming the supermarket refrigeration industry. Although their systems have always been designed with energy efficiency in mind, the supermarket chains have become even more aggressive in their search for new and innovative ways to increase efficiency. Even more admirably, this industry has become committed to decreasing the impact of their operations on the environment through a reduced use of HFC refrigerants, copper, and other products that contribute to global warming. Hopefully, what the supermarket industry is doing today will be adopted in the very near future by other industries that use commercial refrigeration.

SERVICE SCENARIOS

Following are some actual service scenarios. To the best of our ability, we have recreated what the technicians were seeing, thinking, and feeling as they dealt with the service problems, some of which happened years ago. However, through the documentation, pictures, and drawings, it is hoped that you will gain important

insights from these scenarios about servicing supermarket refrigeration. Some of the pictures taken on the job are not as clear as we would hope; however, they still give a better representation of what is being discussed than if there were no pictures at all.

SERVICE SCENARIO #1

ICE CREAM CASE WARM

By Jess Lukin

One of the interesting things about this career is how truth is often stranger than fiction when it comes to refrigeration. It's like a daily occurrence. Just when you think you've seen it all, something new pops up.

Service Call: Ice cream case at +28°F

Observation: No airflow through the case

Diagnosis: Most likely a frozen evaporator

Cause: Not sure yet.

Of course, the dreaded frozen food case emergency always seems to hit on a Friday around 4:00 p.m. A "no airflow" freezer case normally points toward one thing: a frozen evaporator. The hard part is figuring out why it happened. The messy part is defrosting the evaporator (see Figure 10-87).

Ice cream is a very fragile product. If it is allowed to get soft, it is ruined. No case is designed to refreeze warm product and even if a walk-in were able to get it back down to temperature, the product would form crystals. Dissatisfied customers would return it to the store because the taste, consistency, and texture would be off. Therefore, as soon as this store manager realized there was a problem and before the ice cream had a chance to get soft, they unloaded the entire line-up and stored it in the freezer. With no product in the case, it was easy to remove access panels to diagnose the problem and to defrost the frozen evaporators.

FIGURE 10-87 **Example of a glass door freezer lineup.**
Photo by Jess Lukin.

There was so much ice on the evaporator coil that the fan plenum had to be de-iced before the evaporator coil could even be completely accessed. However, the drain pan heater and drain line were clear of ice. This fact is important because a stopped-up drain, bad drain line heater, or bad pan heater will result in exactly the same condition that this freezer is experiencing.

So what do we know up to this point?

1. The fan plenum had to be thawed before we could remove it to see the evaporator.
2. The drain pan heaters were working.
3. No damaged wires were visible.

What do we need to find out?

1. Was it a defrost heater problem?
2. Was it a defrost control problem?
3. Was it something else?

To check the defrost heaters and controls, the cases were put into an emergency defrost cycle. The rack controller has an emergency defrost override feature in the system programming. The override forces the circuit into defrost, bypassing all "Klixon" thermostatic discs in the case. If necessary, defrost will continue until the case temperature reaches about 55°, which is the fail-safe setting. At this point, the controller will put the case back into the freeze cycle.

In this lineup of glass door reach-in freezers there were two 5-door units and one 3-door case. The defrost termination sensor for the rack controller is placed in the "lead case," or the middle 5-door unit, in this instance. Each case has redundant backup controls in the form of Klixon temperature sensing controls for defrost termination, high-limit heater safety, and evaporator fan delay. The word *redundant* as used here means duplicate or more than necessary controls just in case the primary control fails.

After the circuit was forced into an emergency defrost, amperage was taken at the defrost heater contactors for each of the three cases. The results were as follows:

- Circuit case 1—drawing 13 amps at 208 volts (okay for 3-door freezer)
- Circuit case 2—drawing 10 amps at 208 volts (low; should be 15 amps for 5-door freezer)
- Circuit case 3—drawing 15 amps at 208 volts (okay for 5-door freezer)

Case 2 was the iced-up system and the one running at +28 degrees. See Figure 10-88 for an example of a case rating plate showing the amperage of defrost heaters.

FIGURE 10-88 **Case nameplate with circuit volts and amps.**
Photo by Jess Lukin.

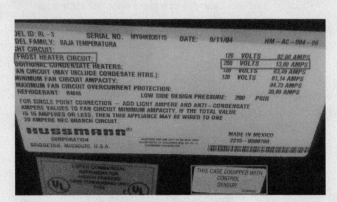

What could cause a defrost heater to draw lower than normal amperage? Usually, the problem is low voltage, but we needed to verify that and to locate exactly where the voltage dropped. The power starts at the control panel to the defrost contactor in the panel. From there, it goes to a terminal block in the case. From there, it goes through the high-limit heater safety in series with the defrost termination Klixon, which is in series with the defrost heater. By checking the voltage and amperage at the control panel, contactors we already know the voltage in and out of the contactor was correct at 208 volts. Any voltage drop had to be in the wiring going to the case, or inside the case.

As soon as we removed the access panels in the case, the problem was apparent. The wire going from the heater wire was melted at the terminal block (see Figure 10-89). This is normally caused by a loose connection, which results in the wire overheating. We attempted to tighten the connection lug, but it would not get any tighter. Further tightening would risk breaking the terminal block into pieces and having the metal screwdriver slip into the high-voltage section—not good.

FIGURE 10-89 **Loose connection of heater wire—caused wire melting.** *Photo by Jess Lukin.*

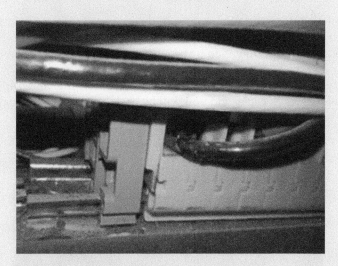

The entire terminal block would have to be ordered from the OEM (Original Equipment Manufacturer). Since the part is a special order with a lead time of 4 weeks, we had to find out what options were available to get the freezer back in service tonight.

The terminal block had some unused terminals, but we were not exactly sure how it was wired internally. By using an ohm meter, we determined that there happened to be an unused terminal on the same side as the bad terminal. From there, it was just a matter of cutting off the damaged portion of the wire, stripping it, and inserting it in the new terminal. We made sure this connection, and all others on the terminal block, were tight.

With the case covers removed, this was a good opportunity to take a look at everything in the case . . . no telling what you may find that could prevent a future service call.

Found Another Problem: When looking over the case after the initial repair, a possible problem was discovered. The defrost termination had burn marks on both wires, which may prevent it from ending defrost (see Figure 10-90). An ohm meter was needed to check the control. The contacts should be closed above 55°F (usually), which signals the controller that all frost is off the evaporator and ready to go back into freeze. The control must also be checked to make sure the contacts are open below 55°F. The actual temperature setting may be stamped on the control (see Figure 10-91).

FIGURE 10-90 **Burn marks on the wires to the defrost termination.** *Photo by Jess Lukin.*

FIGURE 10-91 **This freezer has the defrost termination located on the coil tubing.** *Photo by Jess Lukin.*

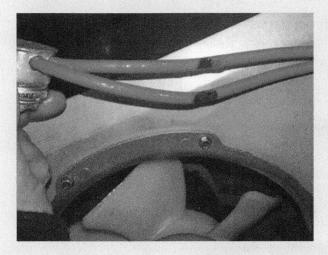

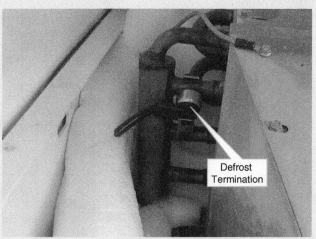

Defrost Termination

The decision was made to order a replacement control. Everything done to this case was noted in detail on the service ticket in case someone else was scheduled to return and replace the control.

Epilogue: Putting the case in defrost did not remove all the ice, especially in the middle of the coil. Remember, defrost heaters are designed only to melt frost, not ice. That is why they are called de-FROST heaters. If the frost on a freezer evaporator is allowed to turn to ice, then the unit is in trouble (see Figure 10-92).

So before putting the unit back into operation, we had to use a water hose to melt every bit of ice in the evaporator. This is usually the long, messy part of the service call, but it is imperative to remove every bit of ice. If not, it will just grow with each cycle, and you will be back soon to defrost it all over again.

FIGURE 10-92 **Drain pan and evaporator ice needs to be melted before putting case into operation.** *Photo by Jess Lukin.*

This service call took a very long time, almost 4 hours. However, it is about average for this type of problem. Taking shortcuts such as not defrosting the coil thoroughly or not checking other wiring often results in a "recall" or "call-back." Remember, take the time to find out why the unit had a problem, then take the time to repair it properly and safely.

SERVICE SCENARIO #2

FIVE-DOOR FREEZER WARM

By Jess Lukin

Service Call: Five-door freezer glass door reach-in running warm at +15°F; customer needs service tonight (see Figure 10-93).

FIGURE 10-93 **Hussmann RL-5.** *Photo by Dick Wirz.*

It is important to follow a procedure or routine when diagnosing problems, especially on emergency calls. Stay calm and think logically. Start with the obvious problem, then do a step-by-step analysis to work backward to the cause.

Initial Findings:

- Frozen food glass door case 1 was +18°F
- No airflow through the case discharge air honeycombs
- Problem case was one of two on the same circuit; second case was OK

Mental Process: Since one case was working properly, it's unlikely there is a major refrigeration problem at the compressor rack. So the problem is most likely in the nonworking case. No airflow in the problem case indicates a blockage of air. Probably a frozen evaporator. If so, what prevented it from automatically defrosting?

To find out, the product had to be removed to access the evaporator area. Once the panels were removed, the evaporators, fans, and controls were checked.

Case Inspection:

- Fans were running
- Front of the evaporator was clear of ice
- Drain and drain pan were clear of ice or blockage
- But the rear of the evaporator was frozen

Conclusion: Ice buildup indicates the heaters were not defrosting. The problem is either the defrost heater or something controlling the defrost heater.

Facts Known by Experience: Most glass door frozen food cases have two defrost heaters, one in the front of the evaporator and one in the back. There are three controls inside the case:

- Heater safety limit—shuts off the heater if the case gets too warm
- Defrost termination—ends defrost and puts it back into freeze
- Evaporator fan delay—fan stays off until the coil drops to 25°F

Rack controllers have a helpful feature called emergency defrost. It performs the following operations:

- Liquid line solenoid de-energized—stops refrigerant flow to cases
- Evaporator fan motors turned off
- Defrost heater contactors energized—heaters come on
- Drain pan heater relay energized—drain pan heaters come on
- Drip time setting—defrost off time verified before start of freeze cycle
- Fan delay On setting verified—temperature fan starts after defrost cycle
- Defrost termination verified—coil temperature reached at end of defrost

The case nameplate gave the defrost heater amps at a specific voltage. In this instance, it stated 14 amps at 208 volts.

The refrigeration circuit pumped down, the evaporator fans shut off, and the defrost heaters came on. Amperage was taken at the circuit's defrost heater contactor inside the electrical panel at the rack. The heaters were drawing only 4.4 amps, not 14 amps. The voltage coming into the contactor at L1 and L2 was only 65 volts, not 208 volts. But the voltage leaving the circuit breaker only a few inches away was 208 volts. Could there be a problem in a wire that was less than a foot long? (See Figure 10-94.)

Then it was noticed that there was a dark area on the electrical solderless connector going into the contactor (see Figure 10-95). After tightening the screw lug about an eighth of a turn, the contactor had 208 volts to it, and the heaters were drawing 14 amps. Notice the arrow pointing to the dark, discolored yellow connector in the figure.

FIGURE 10-94 **Voltage is 208 leaving the circuit breaker, but only 65 at the contactor.** *Photo by Jess Lukin.*

FIGURE 10-95 **Arrow points to discolored insulation caused by loose connection.** *Photo by Jess Lukin.*

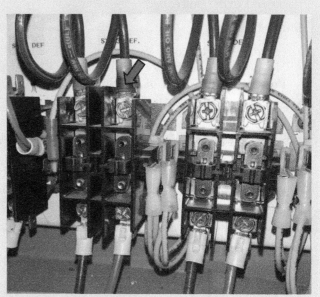

Repairs: There was enough slack in the wire to cut out the existing connector and install a new one. After connecting the wires back up to the contactor, the voltage and amperage were double-checked. In addition, all connections in the control panel were checked to make sure they were tight.

Conclusion: The problem was simply a loose wire connection. However, it caused a major problem with the frozen food case. An inexperienced technician may have condemned the heater, then found the real cause after performing a long and expensive repair. However, experienced techs have learned (often the hard way) to always check voltage before condemning an electrical device.

SERVICE SCENARIO #3

OIL FAILURE

By Jess Lukin

Service Call: Rack 3 down. Losing product. Need immediate help.

On this rack system, the alarms are set off anytime there is a compressor problem or a sensor showing a failure that could result in product loss. The remote EMS (Energy Management System) controller monitoring system alerts the store manager and then contacts the refrigeration contractor. See Figure 10-96 for a picture of the rack.

Verifying What Set Off Alarm: The "lead" or "primary" compressor is off on oil failure.

Before resetting an oil failure control, make sure there is no liquid in the crankcase. Liquid flooding or washout can cause an oil failure control (OFC) to trip. If we just reset the oil failure and there is liquid in the crankcase, the liquid will slug the compressor. Many technicians have made this mistake. But after hearing the horrible sound of a compressor slugging, they don't make the same mistake twice.

FIGURE 10-96 **Rack 3.** *Photo by Jess Lukin.*

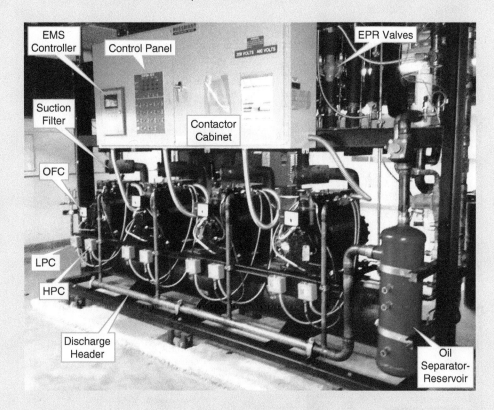

To prevent this, start with isolating the crankcase by attaching a low side gauge to the suction service valve then front-seating it. The front-seated suction service valve will prevent refrigerant from the suction header from flooding the crankcase when the compressor is started. Then do one or more of the following:

1. Hook up the recovery machine and run it until crankcase pressure is down to about 30 psig (or just above the low-pressure control cut-out setting) to be sure no liquid is remaining in the crankcase. Reset the OFC, start it up, and slowly open the suction service valve.
2. Or turn off the control switch or circuit breaker to the compressor. Reset the OFC. Then "bump" the compressor. Or turn on and quickly turn off the compressor a couple times to allow any liquid to vaporize.

We chose option 2. First, we checked the rack's oil separator and the other compressors' sight glasses to confirm adequate oil in the other part of the system. Then we checked the oil level control attached to the crankcase of the primary compressor. It appeared to have no oil in the crankcase. Closing the main oil supply line to the compressor and adding just enough oil with a refrigeration oil pump, we made sure the compressor wouldn't be damaged from lack of oil. Once started, we let it pull down and shut off on the low-pressure control (LPC) to check the control setting. Next, we jumped out the LPC and let the unit run into a vacuum then shut it down to check the valves. This procedure was quick and easy and provided one more factor to help us diagnose this compressor's problem. Satisfied that the compressor reed valves were holding, we restarted the compressor and opened the suction service valve. The unit ran for about 2 minutes and cut out on the OFC.

Diagnosis: If there is oil in the separator and the other three compressors, then the problem is either the oil pump or in the oil circuit in the compressor. Very seldom is there a problem with oil pumps or mechanical oil level controls. But there are two strainers in the compressor's oil circuit. We inspected them first.

In order to do this, we have to first isolate the compressor by doing the following:

1. Shut off power to the compressor.
2. Turn off the oil supply line valve that feeds oil to that compressor's oil level control (Figure 10-97).
3. Valve off the suction line by front-seating the suction service valve.
4. Valve off the discharge line by front-seating the discharge service valve.

See Figure 10-97 for front and rear views of the compressor. Call-outs are used to point out some of the more notable objects in the pictures.

FIGURE 10-97   **Front and rear views of the compressor.** *Photo by Jess Lukin.*

Before doing any repairs, we need to be sure we have the following:

- Oil sump gasket
- Electrical contact cleaner
- Plastic container to catch drained oil
- Oil pump and 1 gallon of oil that rack uses
- Razor-sharp scraper to remove gasket and sand cloth
- An access valve to check crankcase pressure

See Figure 10-98 for preparations necessary for dealing with working on a rack's oil system.

FIGURE 10-98   **Rack 3 preparing to add oil.** *Photo by Jess Lukin.*

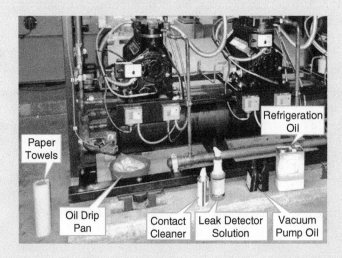

Step 1 – Verify that both main power and control voltage is off to the compressor. If the compressor should unexpectedly come on, it can injure the technician and/or the compressor.

Figure 10-99 shows the main circuit breakers for the compressors and also the control toggle switches.

Figure 10-100 shows how the suction service valve is front-seated on this particular suction cooled compressor.

FIGURE 10-99 **Control circuit toggle switch and main circuit breaker.** *Photo by Jess Lukin.*

FIGURE 10-100 **Suction service valve front seated.** *Photo by Jess Lukin.*

Step 2 – Use the recovery machine to pull the compressor down to 0 psig.

Figure 10-101 shows how the recovery machine pulls refrigerant from the compressor head and crankcase, as well as how the pump discharges the recovered refrigerant into the suction line.

FIGURE 10-101 **Recover refrigerant inside compressor.** *Photo by Jess Lukin.*

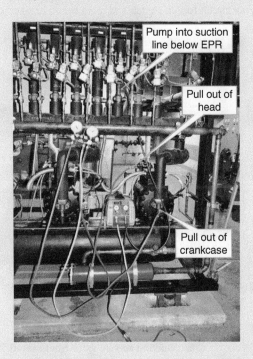

Step 3 – Remove the oil sump strainer plate slowly to prevent making a mess. Use an oil pan to catch the oil.

NOTE: **Loosen the bolts just enough to let the oil drain out. After it is finished draining, remove the plate carefully because it has a spring inside. Figure 10-102 shows removal of the sump plate.**

FIGURE 10-102 **Removing oil sump plate.** *Photo by Jess Lukin.*

Step 4 – Remove all parts from inside.

Figure 10-103 shows how dirty the oil strainer and feed tube were.

FIGURE 10-103   **Dirty oil strainer.** *Photo by Jess Lukin.*

Step 5 – Clean all the parts. The electrical contact cleaner works well and evaporates without leaving any residue. Figure 10-104 shows how well the contract cleaner works on cleaning up the strainer and tube.

FIGURE 10-104   **The cleaner does a very good job on the strainer and tube.** *Photo by Jess Lukin.*

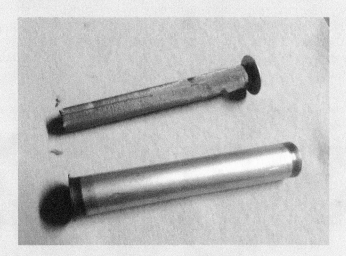

Step 6 – Remove the old gasket from the cover plate surfaces.

Figure 10-105 shows that most of the gasket is removed, but not all of it. Figure 10-106 shows how the surface should look before putting on a new gasket. Prepare the new oil sump gasket by coating it with oil before installing (see Figure 10-107).

FIGURE 10-105   **Remove old gasket. Keep going—it is not clean enough.** *Photo by Jess Lukin.*

FIGURE 10-106   **Okay—it's good to go.** *Photo by Jess Lukin.*

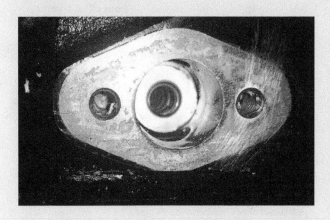
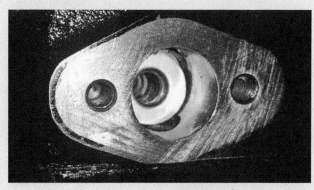

FIGURE 10-107   **New oil sump gasket coated with oil.**
Photo by Jess Lukin.

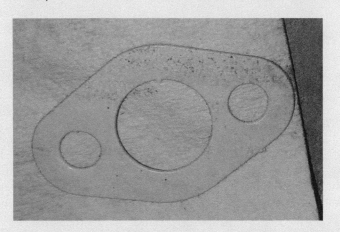

Reassemble the oil strainer parts and insert the strainer into the compressor. There is a slot inside the compressor and a ridge on the oil feed tube to position the tube properly. Figure 10-108 shows insertion of the oil tube and strainer. Don't forget the spring. The spring actually holds the oil feed tube back in place (see Figure 10-109).

Remove and clean the strainer inside the oil level control. Figure 10-110 shows how the strainer looks when it is in place. The strainer looks similar to ones on TEV inlets, but oil strainers have a slightly larger head and more strainer surface. Figure 10-111 is a picture of the strainer removed.

FIGURE 10-108   **Insert cleaned oil tube and strainer.**
Photo by Jess Lukin.

FIGURE 10-109   **Install the spring.** *Photo by Jess Lukin.*

FIGURE 10-110   **Accessing the inlet oil strainer in the oil level control.** *Photo by Jess Lukin.*

FIGURE 10-111   **Inlet oil strainer from oil level control.**
Photo by Jess Lukin.

Step 7 – Add an access fitting to check crankcase pressure.

This is crucial for getting the correct net oil pressure reading and checking the compressor for blow-by. Figure 10-112 shows one of the types and locations of an access valve for checking crankcase pressure.

FIGURE 10-112   **Crankcase pressure access fitting.**
Photo by Jess Lukin.

Step 8 – Disconnect the recovery machine. Hook up your vacuum pump and pull down to at least 500 microns. Figure 10-113 shows how this vacuum pump was hooked up to this compressor.

NOTE: **Always change the vacuum pump oil just before you start evacuating (see Figure 10-114).**

FIGURE 10-113   **Evacuate the compressor.** *Photo by Jess Lukin.*

FIGURE 10-114   **Drain and refill pump with fresh oil.** *Photo by Jess Lukin.*

After evacuation, open the valves, turn on the power, start it up, and check the following:

1. Suction pressure and suction line temperature
2. Discharge pressure and discharge temperature
3. Oil pump pressure at the oil pump
4. Crankcase pressure
5. Compressor body temperatures in the oil sump area, cylinder head temperature, and compressor bell region for excessive cold or hot spots
6. Compressor valves by pumping the compressor down

To determine the net oil pressure reading, use the following formula:

$$\text{oil pump pressure} - \text{crankcase pressure} = \text{net oil pressure}$$

Most technicians take the suction pressure at the suction service valve and subtract that from oil pump pressure. The proper way is to determine the true crankcase pressure in order to calculate the net oil pressure. That is the primary reason for installing the access fitting. But having access to the crankcase pressure will also allow us to check for blow-by, which will be discussed later.

The OFC will start timing out whenever the net oil pressure drops below 10 psig. So we'll check oil pressure first.

Following are the readings taken:

- Oil pump discharge pressure: 98 psig
- Crankcase pressure: 45 psig

$$\text{oil pump pressure} - \text{crankcase pressure} = \text{net oil pressure}$$
$$98 \text{ psig} - 45 \text{ psig} = 53 \text{ psig pump pressure}$$

A properly operating oil pump has between 10 and 65 psig net oil pressure and typically operates between 30 and 55 psig. Most oil pumps have a bleed valve, which prevents net oil pressure from exceeding about 65 psig. Our oil pump is good, and our cleaning of the strainers has the unit operating again, so it should not go off on oil failure.

However, we must also check compressor efficiency and blow-by. We already did a pump-down test, and it did hold a vacuum. That test is fairly accurate but is limited to determining if the valves and the valve plate gaskets are good. It is not a good indicator of piston rings or the condition of the cylinder walls.

As a rule, blow-by is evident if there is more than 3 psig difference between the crankcase pressure and the suction pressure. Blow-by refers to what happens when high discharge pressure during the piston compression stroke is pushed down through the piston rings into the crankcase, "blowing by" the piston rings.

Blow-by is not only a sign of an inefficient compressor, it also causes oil return problems the compressor with blow-by as well as other compressors on that rack. So we check by calculating the following:

$$\text{crankcase pressure} - \text{suction pressure} = \text{blow-by}$$
$$45 \text{ psig} - 35 \text{ psig} = 10 \text{ psig}$$

This unit is far above the maximum of 3 psig. So we have an inefficient compressor with an internal problem of—most likely—damaged cylinder walls or leaking worn rings. The compressor must be replaced.

Is that it? Do we just quote the customer the cost of replacing the compressor and go from there?

We wish it were that simple, but it never is. A bad compressor is usually the result of problems that we have yet to determine and remedy. If we replace the compressor but do not solve the real issues, then we will be back again and again until we finally do. Or the customer gets another company that will. Call-backs are never smart business, nor are they good business.

Epilogue: The compressor failure was only a symptom of many other problems with this rack system. Following is a list of facts as well as assumptions:

1. The blow-by in the compressor may be the result of damaged piston rings due to lack of oil or excessive high heat in the head from high compression ratios or high head pressure.

2. Another possible cause for the compressor failure may be slugging by oil. This unit had a Y-valve (Sporlan Y1236 Oil Pressure Differential Valve), which was fluttering. Instead of maintaining an oil pressure only about 20 psig above the suction header, the valve was allowing high discharge pressure to overpressurize the oil return system. This put too much oil into the compressor crankcase and could have slugged the compressor on start-up. Figure 10-115 is a diagram of a high pressure oil return system with a Y-valve.

FIGURE 10-115 **High-pressure oil system.** *Compliments of Parker Hannifin-Sporlan Division.*

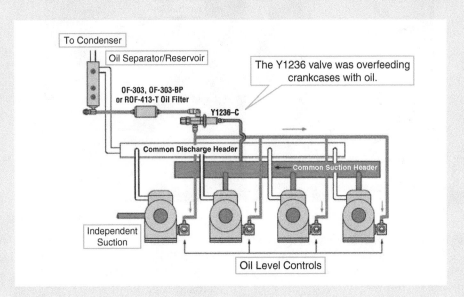

3. Why was the oil system so dirty? The compressors were experiencing superheats of 35°F and higher. This caused high discharge temperatures, which overheated the oil and caused carbon deposits and poor lubrication. The compressor superheat should have been closer to 20°F. Both systems needed to have oil filters and oil replaced.
4. Why were the superheats so high? Because the system was so dirty the TXV (thermostatic expansion valve) inlet screens were probably clogged, causing poor feeding of refrigerant.
5. This rack doesn't have a good load on it. In an attempt to increase energy efficiency, the store added glass doors to all the open-front multideck medium-temperature cases. Therefore, the systems were much bigger than they needed to be and as a result, the compressors were short-cycling. Short-cycling not only means more compressor wear, but the refrigerant velocities are not maintained, which means poor oil return and possible oil-logging in the evaporators. Low load also encourages refrigerant floodback. In addition, the medium-temperature cases have off-cycle or "air defrost." Normally, the recirculating air during defrost is about 35°F, but with the doors on, it remained about 30°F, which was not doing much defrosting.
6. The oil separator is "floating two balls" with some oil in the top sight glass. There should never be oil visible in the top sight glass. This indicates a problem with oil return because the separator is almost full. Figure 10-116 shows two oil separators. One is normal and the other overfilled.

FIGURE 10-116 **Oil separator overfilled (left) compared to normal fill (right).**
Photos by Jess Lukin.

7. The EMS (energy management system) programming needed to be verified on set points and logic, just in case. The condenser fan program was set up for R22 rather than R404, and the fans were not staged properly for head pressure control. The fans closest to the discharge/liquid header should run all the time, but they were shutting off. This caused high head pressure during hot weather, which could have contributed to high discharge temperatures in the compressor heads.

8. This system had two major liquid line leaks within the past four months, each losing about 250 pounds. When that much refrigerant is added, it can kick up some trash in the systems that may block TXVs and other valves in both the refrigerant and oil systems.

Remember, replacing a failed component like a compressor does not necessarily solve the problem. Each of the eight items listed above could have contributed to the compressor failure. One of these items may have been, as the saying goes, "the straw that broke the camel's back." However, lots of "straws" can make up a very heavy bale of hay that causes the breakdown.

SERVICE SCENARIO #4

RACK COMPRESSOR NOISY

By Jess Lukin

Service Call #1: Compressor running but noisy and vibrating badly

Upon entering the mechanical room of the supermarket, it was obvious which one of the compressors was the problem. The rocking of compressor 8 on the medium-temperature rack B was so bad it could have cracked a refrigeration line.

Rocking like that on a 4-cylinder compressor usually means one piston is broken. The compressor was shut down and inspected for internal damage. The easiest method of doing this is to remove the cylinder heads and have a good look. It was not necessary to worry about having a new head gasket because this compressor was not going to be put back into service that day.

⚠ **C A U T I O N**

Safety note: Make sure the discharge valve and suction service valve are closed (front-seated) and that all pressure is released from the crankcase and head. Otherwise, when the head bolts are removed, the pressure inside the compressor could cause the head to come flying off and injure the technician servicing the unit.

With the head removed, it is highly recommended to make sure both the low and high side service valves close tight and hold back the rack's system pressures. If the compressor is shut down overnight with the head removed while waiting for parts, a leaking service valve could mean loss of enough refrigerant to cause the whole rack to go down. Of course, this will happen in the middle of the night, which adds aggravation to the already expensive and embarrassing mistake that happens more often than it should.

The right side head was removed first (see Figure 10-117). Refrigerant oil poured out of the cylinder head as it was removed. This is a very bad sign of oil slugging (see Figure 10-118). When the left head was removed, it was evident that it had "swallowed" a piston. There was no piston in the cylinder; it was apparently laying in pieces down in the crankcase. In Figure 10-119, you can you can see what remains of the lighter-color connecting rod still on the crankshaft.

FIGURE 10-117   **Compressor head with bolts removed prior to inspecting valves and pistons.** *Photo by Jess Lukin.*

FIGURE 10-118   **Oil in the cylinder. Liquid slugging comes from either oil or refrigerant.** *Photo by Jess Lukin.*

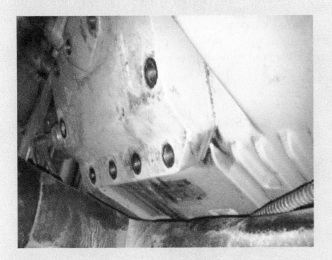

FIGURE 10-119   **Broken connecting rod and crankshaft, piston somewhere in crankcase.** *Photo by Jess Lukin.*

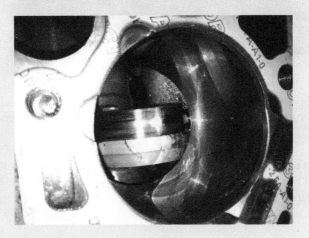

Compressors are designed to compress vapor. When liquid—either refrigerant or oil—enters the cylinder, the piston tries to compress it but can't. The crankshaft keeps turning to the highest point of its rotation, pushing up on the connecting rod. But if the piston cannot travel all the way to the top of the cylinder, something has to give way. Sometimes, it is the head gasket or reed valves. Other times, it is the piston connecting rod. When the motor is rotating at about 1,750 rpm, the damage happens quickly.

Problem: Bad compressor (broken piston)

Recommendation: Replace compressor

Diagnosis: Slugging of compressor.

Liquid slugging damaged the compressor 8 on the medium-temperature rack B. How and why liquid caused the problem will need to be determined, or else the new compressor will suffer the same fate. These compressors were only a couple years old, so old age was not a factor in this situation—there is a saying in the industry: "Compressors don't just die, we kill them!"

Luckily, the Bitzer compressor was still under warranty. Bitzer had excellent customer service and got us a new compressor in just 2 days.

This was the third compressor replaced on this job within a year and a half of the major remodel of the rack systems. Therefore, Bitzer sent us copies of the tear-down reports for the two previous failures. The causes were excessive bearing wash caused by flooding. The current compressor failure had the same type of damage.

After the new compressor was installed, the floodback returned (see Figure 10-120). Although several things can cause this condition, just one look at the case evaporator coils was enough to alert us that cleaning them was the first thing to do. The customer was informed that the evaporators needed attention before we could start checking other causes of flooding. Unless the flooding condition was corrected, the new compressor would soon fail, as would the other compressors on the rack.

FIGURE 10-120 **New compressor has floodback.** *Photo by Jess Lukin.*

The cost of replacing the compressor was approved as an emergency repair. However, the cleaning and further diagnostics was not. Therefore, we had to submit a quote to the corporate office and wait for approval before continuing. This is not unusual, but it can be very frustrating for both the manager and the service company.

After about a month, the rest of the work was finally approved. However, this did not happen before they lost all their product in the meat cases. It is baffling for technicians like us to see our recommendations not taken as a priority until the situation becomes an emergency. A more proactive approach by store management would have prevented a great deal of product loss. However, they are the customer, and we are paid to do what they want us to do in the way they want us to do it.

Service Call #2: Meat case lineup is warm

Initial Inspection: Case and product temperature was 65°F

Diagnosis: No airflow in the case, coils blocked with dirt

Recommendation: Clean the evaporators.

NOTE: The product loss probably would not have occurred if alarms from the CPC (computer process controls) controller had alerted the store manager. However, the previous refrigeration contractor had inserted the temperature sensors in each evaporator's fins rather than in the discharge airstream after the evaporator or the return air entering the evaporator. The CPC was registering the meat case coil temperature of 26°F rather than the case air temperature, which was running at 65°F due to the dirty evaporator coils. Although this prevented the high case temperature alarms from going off, it did not go unnoticed. This customer was very conscientious about checking case temperatures every 2 hours with an infrared temperature gun. Therefore, he called us for service rather than wait for the system alarm.

We were finally given approval to clean all evaporators. Dirt was packed into the evaporators of the meat case. It is easy to see why there was no airflow in those cases (see Figure 10-121).

The produce walk-in evaporator coils were super dirty, and cleaning them was really a chore as we worked on a scissor lift 20 feet up in the air (See Figure 10-122).

An old walk-in freezer had an ice sheet built up on it, which contributed to the rack floodback issue (see Figure 10-123). The Koolgas Defrost system was defrosting the evaporator interior properly, but the pan heater located at the bottom of the evaporator drip tray had failed. Therefore, all the condensate defrost water had no place to drain (see Figure 10-124). So icicles formed when water overflowed the drain pan, and stalagmites grew from the drain pan up the evaporator coil. This produced a solid sheet of ice, which prevented airflow through the coil.

NOTE: Hussmann's Koolgas Defrost is similar to a hot gas defrost. But instead of using hot discharge vapor before it reaches the condenser, it uses cooler vapor off the top of the receiver. This vapor may not be as hot, but it has a higher heat content per pound of vapor than does discharge vapor.

After performing a major cleaning of all the evaporators, we thought our floodback problems were solved and the equipment would return to proper operation. But instead, when the systems were turned

FIGURE 10-121 **Meat case evaporator plugged with dirt.** *Photo by Jess Lukin.*

FIGURE 10-122 **Produce walk-in cooler dirty evaporator.** *Photo by Jess Lukin.*

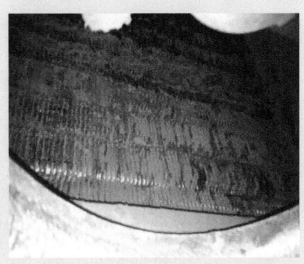

FIGURE 10-123   Walk-in freezer ice in front of coil.
Photo by Jess Lukin.

FIGURE 10-124   Walk-in freezer drain pan frozen.
Photo by Jess Lukin.

on, some were running warm, and others were running too cold. What a nightmare! The customer was upset and insisted we stay on an emergency basis to find and correct the problems.

Service Call #3: All systems erratic and not maintaining proper temperatures

Initial Observations: Units certainly acting very strange

Diagnosis: Not sure why

Recommendations: Go through each system and find out what is going on; call for backup—this is going to be a big job

History of the Refrigeration Equipment: All systems remodeled about two years earlier; following is some information about the original systems:

Hussmann Rack Model CS13111VDM

- Hussmann rack with Copeland scroll compressors installed about 1988.
  - **Significance:** The early scroll compressors had oil return problems. As a result, there were many compressor replacements.
- Hussmann's Koolgas Defrost and Heat Reclaim design was the best at that time and was one of the main reasons for using a Hussmann system.
  - **Significance:** The Koolgas Defrost system was retained in the remodel.
- R-22 with one low-temp suction group and one medium-temp suction group.
  - **Significance:** R22 was the standard in the 1980s. It is used in new systems even though the price of R22 is rising as it is becoming scarce through scheduled EPA phaseout.
- Each receiver holds about 700 pounds of refrigerant at 80 percent fill.
  - **Significance:** There is a lot of refrigerant in the systems, and leaks can become very expensive. In fact, the billing for refrigerant and labor to repair the leaks for this series of service was over $30,000.
- Subcooler with a 50°F feed line.
  - **Significance:** Mechanical subcooling of refrigerant can make systems much more efficient by lowering the temperature of liquid to the TEV (thermostatic expansion valve). However, incorrect settings can cause problems.

In 2012, a remodel was done to retrofit the store's existing Hussmann rack. The original Copeland scrolls, which seemed to fail on a regular basis, were replaced with Bitzer semihermetic reciprocating compressors. The cost of the scroll failures was the justification for spending the money to update the compressor racks. At this time, all SORIT valves (EPRs, or evaporator pressure regulators) were also replaced in conjunction with a remodel of the cooling tower. The cooling tower had repeated refrigerant leaks in the past and was repaired by cutting out those loops. This resulted in decreasing its condensing capacity and inability to maintain condensing temperature in hot weather.

The remodel in 2012 consisted of the following:

* Bitzer Octagon compressors rated at 20 hp each on medium-temperature applications and 12 hp each on low-temp application.
  * **Significance:** The early scrolls had problems. The Bitzers had a long track record of being reliable semihermetic compressors.
* Both medium- and low-temp compressors have a DTC (discharge temperature control) valve for liquid injection.
  * **Significance:** Low-temp and medium-temp systems with R22 have high compression ratios and high discharge temperatures. Injecting refrigerant into the suction line or crankcase will raise the suction pressure. In turn, this lowers the temperature of discharge vapor in the head and prevents heat damage.
* Retention of the existing low-pressure oil return system.
  * **Significance:** The original oil return system was a high-pressure system. The previous service company thought that was part of the problem, so they repiped it and changed to a low-pressure oil return system. The job was not well done.
* Mineral oil removed and Alky 200TD (alkylbenzene) added.
  * **Significance:** Mineral oil tends to slow down and collect in low-temperature evaporators. Alkylbenzene oil flows a little better. This shows the contractor was trying anything he could think of to improve oil return but was missing the real reasons.
* New SORIT valves (large EPRs or evaporator pressure regulating valves) were installed for existing Koolgas Defrost freezer defrosts.
  * **Significance:** They did not need replacing because only the inlet strainers were clogged. Replacing the whole valve only contributed to more debris in the system.

Continue Service Call #3: Erratic case operation meant some evaporators were starved of refrigerant while others were flooded. Figure 10-125 shows excessive liquid floodback occurring at the racks. Liquid refrigerant is returning to the compressors because it is not boiling off in the evaporator. Cleaning the evaporators removed the most obvious cause. The next most common causes are:

* Flooding and starving evaporators—incorrect TXV settings and/or EPR settings
* Defrost problems—incorrect case setting for defrost cycles

But wait—there seemed to be more going on. Here were our observations of rack B:

* Foaming in oil sight glass of compressors
  * **Significance:** Liquid refrigerant floodback washing oil from crankcase
* Starving and flooding evaporators
  * **Significance:** Indicate TEV problems or problems with refrigerant to the TEVs
* Subcooling set points on CPC (control panel) are incorrect
  * **Significance:** Too-low settings can cause TEV starving
* Some circuits show signs of liquid floodback at SORIT (EPR) on the rack
  * **Significance:** Valves not maintaining evaporator pressure
* Refrigerant receiver level gauge dropping to 0 percent
  * **Significance:** No liquid to the expansion valves
* Multiple circuits in hot gas defrost overlapping
  * **Significance:** Could cause the receiver to run out of refrigerant
* Johnson Control P545 oil safety control is short-cycling compressors
  * **Significance:** Could damage compressor and affect oil return

- Low-temp suction group refrigerant leaks
  - **Significance:** Loss of refrigerant
- Liquid line Y-valve flare nuts leaking oil
  - **Significance:** Contractor failed to put in the proper gaskets

FIGURE 10-125 **Floodback still occurring after cleaning evaporators.** *Photo by Jess Lukin.*

All the problems will eventually be addressed, but we needed to determine where to begin. The manager stated that the retrofit contractor was an HVAC company that had little experience with supermarket refrigeration. Therefore, we decided to start by verifying that all major components and controls were doing what they were supposed to do. We began at the rack system and worked our way down the refrigeration circuits to the cases.

We suspected the oil system was dirty, and we were so right. Not only were the oil filters nearly plugged up, but the previous contractor used refrigerant filters rather than oil filters (see Figure 10-126).

FIGURE 10-126 **Refrigerant filter used in place of oil filter.** *Photo by Jess Lukin.*

Comparing Refrigerant and Oil Filters: There is a big difference between the two types of filters. Refrigerant liquid line filter-driers do not catch small enough particles to be effective for filtering oil. Normally, refrigerant circulates particles in the 30- to 50-micron range; therefore, the minimum size of 20-micron filtering is more than adequate for refrigerant filters. However, POE oil suspends and circulates a high concentration of 2- to 20-micron-sized particles, and the larger particles are capable of excessive bearing wear. For that reason, Sporlan designed their oil filters to be 99 percent efficient when filtering out particles down to 3 microns.

In addition, the oil separators on this job had copper particles in one and sludge in the other (see Figures 10-127 and 10-128). This type of trash in a system will clog oil valves, floats, and strainers, which adversely affects the whole oil return system. Moisture indicators in the sight glasses showed moisture in the refrigeration system. However, an oil analysis verified that we did not have any acid. That was good news because acid will eat copper from motor windings and piping and cause all kinds of troubles. Changing the refrigerant filter-driers to high moisture capacity driers will take care of the moisture.

Now that the refrigeration and oil systems were clean, it was time to look at something that could possibly contribute to flooding. Each compressor, both low- and medium-temperature, had a DTC (discharge temperature control) valve. Actually, it was a Sporlan Y1037 TREV (temperature responsive expansion valve). The function of this valve is to inject saturated refrigerant into the suction line or crankcase to lower high discharge temperatures in the cylinder head due to high compression ratios (see Figure 10-129). The Bitzer compressors were being injected into the low-side crankcase through a fitting in the compressor body.

FIGURE 10-127 **Trash in oil separator.** *Photo by Jess Lukin.*

FIGURE 10-128 **Sludge in oil separator.** *Photo by Jess Lukin.*

FIGURE 10-129 **Sporlan Y1037 valve used to lower discharge temperature.** *Compliments of Parker Hannifin-Sporlan Division.*

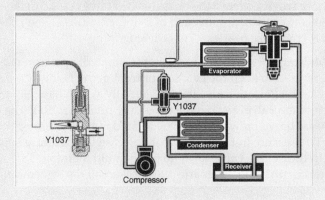

Discharge Temperature Controls: Medium- and low-temperature compressors using R22 refrigerant are prone to high discharge temperatures in the cylinder head. This leads to carbon deposits, burned discharge valves, and oil becoming so hot in the cylinder that it cannot lubricate. Without lubrication, the aluminum pistons wear down against the steel cylinder walls, and discharge vapor blows by into the crankcase.

There is a direct correlation between the compression ratio and discharge temperature. The greater the difference in pressure between the suction vapor entering the cylinder and the high pressure leaving, the more heat of compression is generated in the cylinder.

Compression ratio is the head pressure divided by the suction pressure in absolute pressure. Absolute pressure (PSIA) is PSIG plus 14.7 (usually rounded off to 15). For example, 300 head ÷ 30 PSIA suction = 10:1 ratio.

Now add cool refrigerant vapor to the suction to raise the suction pressure by 30 pounds. 300 head ÷ 60 = 5:1 ratio is half the original, and the discharge temperature will be less than before.

The Y-1037 TREV valves used as discharge temperature controls were rated at 1/2-ton capacity, which was correct for the application. However, the valve's sensing bulb strapped to the discharge line was only rated for 190°F. This means that when the discharge line temperature reaches 190°F, the valve opens to allow refrigerant into the crankcase to lower the discharge temperature. When we checked with Bitzer, they recommended 240°F valves. The original lower temperature valves would inject refrigerant too soon and could cause flooding in the crankcase (see Figure 10-130). After the valves were changed out, the temperature in the crankcase was warmer. We believed this may prevent one of the causes of floodback, or at least some of the foaming in the crankcase.

FIGURE 10-130 **Discharge temperature control (DTC) valves flooding crankcase.** *Photo by Jess Lukin.*

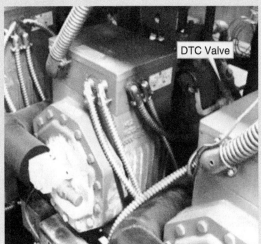

Crankcase Heaters: While working on the DTCs, we noticed some compressors did not have crankcase heaters. On further inspection, we discovered the ones that did have crankcase heaters were wired incorrectly to come on when the compressor did, instead of being energized when the compressor shut off. Rewiring was done, and new crankcase heaters were ordered.

Evaporator Pressure Regulators: Sporlan SORIT (solenoid operated, opens on rise of inlet pressure, with a tee access fitting) valves are used to keep the evaporator from going below a certain pressure. SORIT-PI were also used in this system. The *PI* stands for Piloted Internally, which basically means they do not need high-pressure vapor to operate the valve, as does the externally piloted SORIT. The design engineer determines which valve will be used based on where they are located in the system.

By maintaining a specific pressure in the evaporator, the case temperature stays constant. For example, a system using R22 with the SORIT valve set at 40 psig should maintain 18°F saturation temperature in all the evaporators on that circuit. In this scenario, we checked all the systems and adjusted valves that needed it. There were several valves that would not adjust, so we rebuilt them. Replacing the whole valve is very difficult because the original has to be cut out and a new one brazed in its place. Rebuilding was much easier and less expensive because we left the welded base in place and just changed everything else (see Figures 10-131 and 10-132).

FIGURE 10-131 **SORIT evaporator pressure regulator before rebuild.** *Photo by Jess Lukin.*

FIGURE 10-132 **SORIT after rebuild.** *Photo by Jess Lukin.*

Defrosts: It is important to verify the defrost times to make sure only one Koolgas Defrost system at a time was in defrost. The Hussmann systems are not designed to have two circuits in Koolgas Defrost at the same time on the same rack. Figure 10-133 shows the defrost schedule for rack A. Circuits A2 through A7 and circuit A17 go into defrost four times a day for an average of 20 minutes each. Koolgas Defrosts use refrigerant vapor from the top of the receiver; therefore, more than one defrost at a time will cause a temporary "washout" (uses all the refrigerant in the receiver). This is what was happening when we noticed the receiver at 0 percent during the first inspection of the systems, and then a few minutes later, the receiver was registering 80 percent fill. Balancing defrosts to prevent overlapping is very important.

FIGURE 10-133 **Rack A defrost schedule.** *Photo by Jess Lukin.*

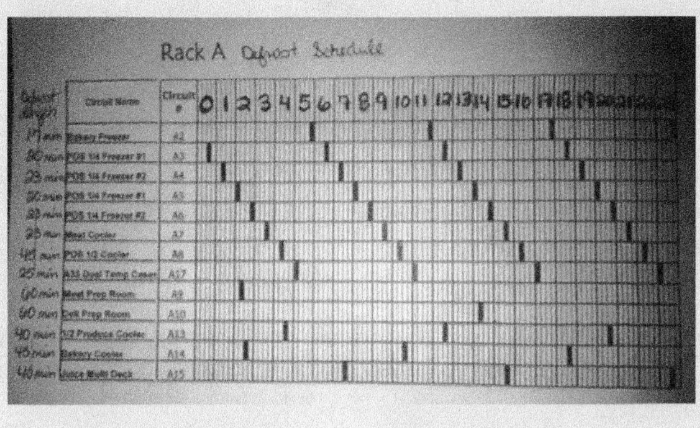

The other circuits are electric defrost, and they range from 40 to 60 minutes. They do not use refrigerant to defrost; however, the expansion valves are wide open when the system goes back into refrigeration, and any oil trapped in the cold coils is returned to the compressor rack. This could cause floodback; however, as long as the TEVs are maintaining superheat, the SORITs are maintaining evaporator pressure, and the oil return system is doing its job, then there should be no floodback problems. And there weren't when we were done.

Continuing our search for problems, we noticed the liquid header and circuit lines were very cold. So the next thing we checked was liquid subcooler.

Mechanical Subcoolers: Subcoolers are installed on rack systems to lower the temperature of liquid before it goes to the liquid header at the rack. The cooler liquid means less liquid in the expansion valve needs to be "flashed off" (vaporized) in order to cool the remaining liquid down to the evaporator temperature. The result is that more of the liquid entering the TEV remains a liquid as it enters the evaporator. Not only does this mean the evaporator is much more efficient, but the TEVs can be smaller (less tonnage) (see Figure 10-134).

FIGURE 10-134   Subcooler. *Photo by Jess Lukin.*

The subcooler was supposed to be set for 45°F to 50°F leaving liquid temperature. However, instead of temperature, the subcooler was set for a pressure of 45 psig. This was an obvious UOM (unit of measure) error. The pressure relates to an R22 saturation temperature of only 22°F, which is way too cold. This causes the case TEVs to be erratic.

After getting the subcooler working properly, we turned our attention to the TEVs in the cases. It was significant that after cleaning the evaporators, many of the superheats went from 2°F to over 40°F. We speculated that the other company tried to prevent floodback by adjusting the valves closed in order get some kind of superheat. With clean coils, the refrigerant boiled off quickly, and the evaporators were starving—as evidenced by the high superheats.

Before adjusting superheat, we checked the valve size. We did the following for each case:

- Determine case Btuh required; **example:** 1,050 Btuh per foot × 8-foot case = 8,400 Btuh
- Determine the tonnage rating of the valve; **example:** 12,000 Btuh is 1 ton
- Determine evaporator temperature; **example:** low temp at −20°F, medium at +20°F
- Choose TEV based on Btuh capacity at the case's evaporator temperature.

The valves used in most supermarkets are "Balanced Port" TEVs. They maintain a more steady flow rate during fluctuating pressures than do standard TEVs. Therefore, they are rated in a tonnage range rather than as a single ton capacity.

Sizing the TEV: A balanced port expansion valve has a tonnage range, for example, 0.75 to 1.5 tons. This valve will be rated at its maximum 1.5 tons at an evaporator temperature of +40°F. However, when operating at a low temperature of −20°F, its maximum is only 1 ton. The liquid temperature coming into the valves is also a factor. The colder the liquid, the higher its tonnage rating based on a correction factor in the TEV chart. For instance, a 1-ton valve is rated at 100°F inlet liquid temperature, but at a +50°F inlet temperature, it would have a 1.25 correction factor, or a rating of 1.25 tons. This is important to know when sizing valves with liquid refrigerant subcoolers. Also, pressure drop across the valve affects valve capacity. The nominal tonnage capacity is usually based on 100-psig drop across the valve between the pressure entering the valve and the pressure entering the evaporator. The lower the pressure drop, the lower the capacity; the higher the pressure drop, the higher the capacity. It is best to refer to valve spec sheets such as Sporlan Bulletin 10-10 for sizing.

The contractor in this scenario used Sporlan BQ valves, which give the user a choice of five different capacities from the same valve body by simply changing the "cartridges." However, in our experience, these particular valves have problems if superheat is adjusted multiple times. We found some TEVs with the wrong size cartridges for the case (see Figure 10-135). In other TEVs, we could not adjust the superheat. And all of them had dirty inlet strainers (see Figure 10-136). It was evident that we were correct in starting with the oil filters to clean up the system. We replaced all the refrigeration filter-driers and then replaced the TEVs and checked for proper superheat.

FIGURE 10-135 **Cartridge of a Sporlan BQ balanced port TEV.** *Photo by Jess Lukin.*

FIGURE 10-136 **Clogged mesh strainer of TEV.** *Photo by Jess Lukin.*

Summary of Problems That Caused Floodback and Killed the Compressor:

- Dirty evaporator coils
- Subcooler set too cold
- TEVs plugged, incorrect, and misadjusted
- Oil return issues
- DTC controls causing floodback
- Crankcase heaters not wired and some missing
- SORIT valves not working
- Koolgas Defrosts occurring at the same time

Epilogue: Thank goodness we don't run into complex jobs like this very often. However, when it does happen, here are a few suggestions:

- Gather all the facts.
- Review service histories.
- Document everything in detail, on every unit.
- Take pictures of equipment, nameplates, and problems.
- Give the customer an estimate for work needed and get a signature.
- Follow a logical sequence of diagnostics and repairs.
- When all repairs are complete, write down all the operational information in detail.

SERVICE SCENARIOS #5

SIMPLE ELECTRICAL PROBLEMS

By Jess Lukin

Sometimes, simple electrical issues cause major system failures. Most can be avoided by following some basic commonsense rules. Following are three examples of things we ran into.

Example #1: Electrical Connection Problems: Before getting into the first service call, it will be helpful if you understand that the bottom of most refrigerated cases serve as the drain pan. The fan plenum and evaporator coil are installed in that drain pan (see Figures 10-137 and 10-138). All too often, the drains clog up, and water backs up into the fan area. In low-temperature cases, there are heaters in the drain line and in the drain pan to keep them clear of ice. If either of them fails, it will cause the defrost condensate water to fill the pan, which soon turns to ice (see Figure 10-139). When water in a drain pan gets deep enough, the fan blades pick it up and throw water in every direction in medium-temp cases. This could create a short circuit because the plug is often located in the fan plenum. In low-temp cases, this water turns to ice, which blocks airflow through the evaporator and will eventually stop the fan blade. When this happens, you can expect problems with the electrical devices in the case.

FIGURE 10-137 **The fan plenum over the evaporator.**
Photo by Jess Lukin.

FIGURE 10-138 **The evaporator coil in the drain pan.**
Photo by Jess Lukin.

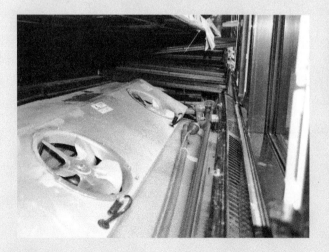

FIGURE 10-139 **Ice in drain pan of bottom evaporator section of a refrigerated case.** *Photo by Jess Lukin.*

Ice in drain pan

Whenever wires in a case have been improperly spliced, you can count on a short circuit when they come in contact with water. Therefore, you must prevent this from happening by making sure all your wiring connections are watertight. Wire splices should be made with heatshrink waterproof connectors (see Figure 10-140); another option is to coat the wire nuts or other connection devices with liquid electrical tape (see Figure 10-141). If possible, make sure the electrical connection is in the dry and covered electrical panel or kick-plate area instead of inside the case.

FIGURE 10-140 **Waterproof electrical connectors.** *Photo by Jess Lukin.*

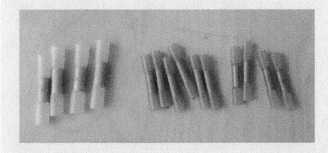

FIGURE 10-141 **Waterproofing with liquid electrical tape.** *Photo by Dick Wirz.*

Service Call: Meat case at 75°F. After-hours emergency with product loss.

The manager showed us the problem case and was very upset about the loss of all the meat. The first thing we checked was case airflow. There was none. Meat cases normally run at product temperatures of about 28°F, which is cold enough that there can be a frozen evaporator. However, not only was there no airflow, but we did not hear any fans running.

When we removed the bottom access panels, it was evident all the fans were off. One of the motors had been replaced, and wire nuts were used to make the connections. Instead of using electrical tape to make the connections even slightly water resistant, the person who did the repair used a plastic bag to wrap the wire nuts. The bag became tangled up in the fan blade, which eventually stopped the fan motor. The exposed connections apparently shorted and tripped the circuit breaker for the case (see Figure 10-142).

FIGURE 10-142 **A plastic bag is not the way to waterproof electrical connections.** *Photo by Jess Lukin.*

It can be a nightmare to find the proper circuit breaker in a supermarket. Most breakers are not labeled correctly, and if any remodels have been done, no one seems to know where the breaker for that system was moved. Supermarkets have many electrical panels because they need power for so many different sections of the store and for things like compressors, condensers, refrigerated cases, store lighting, parking lot lighting, meat cutting room equipment, alarms systems, energy management controls, defrost heater breakers, main computer breakers, cash registers breakers, and so on.

Testing all the breakers in a panel by turning them off and then back on is never a good idea. The following is an actual note found on an electrical panel: "If you like your job don't turn this breaker off or you will be unemployed." These days, we have so many computer-related systems that flipping breakers can really wreak havoc with store operations. To save time and liability, ask the store manager if she knows the location of the correct breaker, or at least which panel it may be in. She may have been there the last time the case was worked on. In addition, always get authorization from the manager prior to turning off any breaker that is not labeled.

In our current scenario, after the breaker was located and verified as tripped, it was labeled for the next technician. The motor was replaced, and correct waterproof wiring connectors were used to join the new motor leads to the original factory male electrical plug.

Epilogue: This 4-hour overtime service call and all the lost product could have easily been avoided if basic electrical principles had been followed. Improper repairs have expensive consequences.

Example #2: Circuit Breakers Do Fail: Don't be afraid to condemn a breaker as being faulty. Given enough time and conditions, any electrical component will need to be replaced. Circuit breakers are no exception to this rule. Breakers typically go unchecked for years at a time, and the refrigeration technician is the first line of defense for getting management to call in a faulty or weak breaker. Because circuit breakers are subject to electricity, heat is always an issue. Following are conditions that should indicate that it is time to replace a circuit breaker:

- It becomes difficult to turn off or on
- It won't reset after tripping (even though the wires have been removed from the load)
- There's any evidence of broken parts or cracks
- There's a measurable voltage drop between power entering and power leaving the breaker
- The breaker is above 130°F (or at least much higher than adjoining breakers when measured with a temperature gun) (see Figure 10-143).

FIGURE 10-143 **Infrared temperature gun or thermal imager can be used to check breakers.**
Reproduced with Permission, Fluke Corporation

Notes about Circuit Breakers for Refrigeration Compressors: Breakers are sized at about 125 percent of the rated load of what they serve, unless the manufacturer has stated differently on the condensing unit nameplate. Compressors have an RLA, or rated load amp draw (i.e., the amp drawn at specific operating pressures). Also on the plate is the LRA or locked rotor amps, which are usually five to seven times higher than RLA. All motors pull LRA for a fraction of a second on each start-up. Circuit breakers and time-delay fuses are designed to handle this high amperage for the short duration needed to get the compressor up to speed.

Circuit breakers can trip from high current, but just as often, they trip from short circuits. The circuit breaker's primary function is to protect the wiring to the compressor. The compressor is protected from overcurrent by its overload. This switch opens from a combination of both heat and high current.

If there is any chance of a short in the electrical circuit or the compressor, it is not wise to reset the breaker without doing some preliminary checks. The worst-case scenario is a short in the compressor windings or a grounded compressor. Resetting the breaker could cause more damage and if shorted to ground, it could electrocute anyone touching the compressor or connected piping. To check for shorts, you need to measure resistances using an ohm meter. For proper readings, the power must be turned off and the power wires to the compressor removed from the load side of the contactor.

Service Call: Fresh meat case down, compressor not running

Fans were running in the meat case, and there was good airflow. We went to the condensing unit on the roof and discovered there was no power to the equipment. We checked the electrical panel, which was 10 feet from the unit, and found the three-pole circuit breaker was tripped (see Figure 10-144). Fearing a possible short in the compressor, we disconnected the power wires going into the line side of the contactor before resetting the breaker. Checking voltage leaving the breaker, we got full voltage on two legs but only 2 volts coming out of the third. Checking the line side, there was full voltage coming into the circuit breaker. We told the customer the breaker had to be replaced. He called his electrician and said he happened to be close by and would be over shortly to replace the breaker.

While waiting for the electrician, we checked the compressor windings with an ohm meter. There were no shorts to ground, and we had continuity between all three windings. Upon checking the compressor contactor, we discovered its coil was shorted out. This may have been a positive thing. The bad coil would

FIGURE 10-144 **Tripped breaker for condensing unit, upper left 50 amp 3-pole.** *Photo by Jess Lukin.*

not let the contactor pull in. It prevented the three-phase compressor from single-phasing, which could have done some real damage to the compressor windings.

The contactor was replaced, and the electrician replaced the breaker. It turned out that the compressor was good, and the system was running. Sometimes you aren't so lucky, but this time we were.

Epilogue: Circuit breakers can and do fail. Don't hesitate to condemn them and have them replaced.

Example #3: Cleanup Wiring: Messy wiring is more than just ugly. It can cause electrical problems and is dangerous for the servicing technicians. In rare cases, it can even cause refrigeration leaks (see Figure 10-145). In this instance, poor wiring by a previous technician caused loose connections that resulted in an electrical short against a refrigerant pressure control. The result was a walk-in freezer system down with product lost.

When you find a rat's nest of wiring, take the time to clean it up. It will protect you and any other technicians who work on it after you. And it will prevent further service calls (see Figure 10-146).

FIGURE 10-145 **A shorted wire that cause a refrigerant leak.** *Photo by Jess Lukin.*

Electrical Short

Bubbles show refrigerant leak

FIGURE 10-146 **Example of wiring that could use some cleaning up.** *Photo by Jess Lukin.*

SERVICE SCENARIO #6

FLOODBACK

By Jess Lukin

What Is Floodback? Floodback is liquid refrigerant returning to a compressor during the running cycle. It may dilute the compressor oil to the point where it cannot properly lubricate the load-bearing surfaces. If liquid enters the cylinders, it can break reed valves, blow head gaskets, and break crankshafts and piston connecting rods. It is ver important we recognize and prevent floodback.

How Do You Know It Is Floodback? Flooding on an air conditioning unit can have a sweating suction line, sweat on the compressor, or even frost on the compressor. A medium- or low-temp commercial refrigeration unit could have frost on the suction line, service valve, or compressor. But a frozen evaporator will show similar signs on all these systems.

The *only* way to be sure it's floodback is to measure superheat. If there is no superheat, then you definitely have floodback. Most evaporators are supposed to have about 10°F of superheat. Most compressors are supposed to have about 20°F of compressor or "system" superheat taken on the suction line about 6 inches before the suction service valve.

How Can You Correct Floodback? First, check for and correct any frost on the evaporator or fan problems. If the coil is clear and fans are running, then turn your attention to the TXV. Checking superheat is easy, but adjusting it calls for patience. A service technician should never rush into adjusting TXV superheat. Too much can be just as bad, or worse, than too little superheat.

Service Call: The four-door section of ice cream lineup is running 0°F rather than −20°F.

The ice cream lineup consists of a five-door, a four-door, and a three-door section of upright glass door reach-in cases (see Figure 10-147 as an example). The manager said that it has been running fine since the last technician came out a week ago because the case wasn't cold enough. The four-door section had more frost on the evaporator than the other cases. Inside the cases, the five-door and three-door suction lines were clear of frost. However, in the four-door section, the suction line had a buildup of frost. In fact, it caused the common suction line to frost all the way back to the compressor body, as seen in Figure 10-148.

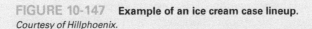

FIGURE 10-147 **Example of an ice cream case lineup.**
Courtesy of Hillphoenix.

Although the frost looks like a problem, we have to remember that the suction line is running well below freezing, and frost will form on it under normal conditions. However, the thick frost on the compressor body is more than we usually see (see Figure 10-149.)

FIGURE 10-148 Top view of flooded compressor.
Photo by Jess Lukin.

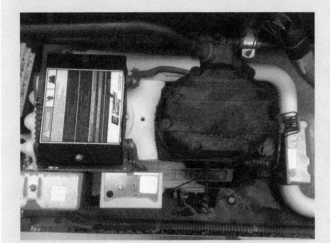

FIGURE 10-149 Side view of flooded compressor.
Photo by Jess Lukin.

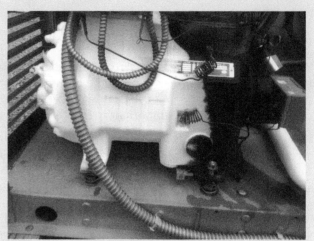

This is a suction-cooled compressor with the suction vapor entering the end bell, cooling the motor windings, and then flowing up to the cylinders at top right. The oil sump and sight glass are in the crankcase section just below the cylinders. Notice the thick frost buildup past the compressor ribs and into the oil sump. This is a red flag suggesting the probability of liquid refrigerant boiling off in that area.

The manager said a technician had been out the previous week because the case temperature was not cold enough. He believes the tech just adjusted the TXV. Apparently, the tech thought lowering the superheat would allow more refrigerant into the evaporator, which would bring the temperature down. This is partially true, but superheat adjustment is used for fine-tuning case operation by getting the maximum efficiency out of the evaporator. By making sure refrigerant is boiling off near the outlet of the evaporator, the proper superheat indicates that the evaporator is neither starving nor flooding. To properly measure the superheat, the case must be within a few degrees of its design temperature. The original call was for a warm case, so it may have been too warm to accurately measure superheat. Adjusting superheat is not like adjusting the thermostat to make the case colder. This case has a pump-down solenoid controlled by an air-sensing thermostat with its remote bulb mounted on the back wall in the discharge airstream about 6 inches from the evaporator. The tech should have first checked to be sure the thermostat was working properly before even thinking about superheat. Adjusting the valve only resulted in more frost on the evaporator and flooding back to the compressor. We had to defrost the case and then let it come down to temperature before checking superheat. This took about 3 hours.

We made an adjustment of 1/4 turn to close the valve (clockwise facing the end of the stem) then let the system stabilize for at least 15 minutes. Yes, this is the hard part because waiting is tough for us service techs. However, being patient just means finding something else to do in the meantime. There are many things that could be productive, like tightening electrical connections, checking amperage, cleaning condensers, or just looking over the job for things that need attention.

A pipe clamp–type temperature probe on the suction line with a meter set to min and max is very good for tracking the progress of any changes in superheat. Figure 10-150 shows how the frost line has started to recede after the first adjustment.

It is obvious we are making progress because the frost pattern is dissipating more with each picture. We were correct to assume that adjusting the TXV would solve the problem. However, we didn't even think about making another adjustment until we were sure it had stabilized. We were patient and waited for the frost pattern to stop receding. One-quarter turn may be all that's needed. Probably not, but it is better to wait and be sure. After about 30 minutes, the TXV stabilized and sure enough, another was needed. So a one-quarter-turn adjustment was done again. Figure 10-151 shows how the frost line continues to recede after each adjustment.

FIGURE 10-150   **Frost recedes after first adjustment.**
Photo by Jess Lukin.

FIGURE 10-151   **After another adjustment more progress is made.** *Photo by Jess Lukin.*

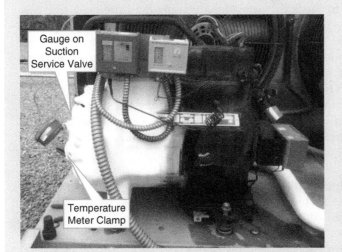

Gauge on
Suction
Service Valve

Temperature
Meter Clamp

Here we are, sitting and waiting for the TXV to stabilize again. But we can see the frost pattern is receding a little more. Since it looks like we're on the right track, we made good use of the time by filling out paperwork documenting what we had done. We took down the compressor model number; unit model number; condensing unit model number, if it differs; and every other model number we could think of. Attention to detail is so important because we were establishing history here for the next technician, so we wrote a history book with plenty of detail and facts of temperatures, adjustments, pressures, and everything else we could think of. Too much is always better that not enough when it comes to documenting a service call.

Look at that compressor now in Figure 10-152, frost free and looking good. Notice the suction line is still frosted, and that's okay because we measured superheat at the compressor, and it was right on at 20°F. But we still need to check superheat at the case. This manufacturer is calling for 8°F superheat at the evaporator when the ice cream case is down to its recommended space temperature of −20°F. However, the superheat was 12°F, and the case temperature was −10°F. Superheat has a tendency to "float" downward as the case cools. So we waited until it came down to temperature and cycled on the thermostat. It shut off at −15°F when set at −20°, and the superheat was 10°F. After adjusting the thermostat and cycling a few times, we determined that the control was erratic, cycling off too low some times and too high other times. We replaced the tstat and just before it cycled off at −20°F, we had an 8°F superheat.

FIGURE 10-152   **Success! No frost on the body, no floodback in the crankcase.** *Photo by Jess Lukin.*

Frost on Suction
Line OK

Epilogue: Apparently, the thermostat was the problem all along, and checking it first would have eliminated a lot of unnecessary work. The previous technician did not allow for the downward trend of superheat as the temperature dropped, nor did he return to check the case after adjusting the superheat.

If the case was not cold enough, the previous service technician should have checked all other possibilities before resorting to superheat adjustment. There could have been many other causes for the warm case temperature. Some of the more obvious include the thermostat setting being too low and freezing the evaporator. Or, if it had a pump-down solenoid, the valve could have been be leaking through, which could cause the evaporator to freeze. Also, too much frost on one section of the evaporator could be caused by product on the shelves or the return blocking airflow or dirty honeycombs blocking the supply airflow. If the problem was high superheat rather than flooding, the TXV could have a blockage of ice or maybe a plugged strainer. For that reason, many techs warm the TXV and check the strainer before adjusting superheat. But the bottom line is that starting case diagnosis by adjusting superheat is just asking for trouble. TXVs do not go out of adjustment by themselves.

In retrospect, we might have saved ourselves time and effort by first adjusting the TXV back to the factory setting by finding the mid-point of the adjustment. Then if the valve was still flooding, we would have replaced it rather than spending a lot of time trying to adjust it.

The amount of evaporator superheat is important for system efficiency and compressor protection. Too much superheat starves the evaporator and overheats the compressor. Too little superheat, as we had on this job, can freeze the evaporator and damage the compressor through flooding. We now have electronic meters with pipe clamp temperature probes that make it easy to measure superheat accurately. This should always be part of the information on our service tickets when servicing refrigeration systems.

According to TXV manufacturers, evaporator superheat measurement is accurate within only 5°F of the design temperature. For instance, if these ice cream cases are designed to run at −15°F, then the superheat reading is not even close to accurate until the case is down to about −10°F. Even then, the superheat will drop some by the time it gets to −15°. At higher temperatures, the valve is almost wide open, and any superheat is due to the heat load on the evaporator and is not based on the TXV adjustment. In this instance, any adjustments about −10°F would have certainly caused flooding when the case finally reached −15°F.

If everything else in the case is operating properly with good airflow and a clear coil and the only thing left to check is superheat, then the following tips may be useful. On frozen food and ice cream case lineups, the coldest case is the one that needs adjustment for low superheat or flooding. Also, check the suction line temperatures of the lineup. The case with the coldest suction line will indicate low superheat.

SERVICE SCENARIO #7

CLEANING UP THE MESS

By Jess Lukin

Supermarket refrigeration works very well when properly maintained. Service and repairs are easy because there are fewer hidden problems, and good records are available for reviewing service histories.

However, a simple service call can become a nightmare when working on a supermarket that has no maintenance program and uses several different companies for service. This is typically known in the field as "Breakdown Maintenance." Following is an example of the problems a service technician can run into when working at one of those types of accounts.

Service Request: This past weekend, the in-house refrigeration company added 300 pounds of refrigerant to the rack in an attempt to resolve refrigeration problems, but all the cases on the rack are still not pulling down to temperature, and 75 percent of the product had spoiled.

Where to Start: It's best to start with the basics and have a logical plan for diagnosing quickly and efficiently. Upon arrival, go directly to the manager. Get all the information you can about what exactly is the problem, which cases are having trouble, and as much recent service history as possible. Spending time with the customer and really listening to what she has to say is both informative and an excellent start to good customer relations.

We were already on the main floor, so we took a look at the cases, noting airflow and anything that could be significant as we headed to the mechanical room. Once there, we tried to determine the following:

- Are all the compressors operating?
- Are any short-cycling or off on a safety switch?
- Is there any evidence of floodback or starving?
- Is there adequate refrigerant in the receiver?
- How do the oil and liquid line sight glasses look?
- How many suction groups are there?
- What types of defrost are being used?
- Does the controller have alarms, and if so, what is the history?

After discussing the situation with the manager, we verified that just about all 34 refrigeration circuits were running warmer than they should. Turning our attention to the mechanical room, we found two old refurbished Hussmann "Super Plus" rack systems—one for low temp and one for medium. We quickly verified that all compressors and condenser fans were running and opened the control cabinets to see what we had (see Figure 10-153).

FIGURE 10-153 **Hussmann "Super Plus" rack system.** *Courtesy of Hussmann.*

Initial Diagnosis:

1. Temperatures in the medium-temperature cases weren't maintaining. It appeared that some TXVs were restricted, while others were flooding (see Figure 10-154).
2. All low-temperature defrost circuits were being affected because of burned wiring on one leg of a terminal block in the rack's defrost electrical cabinet (see Figure 10-155). Without defrost heaters, those cases had no airflow due to frozen evaporators. We decided to repair the low-temp wiring first to get the defrosting started on those cases and then turn our attention to the medium-temp problems.

FIGURE 10-154 **TEV in medium-temperature case. Frost should not be on the valve body.** *Photo by Jess Lukin.*

FIGURE 10-155 **Defrost wires burned at terminal block in control panel.** *Photo by Jess Lukin.*

Start with the Defrost Wiring: We went to the electrical store with the burned terminal block in hand to purchase something similar. We needed to get the heaters working on the low-temp cases so they could defrost properly (see Figure 10-156).

FIGURE 10-156 **Burned terminal block and new replacement.** *Photo by Jess Lukin.*

Other Issues: The initial service call said the in-house refrigeration company added 300 pounds of refrigerant. Both receivers' fill indicators were at 60 percent, which was higher than normal. Usually, we would expect receiver levels in the area of about 30 percent. From experience, we have learned that adding a lot of refrigerant can stir up debris and sediment, which results in clogged valves, especially TXVs. So we decided to change all oil and refrigerant filters and to clean out the TXV strainers and adjust their superheat.

POE oil is known for its solvent capabilities and requires a higher degree of maintenance in a large refrigeration system that contains excessive amounts of debris. POE oil, unlike the other refrigerant oils, keeps the debris suspended and circulating through the system until it clogs something or is filtered out. In older mineral oil systems, we find most of the debris ends up settling in the compressor crankcase rather than continuing to circulate.

We started digging into the compressors operation. One of the two medium-temp compressors that were in override had bad valves. This is probably why we found that the medium-temp suction group was lowered from a +20°F set point to a +14°F set point. The previous tech was trying to get case temperature down by lowering the EPR set point. This erroneous diagnostic was aiding the compressor's demise.

What About the Controls Program? We found what is called a skeleton system program. It monitored suction pressures on both suction groups, discharge pressure, controlled compressor cycling, and defrosts. We noticed the following issues:

- Defrost cycles were too long and too frequent.
- Defrost circuits were not balanced, meaning too many defrosts at certain times and not enough at others; this caused the rack to work harder.
- Two compressors were enabled "On" in override mode, which bypassed the safety controls.
- The medium-temp suction group was running at +14°F instead of +20°F.
- The low-temp suction group was running at −30°F instead of −20°F.

We balanced out the defrost cycles on the rack, labeled the cases with their specific defrost times, and gave copies of the defrost schedule to the manager. We verified defrost contactors worked and took amp draws of heaters for the corresponding circuits, documenting the readings on our service ticket.

We replaced all the refrigerant filter-driers and oil filter-driers then allowed the low-temperature side of the rack to stabilize for 24 hours to allow proper defrosting of the low-temperature circuits.

We tested the refrigeration oil for acid but found none. We prefer to use oil filters with an access tap and internal bypass. The access tap is used to check for pressure drop, which would indicate that the filter was full of debris. If the filter were to become clogged, the internal bypass would allow oil to keep flowing to the compressor. The oil was quite dirty and brown, which indicates high discharge temperatures caused by high compression ratios from high superheat and from running suction pressures too low (see Figure 10-157). We will check out those problems after we get the expansion valves feeding properly.

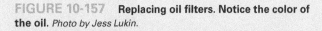

FIGURE 10-157 **Replacing oil filters. Notice the color of the oil.** *Photo by Jess Lukin.*

Safety Reminder: While removing the Schrader valve cap on a liquid line filter-drier, some liquid refrigerant started seeping out from under the cap. It was immediately retightened. The Schrader valve was either missing or damaged.

Evidence of liquid refrigerant is not a good sign and should serve as a reminder that one must be cautious when dealing with a store that isn't regularly maintained by you. Every finger-tight cap, Schrader stem, ball valve, service valve, or refrigeration access point could be an accident waiting to happen.

Replace Filter-Driers, Clean TXV Strainers, and Adjust Superheat: We replaced all the refrigerant filter-driers again and began pulling out the TXV strainers. They were the worst we had ever seen (see Figure 10-158). The black carbon deposits only verified what the oil showed us—that the compressors were having overheating problems.

FIGURE 10-158   **TEV strainers with evidence of high discharge temperatures.**
Photo by Jess Lukin.

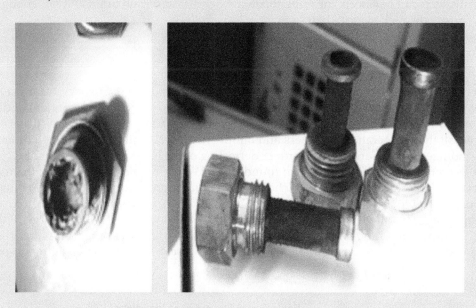

Most of the Sporlan TXVs were adjusted wide open. Apparently, someone was having trouble with starved evaporators and thought adjusting superheat fully open was the answer. We readjusted all the valves to their factory-set superheat. This is done by determining the mid-point of the adjustment. Since they were already adjusted back to the full-open position, all we had to do was count the number of turns until they were fully closed then adjust them in reverse halfway. For example, if from full open you have 20 turns to full closed, then you adjust backward 10 turns to the mid position. Note that these valves are very delicate, so we made sure we stopped turning as soon as we felt resistance when we got to the full open or full closed position.

Return to Check Everything Out: The following day, we came back to see the progress after all the screens were cleaned and defrost cycles were set and balanced, as well as to check temperatures on all the cases.

The first thing we noticed was that the ice cream case's temperature was −24°F. The problem is that the ice cream case should be set for −15°F at about 18 psig. There is no reason to run the case at a temperature as cold as −24°F nor to run the suction that low. It just makes for a higher compression ratio and hotter discharge temperatures. We found the EPR wide open, which allowed the suction pressure to drop too low and the case temperature with it. We had an idea why this was done. Sometimes techs float (open all the way) the EPR as a temporary patch when temperatures aren't coming down. This buys the tech some time to troubleshoot the root of the problem. Sometimes it works, sometimes it doesn't. It depends on what the problems are. In this case, whoever opened the EPR all the way did not find the problem, or he forgot to reset it. After finding this issue with the ice cream EPR, we decided to check every circuit's EPR to verify the set points were correct.

This finding also explained why there were excessive defrost cycles. The lower suction temperatures were frosting the evaporators faster, requiring more defrosts.

Epilogue: The system had about 300 pounds more refrigerant than was needed, EPRs were open, compressors were in manual override, suction pressures were lower than they should be, all low-temperature defrost cycles were offline with excessive defrost cycles, and TXVs were adjusted wide open.

The people working on this supermarket did not understand commercial refrigeration or parallel rack systems. They knew just enough to be dangerous. The root causes of the problems were beyond their understanding, so they tried to respond by first adding refrigerant, then by trying to adjust valves and controllers in order to get the systems to work.

There is a valuable lesson here. A situation that seemed like a lot of different problems was solved by simply cleaning up a system that had not been maintained and then setting everything back to factory specifications.

An inexperienced tech might have come to this job and just started adjusting valves. Experienced and well-trained techs make it look simple because they know how a system is supposed to operate, and they make it so. Diagnostics to them is a matter of asking the following questions:

- What is it doing?
- What is it supposed to be doing?
- Why isn't it doing it?
- What will it take to make it right?

SERVICE SCENARIO #8

ASPHYXIATION ALERTS

By Dick Wirz

Example #1 Service Call: Walk-In Freezer Warm: It was a large walk-in about 16-feet by 16-feet by 10-feet high in the kitchen of a hotel in Fairfax, Virginia. The remote condensing unit was about 50 feet away on the loading dock outside.

Initial Diagnosis: Starving evaporator due to a TEV that was stuck closed.

I told the manager I would go to the supply house and pick up another valve and have it repaired within the next 1 or 2 hours. Upon my return, I pumped the system down and started to replace the valve located in the evaporator. The customer had brought in quite a bit of dry ice in an attempt to prevent his frozen product from thawing out. Although the room was filled with the vapor of the dry ice, I did not think much about it.

Fortunately, the TEV was a flare valve, so the replacement did not take very long. I pressurized the system with nitrogen so I could check leaks at the newly installed valve. I noticed that I was feeling a little dizzy but thought it was maybe because I just hopped up the ladder too quickly. When I went outside to start my vacuum pump, I stumbled as I went through the door. Then I realized what was happening. I was experiencing asphyxiation symptoms.

Dry ice is solid CO_2. In this situation, when it vaporized (sublimated) from a solid directly to a vapor, it filled the room and displaced the oxygen. Apparently, it took me a while to feel the effects because I was working from a ladder near the ceiling. Since CO_2 is heavier than oxygen, there was more oxygen up where I was working than below me. However, it finally had a dangerous effect on me.

Remember, CO_2 displaces oxygen. It is not uncommon for people to use dry ice to prevent frozen foods from thawing in an emergency. We just have to remember that in an enclosed space like a walk-in, CO_2 can cause asphyxiation.

Example #2 Service Call: Meat Cases Warm: This was a butcher shop in Arlington, Virginia. All the cases and the walk-ins were on the first floor, the multicircuited condenser was on the roof, and the remote six-compressor rack was in the basement. Each compressor served one system and had its own receiver.

Initial Diagnosis: Compressor off on low pressure due to refrigerant loss from a loose flare nut at the compressor.

The repair was very simple, but the system had lost about 20 pounds of refrigerant. As I was charging the unit, the customer called down to the basement to ask if I had turned off the hot water heater for any reason. It was next to the compressor rack, so I checked it and found the pilot light was off. Then I started to feel dizzy. I realized immediately that the vaporized refrigerant from the leak had filled the basement and had displaced the oxygen. This was the cause of the pilot light going out and also of my feeling of dizziness. I went outside to get lots of fresh air.

After thoroughly ventilating the basement, I was able to relight the pilot and finish charging the system. As HVACR technicians, we work in confined spaces where refrigerant can displace oxygen. We should keep in mind that anytime we feel dizzy, it could be from a lack of oxygen. We must leave the area immediately and then consider the cause. It may actually be a matter of life or death.

REVIEW QUESTIONS

1. What is the proper temperature of a meat case?

   a. 36°F to 40°F
   b. 34°F to 38°F
   c. 28°F to 32°F

2. What is a parallel rack system?

   a. A system of multiple compressors that share a common source of refrigerant and oil to refrigerate a group of cases
   b. A group of condensing units that are aligned next to each other, rather than facing each other
   c. A rack to support refrigeration pipes that are installed parallel to each other

3. Why are parallel rack systems used in supermarket refrigeration?

   a. They are easier to install than separate condensing units.
   b. They provide excellent temperature control and effective capacity control, and they are more energy efficient than separate condensing units.
   c. There are lots of cases, so they need lots of horsepower.

4. With such long piping runs, how does the rack system ensure oil is returned to the compressors?

   a. By using greater slope than normal on suction piping for better oil return
   b. By using more p-traps to trap the oil and return it to the compressor
   c. By using oil separators to capture oil before it leaves the compressor rack

5. What determines when rack compressors cycle on and off?

   a. Pressure and temperature controls on the rack monitor the refrigeration load and cycle the compressors.
   b. The store personnel monitor case temperatures and cycle the compressors.
   c. The HPR valves block the condenser when system demand is low.

6. How are case temperatures regulated in a parallel rack system?

   a. EPR valves
   b. HPR valves
   c. CPR valves

7. In supermarkets, where are the EPR valves located?

   a. At the evaporator outlet of each case
   b. At the compressor rack suction header
   c. At the condenser on the roof

8. What is a controller, and how is it used on a rack system?

   a. It is an electronic control panel used to control the operation of the system.
   b. It is a temperature control for the condensing unit to cycle the fans.
   c. It is a valve that modulates the flow of refrigerant to the compressors.

9. What methods are used to control head pressure on rack systems?

   a. Condenser fan cycling
   b. Flooded condensers using HPR valves
   c. Split condensers
   d. All of the above

10. Why is energy efficiency so important to supermarkets?

    a. The Environmental Protection Agency (EPA) monitors the activities of supermarkets.
    b. The profit margins of supermarkets are only about 1 percent.
    c. Only energy-efficient equipment will keep product refrigerated properly.

11. List eight ways supermarket refrigeration systems can save energy.

12. For every degree of subcooling at the TEV, how much of an increase in evaporator capacity will there be?

    a. 0.5 percent
    b. 1 percent
    c. 5 percent

13. What is the maximum subcooling used in many modern supermarket refrigeration systems?

    a. 15°F
    b. 25°F
    c. 45°F

14. What is heat reclaim?

    a. It is heat added to the refrigerant for proper condensation.
    b. It is the removal of superheat from the hot gas to heat the building or water.
    c. It is the heat blown out of the condenser with a split coil.

15. If the store humidity is above ___, it will adversely affect the operation of the refrigerated cases.

    a. 50 percent
    b. 60 percent
    c. 70 percent

16. In addition to properly sized A/C, how can the refrigeration system help to maintain store humidity?

    a. Reheating the conditioned air from the heat reclaim system
    b. Splitting the condenser to maintain head pressure
    c. Adjusting the EPRs to maintain proper evaporator temperatures

17. How does the refrigerated air stay inside an open display case?

    a. An air curtain blows down the front opening of the case to prevent outside air from entering the case.
    b. The cold air stays in the case because heat migrates to cold.
    c. It does not stay in the case, which is why so much refrigeration is needed.

18. What are some common causes of airflow problems in refrigerated cases?

19. Why is proper installation of supermarket equipment so important?

    a. Because efficiency can be derived only from good installations
    b. Because the larger the system and the longer the piping runs, the more opportunity for problems and the harder they are to troubleshoot
    c. Because there are few service technicians qualified to service supermarkets

20. How does a good refrigeration maintenance program benefit the grocery store owner?

    a. Regular maintenance can prevent major breakdowns and emergency service calls.
    b. The service technician becomes familiar with what the owner wants.
    c. Lost product and emergency service can be very costly to supermarkets.
    d. All of the above.

21. What distinguishes a high-pressure oil return system from a low-pressure system?

    a. The high-pressure system has a combination oil separator and oil reservoir.
    b. The high-pressure system has an oil separator and separate oil reservoir.
    c. There is an oil separator only on a high-pressure oil return system.

22. Sporlan's Y-valve (Y1236-C) would be used on what type of oil return system?

    a. Both high and low pressure
    b. High pressure
    c. Low pressure

23. What is the function of oil pressure differential valves on rack oil return systems?

    a. To make sure the oil is pulled out of the oil separator
    b. To ensure the oil pressure is correct for lubricating the compressor
    c. To lower the pressure of the separated oil before it enters the crankcase

24. What is blow-by?

    a. Discharge pressure getting into the compressor crankcase
    b. Suction pressure getting by the discharge valves
    c. High-pressure oil entering the crankcase

25. Blow-ny can cause _____?

    a. Oil failure control tripping
    b. Compressor motor overload tripping
    c. Both of the above

26. Why is an EEPR (Electronic Evaporator Pressure Regulator) better for case temperature control than a mechanical EPR or tstat with solenoid pump-down?

    a. It does not pull down too quickly after defrost.
    b. It has less temperature swing.
    c. It limits the amperage draw on the compressor.

27. What is the EEPR sensing in order to control case temperature?

    a. The suction line temperature
    b. The evaporator suction pressure
    c. The evaporator discharge air temperature

28. What is meant by "analog inputs" when talking about controllers?

    a. Signals either "on" or "off" such as a push button switch
    b. Signals that may vary in magnitude or flash intermittently
    c. Signals in a linear range of values such as those from a temperature probe

29. Which best describes an NTC (negative temperature coefficient) temperature probe?

    a. As temperature goes down, resistance remains stable.
    b. As temperature goes down, resistance goes up.
    c. As temperature goes up, resistance goes down.

30. Which best describes a "Dry Contact" in a relay circuit?

    a. Makes primary circuit on one throw but not on the other; example: DPDT relay
    b. Does not make or break the primary voltage; example: auxiliary switch
    c. Both makes and breaks the primary circuit; example: contactor

31. What does "daisy chain" mean when referring to electronic controllers?

    a. Linking controllers in series
    b. Linking controllers in parallel
    c. Linking controllers in both series and parallel

32. R404A has a GWP (Global Warming Potential) of 3922, so what is the GWP of CO_2?

    a. 2,107
    b. 1,430
    c. 1

33. What is the ASHRAE refrigerant designation for CO_2?

    a. R717
    b. R718
    c. R744

34. What is the critical point of a refrigerant?

    a. Temperature above which it cannot be condensed into a liquid
    b. Pressure at which a refrigerant will flash off
    c. Temperature at which it will sublimate

35. What is meant by a *subcritical* refrigerant process?

    a. It takes place above the refrigerant's critical point.
    b. It takes place below the refrigerant's critical point.
    c. It takes place just below the refrigerant's state of volatility.

36. What does *transcritical* mean in a CO_2 refrigeration process?

    a. Heat rejection is above its critical point while heat absorption is below it.
    b. Heat rejection is below its critical point while heat absorption is above it.
    c. The entire process takes place above its critical point.

37. Which of the following is important to remember before opening the canister of a replaceable filter core filter-drier on a CO_2 system?

    a. Pressurize the component to 29.4 psia
    b. Evacuate the component to 500 microns
    c. Pump down the component to atmospheric pressure

38. Because of the difference in volumetric efficiency between R404A and CO_2, the suction lines of a CO_2 system will be _____.

    a. Larger than an R404A system
    b. Smaller than a R404A system
    c. The same for the first 50 feet of run, then increased by 25 percent

39. Which of the following piping materials are currently being used on CO_2 systems?

    a. Stainless steel
    b. Galvanized steel
    c. Type M copper

WALK-IN REFRIGERATORS AND FREEZERS

11

CHAPTER OVERVIEW

Walk-in boxes are one of the most fascinating aspects of commercial refrigeration. They are the largest style of refrigeration equipment and very important to our customers' operations. Walk-ins allow the store owner to save money by making bulk purchases, and to have enough product on hand for the store's operation.

This chapter provides an introduction to how refrigeration systems are selected for walk-ins, how both the box and the refrigeration are installed, and the specific service problems associated with walk-ins. As in any troubleshooting, these systems require a basic knowledge of how a correctly sized and installed walk-in box and refrigeration system are supposed to operate.

OBJECTIVES

After completing this chapter, you should be able to understand

• Walk-in box types and sizes
• Walk-in box installation
• Walk-in door types and adjustments
• Walk-in box applications
• Matching system components
• Refrigeration piping
• Drain piping
• Troubleshooting walk-in problems

WALK-IN BOX TYPES AND SIZES

The insulated panels of walk-ins are normally 4 inches thick and are locked together with special fasteners (see Figure 11-1). Most panels use polyurethane foam that has an insulation value of about R33, equivalent to 10 inches of fiberglass insulation. Unlike fiberglass, closed cell polyurethane does not absorb moisture and therefore maintains its insulating qualities for the life of the walk-in. Panels up to 6 inches thick with R50 insulation value are available and highly recommended for low-temperature freezer applications below −10°F.

FIGURE 11-1 **Walk-in panel locks foamed in place (left) and locks in wood framing (right).** *Courtesy of Jerry Meyer, Hussmann Learning Solutions.*

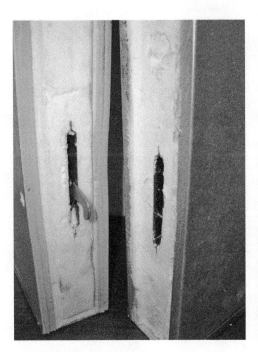

Another type of insulation, polystyrene, is offered by some walk-in manufacturers. Although it is a foam insulation its insulating values are not much better than fiberglass.

Walk-ins are identified by their size and temperature application. The size is given by the outside dimensions of length, width, and height. The narrower dimension of length or width is given first, and height is last. The temperature application is either refrigerator (cooler) or freezer.

FIGURE 11-2 **Installing walk-in panels.** *Courtesy of Imperial Brown Inc.*

EXAMPLE: 1

A walk-in freezer is 10 feet by 12 feet by 8 feet high (10 feet × 12 feet × 8 feet).

Walk-in panels are shipped from the factory and the installing technicians assemble the walk-in on the job site (see Figure 11-2). The smallest walk-ins are called step-ins. Usually they are about 4 feet × 4 feet × 6 feet and have a self-contained refrigeration unit in the top panel (see Figure 11-3).

FIGURE 11-3　**Step-in box.** *Courtesy of Imperial Brown Inc.*

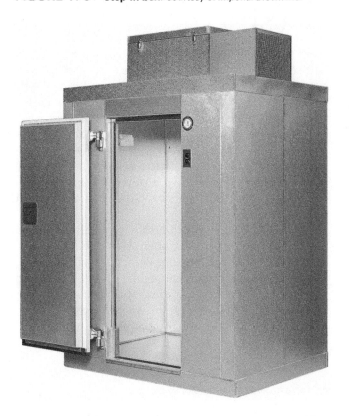

The largest walk-ins are called "refrigerated buildings" or "refrigerated warehouses." They are so big that forklift trucks can be driven inside to load and unload product.

WALK-IN BOX INSTALLATION

Installing a walk-in begins with a level floor. If the floor is not level, then the box will not go together correctly, and the door will not close properly.

The floor of a walk-in refrigerator can be either insulated floor panels or the concrete floor of the building (see Figure 11-4). The building floor can be used if it is on grade (concrete on dirt) because the temperature of the ground under the slab is usually a constant 55°F. The small difference between the ground temperature and a 35°F refrigerator requires little additional refrigeration capacity. If the concrete floor is not on grade, but is over an occupied space, the installation of a walk-in refrigerator without insulated floor panels could cause some very serious problems.

EXAMPLE: 2

A restaurant kitchen is over a parking garage, and the temperature in the garage during the summer is 80°F. A walk-in refrigerator without an insulated floor is installed in the kitchen. The heat from the garage will add to the refrigeration load. Under the walk-in, the ceiling of the garage becomes cold enough to condense water and drip on the cars. The problem could be worse if the area under the walk-in were an office space instead of a garage.

A walk-in freezer cannot be put on a concrete floor on grade because the cold box temperatures would freeze the ground under the slab. The frozen ground would expand, rise, and break up the slab floor. All freezers must have an insulated floor, or have insulation installed under the slab before the concrete floor is poured.

FIGURE 11-4　**Walk-in without floor panels (left) and walk-in floor panels with diamond tread plating (right).** *Photos by Dick Wirz.*

Walk-in installed on concrete floor

Pressure Relief Port

Inside release for exterior door handle

Floor panels with diamond tread plating

If heavy carts or hand trucks are used often on an insulated floor, the floor panels will soon become damaged. This can be prevented by installing 3/16-inch-thick diamond tread aluminum plates on the floor panels in high-use areas (see Figure 11-4).

Concrete floors are seldom perfectly level. Therefore, the installed floor panels will need treated wood or metal shims installed under the panels to ensure the floor of the box is perfectly level. If the walk-in does not have floor panels, the vertical wall panels installed on the concrete base will probably have to be shimmed in order to level the box.

The importance of a level base cannot be emphasized enough. Without a level foundation, the wall and ceiling panels will not latch together, and the door frame will not be aligned. Since the door is the only moving part in the entire walk-in, it is the most noticeable section. If a door does not close or seal properly, it will be a constant source of irritation to the user. This negative impression is often transferred to both the manufacturer of the box and the company that installed it.

WALK-IN DOOR TYPES AND ADJUSTMENTS

Most walk-in doors are flush-mounted, self-closing, and self-sealing. This means the door is even with the wall when it is closed, the hinges are designed to make the door swing closed by itself, and when the door is closed, the gaskets are tight to the frame to seal in the cold.

Some people are very frightened about being locked in a walk-in. However, the door sections are manufactured with an inside-release mechanism. This safety device allows a person to easily open the door, even if the door handle is padlocked (see Figure 11-5).

Cam-lift hinges lift the door vertically about an inch as the door is swung open. When the door is released, the weight of the door, combined with the angle of the hinge insert cams, causes the door to rotate back to its closed position. Optional spring-loaded hinges put even more pressure on the door to ensure it closes (see Figure 11-6).

FIGURE 11-5 **Door releases open walk-ins from inside, even if the door is locked.** *Photos by Dick Wirz.*

Outside handle and inside door release

Outside handle with padlock

EMERGENCY RELEASE
REMOVE KNOB
BY TURNING COUNTERCLOCKWISE.
PUSH DOOR OUTWARD.

FIGURE 11-6 **Door hinges and closer.** *Photos by Dick Wirz.*

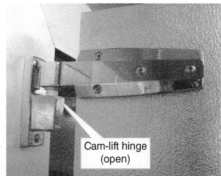

Service problems will definitely occur if the walk-in door does not close fully. A slightly open door will allow heat into the walk-in. To prevent this problem, most doors can be pulled closed by a device simply called a door closer (see Figure 11-6). The closer is a mechanism fastened to the door frame. It uses a hook and a spring or hydraulic piston to engage and pull the door fully closed when the door has swung to within an inch of closing.

Most walk-in door gaskets contain a magnetic strip that is attracted to the metal door frame. This magnet helps both to keep the door closed and to seal out heat. This type of gasket is similar to those used on residential refrigerators.

The bottom of the door has a flap of rubber that covers the opening between the bottom of the door and the threshold. This gasket is called a sweep gasket. The sweep gasket should not be so long that it interferes with the closing of the door. If necessary, the technician can trim or adjust the gasket. Most manufacturers recommend an air space between the bottom of the gasket and the threshold, not to exceed 1/8 inch.

Wire strip heaters are used on the door frame of all freezers to prevent the door from freezing to the frame when it is closed. When the freezer is in operation, these heaters do not seem to make the door frame very warm. However, they are doing the job adequately as long as the gasket does not freeze to the door frame when the door is closed.

If a freezer door is left slightly ajar, the heat from outside the box migrates to the cold freezer interior. Moisture in the air freezes on the unheated portions of the door and frame. The ice buildup will block the door from closing. Technicians should instruct their customers to check daily for ice on the door frame and to gently clear any ice that does appear. If the door is forced closed before the ice is removed, it will damage the door, the frame, the heaters, and the gaskets. This will result in a very expensive and time-consuming repair that could have been prevented if the door had been fully shut every time after opening.

Figure 11-7 shows several things previously discussed, including the door closer, gasket, frame heaters, sweep gasket, and threshold. In addition, the drawing shows a pressure relief port (PRP) that has a two-way flap arrangement. A PRP allows pressure differences between the inside and outside to equalize. Although they are not necessary in medium-temperature units, they should be installed on freezers.

EXAMPLE: 3

When a freezer goes into defrost, the air inside the evaporator cabinet will become 50°F to 70°F warmer than the box temperature. Although the space occupied by the evaporator

FIGURE 11-7   **Side view of freezer door frame.** *Courtesy of Refrigeration Training Services.*

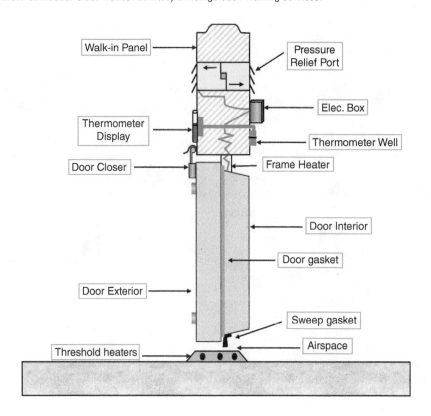

is relatively small, the expansion of the warm air can increase the pressure throughout the box, especially in smaller walk-ins. The increase in pressure inside the walk-in could be enough to force the walk-in door open. If the box has a PRP, the pressure inside will push open the flap and allow the pressure to equalize inside the box.

EXAMPLE: 4

When the door is opened, warm air flows into the box. As the warm air is cooled, it takes up less space, creating a vacuum. If someone were to try to reopen the door within a minute or so, this slight vacuum would tend to hold the door closed, making it very difficult to open. A PRP would relieve the vacuum inside the box by allowing some outside air to enter and equalize the pressures.

All manufacturers mount thermometers on the door frame with the thermometer bulb sensing the air temperature inside the box. Unfortunately, when the door is opened, the air from the outside causes the temperature to rise. After the door is closed, it may take as long as 10 minutes for the thermometer to return to the actual box temperature.

Some technicians use a procedure that helps ensure the door thermometer is less affected by door openings. Figure 11-7 shows a thermometer with its bulb in a "thermometer well." The well is a 6-inch piece made

of 1-inch copper or PVC (polyvinyl chloride) pipe with a cap on the bottom. The pipe is fastened to the inside wall of the walk-in where the thermometer bulb comes through the panel. The bulb is inserted into the pipe, and then covered with food-grade glycol. This arrangement prevents the temperature fluctuations caused by door openings and is accurately indicates the temperature of the product stored in the box.

For the door to seal properly, door frame must be both level and plumb. To check this, place a level across the top of the door frame. If leveling is necessary, loosen the door frame locks and shim the door frame or level the box floor (see Figure 11-8).

To check for plumb, place the level vertically on the face of the door jamb. If adjustment is needed, loosen the panel locks and push the frame in or out. Although the misalignment may only be 1/8 to 1/4 inch, it could greatly affect how the door seals when closed (see Figure 11-9).

Sometimes the frame is as level and plumb as possible, yet one corner of the door may be slightly sticking out from the frame. Shimming the hinge plate opposite the problem corner will twist the door and may bring it back into alignment. Use thin metal shims under the hinge plate that is on the door frame, not on the door itself. As illustrated in Figure 11-10, shimming out the bottom left hinge will push in the top right corner of the door.

FIGURE 11-8 **Leveling a door frame.** *Courtesy of Refrigeration Training Services.*

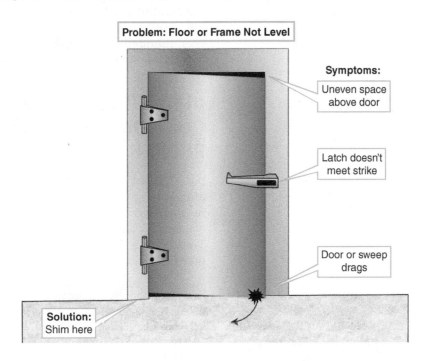

FIGURE 11-9 **Plumbing a door frame.** *Courtesy of Refrigeration Training Services.*

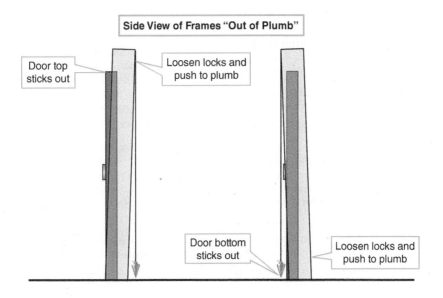

WALK-IN BOX APPLICATIONS

When matching the refrigeration equipment to a walk-in box, it is important to know the walk-in's application. In other words, is the walk-in to be used as a freezer or a refrigerator (cooler)? Is it considered a storage walk-in that needs to store only previously refrigerated product? Or is it a production walk-in that will be subjected to heavy loads and maybe even required to pull down warm product?

Almost all walk-in refrigerators are currently rated at a box temperature of 35°F. Most walk-in freezers are rated for −10°F. However, occasionally a manufacturer will rate its system as a 40°F cooler, or a 0°F freezer. These higher box temperatures require less refrigeration horsepower than the lower temperatures. Therefore, if two manufacturers differ in the refrigeration capacity required for the same-sized walk-in, it could be a difference in the design temperatures of the boxes.

FIGURE 11-10 **Shim hinges to twist the door.** *Courtesy of Refrigeration Training Services.*

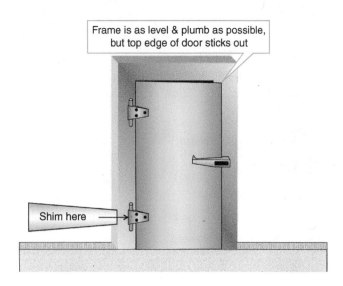

Frame is as level & plumb as possible, but top edge of door sticks out

Shim here

When a manufacturer supplies a chart that gives the size of the refrigeration equipment based on the physical dimensions of the box, then the walk-in is rated for storage only. Storage-only means the listed refrigeration unit will maintain the rated temperature inside the walk-in as long as the ambient temperature around the box is not over 95°F and there are no more than two door openings per hour. Also, and this is the important part, the product to be refrigerated must enter the box within a few degrees of the walk-in's rated temperature.

In other words, a walk-in rated at 35°F will maintain the product at 35°F, as long as the temperature of the product entering the box is near 35°F. At these conditions the compressor will not have to run more than 16 hours per 24-hour period.

NOTE: **Only 16 hours of run time may seem to allow for plenty of reserve capacity, if needed. However, refrigerators need to off-cycle defrost. For instance, a cycle of 10 minutes of run time and 5 minutes of off-cycle would allow only a total run time of 16 hours per day.**

Figure 11-11 is an example of a walk-in equipment sizing chart. All the charts in this chapter for the walk-ins, condensing units, and evaporators are to be used

only for the examples in the following sections and for the questions at the end of the chapter. The Btuh loads are close to actual factory ratings, but the figures have been rounded off and adjusted to make the examples and calculations easier to understand.

MATCHING SYSTEM COMPONENTS FOR PROPER OPERATION

For an 8-foot × 12-foot × 8-foot walk-in 35°F refrigerator, see Figure 11-11 for the amount of refrigeration needed. Next, see Figure 11-12 to determine which condensing unit would provide the necessary capacity. Then skip to Figure 11-14 to pick the evaporator that matches the condensing unit.

EXAMPLE: 5

An 8-foot × 12-foot × 8-foot refrigerator (35°F) needs 8,000 Btuh of refrigeration (Figure 11-11). The MT-100, a medium-temperature 1-hp condensing unit, provides 8,000 Btuh at a 25°F suction temperature (Figure 11-12). The MEV-080 evaporator (Figure 11-14) exactly matches the 8,000 Btuh output of the MT-100 unit.

T . R . O . T

MATCHING COMPRESSOR HORSEPOWER TO BTUH
- 1 hp = 12,000-Btuh air conditioning (75°F room, 40°F evaporator, 10 SEER and below)
- 1 hp = 8,000-Btuh medium-temperature refrigeration (35°F box, 25°F suction temperature, 95°F ambient)
- 1 hp = 4,000-Btuh low-temperature refrigeration (−10°F box, −20°F suction temperature, 95°F ambient)

For example, assume the manufacturer states that an 8-foot × 28-foot × 8-foot walk-in refrigerator requires about 16,000 Btuh. Based on TROT, the walk-in should have about a 2-hp condensing unit (16,000 ÷ 8,000 = 2 hp). However, if the walk-in were a −10°F freezer, it should have about a 4-hp low-temperature unit (16,000 ÷ 4,000 = 4 hp).

NOTE: **These are close approximations, but always use factory specs if the calculations must be accurate.**

FIGURE 11-11 **Walk-in box refrigeration equipment sizing chart for storage only.** *Courtesy of Refrigeration Training Services.*

WALK-IN BOX EQUIPMENT SIZING		
Box Dimensions	**35°F Box (25°F Evap) Btuh Load**	**−10°F Box (−20°F Evap) Btuh Load**
8-foot × 8-foot × 8-foot	6,000	6,000
8-foot × 12-foot × 8-foot	8,000	8,000
8-foot × 20-foot × 8-foot	12,000	12,000
8-foot × 28-foot × 8-foot	16,000	16,000

FIGURE 11-12 **Medium-temperature condensing unit chart.** *Courtesy of Refrigeration Training Services.*

MEDIUM-TEMPERATURE CONDENSING UNITS @ 25°F SUCTION TEMPERATURE		
Unit Model	**Horsepower**	**Btuh**
MT-075	3/4	6,000
MT-100	1	8,000
MT-150	1 1/2	12,000
MT-200	2	16,000
MT-300	3	24,000

FIGURE 11-13 **Low-temperature condensing unit chart.**
Courtesy of Refrigeration Training Services.

LOW-TEMPERATURE CONDENSING UNITS			
Unit Model	**Horsepower**	**Btuh @ − 10°F Suction**	**Btuh @ − 20°F Suction**
LT-075	3/4	4,000	3,000
LT-100	1	5,000	4,000
LT-150	1 1/2	7,500	6,000
LT-200	2	10,000	8,000
LT-300	3	15,000	12,000
LT-400	4	20,000	16,000

If the 8-foot × 12-foot × 8-foot walk-in was a freezer, it would also require 8,000 Btuh. However, it would require a low-temperature condensing unit instead of the medium-temperature unit.

EXAMPLE: 6

According to the chart in Figure 11-13, the low-temperature 2-hp condensing unit, LT-200, will provide the necessary 8,000 Btuh at a −20°F suction temperature. The low-temperature evaporator LEV-080, in Figure 11-14, will match it.

FIGURE 11-14 **Evaporator rating charts for refrigerators and freezers.** *Courtesy of Refrigeration Training Services.*

EVAPORATORS			
Medium Temp. 25°F Evap. @ 10°F TD		**Low Temperature −20°F Evap. @ 10°F TD**	
Model	**Btuh**	**Model**	**Btuh**
MEV-040	4,000	LEV-040	4,000
MEV-060	6,000	LEV-060	6,000
MEV-080	8,000	LEV-080	8,000
MEV-100	10,000	LEV-100	10,000
MEV-120	12,000	LEV-120	12,000
MEV-140	14,000	LEV-140	14,000
MEV-160	16,000	LEV-160	16,000
MEV-180	18,000	LEV-180	18,000
MEV-200	20,000	LEV-200	20,000
MEV-220	22,000	LEV-220	22,000

The previous two examples show the basics of matching refrigeration equipment to the size of a walk-in used for storage-only. Following is a summary of the process:

1. Determine the Btuh requirements of the walk-in box.
2. Determine which condensing unit will provide the required Btuh.
3. Match the evaporator to the condensing unit with the same Btuh.

NOTE: Most walk-ins operate at a 10°F TD. Therefore, choose equipment that operates at a suction temperature (evaporator temperature) that is 10°F below the box temperature.

The −10°F walk-in freezer condensing unit in the previous example had to be picked from the column "Btuh at −20°F Suction" in Figure 11-13. If the −10°F column had been incorrectly chosen, the condensing unit would be the LT-150, which produces 7,500 Btuh at −10°F suction, but only 6,000 Btuh at the required −20°F suction. The LT-150 refrigeration system could not handle the 8,000 Btuh load at −20°F suction for an 8-foot × 12-foot × 8-foot walk-in freezer operating at a box temperature of −10°F.

Matching Compressor Horsepower to Btuh

A condensing unit has a Btuh capacity and a rated horsepower. For a given horsepower, the Btuh capacity will vary, depending on the compressor type and condenser size used on the condensing unit. Therefore, factory literature has pages and pages of different ratings for the same horsepower units. However, when on the job without the factory manuals, remembering the rules of thumb for determining condensing unit capacity can be very useful.

Matching Evaporators to Condensing Units

As discussed in Chapter 2, "Evaporators," the standard 10°F TD system is good for most food service applications where high humidity (85% relative humidity [RH]) is needed to prevent fruits, vegetables, and other moisture-sensitive foods from drying out. A 10°F TD commercial refrigeration system is obtained by simply matching the Btuh of the condensing unit with the evaporator. A 16,000-Btuh condensing unit matched with a 16,000-Btuh evaporator would maintain a 10°F TD system. Following is a formula to calculate system TD:

system TD = condensing unit rating ÷ evaporator rating (at a 1°F TD)

NOTE: Most evaporators are rated at a 10°F TD. Therefore, 1°F TD is 10 percent of the original rating.

EXAMPLE: 7

A condensing unit rated at 16,000 Btuh is matched to an evaporator with a 16,000 Btuh capacity at a 10°F TD (or 1,600 Btuh at a 1°F TD).

16,000 ÷ 1,600 = 10°F system TD (maintains about an 85% RH)

What would happen to the box temperature and humidity if the condensing unit and the evaporator had different ratings?

EXAMPLE: 8

The condensing unit rating is 16,000 and the evaporator is rated at 10,000.

16,000 ÷ 1,000 = 16°F system TD
(maintains about a 75% RH)

The greater the system TD, the colder the evaporator becomes. The colder the coil surface, the more moisture is condensed on the evaporator and removed from the space. Therefore, the humidity falls. A 16°F TD may dry out unpackaged meat, vegetables, and other moisture sensitive products. However, it would not be a problem with wrapped product or bottles. If a similar match were encountered on the job, it may have been intentional because available space in the walk-in limited the size of the evaporator. As a matter of fact, reach-ins normally have 15°F to 20°F TD systems because the lack of space in the cabinets necessitates small evaporators.

Using a smaller evaporator can reduce equipment costs, but it will also reduce the overall system Btuh capacity. System capacity is addressed in the next section.

If the system TD is less, the evaporator temperature is higher; less moisture will condense on the evaporator. Therefore, the humidity in the box increases. High humidity is important for floral and fresh-meat walk-ins. In addition, low airflow evaporators are often used to reduce the chances of air currents damaging delicate flowers or turning meat surfaces darker.

EXAMPLE: 9

The condensing unit rating is 16,000, and the evaporator is rated at 20,000.

16,000 ÷ 2,000 = 8°F system TD
(maintains about a 90% RH)

System Capacity Based on Condensing Unit and Evaporator Matching

Total system capacity is determined by how the condensing unit is matched to the evaporator. The system capacity is the average of the two components, as long as the evaporator capacity is less than or equal to the condensing unit capacity. Following is the formula for system capacity:

system capacity = (condensing unit capacity
+ evaporator capacity) ÷ 2

EXAMPLE: 10

The condensing unit rating is 16,000, and the evaporator is rated at 16,000.

system capacity = (16,000 + 16,000) ÷ 2 = 32,000 ÷ 2
= 16,000 Btuh

EXAMPLE: 11

The condensing unit rating is 16,000 and the evaporator is rated at 10,000.

system capacity = (16,000 + 10,000) ÷ 2 = 26,000 ÷ 2
= 13,000 Btuh

The previous example shows how a smaller evaporator lowers total system capacity.

There is an exception to determining total system capacity by averaging the two components. That exception is when the evaporator capacity is more than the capacity of the condensing unit.

EXAMPLE: 12

The condensing unit rating is 16,000 and the evaporator is rated at 20,000.

If the evaporator is larger than the condensing unit, the system capacity is basically limited to the capacity of the condensing unit, in this case 16,000 Btuh. The reasoning is that the evaporator cannot absorb any more heat from the space than the condensing unit can reject.

Matching the TEV to the Evaporator

Chapter 5, "Metering Devices," states that the capacity of a TEV depends on the pressure drop across the valve and the temperature of the liquid refrigerant entering the valve. Use the manufacturer's charts to properly determine the capacity of a TEV based on the expected operating conditions. Select the valve that has a capacity closest to the Btuh rating of the evaporator on which it is to be installed.

NOTE: Sizing the TEV to the evaporator is correct only under the following conditions:

1. When the evaporator and the condensing unit are the same capacity
2. For medium-temperature applications (freezers will be discussed later)

Matching by nominal tonnage (the rating printed on the valve) is pretty close to the rating tables on medium-temperature refrigeration. For example, a nominally rated 1-ton TEV will provide about 12,000 Btuh of refrigeration.

EXAMPLE: 13

Evaporator capacity is 12,000 Btuh; TEV capacity should be 1 ton.

EXAMPLE: 14

Evaporator capacity is 9,000 Btuh; TEV capacity should be 3/4 ton.

T . R . O . T

When between valve capacities, choose the higher rated TEV.

EXAMPLE: 15

The evaporator is rated at 9,000 Btuh, but the TEV comes only in 1/2-ton (6,000 Btuh) and 1-ton (12,000 Btuh) capacities. Which would you choose?

The smaller 1/2-ton valve will probably starve the evaporator, especially during start-up or hot pull-down. However, the larger 1-ton valve will provide plenty of refrigerant, and it will do so without flooding the compressor. An expansion valve is designed to properly maintain superheat even when it is slightly oversized.

Special TEV Sizing

Proper sizing of a TEV for low-temperature systems requires looking up the TEV capacity on the manufacturer's chart. For example, an F-type Sporlan valve rated for 2 tons (24,000 Btuh) at a +20°F evaporator temperature would produce only 1.68 tons at a −20°F (minus 20°F) evaporator. Therefore, the next available size is 3 tons, which would provide a sufficient capacity of 2.10 tons at the same low-temperature conditions.

TEV sizing is easy on medium-temperature systems that have matched condensing unit and evaporator capacities. However, the capacity of an evaporator is primarily a function of the size of the condensing unit connected to its outlet and the tonnage of the TEV feeding it. Earlier in the chapter it was shown that the TD of a system is the condensing unit capacity divided by the evaporator capacity at a 1°F TD. For example, a 12,000-Btuh evaporator rated at 10°F TD will have a 1,200 Btuh rating at 1°F TD. As the TD is increased, so is the evaporator's capacity. The evaporator in the example will provide 12,000 Btuh at 10°F TD; 24,000 Btuh at 20°F TD; 36,000 Btuh at 30°F TD; and so on.

REALITY CHECK 1

Sometimes the factory charts are not available and the decision for TEV sizing must be made based only on the nominal tonnage of the valve. Therefore, the following examples show how to match the TEV to the evaporator capacity.

Assume a customer needs a 35°F walk-in to handle a load of 24,000 Btuh, but wants a drier box of 65 percent maximum humidity rather than the 85 percent humidity of a 10°F TD system. A 20°F TD will give the customer about 65 percent, but what size evaporator would we use? Earlier in this chapter we calculated the system TD by the following equation:

condensing unit capacity ÷ evaporator capacity at 1°F TD = system TD

We can use the same equation to solve for the evaporator size:

condensing unit capacity ÷ system TD = evaporator capacity at 1°F TD

24,000 Btuh ÷ 20°F TD = 1,200 Btuh at 1°F TD

= 12,000 Btuh at 10°F TD

= 24,000 Btuh at 20°F TD

Since the evaporator is now expected to handle 24,000 Btuh, it will need to be fed by a 2-ton-capacity TEV. In addition, the condensing unit will need to have a matched rating of 24,000 Btuh at an evaporator temperature of 15°F, which is 20°F below the 35°F box temperature (20°F TD). This is one example of why some TEVs may seem to be mismatched to the evaporator, but in fact are perfectly correct. It also shows why some technicians believe the TEV should be sized to the condensing unit capacity.

Another method of determining system capacity and TEV sizing is the 3:1 rule. This method uses the average of the condensing unit capacity plus the evaporator capacity, but it puts three times more emphasis on the capacity of the condensing unit.

First, multiply the condensing unit capacity by three, then add the capacity of the evaporator at the system TD. Finally, divide by four to get the average. In the previous example the equation would be:

[(condensing unit capacity × 3) + evaporator capacity at system TD] ÷ 4 = TEV capacity
[(24,000 × 3) + 24,000] ÷ 4 = 24,000 Btuh

Assume the design calls for the box temperature to be 35°F as before, but the condensing unit is to operate at a 25°F suction temperature, which means a 10°F TD. The evaporator will be at its original 12,000-Btuh rating at a 10°F TD, and the condensing unit's rating at the higher suction temperature is now only 20,000 Btuh.

Applying the 3:1 rule:

[(20,000 × 3) + 12,000] ÷ 4 = 18,000 Btuh TEV capacity

Although the 3:1 rule works pretty well in the field, it is not the most accurate method. The walk-in box and expansion valve manufacturers use charts to determine the correct TEV for the application. These charts allow them to plot the total capacity of a system based on the condensing unit and evaporator performance data. Whenever possible, it would be best to consult the manufacturer when sizing an expansion valve on a system where the condensing unit and the evaporator are not exactly matched.

Production, or Pull-down, Boxes

Storage walk-ins are designed to maintain previously refrigerated product at approximately the same temperature it was when brought into the box. Production, or pull-down, boxes are used to reduce the temperature of specific products in a certain amount of time. Because of the warm product, the fast pull-down time, and other factors, the refrigeration requirements are quite different from a storage-only walk-in. Equipment sizing for these systems should be done by the walk-in manufacturer's application engineers. They are experienced in calculating refrigeration loads based on specific requirements.

The aforementioned form is an example of a load estimate for a 35°F convenience store cooler with glass doors. Additional calculations are made for an uninsulated concrete floor and the product loads for sodas and dairy products with an entering temperature higher than the box. This form is adopted from the one used by Heatcraft, a refrigeration-equipment manufacturer.

Based on the sizing chart in Figure 11-11, a 35°F storage-only 8-foot × 28-foot walk-in refrigerator would require a 16,000-Btuh refrigeration system, which is about a 2-hp unit. However, the same-sized box in Figure 11-15 has some additional

FIGURE 11-15 **Walk-in load estimate form.** *Adapted from Heatcraft, Inc.*

Refrigeration Load Estimate Form (adapted from Heatcraft, Inc.)
35° Convenience Store Cooler with Glass Doors

Basis for Estimate
Room Dimensions: Width __8__ ft. × Length __28__ ft. × Height __8__ ft. Volume: __1,792__ cubic feet
Ambient Temp. __85__ °F. (corrected for sun load) - Room Temp. __35__ °F. = __50__ °F. T.D.
Insulation: CEILING 4" Styrene   WALLS 4" Styrene   FLOOR 6" Concrete

Product Load
  (a) __2,000__ lbs./day of __Soda__ to be reduced from entering temp. of __85__°F. to __35__°F. Temp. Drop __50°F__
  (b) __200__ lbs./day of __Dairy__ to be reduced from entering temp. of __40__°F. to __35__°F. Temp. Drop __5°F__

Miscellaneous
Fan motors __0.2__ HP   Ground Temp. __60__   Lights (1 watt/sq. ft.) __224__ Watts   No. of people _____

1. Transmission Loads

Ceiling: (L)	28	× (W)	8	× Heat Load	72	(Table 1)	=	16,128
North Wall: (L)	28	× (H)	8	× Heat Load	72	(Table 1)	=	16,128
South Wall: (L)	28	× (H)	8	× Heat Load	72	(Table 1)	=	16,128
East Wall: (W)	8	× (H)	8	× Heat Load	72	(Table 1)	=	4,608
West Wall: (W)	8	× (H)	8	× Heat Load	72	(Table 1)	=	4,608
Floor: (L)	28	× (W)	8	× Heat Load	125	(Table 1)	=	28,000

2. Air Change Load
  Volume: __1,792__ cu ft. × __19.5__ Factor (Table 4) × __1.86__ Factor (Table 6)   = __64,996__

3. Additional Loads

Electrical Motors:	0.2	HP × 75,000 BTU/HP/24 hr.	=	15,000
Electrical Lights:	224	Watts × 82	=	18,368
People Load: _____	People × _____	BTU/24 hrs. (Table 12)	=	
Glass Door Load:	10	Doors × 19,200 BTU/Door/24 hr.	=	192,000

4. Product Load: Sensible   (Product Load Figured @ 24 hr. Pulldown)*
  (a) __2,000__ lbs./day × __0.9__ Spec. Heat (Table 7) × __50__ °F Temp. Drop = __90,000__
  (b) __200__ lbs./day × __0.7__ Spec. Heat (Table 7) × __5__ °F Temp. Drop = __700__
  *For product pulldown time other than 24 hrs. figure 24 hr. load × (24/Pulldown Time)

5. Product Load: Respiration
  _____ lbs. stored × _____ BTU/lb./24 hrs. (Table 8)   = _____

Total Refrigeration Load (1+ 2 + 3 + 4 + 5) BTU./24 hrs.	466,664
Add 10% Safety Factor	46,666
Total with Safety Factor BTU./24 hrs.	513,330

Divide by No. of Operating Hrs. (16) to obtain BTUH Cooling Requirement	**32,083**

loads that the storage-only unit does not. Without going into the specifics of each load calculation, the important point is that the floor, product loads, and glass doors add another 16,000 Btuh to the refrigeration requirements of this walk-in. Therefore, to properly refrigerate a same-sized walk-in but with a higher load would require a 32,000-Btuh, or 4-hp, refrigeration system.

Application engineers will need the following information to properly calculate product loads for a proposed walk-in:

1. Type of product (examples: dairy, meat, type of vegetables)
2. Amount of product
3. Entering temperature of product
4. Pull-down time (standard is 24 hours)

Refrigeration Piping

Once the refrigeration system is selected, the next step is to determine the proper size of liquid and suction lines between the condensing unit and the evaporator. The primary concerns for refrigeration piping are to keep pressure drop to a minimum, to prevent oil from lying in the suction line, to return oil to the compressor, and to make sure the suction line does not pick up ambient heat or cause water damage due to sweating.

Following are a few rules for proper refrigeration piping.

NOTE: **Always follow the recommendations of the manufacturer. However, the following basic guidelines are usually adequate for systems up to 3 hp.**

1. Use a piping chart for sizing suction and liquid lines (see Figure 11-16).
2. Determine equivalent length of pipe by one of the following two methods:
   a. Length of pipe plus 3 feet per fitting
   b. Length of pipe multiplied by 1.5 (the quick method)
3. Slope suction lines downward 1/16 inch per foot in the direction of refrigerant flow. (This is a conservative figure. Most manufacturers recommend a slope between 1/2 and 1 inch for every 20 feet of horizontal run.)
4. Trap the suction line at the evaporator whenever the compressor is more than 3 feet above the evaporator.
5. Trap the suction line every 20 feet of vertical riser when the compressor is above the evaporator.
6. Install an inverted p-trap at the top of a vertical suction-line riser.
7. Insulate medium-temperature suction lines with 1/2-inch pipe insulation, and low-temperature suction lines with 1-inch pipe insulation.

When the condensing unit is located on a roof, too often the piping becomes a large oil trap. The dotted line in Figure 11-17 illustrates how this can happen. When the suction line exits the roof pipe chase (or gooseneck) it drops down to the roof and runs horizontally until it reaches the condensing unit. The horizontal section of the suction line traps the oil and restricts the suction line as effectively as if someone had flattened the pipe with a hammer. Figure 11-17 shows an alternate piping design for an application where the compressor is above the evaporator. In the diagram the suction line exits the gooseneck and is sloped downward to the condensing unit. This method prevents unnecessary traps. The picture in Figure 11-18 shows an example of good suction-line piping. At the top of the picture you can see an example of poor piping when the suction line is looped to form several oil traps.

For the following example, use the piping chart in Figure 11-16 to determine the line sizing for the pipe in a system similar to the one in Figure 11-17.

EXAMPLE: 16

A low-temperature system is 1 1/2 hp using R404A. The total piping run is 40 feet, and there are seven fittings. Determine the correct size of liquid and suction line.

First, figure the equivalent length of pipe. This is the actual length of pipe plus an additional length of straight pipe that would be equivalent to the pressure drop in the fittings. The following two methods show how to calculate the length of pipe in terms of equivalent feet. The first one actually accounts for every fitting; the second is a quick method used by many installers.

1. Equivalent feet = 40 feet + (7 fittings × 3 feet per fitting) = 40 feet + 21 feet = 61 feet
2. Equivalent feet = 40 feet × 1.5 = 60 feet

Refer to the low-temperature R404A piping chart at the bottom of Figure 11-16. Since the equivalent run is over 50 feet, it is necessary to use the line sizing under the next higher column, or 100 feet. Locate 1 1/2 hp on the chart and follow the row to the right until you intersect the column under 100 feet. The suction-line size for all horizontal piping is 7/8 inch to reduce pressure drop. However, under the Riser column, the maximum suction-line size for the vertical sections of piping is 5/8 inch. The vertical risers must maintain a sufficient velocity in order for the refrigerant to push the oil up the pipe.

The liquid-line sizing is on the same row, at the far right. According to the chart, the liquid line should be 3/8-inch pipe size. Because oil mixes well with the liquid refrigerant and the liquid is traveling at a high pressure, line slope and p-traps are not a concern when piping the liquid line.

FIGURE 11-16   **Piping chart.** *Courtesy of Refrigeration Training Services.*

REFRIGERATION PIPE SIZING CHART								
R134a R401A (MP39) R12 MEDIUM-TEMP 35° BOX (25° Suction)								
HP	BTUH	SUCTION LINE (Equivalent Feet)				MAX. SUCTION RISER	LIQUID LINE (feet)	
		25	50	100	150		50	100
1/4	2,000	3/8	3/8	1/2	1/2	1/2	1/4	3/8
1/3	3,000	1/2	1/2	5/8	5/8	1/2	3/8	3/8
1/2	4,000	1/2	5/8	5/8	7/8	5/8	3/8	3/8
3/4	6,000	1/2	5/8	7/8	7/8	5/8	3/8	3/8
1	8,000	5/8	7/8	7/8	7/8	7/8	3/8	3/8
1 1/2	12,000	7/8	7/8	1 1/8	1 1/8	7/8	3/8	3/8
2	16,000	7/8	7/8	1 1/8	1 1/8	1 1/8	3/8	1/2
3	24,000	7/8	1 1/8	1 1/8	1 3/8	1 1/8	1/2	1/2

R22 R404A MEDIUM-TEMP 35°F BOX (25°F SUCTION)								
HP	BTUH	SUCTION LINE (Equivalent Feet)				MAX. SUCTION RISER	LIQUID LINE (feet)	
		25	50	100	150		50	100
1/4	2,000	3/8	3/8	1/2	1/2	1/2	1/4	3/8
1/3	3,000	3/8	1/2	1/2	5/8	1/2	1/4	3/8
1/2	4,000	1/2	1/2	1/2	5/8	1/2	1/4	3/8
3/4	6,000	1/2	1/2	5/8	7/8	5/8	3/8	3/8
1	8,000	5/8	5/8	5/8	7/8	7/8	3/8	3/8
1 1/2	12,000	5/8	5/8	7/8	7/8	7/8	3/8	3/8
2	16,000	7/8	7/8	7/8	7/8	7/8	3/8	3/8
3	24,000	7/8	7/8	7/8	1 1/8	1 1/8	3/8	1/2

R404A LOW TEMP −10°F BOX (−20°F SUCTION)								
HP	BTUH	SUCTION LINE (Equivalent Feet)				MAX. RISER	LIQUID (feet)	
		25	50	100	150		50	100
1/4	1,000	3/8	3/8	1/2	1/2	1/2	1/4	1/4
1/3	1,500	3/8	3/8	1/2	1/2	1/2	1/4	1/4
1/2	2,000	3/8	1/2	1/2	5/8	1/2	1/4	3/8
3/4	3,000	1/2	1/2	5/8	7/8	1/2	3/8	3/8
1	4,000	1/2	5/8	7/8	7/8	5/8	3/8	3/8
1 1/2	6,000	5/8	5/8	7/8	7/8	5/8	3/8	3/8
2	8,000	5/8	7/8	7/8	1 1/8	7/8	3/8	3/8
3	12,000	7/8	7/8	11/8	1 1/8	7/8	3/8	3/8

The recommended insulation thickness on the suction line of this example is 1 inch.

In a large walk-in it is likely that there will be more than one evaporator on a single compressor. Multiple evaporators provide a more even airflow throughout the box. Figure 11-19 shows three methods of running the suction lines on a multiple evaporator installation where the compressor is above the evaporator.

Discharge Line Piping and Suction Line Double Risers

Discharge line piping from a compressor to a remote condenser follows most of the same rules of slope and p-traps as a suction line. Similar to cool suction vapor, hot discharge vapor moves oil through the discharge lines (see Figure 11-20). Discharge line piping is usually

FIGURE 11-17 **Refrigeration piping when the compressor is above the evaporator.** *Courtesy of Refrigeration Training Services.*

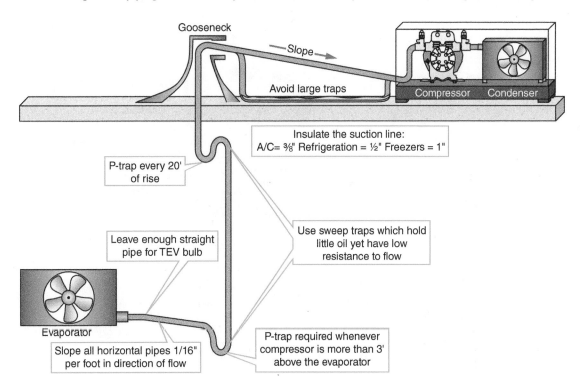

Gooseneck

—Slope—→

Avoid large traps

Compressor   Condenser

Insulate the suction line:
A/C= ⅜" Refrigeration = ½" Freezers = 1"

P-trap every 20'
of rise

Leave enough straight
pipe for TEV bulb

Use sweep traps which hold
little oil yet have low
resistance to flow

Evaporator

Slope all horizontal pipes 1/16"
per foot in direction of flow

P-trap required whenever
compressor is more than 3'
above the evaporator

FIGURE 11-18 **Examples of piping condensing units above the evaporator.** *Photo by Dick Wirz.*

Looped piping
traps oil

Proper piping for
good oil return

one pipe size larger than the required liquid line size. However, it is best to consult factory recommendations before deciding what size discharge line to run to a remote condenser.

Approximate minimum velocities of 700 FPM (feet per minute) in horizontal suction lines and 1,500 FPM in vertical lines have been recommended and used successfully for many years for standard suction-line sizing. On large systems with capacity-control compressors, or where multiple compressors are cycled for capacity control, single suction-line risers may have insufficient suction velocity during part load conditions to adequately move oil up the suction riser. If the riser was sized small enough for proper oil return during minimum load conditions, the pressure drop would be too great in the small riser during maximum loading. Air conditioning applications can tolerate somewhat higher pressure drops without a major penalty in system performance. However, on medium- and low-temperature systems, pressure drop is more critical. Where separate risers for individual evaporators are not possible, a double riser may be necessary to avoid an excessive loss of capacity (see Figure 11-21).

A typical double-riser configuration is shown in Figure 11-22. The two lines will have approximately the same cross-sectional area of a single riser sized for maximum-load conditions. The two lines are normally of different sizes. The larger line is trapped, while the smaller line is sized for the reduced load with adequate velocity to move oil vertically under partial-load conditions.

During operation at maximum load the vapor and entrained oil will be flowing through both risers. At minimum load conditions, the vapor will not be at sufficient velocity to carry oil up both risers. The entrained oil will drop out of the refrigerant vapor and accumulate in the p-trap, forming a vapor seal. As a result, the flow of vapor will be forced up the smaller riser at a velocity sufficient to circulate oil through the system.

FIGURE 11-19 **Three options of suction-line piping for multiple evaporators.** *Courtesy of Refrigeration Training Services.*

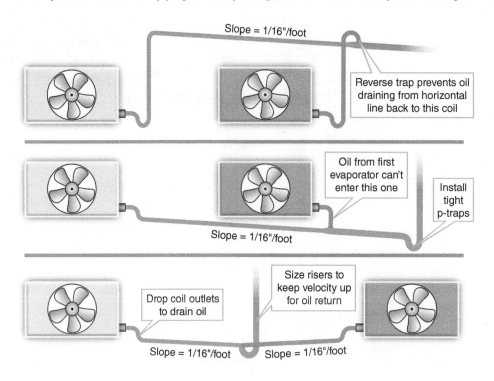

FIGURE 11-20 **Comparison of discharge line and suction-line riser piping.** *Courtesy of Refrigeration Training Services.*

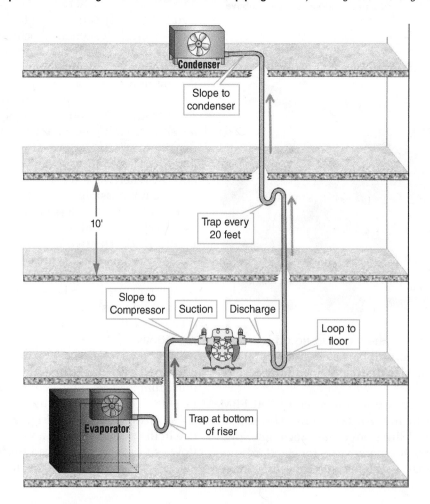

FIGURE 11-21   **Single- and multiple-evaporator piping when capacity control is used.** *Courtesy of Refrigeration Training Services.*

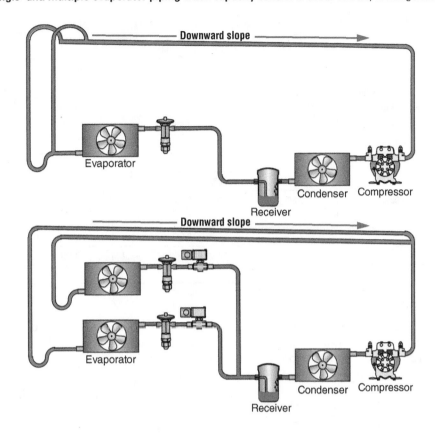

FIGURE 11-22   **Suction line double riser.** *Courtesy of Refrigeration Training Services.*

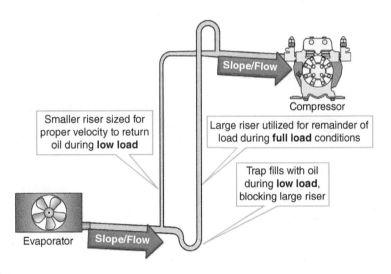

Proper sizing of vertical risers requires the use of charts where one can determine the line size based on compressor capacity and evaporator temperature. For double-line risers the smaller of the two lines must be sized according to the minimum capacity operation while the sum of the two risers will handle maximum-load conditions. Obviously this is best left to factory engineers to determine, but they may utilize charts similar to the one in Figure 11-23. The figure is a portion of a chart reproduced from Emerson Climate Technologies' *Refrigeration Manual Part 4–System Design.*

Assume a 50-hp compressor with unloaders capable of 50 percent load-reduction capacity. The system is capable of 100,000 Btuh at a $-20°F$ evaporator

FIGURE 11-23 Suction-line sizing for horizontal piping and two-pipe vertical risers. *Courtesy of Emerson Technology.*

RECOMMENDED SUCTION LINE SIZES							
		R-502 — 20° F Evaporating Temperature					
		Equivalent Length (feet)					
		50			100		
Capacity (BTU/hr)	Light Load Capacity Reduction	Horizontal	Vertical		Horizontal	Vertical	
60,000	0 to 33%	1 5/8	1 5/8		2 1/8	1 5/8	
75,000	0 to 33%	2 1/8	1 5/8		2 5/8	1 5/8	
100,000	0 to 33%	2 1/8	2 1/8		2 5/8	2 1/8	
	50%	2 1/8	*1 3/8	*1 5/8	2 5/8	*1 3/8	*1 5/8
150,000	0 to 50%	2 5/8	2 1/8		2 5/8	2 1/8	
	66%	2 5/8	2 1/8		2 5/8	*1 3/8	*1 5/8
200,000	0 to 50%	2 5/8	2 5/8		3 1/8	2 5/8	
	66%	2 5/8	*1 5/8	*2 1/8	3 1/8	*1 5/8	*2 1/8
					*Double Riser		

temperature. The total piping run is equivalent to 100 feet. According to the chart in Figure 11-23, the horizontal pipe should be 2 5/8 inch. If the compressor did not have unloaders, the suction riser would be only slightly reduced to 2 1/8 inch for proper oil return.

Assume the compressor in our example has the capability to unload 50 percent of its capacity. The chart shows that a two-pipe suction riser of 1 3/8 and 1 5/8 inch will be required. The cross-sectional area of the two pipes is about equal to the area of the original 2-1/8-inch riser. When unloaded by 50 percent of its capacity the unit will utilize only the smaller 1-3/8-inch

riser for suction vapor and oil return. However, between 33 percent and full-load conditions, both risers will be put into use to return suction vapor and oil to the compressor.

DRAIN PIPING

In a medium-temperature walk-in, water is condensed on the cold evaporator fins as the humid air is blown through the coil (Figure 11-24). The *condensate* collects in the evaporator drain pan, exiting the box through a PVC

FIGURE 11-24 Drain piping for walk-ins. *Courtesy of Refrigeration Training Services.*

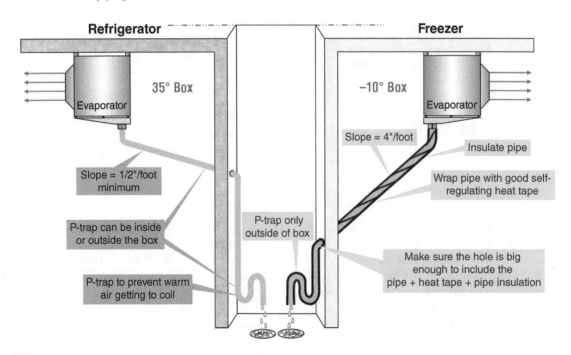

or copper drain line. The size of the drain line should be at least the same size as the drain pan outlet, and should be sloped at a minimum of 1/2 inch per foot of run. A p-trap must be installed in the drain line. The water seals the trap so that no warm air outside the walk-in can migrate up the drain and into the evaporator.

The moisture in a walk-in freezer sticks to the evaporator in the form of frost. During the defrost cycle, water collects in the evaporator drain pan. It is important for all water to be removed quickly from both the pan and the drain line during defrost. Any remaining water would refreeze and block the drain or burst the drain pipe. Following are requirements for running drains in walk-in freezers:

1. Slope the drain line 4 inch per foot.
2. Use copper pipe, not PVC.
3. Install self-regulating heat tape on the drain line inside the walk-in.
4. Insulate the drain line.
5. Install the p-trap outside of the box.

The steep slope helps remove the condensate quickly during the relatively short defrost cycle. The copper pipe is warmed evenly by the heat tape, and the insulation keeps it warm. Any water left in the drain line after defrost may freeze, even with heat tape. Therefore, the p-trap must be installed outside the freezer.

TROUBLESHOOTING WALK-IN PROBLEMS

Chapter 7 covered troubleshooting the refrigeration system. However, there are many problems in a walk-in that are not associated with the refrigerant, compressor, or metering device. Therefore, this section explores the service issues unique to walk-ins, such as evaporator freezing, product loading problems, door problems, and drain problems. It will become apparent that these problems are often related.

As pointed out in prior chapters, coil frosting is a normal result of the cold evaporator temperatures in walk-in refrigerators. Also, the frost should melt during the off cycle. However, if there is too much frost or the off cycle is not long enough, the frost can build up until it causes a problem. Following is a list of common causes for evaporator freezing:

- Refrigeration system problems (covered in Chapter 7) (see Figure 11-25, left side picture).
- Evaporator airflow problems
- Door problems
- System sizing problems or improper product loading
- Drain problems
- Freezer defrost problems (see Figure 11-25, right side picture)

In Figure 11-25, an evaporator with refrigeration problems will have uneven frost buildup (left), and a freezer evaporator that is overdue for its defrost has even frost buildup (right).

Evaporator Airflow Problems

If the evaporator is dirty, if the air inlet is blocked by product packaging, or if there are fan motor problems, the evaporator temperature will fall and excessive frost will form. The problem is usually quite evident, except in the case of a bad fan motor on a multifan evaporator. In this situation, the fan motor stops but the fan blade is free to rotate. The air is short-circuited through the opening of the bad fan motor. The reverse airflow turns the fan blade, but in the opposite direction of the good fan motors (see Figure 11-26).

FIGURE 11-25 **Evaporator with refrigeration problems will have uneven frost buildup (left) and a freezer evaporator that is overdue for its defrost has even frost buildup (right).** *Photos by Dick Wirz.*

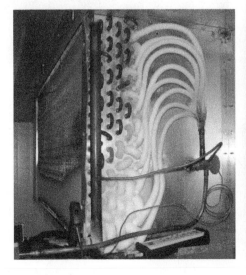

FIGURE 11-26   **Bad fan motor on Evaporator B.** *Courtesy of Refrigeration Training Services.*

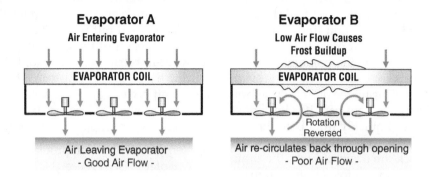

Very little air is pulled through the evaporator behind the bad fan motor, so frost and ice develop. The thickest ice is usually in the area of the bad fan motor. To find out what is actually going on, shut off the fans and watch them. As they slow down, the reversed rotation of the fan blade on the bad motor will be very noticeable.

Door Problems

If the door is not closed or the gaskets are bad, warm air from the outside will enter the walk-in, adding heat to the box. The box temperature will rise, the refrigeration unit will run too long, the evaporator will not self-defrost, and the coil will eventually freeze up.

Assuming the door was originally installed correctly, there are a number of remedies for door closing problems. Some of these solutions include installing door closers, installing plastic strip curtains in the doorway (Figure 11-27), and replacing worn-out door gaskets. Sometimes the simplest solution is the best: Remind the people using the walk-in to make sure the door is closed after they exit the box.

System Sizing Problems or Improper Product Loading

If the refrigeration system is sized for storage only, but the walk-in is being used as a production box, the compressor will run continuously and the evaporator will freeze. To determine if this could be the problem, compare the existing equipment size to the factory recommendations. Also, ask the customer about how much product is being loaded into the walk-in each day and what temperature it is when entering. Doing some research like this will help determine if the existing unit is being overworked.

FIGURE 11-27   **Walk-in door with strip curtains.**
Photo by Dick Wirz.

In walk-in freezers, how the product is loaded can be as important as what is stored. In Figure 11-28, the right side of the box has good airflow around the product, so it remains frozen at the box temperature of −10°F. On the left side of the box, the ice cream is left on the floor and is pushed tight to the wall. The higher heat outside the box has a direct path to the cold

FIGURE 11-28 **The importance of airflow around frozen product.** *Courtesy of Refrigeration Training Services.*

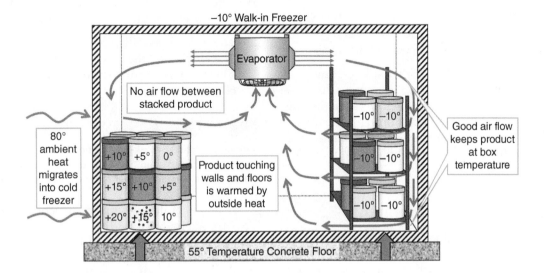

product inside the freezer. Insulation does not stop heat transfer; it only slows it down. If there is no air space between the product and the insulated panels, the product will eventually be warmed. Soft ice cream is a sign that it is too warm, and crystal formation in the product is an indication of thawing and refreezing.

Drain Problems

A plugged drain in a refrigerator will cause the pan to overflow, and the customer will notice water on the floor. The repair is easily accomplished by blowing out the drain line.

In a freezer, water backing up in the drain line will turn to ice and split the pipe. In addition, the drain pan will fill with water, freeze, and produce icicles hanging from the pan. The ice in the drain pan will also cause ice to build up into the evaporator. The ice in the evaporator will block the airflow, which will cause the walk-in temperature to rise and eventually thaw the frozen product in the box. Not only can it be costly to the customer in terms of lost product, but the customer will incur a repair bill for the services of a technician for at least four hours and the cost of materials.

The most common causes for freezer drain problems are improper drain installation and heat tape failure. The use of self-regulating heat tape is recommended because it has the ability to add more heat as the pipe gets colder, and it does not burn out as easily as the standard heat tape. Although the cost of the self-regulating type is 5 to 10 times more than the standard heat tape, the difference is insignificant compared with the high cost of repairs and product loss from a frozen drain line.

Freezer Defrost Problems

The number of freezer defrosts per day depends on ambient conditions and how the walk-in is used.

A walk-in freezer in Phoenix, Arizona, may require only one or two defrosts per day (i.e., every 24 hours), while a freezer in Washington, D.C., may require four defrosts. The primary factor that affects the number of defrosts is the amount of moisture in the air. Although Phoenix may get hotter than Washington, the humidity in D.C. is much higher than in Arizona.

Following is the defrost sequence for a standard walk-in freezer:

1. Defrost is time initiated (started at a certain time of day).
2. The fans shut off, the heaters come on, and the frost melts.
3. Defrost is temperature terminated (i.e., defrost ends when the coil reaches 55°F).
4. The heaters shut off and the compressor starts.
5. The evaporator fans are delayed until the coil temperature drops to about 25°F.

A problem with any one of the sequences will require a service call. Following are some of those problems and what to look for:

- Evaporator does not go into defrost. Check for a bad clock or defrost control.
- Heater does not heat. If there is voltage to the heater but no amperage draw, the heater is bad. To verify, an infinite ohm reading means the heater is open and must be replaced.
- Long defrosts. The primary defrost termination is based on coil temperature. If that control fails, the defrost cycle will continue until it comes out on the timed fail-safe or backup defrost termination. The box temperature will rise as a result of longer defrosts because the off time is too long and there may be less freeze time.

When defrost is terminated, the compressor starts, but the fan delay keeps the fans off until the evaporator has cooled to about 25°F. This ensures that any remaining heat from the defrost heaters is removed from the coil area. Also, any droplets of water remaining on the coil fins will be refrozen. The next problem illustrates what happens if the fan delay fails with its contacts stuck together:

> Icicles and frost form on the walk-in ceiling near the fans, and ice develops on the fan blades. The problem is a bad fan delay switch, because the fans usually fail in the closed position. This means the fans do not delay; rather, they start immediately after defrost, throwing water droplets and warm air out into the box.

Glass Door Sweating Problems

A sweating glass display door is a special area of troubleshooting walk-in problems (Figure 11-29). Although sometimes perplexing, finding the cause of sweating glass and frames is fairly easy once you understand what conditions cause the water formation.

Following are the three basic types of glass doors available for walk-ins:

1. Double-pane glass doors
2. Double-pane glass doors with heaters in the glass
3. Triple-pane glass doors with heaters in the glass (for freezers)

All door frames have heaters to prevent condensation on the frames. If there is moisture only on the door frames, check for power to the heaters.

Although most food service walk-in coolers are designed for 35°F, some applications allow for the box

to be a little warmer. If the box temperature is above 38°F, nonheated glass doors can be used. However, they will sweat for the following three reasons:

1. The inside glass is below 38°F.
2. The air outside the box is over 80°F.
3. The relative humidity outside the box is over 60 percent.

A recently installed walk-in with nonheated glass doors is in a newly constructed building. The customer is concerned because the doors are sweating. What are the most likely causes?

In a new building with construction still going on, it is very likely the humidity is above 60 percent or the room temperature is above 80°F. Also, the cold air coming out of the evaporator may be hitting the back of the glass, cooling it well below 38°F. Stocking the shelves behind the glass door may solve the problem simply by blocking the cold air from the evaporator. If not, an air baffle may need to be installed on the evaporator to direct the air away from the glass doors.

A customer is concerned because the glass doors in her walk-in are sweating when she opens it up for business in the morning, but then the doors clear up within about an hour. What is causing the problem?

There are two likely causes for this problem. First, the air conditioning in the store is being turned off at night, and the increase in heat and humidity causes the

FIGURE 11-29   **Conditions for sweating of unheated glass display doors.** *Courtesy of Refrigeration Training Services.*

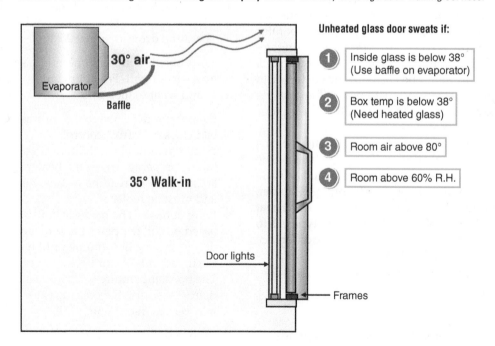

Unheated glass door sweats if:

1. Inside glass is below 38° (Use baffle on evaporator)
2. Box temp is below 38° (Need heated glass)
3. Room air above 80°
4. Room above 60% R.H.

doors to sweat. When the AC is turned on in the morning, the sweating disappears as the temperature and the humidity drop.

The second cause would apply to heated glass doors. The door heaters are often wired to the same circuit as the door lights. If the customer is using the circuit breaker to shut off the walk-in lights, she is also shutting off the heaters. The doors will clear up when the lights are again turned on in the morning because the door heaters will be energized at the same time.

SUMMARY

Walk-ins are usually either refrigerators at 35°F or freezers at −10°F. However, the customer has a wide range of sizes and temperatures to choose from.

The floor must be level if the box is to go together correctly and the doors are to close properly. Some leveling and adjusting of the door frame can be done after the box is installed. The customer uses only the door, so it should work perfectly.

Walk-in refrigeration systems are sized according to how big the box is and how it will be used. The refrigeration equipment for a storage-only walk-in can be determined from a chart furnished by the manufacturer. However, it is important for service technicians to understand the basics of equipment sizing in order to properly service walk-ins.

Correct refrigeration suction piping prevents excessive pressure drop and returns oil to the compressor. Proper evaporator drain piping in walk-ins is also important, especially for freezers.

There are many walk-in service problems that are not directly associated with the refrigeration system. Understanding the other causes for walk-in service problems is essential for proper troubleshooting.

Following is an actual service scenario related to the author. It relates to some of the material in this chapter.

SERVICE SCENARIO:

OIL LOGGING PROBLEM

By Gary Perdue

Service Call: Walk-in freezer warm in a high school cafeteria

Initial Diagnosis: Compressor locked up

Repairs: Replace the compressor

The unit was a walk-in freezer with a rooftop condensing unit installed in 1974. The evaporator had no p-trap on the suction line riser coming off the evaporator. Also, the freezer had been operating below its normal box temperature of −10°F. The freezer's temperature recordings showed that it had been running from −18°F to −20°F. This indicated a problem with the thermostat setting, or that the tstat contacts were stuck closed. The recordings also showed that for 2 months prior to the compressor failure the box temperatures were slowly rising as the compressor deteriorated. Eventually the compressor failed. Inspection of the compressor revealed there was no oil in the crankcase and the compressor was locked up.

Examining as much of the unit history we could find showed this would be the third compressor replacement. The scary thing is that I was sure I was involved with at least one of those replacements because back in the day I was a compressor replacement fool. It just so happened that this time we were following a much better and detailed diagnostic process. This process uncovered the obvious piping mistake of no trap on the suction line riser to the condensing unit on the roof. This situation definitely leads to evaporator oil logging, which effectively blocks off part of the evaporator coil. This not only decreased evaporator efficiency, it also restricted refrigerant flow, which lowered suction pressures, increased the compression ratio, and resulted in high-heat problems in the compressor cylinder. In addition, any oil remaining in the evaporator was starving the compressor of the oil it needed for lubrication.

We installed a suction line p-trap, replaced the temperature control, properly set defrosts and terminations, and disconnected the evaporator piping because we knew there had to be substantial oil there. We flushed it at least three times and removed more than a quart of oil that had been trapped in the evaporator.

That was about 4 years ago. We did some follow-ups afterward, and oil levels seemed consistent. Then last summer we did our scheduled preventive maintenance on all the rooftops, and as part of the inspection we checked the oil sight glasses. The oil level had remained good.

Epilogue:

Compressor failures often occur due to more than one problem. This unit ran for over 30 years, which means it took more than 10 years to kill each compressor. The unit had been running below its design temperature. Could this have caused high compression ratios and overheating of the compressor? Replacing the thermostat and making sure it was cutting off at −10°F should take care of that.

The lack of a p-trap was definitely a contributing factor to oil logging. But so are low flow rates in the vertical suction line piping. Was it sized to keep refrigerant velocities high enough to properly return the oil to the compressor? Also, was the oil being moved by the refrigerant? The original unit used R502 with mineral oil. It was changed to R404A sometime in the past 10 years. Was the oil changed to POE (polyolester) at that time? At least we know that the last compressor came with a full charge of POE and was rated for R404A.

Have we solved all the possible problems that could have killed the compressor? Maybe we will know in 10 years.

REVIEW QUESTIONS

1. Four inches of urethane insulation is equivalent to how many inches of fiberglass insulation?

   a. 8 inches
   b. 10 inches
   c. 12 inches

2. Why is urethane insulation better than fiberglass insulation?

   a. It is thicker than fiberglass.
   b. It does not absorb moisture like fiberglass, and it maintains its insulating properties longer than fiberglass.
   c. Urethane can be used where fiberglass is not allowed.

3. When installing a walk-in with prefabricated floor panels on a concrete floor, is it necessary to level the floor panels with shims?

   a. No, because when the floor panels are locked in place, the box will be level.
   b. Yes, because the floor panels are not locked together.
   c. Yes, because even locked floor panels will not keep the box level on an uneven floor.

4. It is important that the walk-in floor be level. What happens when it is not?

   a. The wall and ceiling panels will not latch properly, and the door will not seal completely.
   b. The insulation value of the panels will be affected.
   c. The product will fall off the shelving inside the walk-in.

5. What device can be added to a walk-in door to help it close fully?

   a. A door closer
   b. Magnetic gaskets
   c. A kick plate

6. What is the name of the gasket at the bottom of the door?

   a. Balloon gasket
   b. Blade gasket
   c. Sweep gasket

7. Door frames must be both _____ for a door to seal properly.

   a. Upright and hinged
   b. Level and plumb
   c. Rigid and locked in place

8. When refrigeration equipment is sized for storage only, and what does that mean in terms of product load?

   a. The entering product is at about the same temperature as the walk-in.
   b. The product will be pulled down to the storage temperature within 24 hours.
   c. The walk-in can handle only a certain amount of product per day.

9. What are the three basic steps for matching storage-only refrigeration to the size of a walk-in?

For questions 10 through 14, use information in Figures 11-9, 11-10, 11-11, and 11-12.

10. What condensing unit and evaporator should be used on a 35°F 8-foot × 8-foot × 8-foot walk-in?

    a. MT-075 and MEV-060
    b. MT-100 and MEV-100
    c. MT-150 and MEV-120

11. What nominal tonnage TEV should be used on an MEV-060 evaporator?

    a. 1/2 ton
    b. 3/4 ton
    c. 1 ton

12. What are the system TD and the total system Btuh if an MT-100 condensing unit is matched with an MEV-080 evaporator?

    a. 8°F TD and 6,000 Btuh
    b. 10°F TD and 8,000 Btuh
    c. 10°F TD and 10,000 Btuh

13. What are the system TD and the total system Btuh if an MT-150 condensing unit is matched with an MEV-060 evaporator?

    a. 10°F TD and 6,000 Btuh
    b. 15°F TD and 8,000 Btuh
    c. 20°F TD and 9,000 Btuh

14. What condensing unit and evaporator should be used on a −10°F walk-in that is 8 feet × 20 feet × 8 feet?

    a. LT-300 and LEV-120
    b. LT-150 and LEV-200
    c. LT-400 and LEV-220

15. Using TROT, what is the approximate Btuh of a 2 1/2-hp condensing unit at a 25°F suction temperature?

    a. 10,000 Btuh
    b. 16,000 Btuh
    c. 20,000 Btuh

16. Using TROT, what is the approximate Btuh of a 2 1/2-hp condensing unit at a 220°F suction temperature?

    a. 10,000 Btuh
    b. 16,000 Btuh
    c. 20,000 Btuh

17. What are the seven rules given in this book for proper refrigeration piping?

18. Refer to the piping chart in Figure 11-15. Assume you must pipe an R404A freezer with a 1 1/2-hp condensing unit. It has a 30-feet piping run with five fittings. What are the recommended suction and liquid line sizes?

    a. 1/2-inch suction line and 3/8-inch liquid line
    b. 5/8-inch suction line and 3/8-inch liquid line
    c. 5/8-inch suction line and 3/8-inch liquid line

19. What is the slope of the drain line in a walk-in refrigerator?

    a. 1/8 inches per foot of run
    b. 1/2 inch per foot of run
    c. 4 inches per foot of run

20. What is the slope of the drain line in a walk-in freezer?

    a. 1/8 inch per foot of run
    b. 1/2 inch per foot of run
    c. 4 inches per foot of run

21. Why is a p-trap used in a walk-in drain line?

    a. To prevent bugs and rodents from getting into the box
    b. So the water flows better out of the drain
    c. So warm air does not get into the box through the drainpipe

22. Walk-in freezer drain p-traps must never be located inside the box. Why?

    a. Because even with heat tape, the water will freeze in the trap
    b. Because the heat tape shuts off during the freeze cycle
    c. Because with the steep slope of a freezer drain, there is no room for a p-trap inside the box

23. What are the six common causes of evaporator freezing?

24. If a door is left slightly open on a walk-in refrigerator, why does this cause a frozen evaporator?

25. Why does an evaporator freeze up if the load in the box is more than the refrigeration system can handle?

26. What happens in a walk-in freezer when the drain freezes?

27. What are the five steps in the basic defrost sequence of a walk-in freezer?

12 ICE MACHINES

OBJECTIVES

After completing this chapter, you should be able to
- Describe ice machine types and applications
- Explain basic ice machine operation
- Understand installation-related service
- Undertake ice machine maintenance and cleaning
- Describe ice machine warranties
- Troubleshoot ice machines

CHAPTER OVERVIEW

Ice machines are very specialized pieces of equipment. This chapter is an introduction to the basics of ice maker operation, maintenance, and trouble-shooting. To learn the specifics of a particular machine, a technician would need to attend a factory school or seminar that provides comprehensive training on that brand of equipment. However, a thorough understanding of ice machines may take years of hands-on experience.

Just as understanding airflow is critical to air conditioning service, a technician needs a thorough knowledge of how water affects an ice machine. Water, the cause of most ice maker service problems, is discussed frequently in this chapter.

ICE MACHINE TYPES AND APPLICATIONS

The two main types of ice machines are cube ice makers and flake ice makers (Figure 12-1). Cubes are harder than flake ice and last longer. Also, cube ice machines produce "pure ice." Cubes are formed by water flowing over a cold evaporator. Only the water freezes; minerals and other impurities remain in the solution and are flushed out of the machine during the harvest cycle. Also, air and chlorine entrained in the water supply tend to vaporize out of the water as it freezes. To sample pure ice, fill a cup with commercial ice and let it melt. Drink the water from the melted ice and what do you taste? Nothing; in fact, it tastes "flat." That is because there is no chlorine, air, odor, or minerals in the water. Drink manufacturers are counting on this purity. When a soft drink is dispensed into a cup of commercial ice, the cold beverage should have only the taste of the soft drink itself.

NOTE: **This chapter has quite a few new terms, or jargon, that are used when discussing ice machines. The first time they appear, the words will have quotation marks to identify them.**

Compare commercial ice cubes to those made in the freezer compartment of a residential refrigerator. In the domestic unit, the water is simply poured into a tray to be frozen into cubes. The ice will contain everything in the water: air, chlorine, minerals, plus odors from foods in the freezer (like onions). The air in the water is what makes domestic ice white in color.

Commercial flake ice is similar to residential cube ice; it contains everything in the water. Flake ice is used primarily for food displays of everything from fish to salad bars. The flake ice forms well to the shape of the product pressed into it, providing excellent refrigeration of the merchandise. However, flake ice melts much faster than cube ice due to the air and other impurities trapped in the flake ice when it was formed.

A type of flake ice machine called nugget ice has been developed to partially overcome the drawbacks of flake ice. The chunks of ice are about 1/2-inch long and the thickness of a wooden pencil (1/4 inch). Nugget ice is used primarily in self-serve soft drink dispensers in cafeteria-style restaurants and convenience stores. Although they have some very important uses, flake and nugget ice machines account for only 10 percent of ice machine sales.

FIGURE 12-1 **Types of ice: cubes, flakes, and nuggets.** *Courtesy of Hoshizaki America, Inc. and Manitowoc Ice, Inc.*

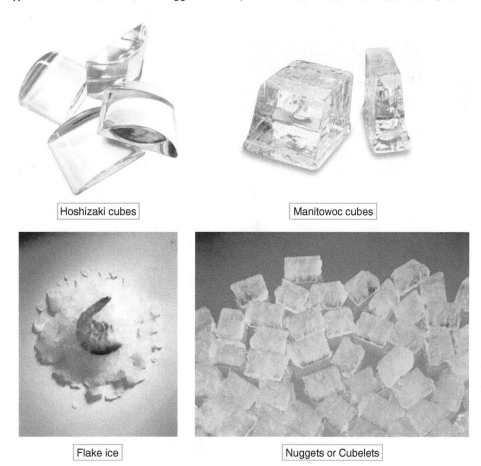

Hoshizaki cubes

Manitowoc cubes

Flake ice

Nuggets or Cubelets

BASIC ICE MACHINE OPERATION

Cube ice machines make "batches" of ice. Water fills a sump (water tray). A pump circulates the water over the surface of a cold evaporator. When the ice is thick enough, the machine switches to the harvest cycle, which is a hot gas defrost. As the evaporator warms, it releases the ice, which falls into the storage bin. The water that is left over from the batch of ice is very concentrated with minerals and must be flushed from the sump. In preparation for the next cycle, the sump is refilled with fresh water (see Figures 12-2 and 12-3).

FIGURE 12-2 **A batch of cube ice falls into the ice storage bin.** *Courtesy of Manitowoc Ice, Inc.*

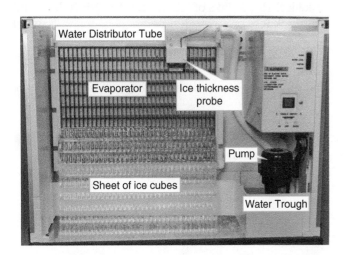

FIGURE 12-3 **Hoshizaki cube ice forming on the evaporator.** *Courtesy of Hoshizaki America, Inc.*

The ice-making cycle is basically the same for all cube ice machines. The main differences between the brands are the evaporator (cube shapes), the mechanism by which the machine starts to harvest, and how the machine senses when the bin is full.

A flake ice machine has a long cylinder filled with water. The evaporator is the wall of the cylinder. A stainless steel auger rotates in the center of the cylinder. The auger flights (fins) gather and compress the ice flakes forming in the water, move the ice up the cylinder, and push it out through a chute into an ice storage bin.

Producing nugget ice (also known as chiplet ice) is similar to the flake ice process, except that it takes an extra step. As flake ice leaves the freezing chamber it is forced through a thick metal plate with 1/4-inch holes in it. The plate, or "extruder," squeezes out excess water as it compresses the ice into hard chunks before they drop down the ice chute to the ice storage bin (see Figure 12-4).

The Vogt Tube-Ice machine is one example of a piece of equipment that offers the features of both cube ice makers and flakers in one machine. The ice is formed by circulating water through a cold vertical tube. When the ice is thick enough, the harvest (defrost) cycle starts and the ice is released from the evaporator. As the ice drops, a rotating blade cuts the ice into 1- to 2-inch-long cylindrical tubes. If the owner wants a flake-type ice, he simply moves a switch and the cutter rotates faster to produce "crushed" ice. The crushed ice is in slivers like flake ice, but it is the hard pure ice of a cube machine (see Figure 12-5).

When deciding which ice machine to recommend to a customer, price is not the only consideration.

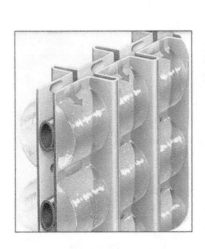

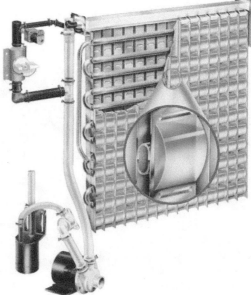

The equipment is sold only once, but it will have to be serviced for many years. Therefore, the customer should be offered a quality machine that has good factory technical support, training, and parts availability.

NOTE: **Price is what you pay, value is what you get.**

FIGURE 12-4   **Manitowoc flake ice assembly.** *Courtesy of Manitowoc Ice, Inc.*

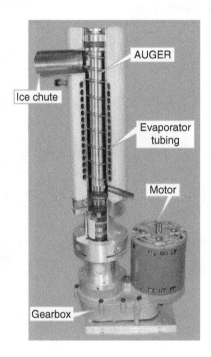

FIGURE 12-5   **Vogt Tube ice machine 2,000- to 4,000-pound capacity.** *Courtesy of Vogt Ice, LLC.*

INSTALLATION-RELATED SERVICE

Proper ice machine installation is simply a matter of following the manufacturer's instructions. However, many units are installed incorrectly, contributing to service problems. This section describes some of the common installation mistakes that a service technician should look for when troubleshooting an ice machine.

The first thing to consider is the location. Is there enough space around the machine for condenser air as well as service access? Most ice machines need at least 6 inches of clearance on all sides and the top.

The next consideration is the drainage (see Figure 12-6). An ice machine has several drains: an ice-storage-bin drain, a "dump drain" to remove the water left over from each batch of ice, and a condenser drain if the unit is water cooled. All drain lines should be run separately, be sized large enough, and maintain a 1-inch air gap (open space) above the floor drain (per plumbing code). If the bin drain is too small, it will easily clog up, filling the storage bin with melted ice water. If the dump drain is too small, the machine will not be able to get rid of all the old water from the last batch of ice. This will cause scale buildup in the machine and poor ice quality. The dump drain and the condenser drain should not be connected to, or "teed" into, the bin drain line. The water in these lines

FIGURE 12-6   **Ice machine water and drain piping.** *Courtesy of Manitowoc Ice, Inc.*

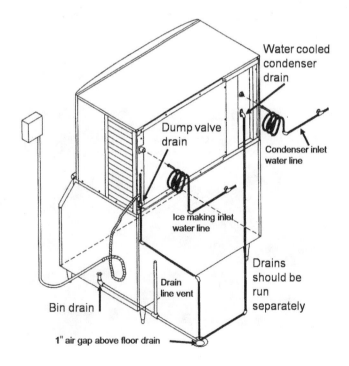

is under pressure and could force its way back up the bin drain and into the storage bin. This would cause some of the ice at the bottom of the bin to melt.

The hidden loss of ice due to drain problems is difficult to troubleshoot and can be very frustrating for both customer and technician. Therefore, ice machine service procedures should always include inspection of the drain piping for possible defects. Normal ice melting is usually indicated by a steady drip of water from the bin drain. However, no dripping water may indicate a stopped-up drain. On the other hand, a steady stream of water may be a sign that there is a problem in the ice-making section and water is leaking down into the ice bin.

The dump valve is supposed to be open only during certain parts of the ice machine's cycle. A continuous flow of water from the dump drain line would indicate the dump valve is stuck open (see Figure 12-7). On water-cooled ice machines any water flowing from the condenser drain when the machine is off means the water pressure regulating valve (WRV) is not closing down completely. This not only is a waste of water, but during harvest any amount of water going through the condenser will cool the discharge vapor needed for defrost and prolong the harvest cycle. Inlet water-line size and water pressure are also important. If the inlet water line is too small, it may reduce production. If the water pressure is too low (below 20 pounds per square inch [psi]), the ice machine will not produce to capacity. If there is too much pressure (above 80 psi), the inlet water floats or electronic fill valves may be damaged. Water-regulating valves on the inlet of the ice machine can protect it from excessive water pressure.

NOTE: Many commercial buildings have water pressure well above 80 psi.

FIGURE 12-7 Manitowoc ice machine dump valves.
Courtesy of Manitowoc Ice, Inc.

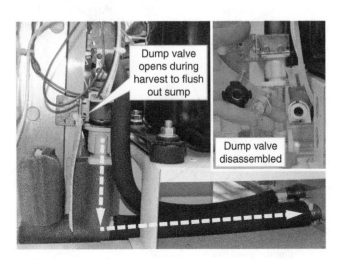

Dump valve opens during harvest to flush out sump

Dump valve disassembled

Ice machines with remote condensers have line sets available in several lengths. However, sometimes the distance between the machine and the condensing unit is less than expected, resulting in excess tubing. The unnecessary loops of tubing can become oil traps and are also easily kinked. Even though the factories have recommendations for coiling excess tubing, the best solution is to cut out the extra pipe, braze in a coupling, evacuate the tubing, and then connect it to the units (see Figure 12-8).

FIGURE 12-8 **Excess line set (should have been cut out).**
Courtesy of Manitowoc Ice, Inc.

This procedure should be performed at the time of installation, before the tubing is connected to the ice machine. Initially, the tubing only has a vapor charge, whereas the full refrigerant charge of the system is stored in the ice machine. Only after the tubing is connected between the ice machine and condenser is the charge released into the system. Once released, all of the refrigerant would have to be recovered before cutting out the excess tubing.

NOTE: It is best to verify the manufacturer's recommendations for handling excess line sets for the specific brand and model.

ICE MACHINE MAINTENANCE AND CLEANING

The most important part of ice machine maintenance is to clean the water circuit. Everywhere the water flows in an ice machine, a scale buildup of mineral deposits will eventually collect. Removal of the accumulated scale requires regular service with ice machine cleaner and a brush (see Figure 12-9). The critical areas are the

FIGURE 12-9   **Use cleaner and a brush to clean ice machine evaporators.** *Courtesy of Manitowoc Ice, Inc.*

evaporator and water spray tubes. Scale buildup on the evaporator will prevent the ice from releasing during harvest. Stoppage of the spray tubes will result in an incomplete batch of ice. In some ice machines, slime is also a cause of spray tube restriction. A regular program of running sanitizer through the machine water circuit can prevent this problem by killing the bacteria growth that causes slime (Figure 12-10).

Many ice machines have a self-cleaning feature. This usually means that after adding cleaner or sanitizer, the technician just moves the selector switch to the clean position. The machine shuts off the compressor and circulates the cleaner or sanitizer for a preset amount of time. Then, the machine goes through several flushing sequences before returning to the

FIGURE 12-10   **Ice machine cleaners and sanitizers.** *Courtesy of Manitowoc Ice, Inc. and (right) photo by Dick Wirz.*

| Manitowoc cleaner and sanitizer | Nu-Calgon ice machine cleaners |

ice-making mode. Manitowoc ice machines not only have a timed clean and flush cycle, but also offer an optional AuCS (automatic cleaning system) that performs preprogrammed cleaning cycles with cleaner or sanitizer.

The quality of water used to make ice determines the amount of maintenance an ice machine requires. To greatly reduce the need for ice machine cleaning, the manufacturers recommend using water filters to trap sediment and large particles in the water before they enter the machine. Unfortunately, the minerals dissolved in the water cannot be filtered out. Therefore, some ice machine water filters contain phosphates that bond to the minerals to prevent them from clinging to the evaporator and other parts inside the ice machine. The dissolved minerals are flushed out at the end of the ice-making cycle. Under normal use the phosphates in a properly sized water filter should last about one year (see Figure 12-11).

NOTE: **If the ice machine is water cooled, hook up the ice machine filter so that it only serves the water inlet for ice making.**

If the filter is installed on the main water line, it will condition the water for both the condenser and for making ice. However, the large amount of water used by the water-cooled condenser would quickly deplete the phosphates in the filter, requiring frequent filter changes. It is more cost effective to simply clean a water-cooled condenser when needed.

An important part of ice machine maintenance is making sure the condenser is clean. Most air-cooled ice machines have some type of cleanable air filter on the condenser inlet (see Figure 12-12). However, checking water-cooled condensers is a little more

FIGURE 12-11   **Ice machine water filters.** *Courtesy of Hoshizaki America, Inc. and Manitowoc Ice, Inc.*

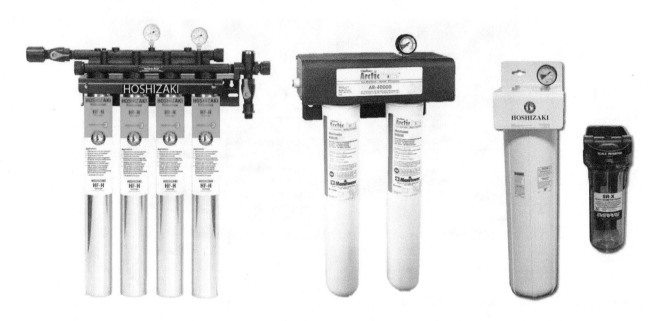

FIGURE 12-12   **Ice machine air-cooled condenser filter.** *Courtesy of Hoshizaki America, Inc. and (left) photo by Dick Wirz.*

challenging than simply checking filters (see Figure 12-13). The leaving condenser water temperature should be about body temperature (between 95°F and 105°F) when the unit is operating at the factory-recommended head pressure. Although an excessive flow rate of water can be an indication of scale buildup, it is not always an accurate measurement of the condition of the condenser on ice machines. A fairly effective way to determine mineral buildup in a water-cooled condenser is to measure the temperature of the water

FIGURE 12-13 **Water-cooled ice machine.** *Photo by Dick Wirz.*

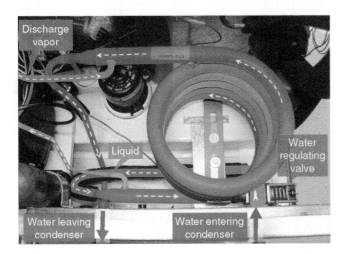

Discharge vapor

Liquid

Water regulating valve

Water leaving condenser

Water entering condenser

leaving the condenser. Following is a summary of the water temperature diagnostic procedures:

1. Measure the water temperature leaving the condenser.
2. If it is lower than body temperature (about 95°F), check the head pressure.
3. If the head pressure is within factory recommendations, the condenser probably only has a thin layer of scale buildup.

EXAMPLE: 1

The temperature of the water leaving the condenser is 85°F, lower than the usual 95°F–105°F. The condenser head pressure is within the range specified in the ice machine service handbook. This condition may signify a small mineral buildup in the condenser.

In this example, the scale problem is not bad enough to cause the head pressure to rise. By allowing a greater flow of water to transfer heat, the valve is able to keep the head pressure down to what it was originally adjusted to maintain. However, the scale prevents efficient transfer of heat, resulting in the lower water temperature. There is an easy way to verify the condition. Raise the condenser water temperature by adjusting the valve to reduce the water flow rate. If the head pressure then rises above the factory recommendations, the likely cause is a thin coating of minerals inside the condenser. In this case, a little scale remover flushed through the condenser should return the water temperature to its original level.

4. If the head pressure is high, the condenser probably has a thick layer of scale buildup.

EXAMPLE: 2

If the head pressure is high and the leaving water temperature is low, the water-cooled condenser most likely has a thick layer of mineral buildup.

In this example, even with the valve wide open, the flow rate is not enough to maintain the proper head pressure for which it was adjusted. Therefore, the scale buildup is probably quite thick and will definitely need acid cleaning.

Sometimes the layer of minerals is so thick that the cleaning solution cannot completely remove it. In other instances, the scale breaks off in big enough chunks to totally stop up the condenser. If the condenser cannot be cleaned, then it must be replaced. Until a new condenser is installed, the customer's ice machine is out of service. Therefore, an experienced technician will check on the availability of a replacement condenser before beginning a water-cooled condenser cleaning process.

NOTE: **If the mineral buildup is excessive, cleaning the condenser may not always be successful.**

5. If the head pressure is low, adjust the water valve to raise the head pressure. If both the head pressure and leaving water temperature rise to normal levels, the condenser is clean.

ICE MACHINE WARRANTIES

Before covering ice machine troubleshooting, it is a good idea to discuss warranties. Following is a list of the basic factory warranties for most ice machines:

- Three-year parts and labor warranty on everything in the machine
- An additional two-year parts-only warranty on the compressor
- Five-year parts warranty and labor on the evaporator

To perform warranty work and receive factory reimbursement for it, the person servicing the equipment must be a dealer or certified technician for that brand of ice machine. This is understandable because the factory wants only properly trained technicians performing in-warranty repairs.

The owner of the machine is responsible for normal maintenance. In addition, the factory will not cover problems caused by water or electrical power external to the machine, problems due to ambient conditions beyond system design limits, or problems resulting from an improper installation.

 REALITY CHECK 1

Lack of maintenance is the primary cause of service calls for ice machines. The need for factory-warranty parts replacement is not that common.

TROUBLESHOOTING ICE MACHINES

This section covers some general troubleshooting guidelines. However, an effective ice machine service technician must have factory training as well as the service handbook for the ice machine model he is working on.

The first thing a technician must remember about ice machine service is to leave his gauges in the truck. That way he will not be tempted to check pressures every time he troubleshoots a unit. Ice machines are critically charged; checking pressures every time it is serviced is a good way to end up with a low charge. Also, 50 to 75 percent of ice machine problems are water related— dirty water circuit, too much or too little water pressure, drain problems, and so on. However, when the pressures must be checked, the technician should use a gauge set that will not trap refrigerant inside the gauge hoses. Figure 12-14 is an example of gauges without hoses.

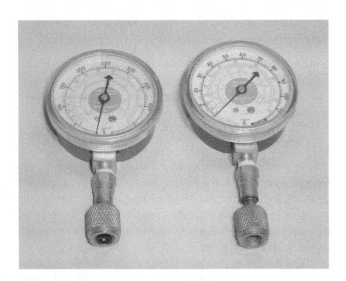

FIGURE 12-14 **Gauges used on ice machines.** *Photo by Dick Wirz.*

Troubleshooting a "No Ice" Service Call

When an ice machine is not running, it is fairly easy to trace the problem. If an ice machine has power and water, but is not running, a safety control has probably shut off the unit.

The most common safety control is a high-pressure switch with a manual reset button. If resetting the control starts the machine, then the technician must find out what made it trip.

NOTE: Occasionally high-pressure controls are condemned because the technician cannot restart the machine right after the control trips. Keep in mind that if a machine trips on a high-pressure switch, it needs time for the pressure to fall before it can be reset.

Other safety controls turn off the ice machine and turn on lights on the circuit board. These lights are a diagnostic key for the service technician. To diagnose the problem using the circuit board lights, the technician will need the service handbook for that model.

Manitowoc's high-pressure controls on its R404A machines will trip at about 450 psig, but will not reset until the pressure drops to 300 psig.

Troubleshooting a "Slow Ice" or "Not Enough Ice" Service Call

When an ice machine is not making enough ice for the customer, the technician must first do a "capacity check":

1. Take the temperatures of the ambient air and of the water to the machine.
2. Determine total cycle time (freeze cycle + harvest cycle).
3. Weigh the batch of ice.

NOTE: Some handbooks give the average weight of a properly formed batch of ice.

4. Calculate daily production as follows:

$$[1{,}440 \text{ (minutes in a day)} \div \text{total cycle time}] \times \text{weight of ice} = \text{production}$$

Freeze cycle is 14 minutes, harvest is 1 minute, and total cycle time = 15 minutes.

The weight of the batch of ice is 5 pounds. Calculate the total production in 24 hours:

[1,440 minutes per day ÷ 15 minutes per cycle] × 5 pounds per cycle = 96 cycles × 5 pounds = 480 pounds produced in 24 hours

Each model ice machine has a chart showing its average 24-hour ice production based on ambient air and incoming water temperatures (see Figure 12-15). The machine is considered to be operating properly if its production is within 10 percent of its rated capacity. If the machine in Example 4 is rated for 520 pounds per day, the production of 480 pounds would be about 8 percent below the rating, but it would be acceptable. Almost as often, ice machines are found to be producing ice above their rated capacity.

If the customer regularly needs to buy more ice than the machine is producing daily, then a larger machine, or an additional machine, may be the answer. However, if the machine fills up and sits idle much of the week, but the customer still needs more ice only on busy weekends, the answer may just be a larger bin. In this case, the ice machine can run longer during the week to make more ice available for the weekend.

FIGURE 12-15 **Ice machine production charts.** *Adaptation of Manitowoc Ice charts by Refrigeration Training Service.*

B450 SERIES WATER-COOLED ICE MACHINE 24 HOUR ICE PRODUCTION (POUNDS)			
Ambient Air Temp °F	**Production at Water Temperature of:**		
	50°	**70°**	**90°**
70°	440	400	355
80°	435	395	350
90°	430	390	345
100°	420	380	335
Based on average ice slab weight of 4.12 lb. to 4.75 lb.			

CYCLE TIMES (MINUTES)			
Ambient Air Temp °F	**Freeze Time at Water Temperature of:**		**Harvest Time**
	50°	**70°**	**90°**	
70°	12.0–14.1	13.3–15.6	15.2–17.8	1–2.5 min.
80°	12.1–14.2	13.5–15.8	15.5–18.1	
90°	12.3–14.4	13.7–16.0	15.7–18.3	
100°	12.6–14.8	14.1–16.5	16.2–18.9	
freeze time ⌐ harvest time = total cycle time			

Another solution is for the customer to bag some ice during the week and store it in the walk-in freezer for the weekend's heavy demands.

Remember to check the drains. The melting of ice in the bin from drain problems will be evident only when the ice is removed from the bin. Look in the empty ice bin occasionally during the ice-making cycle and especially during the harvest cycle. If any water is backing up in the bin, the drain piping will have to be corrected.

During the capacity check is an excellent opportunity to observe every detail of the ice-making and -harvest operation. Many ice machines can go through a complete cycle in 10 to 20 minutes. Experienced technicians can determine most ice machine problems within two complete cycles. Even ice machines like Hoshizaki, which have relatively long cycles, have a procedure for checking out most of the machine's operation in about 7 minutes (Figure 12-16).

Additional Troubleshooting Tips

The way the ice is formed can be a valuable piece of information. If there is ice in the bin, look for irregularities in its shape. Typical indicators include too thick, too thin, or "burned." If cubes stick on the evaporator, they "burn," or melt, into irregular shapes during the defrost process (see Figure 12-17). Cube hang-up can usually be cured by removing mineral deposits from the evaporator with ice machine cleaner and a stiff brush.

The water regulating valve (WRV) on a water-cooled machine should shut off the water flow through the condenser when the machine is off or during the harvest

FIGURE 12-16 **Hoshizaki cube ice machine.** *Courtesy of Hoshizaki America, Inc.*

cycle. If not, there is a great waste of water during the off cycle. In addition, a leaking water valve can cool the discharge gas during harvest, resulting in longer harvest times and lower overall production.

FIGURE 12-17　**Check the ice cubes when troubleshooting.** *Courtesy of Manitowoc Ice, Inc.*

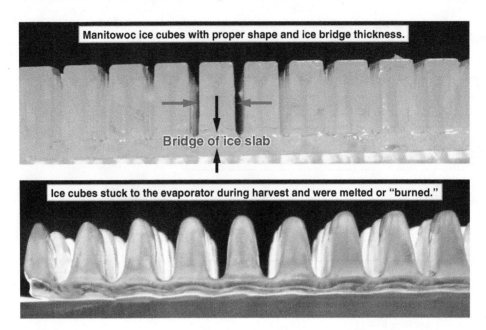

An ice machine leaking water on the floor can cause someone to slip and fall, as well as cause water damage to the building. This problem should be corrected promptly. However, sometimes what looks like an ice machine leaking may just be the sweating of copper drain lines.

The water from the melting bin makes the drain line cold, providing a place for moisture in the air to condense. On warm and humid days the drains may sweat so much the customer thinks the ice machine has a water leak. Insulating the drain lines will solve the problem.

SUMMARY

Commercial ice machines make two types of ice: cubes and flakes. Cube ice machines are the most popular, but flake ice machines are preferred for some applications. Although ice machines are similar, each has specific operating characteristics. The best way for someone to become an effective ice machine technician is to take advantage of all the factory training available for the brands he or she services.

A properly installed ice machine allows the machine to produce ice to its rated capacity and eliminates most unnecessary service calls. In addition,

maintenance is just as important to ice machines as any other piece of refrigeration equipment. The main difference is that the water circuit requires special attention. About 50 to 75 percent of ice machine problems are water related. Water filters can help keep maintenance and service problems to a minimum.

A quick capacity check will help determine if the problem is in the machine or the customer's ice usage. It also provides the technician with an excellent opportunity to view every function of the ice machine.

REVIEW QUESTIONS

1. What are the two main types of ice machines?

2. Why are commercial ice cubes considered "pure ice"?

   a. They have different shapes.
   b. They do not have impurities in them.
   c. They are not frozen in a tray.

3. What is the basic process for flake ice?

4. What is the basic process for cube ice?

5. Why do manufacturers require a minimum space around ice machines?

   a. For service access and airflow
   b. For storage and to look nice
   c. For safety and drainage

6. What problems occur if the ice machine bin drain is too small?

   a. The ice cubes cannot form properly.
   b. The machine will not produce enough ice.
   c. The drain could become clogged, back up water, and melt the ice in the bin.

7. What problems occur if the "dump drain" is too small?

   a. The ice-making cycle will be too long.
   b. The machine will not produce enough ice.
   c. An excessive buildup of minerals will occur inside the machine.

8. What problems occur if the dump drain or the water-cooled condenser drain is piped into the bin drain?

   a. Water could back up into the ice bin, melting the ice.
   b. The ice machine will make less ice.
   c. Excessive minerals will build up in the bin.

9. What happens if the water to the ice machine has too little water pressure?

   a. It will not be able to produce its rated capacity of ice.
   b. The inlet water valve, or float valve, can be damaged.
   c. The machine will make too much ice and overflow the bin.

10. What happens if the water to the ice machine has too much water pressure?

    a. It will not be able to produce its rated capacity of ice.
    b. The inlet water valve, or float valve, can be damaged.
    c. The machine will make too much ice and overflow the bin.

11. What should be done with the excess tubing on remote condensers?

    a. Cut it out or check with the manufacturer's recommendations.
    b. Loop the excess tubing vertically, like a Ferris wheel.
    c. Do not worry about it because it is only discharge gas.

12. What is the most important part of proper ice machine maintenance?

    a. A clean bin drain
    b. A clean water circuit
    c. A fast harvest cycle

13. On water-cooled condensers, what is the proper leaving water temperature?

    a. 20°F to 30°F above ambient
    b. 115°F to 125°F water
    c. Body temperature (95°F to 105°F) water

14. What is the basic parts and labor warranty on most ice machines?

    a. One year, parts and labor
    b. Three years, parts and labor
    c. Five years, parts and labor

15. What is usually not covered under an ice machine factory warranty?

16. Why should ice machine service technicians leave their gauges in the truck?

    a. Because most problems are water related
    b. Because there are no gauge ports on an ice machine
    c. Because ice machines do not lose refrigerant

17. Describe the procedure for taking an ice machine capacity check.

18. If an ice machine is producing within _____ of its rated production, it is considered to be operating properly.

    a. 10 pounds
    b. 10 percent
    c. 20 percent

19. Why is checking the bin drain part of the troubleshooting process?

    a. To see if the bin drain has a vent in it
    b. To see if any water is backing up into the bin
    c. To see if the bin drain is sloped properly

20. If the ice cubes are "burned," or show signs of sticking to the evaporator in harvest, what is the first thing a technician can do to correct the problem?

    a. Clean the ice machine's water circuit and evaporator.
    b. Clear the bin drain.
    c. Add some refrigerant.

21. When should there be no water flow through the water regulating valve on a water-cooled ice machine?

    a. During start-up of the machine
    b. When the inlet water temperature is below 50°F
    c. When the machine is off and during the harvest cycle

22. Why is bin drain sweating a problem?

    a. It may cause water damage to the building or may cause someone to slip and fall.
    b. It will cause a decrease in the ice production.
    c. It will cause the bin drain to clog and back up into the bin.

13 PRODUCT TEMPERATURES FOR PRESERVATION AND HEALTH

OBJECTIVES

After completing this chapter, you should be able to understand, explain, and describe the following topics as they apply to product preservation and consumer health:

- Minimum temperatures
- What the health inspector looks for
- Problem areas and solutions
- Inspecting ice machines
- Hinges, handles, and gaskets
- Refrigeration maintenance program
- Interesting temperature and health facts

CHAPTER OVERVIEW

Restaurant managers typically dread visits from their local health inspectors. The inspections are unannounced and often occur at the restaurant's busiest time. The timing is intentional; it provides the health department with a good opportunity to see how well the facility adheres to health regulations under full operating conditions. The local health department has the authority to impose fines or even close the restaurant if there are substantial violations of the health or food codes.

Health inspectors check everything in a restaurant's operation, from pest control to the drain in the kitchen sink. However, the purpose of this chapter is to make technicians aware of the inspector's health concerns as they relate to refrigeration. The more technicians know about health regulations, the better they can serve their customers by ensuring the refrigeration equipment is operating according to code. Just as important is their understanding of how their customers are using their refrigeration equipment in an attempt to comply with that code.

NOTE: **Local codes may vary from region to region.**

MINIMUM TEMPERATURES

There are basically three levels of requirements for food product temperatures. The minimum level for the entire United States is set by the Food and Drug Administration (FDA). Next, the local health departments can enforce tougher regulations. Finally, the policies of the restaurant, especially large restaurant chains, may be even stricter than the local requirements.

NOTE: **For information, go to www.fda.gov and search the website under "Food Code."**

For example, the maximum temperature for a commercial refrigerator is 41°F according to the FDA. However, some restaurants want 35°F in the walk-in and 38°F in the reach-ins. The lower temperatures keep the product fresher longer. Plus, in the event of a surprise visit by the health department, the equipment can be a few degrees above the restaurant's requirements and still pass inspection.

The FDA's standard for frozen food is simply that the product is frozen solid. The degree of hardness depends on the type of food being frozen. For example, water can be frozen solid at about 32°F, but ice cream will become soft above 5°F. Most reach-in freezers are designed for only 0°F to 5°F, and most walk-ins for −10°F to 0°F.

WHAT THE HEALTH INSPECTOR LOOKS FOR

In an attempt to make the inspector think the unit is running even colder than necessary, some kitchen employees place the thermometer bulb in the evaporator outlet air stream rather than in its inlet.

During an inspection, the food temperature is checked with an electronic thermometer that is usually equipped with a stainless steel probe. The probe enables the inspector to check the temperature inside the product. Some kitchen managers use an infrared thermometer to "shoot" the surface temperature of the product several times throughout the day. Although not as accurate as the probe, monitoring temperatures

REALITY CHECK 1

Health inspectors use their own thermometers to check box temperature, and they record the temperatures in the warmest part of the cabinet. More important, they often check the temperature of the product to make sure it is also down to the proper temperature.

with the infrared unit is easier than with a probe, and it still gives a fairly good indication of the food's temperature.

The temperature of the product is more important than the temperature inside the refrigerator. For instance, if the box temperature is up to 54°F but the product temperature is below 41°F, the inspector will tell the customer to get the refrigerator checked. She will assume the box needs repairs, and the product was recently transferred from the walk-in. However, if the product is 54°F and the refrigerator is 54°F, she has no choice but to cite the kitchen for the violation.

PROBLEM AREAS AND SOLUTIONS

Prep units like those in Figures 13-1 and 13-2 are prime targets for the inspector's temperature probe. The food pans are refrigerated from the bottom whereas the top is exposed to ambient air. The storage area of the prep unit is often adjusted to about 35°F in an attempt to keep the product in the pans under 40°F. Most prep units have a lid to cover the pans when not in use, but the units are often left open. At least one manufacturer has a prep unit that blows refrigerated air over the top of the pans. According to the company, this design keeps product temperatures from rising when the cover is removed.

FIGURE 13-1 **Example of a prep table.** *Courtesy of Master-Bilt Products.*

If the inspector discovers the product temperature is higher than it should be, kitchen managers often use the explanation that the food was recently put in the refrigerator. The inspector has to determine whether the claim is valid. According to the FDA's Food Code, cooked food is allowed about 4 to 6 hours (depending on the food) to be refrigerated down to 41°F. In addition, refrigerated food that has been removed and mixed with other food should drop to at least 41°F within an hour of being returned to the refrigerator.

Standard refrigeration equipment (reach-ins, walk-ins, and prep units) is not designed to quickly lower the temperature of warm food to 41°F. These units are for storage only. Introducing too much warm product would raise the box temperature, causing the compressor to run continuously without its normal off-cycle defrost. As discussed in Chapter 2, "Evaporators," without a significant off cycle the coil will not have a chance to defrost, resulting in a frozen evaporator. One solution is to install a thermostat (tstat) with a sensing bulb mounted inside the coil fins. This will shut the compressor off when the evaporator becomes too cold; the compressor will not restart until the evaporator temperature is above freezing and the frost has melted. Another preventive measure is to use a time clock to provide a planned off-cycle defrost. Both these suggestions will prevent the evaporator from frosting, but will not speed up the refrigeration process.

The best way to quickly lower prepared food temperatures is to use blast chillers. These refrigerated cabinets are specially equipped with large refrigeration units and increased airflow to quickly bring down the temperatures of large quantities of food.

EXAMPLE: 1

A standard single-door reach-in usually has a 1/4-horsepower (hp) compressor. However, a similar size blast chiller has a 4-hp condensing unit, or about 16 times the refrigeration capacity of a storage refrigerator. This specialized refrigerator can lower the temperature of up to 200 pounds of cooked product, from 170°F down to 40°F in about 1 1/2 hours.

The previous example illustrates how much additional refrigeration capacity is required to lower product temperatures. Therefore, it should be easier to understand why even large walk-ins, designed for storage only, may have difficulty pulling down the temperature of warm food.

Proper airflow is important around the product inside all refrigeration equipment, not just blast chillers and freezers. The food must be placed on shelves or trays with adequate space for the cold air to circulate. Reach-in manufacturers often use shelves with "product stops" in the back to prevent packages from contacting the box wall and blocking the circulation of cold air. Even with these safeguards, some users overload the reach-in with product that the refrigerated airflow is restricted. Many people who stock reach-ins are not aware of this problem. Therefore, it is often up to the technician to instruct the customer about proper loading and the need for good airflow. Occasionally, a technician will have to run a nuisance service call where the warm product temperatures are the result of overstocking, not an equipment failure.

INSPECTING ICE MACHINES

Airborne yeast forms a clear slimy residue on the dark wet interior of an ice machine. This fungal growth looks disgusting and can interfere with ice production. Yeast is present in places where dough is made and where beer is served. Regular ice machine cleaning with an approved sanitizer will take care of this problem.

Most restaurants have an ice scoop holder on the side of the ice machine where the scoop is kept when not in use. Because ice is part of the Food Code, the inspector does not want the employees to use their hands to search for the ice scoop under the ice in the bin (see Figure 13-3).

HINGES, HANDLES, AND GASKETS

According to the FDA's Food Code, the hardware of refrigeration equipment must be in good order.

FIGURE 13-3   **Ice scoops must be stored outside the ice bin.** *Photos by Dick Wirz.*

Door gaskets should be cleaned and sanitized at least once a day. Food handlers often have animal fat on their hands when they touch the gaskets while opening and closing the doors. This leaves a residue that will quickly deteriorate rubber and plastic gaskets (Figure 13-4). Torn or sagging door gaskets allow outside air into refrigerated space, which not only results in higher product temperatures but also increases the chance of a frozen evaporator. For the same reason, loose hinges and handles should be tightened regularly, and broken hardware must be replaced (Figure 13-5).

Sliding doors are often troublesome because the rollers and door tracks seem to wear out relatively quickly from normal opening and closing. In addition, if food is dropped into the door tracks, it not only contributes to sliding problems but causes health code issues as well. When the door does not close easily, employees often leave it slightly open, allowing warm ambient air to enter the case. Therefore, frequent cleaning of the door tracks is important for proper door operation, to maintain refrigeration, and for proper sanitation.

FIGURE 13-4   **Examples of gasket problems.** *Photos by Dick Wirz.*

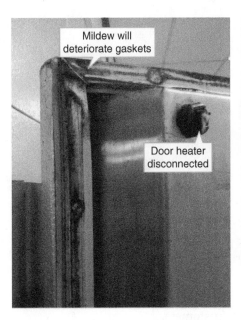

FIGURE 13-5 **Door hardware problems.** *Photos by Dick Wirz.*

Sweating caused by gasket or latch problems

Part of normal maintenance is tightening hinges

REFRIGERATION MAINTENANCE PROGRAM

The first line of defense against health department violations is a good refrigeration maintenance program. Whether the maintenance is performed monthly, quarterly, or semiannually depends on the conditions in which the equipment must operate. However, the important point is that the equipment checkups are scheduled at regular intervals. Following are some of the more important maintenance procedures:

- Checking and adjusting equipment temperatures
- Cleaning air-cooled condensers (or replacing condenser filter material)
- Checking water-cooled condensers for water use
- Checking evaporators and condensate drains
- Checking door handles, gaskets, and hinges
- Checking ice production and cleaning the water circuit
- Asking the customer about any problems or equipment concerns he may have

If the above items are part of a comprehensive program, the customer will have few refrigeration-related issues to be concerned about when the health department makes an inspection.

Be very careful not to get chemicals or dirt into the food when cleaning evaporators and condensers (Figure 13-6). Prevention is the best protection; either

FIGURE 13-6 **Dirty condenser on top of a reach-in refrigerator.** *Photo by Dick Wirz.*

cover the food or remove it from the area. If there is an accidental contamination of the food, make sure the manager is notified immediately. It is much better to apologize and risk his anger than to poison one of his customers.

INTERESTING TEMPERATURE AND HEALTH FACTS

Following is some information that should be both interesting and helpful to commercial refrigeration technicians.

Food service operations use a sanitizer to kill germs, to remove light grease from food preparation surfaces, and as a disinfectant in the dishwashing equipment. Sanitizer is a very mild solution of chlorine bleach and water; it requires only a teaspoon of bleach to a gallon of water. Too much bleach leaves a toxic residue on all the surfaces it touches. The health inspector uses a piece of litmus paper to make sure there is not too much bleach in the sanitizing solution.

Fast freezing is important to maintaining the taste of most foods, especially meat. In a commercial freezer, meat is quickly blast frozen to subzero temperatures in a matter of minutes, rather than hours. The speed of the freezing process locks in the natural moisture of the meat and produces very small ice crystals. When the meat is cooked, it retains its original juices and, therefore, its original taste. However, when meat is put in a regular freezer, the slow freezing process creates relatively large ice crystals, which tear the cells of the product being frozen. When the food is cooked, the juices drain out, leaving the meat tasteless and tough.

NOTE: This problem can also be created when properly frozen meat is thawed, then slowly refrozen.

Some foods fry best when they are kept frozen until placed in the fryer. For instance, fried potatoes that have thawed will absorb the hot frying grease and come out limp and chewy, rather than crisp. For this reason, restaurants usually have a reach-in freezer close to the frying station.

Ice cream quality varies greatly; therefore, the temperature at which it freezes solid also varies. Less expensive ice cream freezes hard at about −5°F. However, the higher-quality ice creams contain more cream, which requires a lower temperature of −10°F to −20°F before it is considered "hard" enough for storage. In ice cream parlors, the ice cream is taken out of the storage freezer and placed in a "thawing cabinet" to raise its temperature. When the ice cream has risen to about 0°F to 5°F, it can be placed in the ice cream dipping cabinet, ready for easy scooping.

Refrigeration unit temperatures should be checked each day before the staff begins using the equipment. If the units start out at the proper temperature, yet the temperature rises excessively during the day, it may be the way the units are being used rather than an equipment problem. There is another important reason for checking equipment first thing in the morning. If a unit needs service, the sooner the refrigeration company is notified, the sooner the problem can be corrected.

Because health departments usually have a limited number of qualified inspectors, they would like to concentrate their efforts on places that are most likely to have food code violations. Some health inspectors suggest that refrigeration technicians help them. Since technicians often see the condition of a kitchen, they should turn in food service establishments they believe pose a dangerous health risk to the public. This may seem like "biting the hand that feeds you." However, technicians need to ask themselves how they would feel if someone they knew became very sick, or even died, as a result of eating at a restaurant that a technician should have turned in to the health department. It certainly is something to consider.

SUMMARY

Although a health inspection will never be a joyful occasion for a food manager, it can be much less stressful if the refrigeration temperatures are correct. The FDA requires a minimum refrigerated food temperature of 41°F and that frozen food is frozen solid. All the refrigeration equipment manufactured for the food industry is capable of maintaining the required Food Code temperatures. However, equipment must be used according to its designed conditions, and it must be maintained properly.

The primary reason some refrigeration equipment does not maintain product temperature is user misuse and neglect. Standard refrigeration equipment is designed for storing already refrigerated product, not to pull down the temperature of warm product. Special blast chillers and blast freezers are designed for this type of application. Also, there must be enough space around the product for the cold air to circulate. And finally, the equipment must be cleaned and checked on a regular basis.

Ice machines provide a dark and wet environment that is a prime breeding ground for mold and bacteria problems. Regular sanitizing will prevent ice machines from becoming a health issue.

A refrigeration maintenance program is essential to keep refrigeration in compliance with health department requirements. Maintenance inspections help ensure that the temperatures will be correct and that the equipment hardware is in good working order.

REVIEW QUESTIONS

1. What are the three levels of requirements for food product temperatures?

2. Where does the health inspector take the refrigeration temperature?

   a. In the warmest part of the refrigerator
   b. In the coldest part of the refrigerator
   c. In the first place he can

3. If the thermometer on a refrigerator is reading 35°F, will the inspector still check the temperature of the product inside the refrigerator?

   a. Yes, because the product may be warmer than the box temperature
   b. Yes, because the thermometer may be adjusted to read colder than it should
   c. Yes, because the thermometer bulb may be reading the colder evaporator outlet temperature rather than the box or return temperature
   d. All of the above

4. According to the FDA, what is the maximum temperature allowed for a commercial refrigerator?

   a. 35°F
   b. 41°F
   c. 54°F

5. Should the customer expect her reach-in refrigerator to quickly pull down hot product?

   a. Yes, because all refrigerators are designed to refrigerate food
   b. No, because most refrigerators are storage only, and are not designed to pull down warm product

6. What harm is there in trying to pull down warm product in a walk-in or reach-in?

   a. The coil could freeze up and cause the box temperature to rise.
   b. None, because the temperature of the box will just take a little longer to come down.
   c. The product will dry out too quickly.

7. What is main difference between a blast chiller and a regular reach-in refrigerator?

   a. The blast chiller has more air movement.
   b. The blast chiller has much more refrigeration and airflow.
   c. The blast chiller is bigger than a regular refrigerator.

8. What is the best way to get rid of yeast mold inside an ice machine?

   a. Use ice machine cleaner only
   b. Use ice machine cleaner and sanitizer
   c. Air out the ice-making cabinet regularly

9. What is the best way to keep door gaskets in good condition?

   a. Coat them with cooking grease to keep them pliable
   b. Clean and sanitize them about once a week
   c. Clean and sanitize them at least once a day

10. How can technicians help their customers keep their refrigeration equipment in good condition and prevent health department violations?

    a. By offering a good refrigeration maintenance program
    b. By selling the customer only good quality refrigeration equipment
    c. By teaching the customer the basics of refrigeration

11. What should a refrigeration technician do if she accidentally contaminates the customer's food while servicing the unit?

    a. Sneak out and not tell anyone
    b. Tell the customer immediately
    c. Clean off the food as best as possible, and then forget about it

12. What is sanitizer?

    a. A solution of ammonia and water
    b. A solution of soap and water
    c. A solution of chlorine bleach and water

13. Should the sanitizer solution be very strong?

    a. Yes, the stronger it is the better.
    b. No, it should be a mild solution. Too much bleach can be toxic.
    c. No, sanitizer is expensive. Too much erodes profits.

14. Why does commercially frozen meat seem to taste better than meat frozen in a residential freezer?

    a. Commercial freezing is done very quickly and does not damage the cells in the meat.
    b. The tastes and odors from other food in a domestic freezer affect the taste of the meat.
    c. It is just your imagination; as long as it gets down to 0°F, meat tastes the same.

15. Why do frozen potatoes fry better than thawed potatoes?

    a. A frozen potato will keep the hot oil from getting inside the fried potato.
    b. The crystals of ice on frozen potatoes react with the hot oil for a more tender fried potato.
    c. Frozen potatoes fry up faster than thawed potatoes.

REFRIGERATION BUSINESS TIPS

CHAPTER OVERVIEW

This chapter provides some tips on running a refrigeration business, a little about starting a business, and a whole lot about what it takes to stay in business. It may seem strange to have a chapter on business strategies in a refrigeration textbook. However, technicians are much more accepting of company policies if they understand the concerns their employers have about running a business. Who knows, maybe someday these same technicians may be making similar business decisions.

When someone is in the planning stage of starting her own small business, a good book to read is *The E Myth Revisited*, by Michael E. Gerber, available in paperback and as an e-book, as well as on CD and audiotape. It is an excellent analysis of why people want to go into business and how they need to prepare for it mentally.

OBJECTIVES

Upon completing this chapter, you will have a good understanding of what the following topics have to do with the operation of a refrigeration business:

- Starting and staying in business
- Records and paperwork
- Cash is king
- Pricing and costs
- Customers
- Employees
- Expanding
- Technicians as salesmen
- Maintenance contracts
- Exit strategy

STARTING AND STAYING IN BUSINESS

According to statistics from the U.S. Department of Commerce, more than 1 million new businesses are started in this country each year. Nearly 50 percent of them fail in the first year. After five years, only 20 percent of the original companies remain. At the end of 10 years, about 5 percent of those in the original group are still in business.

New companies spend much of their energy just on surviving. Even old established companies have their good times and bad times, but they have experience in doing what it takes to stay in business.

Many refrigeration businesses are started by technicians who are very good at what they do, but believe they can make more money on their own than working for someone else. When they start their one-person operation, the first shock is the cost of doing business. Each time the technician collects from a customer, it seems like there are so many people waiting for some of his revenue:

- The distributor needs to be paid for parts and materials.
- There are taxes (federal, state, and FICA) on both his personal income and the company's.
- There is the rent on the company space (even if it is his home), utilities, maintenance, truck payments, more taxes on personal property, gas expenses, and repairs.
- There are insurance costs for health, truck, and building, and for the liability insurance and license bonds for the company.

Although this is only a partial list of business expenses, it illustrates how little is left for the technician from each sale.

Knowing a company's cost of business is the first step to setting the prices that a company must charge to be successful. Pricing should be based on the cost of doing business plus a reasonable profit, not necessarily on what competitors are charging.

EXAMPLE: 1

Company A is charging higher prices than its competitors. The management believes they can charge more because they give more than other companies in terms of good service and quality of work.

EXAMPLE: 2

Company B tries to charge less than its competition. The owner of this company may believe they will attract more customers by offering a lower price.

Both the companies may be successful, or not. In any case, there are many pathways to business success as well as failure. The direction the company takes is up to the owner and his management team.

Just about every decision an owner or manager makes involves costs in some way. The following are some examples of a few things a manager may have to consider before making a decision:

- To work the technicians overtime to get a job done, even though it will lower profits
- To discount a sale in order to get the customer's business, thus reducing profit
- To go to a higher-priced supply house that is closer to the job, rather than spend more time and gas to travel farther to a cheaper supply house
- To buy a new truck, or repair an old one
- To hire a new technician, or have the other technicians work more overtime
- To lay someone off because there is not enough work, or to keep them and hope that more work will come
- To pay technicians for a full week's pay during slow times, or send them home when there is no work
- To add employee benefits to keep them from going somewhere else, or to lower employee benefits in order to keep from having to raise prices

Technicians should be aware of the many concerns management must consider before making decisions. This knowledge may help the technicians understand the choices their employers have to make.

With so many things to consider, owners and managers cannot be expected to please everyone all the time. The first rule of management decision making is to do what is best for the company. If the business fails, everyone loses: owners, employees, and customers. In addition, the owners must benefit because they have invested their own money and energy into building the company. Then there are the employees who take care of the customers. And finally the customers, the sole reason the company is in business.

RECORDS AND PAPERWORK

Few people actually enjoy paperwork. Most technicians view it as something that prevents them from doing their "real" jobs. Although it may be a pain in the neck, paperwork is part of record keeping that is a necessary part of business. Imagine what business would be like without it:

EXAMPLE: 3

A customer calls in and the dispatcher takes the information but does not write it down. The technician is dispatched, but it is to the wrong address. He finally gets there but has to ask the customer what the problem is, because the dispatcher did not write that down either. When the work is done, the technician does not fill out a service invoice, so the customer has no record of what was done. The customer will not pay until he has a bill. The technician tells the customer that a bill will be sent from the office. The technician then calls the office and tells the

dispatcher what work he performed. The dispatcher does not write it down but tells the billing department what he thinks the technician said. Unfortunately, someone forgot to mention the $500 compressor and $200 worth of other parts the technician put on the customer's unit. Because there is no paperwork, the billing department forgot to bill it; therefore, the customer never pays. The supply house charges the company for parts that were never billed to the customer. As a result, there is not enough money available to pay the technician for his work. The technician is upset, so he leaves the company. But wait, the damage caused by the lack of written information does not end there.

The customer calls a few days later because the unit just worked on is not running. The dispatcher tells the customer he must pay for the upcoming service call, because there is no record of the company ever servicing that unit before. Instead of arguing about the error, the customer calls another company, hoping the new company does better record keeping.

This example may seem a little extreme, but which part of the paperwork was not necessary? It is pretty obvious that all the paperwork was needed. At this point, most technicians would be saying, "OK, so filling out paperwork may be necessary, but do I have to write a book?" The next example may help answer that question.

EXAMPLE: 4

The dispatcher did not indicate which unit needed service. The technician forgot to write down some of the materials used and did not enter enough time on his invoice or time sheet. When he picked up materials, he forgot to check the order, only to discover later that he did not receive everything he was charged for. The service company invoiced the wrong customer for the wrong amount because they could not read the technician's handwriting. The technician's paycheck was short because the payroll person was not able to match the technician's time sheet to his invoices.

Anyone would really notice that last item. The response would probably be something like, "Hey, they messed up my paycheck." As a matter of fact, all of the mistakes were equally bad and affect the company's costs and profits. However, the payroll error got the technician's attention because it hit his wallet, not the company's. Mistakes always cost someone, so it is everyone's responsibility to minimize costly errors by doing paperwork properly.

CASH IS KING

One way to cut out a lot of paperwork is to collect cash on delivery (COD) as often as possible. The following are a few of the many benefits of COD service:

- Eliminates the costs of billing invoices
- Provides cash for making payroll
- Provides cash to discount bills (the amount on the supplier's invoice is reduced if paid early)
- Few bad debts (a bad debt is when a customer doesn't pay her bill)

It is a common problem in business that customers will take a long time to pay, and occasionally there are those who never send in the money (bad debt). If there is not enough cash to pay the company's own bills, or to cover payroll, the owner will have to borrow the money. Often the interest rate on borrowed money is about the same as the profit a company is trying to make from its operation. Therefore, slow cash flow strips profits from the company and gives them to the lender.

NOTE: Many companies go out of business not because they are unprofitable but because they do not have the cash flow necessary to pay their bills.

PRICING AND COSTS

The cash a technician collects has to be enough to cover expenses and leave a little extra for profit. For some reason, many people believe *profit* is an offensive word. However, in the business world, every company must make a profit if it is to survive. Some technicians think profit is the excess cash that goes into the owner's pocket. In fact, it is money that is put away to cover expenses during periods of low cash flow, and also to expand the business.

Cost is what was paid for what was sold. When a company sells something, it must cover all its costs, or the organization will soon be out of business. In the refrigeration industry, the cost of a service technician's time is much more than what his hourly wage is. Here is a list of the different types of costs and profits that make up a sale:

- *Direct costs* are usually considered the hourly pay of the technician and what the company paid for the parts.
- *Gross profit* is the amount of money remaining after the direct costs are subtracted from the sale price.
- *Indirect costs* or *variable costs* are those that are required to help the technician get to the job but would not be paid if the technician did not work that day (e.g., gas for the truck).
- *Overhead* is the term we use for all other expenses.
- *Net profit* is what remains after all expenses are deducted from sales income.

EXAMPLE: 5

The direct cost for a labor-only service call (no parts were used) is usually considered to be the technician's wages for the time on the job. There are also indirect or variable costs such as payroll taxes, insurance, and truck expenses that could be added for every hour the technician works. Finally, overhead is the cost of the office staff, accountant, rent, truck payments, utilities, phone, and everything else that must be paid whether the technician is working or not.

Table 14-1 is an example of the costs that may be associated with a labor-only service call for a medium-size refrigeration company. The indirect and overhead costs are listed together.

TABLE 14-1   Costs of a labor-only service sale

	EXPENSES	REVENUE
$100 labor-only sale		$100
Technician's pay	$20	
Direct cost		−$20
Gross profit		**$80**
Indirect costs and overhead		
Office and owner salaries	$20	
Truck expense	$8	
Insurance (all)	$6	
Payroll taxes	$4	
Rent and utilities	$5	
Office expenses such as telephone and the like	$5	
Miscellaneous (instruments, tools, etc.)	$4	
Advertising	$2	
CPA and lawyers	$1	
Total overhead		−$55
Net profit		**$25**

A true net profit of 25 percent ($25 in the example) would be considered great. However, in the real world, it does not quite work out that way because of what we call *unproductive time*. A technician is 100 percent productive only if he can charge a customer for every hour the company pays his wage. The following example illustrates the productivity of a typical service technician.

EXAMPLE: 6

Assume a technician runs an average of four service calls in an 8-hour day.

- **There is 30 minutes of travel time (nonchargeable) for each call (total 2 hours)**
- **Deduct time for paid vacations, holidays, meetings, and training (1 hour)**
- **Deduct callbacks, nonchargeable service, and parts pick up (1 hour)**

Based on these estimates, the average technician charges for 4 hours out of 8, or has 50 percent productivity. In Table 14-2, the real cost of time is recalculated to show that for every hour the technician charges a customer, the company is paying him for 2 hours.

The net profit of $5 (or 5% of total labor charges) is closer to a realistic but still acceptable profit for the average service company.

TABLE 14-2   Cost of labor sale at 50 percent productivity

	EXPENSES	REVENUE
$100 labor-only sale		$100
Technician's pay	$20	
50 percent unproductive time	$20	
Direct cost		−$40
Gross profit		**$60**
Indirect costs and overhead		−$55
Net profit		**$5**

REALITY CHECK 1

Most small air-conditioning and refrigeration companies are operating at only a 1 to 3 percent annual net profit. Ideally, they should be operating at least between 5 and 10 percent.

NOTE: Profit is usually referred to as a percentage. In Table 14-1, the gross profit is 80 percent and the net profit is 25 percent. In Table 14-2, the gross profit is 60 percent and the net profit is 5 percent. There will be more about gross profit in the section on maintenance contracts later in this chapter.

In order to increase profits and customer satisfaction and make service pricing easier, many air-conditioning and commercial refrigeration contractors are now using flat rate pricing (FRP). With FRP, the fee for a specific repair is quoted from a pricing book rather than charged on a time and material (T&M) basis. The problem with T&M is that the customer does not really know how much the repair will be until the job is completed and the totals of T&Ms are added up. With FRP the technician uses a pricing book to determine the repair quote. Some of the more important reasons for using FRP are the following:

- The customer knows exactly what the charge will be before the work starts.
- The customer is not charged for any additional labor if the job takes longer.
- Repair quotes using FRP are easier for the technician to make on the job.
- A service company using FRP does not have to worry about competing with another service company based on service labor rates.
- The customer's repair quote will be the same no matter which technician the service company uses for the job.

The following is an illustration of how a service company initially determines the labor portion of a particular repair before printing it in its FRP book.

A certain type of compressor change takes 3 hours for Technician A, 4 hours for Technician B, and 5 hours for Technician C. Technician A is very experienced and makes $25 per hour. Technician B is pretty experienced and makes $20 per hour. Technician C is the least experienced and makes $15 per hour. Let us see what the difference is between what the technicians cost the company and what the company would have to charge the customer at a service rate of $100 per hour.

Table 14-3 shows that the labor costs for the company are about the same for any one of the technicians to replace the compressor. The customer would pay the least if Technician A did the job, but the company would make the least. If Technician C did the job, the company would make more money, but the customer would have to pay more. Technician B's charges represent the average, or middle, of the three technicians.

TABLE 14-3 Comparing labor costs with customer charges for the same job—compressor replacement

	COMPANY COST (HOURS × PAY RATE)	CUSTOMER CHARGE (HOURS × $100 PER HOUR)
Technician A	3 × $25 = $75	3 × $100 = $300
Technician B	4 × $20 = $80	4 × $100 = $400
Technician C	5 × $15 = $75	5 × $100 = $500

The idea of FRP is to make it equally fair for both the customer and the company by charging based on the average time it takes technicians to make a repair. In the example, the FRP quote would be based on the average time to change the compressor, or 4 hours. The customer's quote would include total labor at $400 no matter which technician the company sent to do the job.

Some companies have produced their own FRP systems, whereas others use a method developed by an organization that specializes in FRP. Although the programs may be different, the basic FRP concept is gaining popularity in commercial refrigeration and is proving very beneficial for both the service companies and their customers alike.

CUSTOMERS

It costs a lot more in time and money to bring in a new customer than it does to keep an existing one. That is why a company should make sure its current customers are satisfied.

What does a customer want from a company? The best way to answer that question is for the technician to put himself in the customer's place. The following is a list of things almost anyone would expect of a company servicing his or her home or business:

- The person answering the phone should be pleasant and professional.
- The person taking the call should be able to give a fairly accurate idea of when to expect the technician.
- The service technician should:
  - Arrive during the time promised, or call in advance if there is a problem.
  - Be clean and neatly dressed.
  - Go over the service request with the customer to ensure that he knows what needs to be done.
  - Explain the problem to the customer in simple terms after troubleshooting the unit.
  - Give an estimate of the costs and the time necessary to gather the parts and complete the work.
  - Provide alternatives, when possible, in order for the customer to make a more informed decision.
  - Perform the repairs promptly once given the approval.
  - Produce a neat and legible invoice upon completion.
  - Explain what he has done and what labor and parts warranties may apply.

The abovementioned items are just some of the basic courtesies all customers should be able to expect from their service companies. The care and treatment their customers receive will determine the future of those organizations. As you can see, of everyone in the company, the technician has the best opportunity to make a good impression with the customer.

When a customer's equipment breaks down, he is not only inconvenienced but may have lost some costly product. Therefore, it is understandable that some customers are upset when they call in for service. The faster the technician can respond, the sooner the customer's problem will be resolved.

When presenting an estimate to the customer, the technician should try to understand that the customer may not be very happy about spending money on repairs. However, if the technician shows up on time, looks and acts like a professional, and efficiently solves the problem, the customer will at least be satisfied with the company he chose to service his equipment.

Occasionally, the technician or someone in his organization will make a mistake: The technician does not make it to the customer on time; a part previously installed has failed; or the dispatcher was impatient with the customer. If the customer complains to the technician, the first thing the technician may have to do is apologize. Even if it was not the technician's fault, he represents everyone in the company. By showing respect and understanding for the customer's feelings, the technician is demonstrating the desire of his organization to care for its customers. It is surprising how admitting a mistake, and making a simple apology, can calm an angry client. Next, the technician should correct the problem to the best of his ability and within his company's policies.

Even though companies try to please all their customers, there are occasionally some that seem unreasonable, demanding, or even abusive. It is best to assume the customer is just having a bad day and that he is not usually this negative. There is no way of knowing what pressures others are under in their professional and private lives.

However, if the customer always seems to be negative, tries to blame the technician for things not his fault, or often tries to get the technician to reduce the bill, then the technician may want to discuss the situation with his supervisor. In this situation, it is not unusual for management to decide to "fire the customer." A service company sometimes has to determine whether the emotional and financial cost of trying to keep a negative customer is worth the effort.

EMPLOYEES

A company is known by its employees, from the person who answers the phone to the technician who services the equipment. The owner is placing his company's reputation and future in the hands of his employees. He hopes that the employees will maintain the company image and will take the organization to a higher level. The day a one-person operation hires its first employee, the owner has begun to put the reputation and future of the company in the hands of others. That is why employees must be chosen with care and the company philosophy perfectly understood by all.

A company is made up of people with different strengths and weaknesses. The combination of skills helps make a strong company team. Like a good team, however, all members must conform to the same codes, rules, and regulations.

If one person is allowed to ignore the rules, then other employees may feel they have the right to ignore those rules, as well. If this happens, the owner must make a decision to enforce the rules or, possibly, to change the rules. Sometimes, the offender is one of the most experienced and qualified employees, or maybe even a relative. Although all employees are important to a company, no one is indispensable. The owner has the responsibility as leader of the company to make sure that all employees are working together with mutual respect. It is a loss for everyone when talented and otherwise good members of a company team are asked to leave because they were unable to abide by company policies.

Some company owners hire family members in the belief that they will be more honest and loyal than others, that family will put extra effort because the owner is related, and that they will want to take over the business when the owner retires. Although this arrangement may work in some cases, hiring family members may have negative effects on employee teamwork and incentive. Employees who are not related to the owner see their opportunities for advancement as very limited. They often believe the owner's family members stand to inherit the best positions as well as the company.

When hiring and promoting employees, the owner should make sure relatives do not receive favorable treatment. Family members should go through the same process as any other applicant and earn their income and position based on their qualifications, not their relationships. When companies hold family members to the same standard as others, the management and family members are more respected. Also, it sends a message to everyone in the company that no matter who you are, you are responsible for an acceptable level of conduct and performance.

EXPANDING THE BUSINESS

Both owners and employees enjoy seeing their company grow. However, expanding a business is very similar to starting a business—it takes money. Capital, sometimes called *working capital*, is money that is used to bring in more money. These funds can come from the company's net profits or by borrowing.

Usually a company can grow by about 10 percent per year without causing too much of a cash flow problem. Larger expansion may take some careful planning.

Hiring even one additional technician is not a decision to be taken lightly. For every technician hired, there is the added expense of a truck, inventory, and, possibly, some company-supplied tools, not to mention insurance costs and taxes. If several technicians are hired, more office staff may be needed to support the additional technicians.

EXAMPLE: 8

Assume that for every new technician in a certain company, the total cost of a new truck, shelving package, truck lettering, tools, and parts inventory will be about $50,000. Also assume an experienced technician in this example makes $50,000 a year and that the annual expected sales of labor for each technician is $100,000.

Although the truck payments are spread over three years, the additional costs from hiring a new employee may still use up the net profits he generates during the first year. Therefore, even though the company is doing more work and bringing in more sales, profits may not increase for a while.

This example illustrates only part of the complex decision making it takes to justify expanding a business. Just because a company is growing does not necessarily mean it is making money. Like starting a new company, growth also takes a while to settle down and become profitable.

MAINTENANCE CONTRACTS

In the commercial refrigeration business, there are some very busy times and then some slow times. During the busy times, the hourly employees enjoy the extra income overtime provides. However, during the slow times, the company may not have 40 hours of work for every technician.

Maintenance contracts, or service contracts, originated as a way to keep technicians busy during slack times. However, as the following list of benefits shows, maintenance contracts have come to mean much more to the service companies that use them:

- Keep technicians busy performing inspections during slow periods.
    - These inspections prevent equipment breakdowns in peak periods.
    - This frees up service technicians to perform lucrative service on noncontract customers during peak periods.
- Provide a good training ground for apprentices, with equipment maintenance.
- Build a large and loyal customer base.
- Provide more opportunities to sell replacement equipment.
- Provide steady income.

Not only are maintenance contracts good for the service company, but they are of great benefit to the customer as well:

- Regular maintenance keeps the refrigeration equipment operating at peak efficiency, at lower operating costs, and with many fewer breakdowns.
- Contract customers get priority service over customers who do not have a contract.
- Contract customers are eligible for discounts for new equipment, installation, and other services.
- Technicians become more familiar with the contract customers' equipment and operation.

Many companies are uncomfortable with starting a maintenance contract program. They are not sure how to set up pricing, what should be included or excluded, how big a program is needed to be profitable, and how to keep track of it. However, maintenance contracts are so important to a profitable commercial refrigeration business that the rest of this section will be devoted to discussing them in some detail.

There are two basic types of contracts: full coverage (FC) and preventive maintenance (PM). The FC contract covers labor, most parts, and all maintenance inspections. The following exclusions are usually noted in the contract but are typically performed by the service company as an extra to the contract:

- Water supply, drains, and electricity outside the equipment
- Doors and accessories (gaskets, hinges, handles, springs, closers, heaters, or glass)
- Lighting
- Thermometers
- Out-of-warranty compressors
- Evaporators and condensers (air- and water-cooled) over 5 years old

The last few extra charge items (compressors, evaporators, and condensers) are important for preventing the service company from having to work on equipment that is so old, and in such bad condition, that it is not worth repairing. As a rule, when a customer has to pay to repair a major component, he often decides to replace the whole unit.

It is up to each individual service company to set its own FC contract rates and to determine what type of equipment it will cover. The following are some basic pricing guidelines for an FC contract on small to medium food service operations, with a combination of reach-ins, walk-ins, and ice machines.

- Reach-in refrigerators and self-contained display cases = 3 hours
    - Prep units, add 25 percent
    - Freezers, add 50 percent
    - Remote units, add 50 percent
- Walk-in refrigerators:
    - Up to 2 hp = 4 hours
    - Over 2 hp and up to 3 hp = 6 hours
    - Over 3 hp and up to 5 hp = 9 hours
    - Over 5 hp and up to 7.5 hp = 12 hours
    - Freezers, add 50 percent
- Cube ice machines (a brand on which the company technicians are trained):
    - 100–400 Series = 4 hours
    - 600–800 Series = 6 hours
    - 1,000–1,200 Series = 8 hours
    - 1,700–2,000 Series = 9 hours
    - Remotes, add 25 percent
    - Ice dispenser (soda work not included), add 50 percent

NOTE: **These pricing guidelines are estimates and may not work for every service operation. Although all readers of this book are welcome to use them, please be sure to monitor profitability regularly to make sure the program is working satisfactorily.**

The contract prices are based on an average number of hours per year a technician spends on certain types of equipment. To determine the annual service contract price for each unit, multiply the hours per unit by the company's regular service rate. This total will be the amount to charge the customer per year for maintenance, service, and parts on an FC service agreement.

The following are examples of FC contract pricing for a company whose normal service rate is $100 per hour:

1. Reach-in refrigerator = 3 hours × $100 = $300 per year
2. Walk-in freezer, 2 hp = (4 hours + 50%) × $100 = $600 per year
3. 400 Series ice machine = 4 hours × $100 = $400 per year

If the customer had one of each, her annual contract would be $1,300. Usually, both service companies and customers prefer monthly billing.

The above prices include:

1. Two PM inspections per year
2. All parts and materials, except the exclusions noted previously
3. All labor, including overtime

Overtime labor coverage varies between service companies. Some companies will run emergency service at no additional charge; some will have an extra charge. There are companies that will run emergency service only before 10:00 P.M. Their reasoning is that running calls in the middle of the night is just not good business. One midnight call can make the technician very unproductive on all his calls the next day.

FC maintenance contracts are similar to insurance policies and will be profitable only if the service company has enough units under contract. Usually, if there are enough contract customers to cover about 50 pieces of equipment, the company is on its way to a successful maintenance contract program.

The first maintenance inspection of a new customer requires a very thorough cleaning and equipment assessment. As a rule, the customer is responsible for any parts replaced during the initial inspection. Even with a good cleaning, the first year for a new contract is typically not very profitable. During the initial 12 months, a service company usually has to spend extra time getting its new customer's equipment into shape and becoming familiar with that customer's business operation.

NOTE: Some customers want the service company to cover only the equipment that gives them the most trouble. Don't do it. The contract must cover all of their equipment.

A successful FC maintenance contract program requires accurate costing. Each contract is checked annually to make sure the required profit is being made. Depending on how the service company figures direct costs and productive time, the gross profit for an FC service contract should be somewhere between 60 and 80 percent. If a job is not profitable, the company should first ask the following questions:

*   Was it the service company's fault for not doing the maintenance properly?
*   Was there a problem with the service work performed?
*   Were there only a few pieces of equipment that gave most of the trouble?
*   Was all the equipment causing problems because it is too old?
*   Was there excess overtime costs, and if so, why?

Finding answers to these questions is the only way the service company can honestly determine whether the contract price should be increased or not.

NOTE: Most service companies have a minimum contract increase between 2 and 5 percent each year. This helps cover inflation and the increased cost of servicing aging equipment.

Companies that are not quite ready to jump into FC contracts can ease into maintenance work by starting with a PM program. With PM contracts, the customer is quoted a price for regularly scheduled maintenance inspections only. Service calls for repairs will be an extra charge for both labor and parts.

The advantage to the PM customer is that his equipment will be inspected and maintained regularly for top performance. In addition, the PM customer will get priority service over noncontract customers. Some companies even give their PM customers a discounted labor rate.

When a service business has had PM contracts for a few years, it can use its service records on those contract customers to determine how to price an FC maintenance contract.

TECHNICIANS AS SALESPEOPLE

Technicians "selling" is a topic of much discussion among HVAC/R owners, managers, and the technicians themselves. Some owners believe their technicians should sell, whereas others believe it is a job only for a trained salesperson. Some technicians believe their only job is to repair or install, but not to sell. Most technicians do not believe they have the ability to sell anything. Secure in their technical ability, these technicians lack the confidence to approach a customer and suggest additional service, a maintenance contract, or an equipment upgrade.

Unfortunately, there is a common misconception of salesmen as peddlers who knock on people's doors and try to convince the homeowner that they need whatever the peddler is selling. Or they envision a high-pressure salesman who tries to force people to buy something they do not really need. Actually, these examples are far from the type of selling that we are talking about.

The fact is that every technician is a salesman, to some degree. When we arrive on a job to do repairs or an installation, we try to make a good impression and to show the customer that we are professional and good at what we do. In essence, our first task is to sell ourselves to the customer.

When we find a problem with the equipment, we generally contact our office to get an estimate so that we can quote the customer. After telling the customer what the repairs will cost, we answer her questions and assure her that we are qualified to do the work. Basically, that is what selling is—determining a customer's need and fulfilling that need.

Technician selling is nothing more than telling our customers what it will cost to take care of their problems. It is not hard to visualize yourself pointing out things that a customer might need, but is not aware of yet. For instance, is it selling to inform customers that a service contract would be in their best interest and will actually save them money in the long run? Do you have to be a salesman to suggest that the customer may be better off replacing an old unit rather than having it repaired? Maybe you are working on one unit, but see that the evaporator of another unit is frozen. Should you ask your customer if she would like you to take care of that problem as well? These are examples of serving your customers' needs, not pressuring them to buy something they do not need. The best type of technician selling is trying to satisfy customer's wishes and desires. If you look out for their interests, your customers will be pleased, your employer will be pleased, and you will be more valuable to both of them.

SUMMARY

Hopefully, this chapter helps technicians understand why owners and managers are so concerned about billable time, costs, collections, and company growth. Also, it would be beneficial if these insights are shared with other technicians, who may not have had a chance to learn about running a refrigeration business. This sharing of knowledge could prove to be an invaluable service to both one's employer and fellow employees.

Starting a business is tough, and anyone who has been through it has a great appreciation for those who are still in business after 10 years. However, there is more to being in business than just surviving. Owners are proud of their business, and although some do not show it often enough, they are usually very concerned about their employees and their continued success with the organization.

EXIT STRATEGY

Someone once said, "Begin with the end in mind." This is so true when starting a business. Eventually the owner will want to retire or move on to something else and needs to plan far in advance for this event. Planning of this nature is known as an exit plan or *exit strategy*.

Most small refrigeration company owners are so focused on day-to-day activities just trying to survive that they fail to plan for the future of the company without them. As a result, most of these companies are sold for less than the full value of their assets (vehicles, equipment, parts, etc.). Selling a company for just its assets may not result in a suitable return for the owner's initial investment and years of management. Moreover, if the owner did not plan for the continuation of the company, the employees will have to look for another job.

With some planning, an owner can retire with a good income and the assurance that both the company and its employees will remain in business. There are more options than just selling the business to another company. In some cases, key employees can buy out the retiring owner in a manner that will not require them to take large loans. Often, the financing arrangements can be made through the normal operations of the business.

If the business is turned over to existing employees, the transition should be made while the owner is still present. Gradually, the managing employees should be given more authority and opportunity to run the organization under the owner's guidance. In this manner, the eventual transition of ownership and management will be smooth and effective.

Technicians who advance into company management, or even open their own businesses, should take advantage of as many seminars and schools on management and basic business practices as possible. As important as it is for technicians to receive technical training, properly managing people and running a company also requires training.

I appreciate the dedication and effort you have shown in completing this book about commercial refrigeration. I wish you the best in whatever you choose to do in the refrigeration industry. Hopefully, what you have learned in this book will make you not only a better technician but a more confident person, and a credit to the trade. I look forward to hearing from you. Please contact me at teacherwirz@cox.net

Good luck,
Dick Wirz

REVIEW QUESTIONS

1. Out of 1 million businesses started each year, what percentage is still in operation by the end of their first year?

   a. 95 percent
   b. 75 percent
   c. 50 percent

2. Of the original 1 million businesses, what percentage is left after 10 years?

   a. 50 percent
   b. 20 percent
   c. 5 percent

3. How should service prices be determined?

   a. Based on the cost of doing business, plus a profit
   b. Based on what the competition is charging
   c. Based on the highest price a company thinks it can charge

4. Which factor most influences management decisions?

   a. Employee satisfaction
   b. Company image
   c. Costs

5. What is the first rule of management decision making?

   a. Do what the customers want.
   b. Do what is best for the company.
   c. Do what the employees want.

6. Why is doing paperwork correctly an important part of everyone's job?

   a. It helps eliminate costly mistakes.
   b. It is a part of the company's record keeping.
   c. It is an effective means of company communication.
   d. All of the above.

7. How does collecting cash on delivery help a company's operation?

   a. Eliminates the costs of billing invoices
   b. Provides cash for making payroll and discounting bills
   c. Reduces bad debts
   d. All of the above

8. What is net profit?

   a. What the owner of the company keeps for herself
   b. What is left after all expenses; it is saved for hard times and for growth
   c. What the customer is charged for the service call

9. What are considered direct costs?

   a. Technician's pay + cost of parts

   b. Sale price − direct costs
   c. Sale price − cost of parts

10. What is the average productivity of a service technician?

    a. 95 percent
    b. 75 percent
    c. 50 percent

11. What is the average net profit of small air-conditioning and refrigeration companies?

    a. 15 to 20 percent
    b. 5 to 10 percent
    c. 1 to 3 percent

12. What is flat rate pricing?

    a. The customer is charged a set hourly fee.
    b. A specific repair is quoted according to a price book.
    c. The customer is charged based on total labor and materials.

13. Why is flat rate pricing fair for both the company and the customer?

    a. The customer is charged only the average time it takes a technician to make that type of repair.
    b. The customer pays only what he wants to pay.
    c. The customer pays only the time the fastest technician would take to fix the problem.

14. Why do businesses try so hard to keep their current customers satisfied?

    a. It costs much more to bring in a new customer than it does to keep an existing customer.
    b. Because current customers are easy to please.
    c. Because new customers are harder to break in.

15. Why is it important to hire the right kind of employees for a company?

    a. Because they are easier to train
    b. Because the wrong kind of employees will not stay long with the company
    c. Because the reputation and future of the company are in the hands of employees

16. What is working capital?

    a. Money used to generate more money
    b. All profits generated by the company
    c. The borrowing of money from a bank

17. Describe the benefits of a maintenance contract program for the company, and for the customer.

18. How is a preventive maintenance contract different from a full coverage contract?

APPENDIX

SUMMARY OF TROT

Evaporator TD:

- A/C = 35°F
- Reach-ins = 20°F
- Walk-ins = 10°F

(Refer to Chapter 2 for more information.)

Evaporator TD and Space RH:

- A/C = 35°F TD at 50% RH
- Reach-ins = 20°F TD at 65% RH
- Walk-ins = 10°F TD at 85% RH
- Walk-ins = 8°F TD at 90+% RH

(Refer to Chapter 2 for more information.)

Defrost methods for medium-temperature walk-in refrigerators:

- Box temperature 37°F and above = use normal tstat "off-cycle" defrost
- Box temperature 35°F = use a time clock for "planned" defrost
- Box temperature below 33°F = use time clock and heat for defrost

(Refer to Chapter 2 for more information.)

Condenser splits:

- A/C (10 SEER and below) = 30°F
- Medium-temperature refrigeration = 30°F
- Low-temperature refrigeration = 25°F
- High-efficiency condensers = 20°F and below

(Refer to Chapter 3 for more information.)

Subcooling (average):

- A/C units = 15°F
- Standard refrigeration systems = 10°F
  - Minimum subcooling = 5°F
  - Maximum subcooling = 20°F

Note: Ambient temperatures below 70°F will increase subcooling.

(Refer to Chapter 3 for more information.)

Superheat (at the evaporator) average for a TEV system:

- A/C = 15°F
- Medium-temperature refrigeration = 10°F
- Low-temperature refrigeration = 5°F

Note: Take superheat readings within 5°F of design conditions.

- Minimum superheat = 5°F
- Maximum superheat = 20°F

(Refer to Chapter 5 for more information.)

Short cycling:

- Less than a 2-minute compressor run time
- More than six complete cycles per hour

(Refer to Chapter 6 for more information.)

Liquid-line filter drier replacement:

- When temperature drop across the filter drier is more than 3°F
- Whenever the system is opened for repairs
- When the sight glass indicates moisture in the system
- When in doubt, change it out

(Refer to Chapter 6 for more information.)

Suction-line filter drier removal:

- Medium-temperature units = more than 2 psig pressure drop across the filter drier
- Low-temperature units = more than 1 psig pressure drop across the filter drier
- Any suction-line filter drier with 3°F or more temperature drop
- Any suction-line filter drier that has been installed for more than three days

(Refer to Chapter 6 for more information.)

Charging an R12 cap tube system with a retrofit HCFC refrigerant:

- Charge the unit to a head pressure equal to the original R12 refrigerant

(Refer to Chapter 9 for more information.)

Compressor hp and Btuh (very approximate):

- 1 hp = 12,000 Btuh for A/C
- 1 hp = 8,000 Btuh for medium-temperature refrigeration
- 1 hp = 4,000 Btuh for low-temperature refrigeration

(Refer to Chapter 11 for more information.)

PRESSURE TEMPERATURE (P/T) CHART

| PRESSURE TEMPERATURE (P/T) CHART | | | | | | | | | | | | |
| LOW-PRESSURE REFRIGERANTS (PSIG) | | | | | | | LOW-PRESSURE REFRIGERANTS (PSIG) | | | | | |
°F	CFC 12	HCFC 401A	HCFC 409A	HCFC 414B	HCFC 416A	HFC 134a	°F	CFC 12	HCFC 401A	HCFC 409A	HCFC 414B	HCFC 416A	HFC 134a
−30	6	9	5	6		10	36	34	33	31	38	24	31
−28	4	8	4	4		9	37	34	34	31	39	25	32
−26	3	6	4	2		7	38	35	35	32	40	26	33
−24	2	5	3	1		6	39	36	36	33	41	27	34
−22	0	4	3	0		5	40	37	37	34	42	28	35
−20	1	2	2	1		4	42	39	39	36	44	30	37
−18	1	1	1	2		2	44	41	41	37	46	31	39
−16	2	0	1	3		1	46	43	43	39	48	33	41
−14	3	1	0	4		0	48	45	45	42	50	35	43
−12	4	2	1	4		1	50	47	47	44	53	37	45
−10	5	3	2	5		2	52	49	60	60*BP	55	39	48
−8	5	4	3	6	0	3	54	51	63	66	58	42	50
−6	6	5	4	7	1	4	56	53	66	68	60	46	52
−4	7	5	4	8	2	5	58	56	68	72	63	50	55
−2	8	6	5	9	2	6	60	58	71	74	65	53	57
0	9	7	6	10	3	7	62	60	74	77	68	56	60
1	10	8	6	11	3	7	64	63	77	80	71	59	63
2	10	9	7	12	4	8	66	65	80	83	74	62	65
3	11	9	7	12	4	8	68	68	82	87	77	65	68
4	11	10	8	13	5	9	70	70	86	90	80	68	71
5	12	10	8	13	5	9	72	73	90	93	83	70	74
6	12	11	9	14	6	10	74	76	93	97	86	73	77
7	13	11	10	14	6	10	76	78	97	100	89	75	80
8	14	12	10	15	7	11	78	81	100	103	93	78	83
9	14	12	11	16	7	11	80	84	102	107	96	80	87
10	15	13	11	17	8	12	82	87	107	110	99	82	90
11	15	13	12	17	8	13	84	90	111	115	103	87	94
12	16	14	12	18	8	13	86	93	114	118	106	90	97
13	17	14	13	18	9	14	88	97	118	123	110	93	101
14	17	15	14	19	9	14	90	100	122	127	114	95	104
15	18	16	14	20	10	15	92	103	126	130	118	100	108
16	18	17	15	21	10	16	94	106	130	135	122	103	112
17	19	17	15	21	11	16	96	110	135	140	126	107	116
18	20	18	16	22	12	17	98	113	139	143	130	111	120
19	20	18	17	23	12	18	100	117	142	147	134	114	124
20	21	19	17	24	13	18	102	121	148	152	139	118	128
21	22	20	18	24	14	19	104	125	152	157	143	123	133
22	23	21	19	25	14	20	106	128	157	162	148	127	137
23	23	21	20	26	15	21	108	132	162	167	152	130	142
24	24	22	20	27	16	21	110	136	167	172	157	135	146
25	25	23	21	28	16	22	112	140	172	177	161	140	151
26	25	24	22	29	17	23	114	144	177	182	166	145	156
27	26	25	23	29	18	24	116	149	182	187	171	148	161
28	27	26	23	30	18	25	118	153	187	192	176	153	166
29	28	26	24	31	19	25	120	157	193	198	181	158	171
30	29	27	25	32	20	26	125	169	207	213	196	168	185
31	29	28	26	33	21	27	130	181	221	227	210	182	199
32	30	29	27	34	22	28	135	193	235	243	225	198	214
33	31	30	28	35	23	29	140	206	253	257	239		229
34	32	31	29	36	23	30	145	220	270	275			246
35	33	32	30	37	24	30	150	234	287	295			263

*BP designates start of bubble point values used to determine subcooling.

The figures with italics represent inches of mercury. It is a measure of vacuum (a kind of negative pressure) as compared to the nonitalicized figures, which represent a positive pressure in psig.

PRESSURE TEMPERATURE (P/T) CHART

HIGH-PRESSURE REFRIGERANTS (PSIG)							HIGH-PRESSURE REFRIGERANTS (PSIG)					
°F	CFC 502	HCFC 408A	HCFC 22	HFC 404A	HFC 507	HFC 410A	°F	CFC 502	HCFC 408A	HCFC 22	HFC 404A	HFC 507	HFC 410A
−30	*9*	*7*	*5*	*10*	11	*18*	36	74	72	63	79	83	109
−28	*10*	*8*	*6*	*11*	12	*19*	37	76	74	64	81	85	111
−26	*12*	*10*	*7*	*12*	14	*21*	38	77	76	66	82	86	114
−24	*13*	*11*	*8*	*14*	15	*23*	39	79	77	67	84	88	116
−22	14	*12*	*9*	15	16	*25*	40	81	78	69	86	90	118
−20	15	*13*	*10*	16	18	*26*	42	84	82	72	89	93	123
−18	17	*15*	*11*	18	19	*28*	44	87	85	75	93	97	127
−16	18	16	*13*	19	21	30	46	90	88	78	96	101	132
−14	20	17	14	21	23	32	48	94	92	81	100	105	137
−12	21	19	15	22	24	34	50	97	95	84	104	109	142
−10	23	20	17	24	26	36	52	101	100	87	110	113	147
−8	24	22	18	26	28	39	54	105	104	91	114	117	153
−6	26	23	19	27	29	41	56	109	107	94	118	121	158
−4	28	25	21	29	31	43	58	112	112	98	122	125	164
−2	29	27	22	31	33	46	60	116	115	102	126	130	170
0	31	29	24	33	35	48	62	121	120	105	131	134	175
1	32	30	25	34	36	50	64	125	123	109	135	139	182
2	33	30	26	35	37	51	66	129	127	113	140	144	188
3	34	31	27	36	38	52	68	133	132	117	145	149	194
4	35	32	27	37	39	54	70	138	137	122	150	154	200
5	36	33	28	38	41	55	72	142	143	126	155	159	207
6	37	34	29	39	42	56	74	147	147	130	160	164	214
7	38	35	30	40	43	58	76	152	150	135	165	169	221
8	39	36	31	41	44	59	78	156	155	139	170	175	228
9	40	37	32	42	45	61	80	161	160	144	176	180	235
10	41	38	33	44	46	62	82	166	167	148	181	186	242
11	42	39	34	45	47	64	84	171	173	153	187	192	250
12	43	40	35	46	49	65	86	177	178	158	193	198	258
13	44	42	36	47	50	67	88	182	183	163	199	204	265
14	45	43	37	48	51	68	90	187	188	168	205	210	274
15	47	44	38	49	52	70	92	193	195	174	211	217	282
16	48	45	39	51	54	72	94	199	200	179	217	223	290
17	49	46	40	52	55	73	96	204	205	185	224	230	299
18	50	47	41	53	56	75	98	210	212	190	230	237	308
19	51	48	42	54	57	77	100	216	220	196	237	244	316
20	53	50	43	56	59	78	102	222	225	202	244	251	326
21	54	51	44	57	60	80	104	229	230	208	251	258	335
22	55	52	45	58	62	82	106	235	240	214	258	265	344
23	56	53	47	60	63	84	108	241	246	220	265	273	354
24	58	54	48	61	64	85	110	248	253	226	273	281	364
25	59	56	49	62	66	87	112	255	260	233	280	289	374
26	60	57	50	64	67	89	114	262	268	240	288	297	385
27	62	58	51	65	69	91	116	268	275	246	296	305	395
28	63	60	52	67	70	93	118	276	283	253	304	313	406
29	64	61	54	68	72	95	120	183	290	260	312	322	417
30	66	62	55	70	73	97	125	301	309	278	333	344	445
31	67	64	56	71	75	99	130	321	330	297	355	368	475
32	68	66	58	73	76	101	135	341	350	317	378	393	507
33	70	68	59	74	78	103	140	363	368	338	403	419	540
34	71	70	60	76	80	105	145	385		359	428	446	575
35	73	71	62	77	81	107	150	408		382	455	475	612

The figures with italics represent inches of mercury. It is a measure of vacuum (a kind of negative pressure) as compared to the nonitalicized figures, which represent a positive pressure in psig.

TROUBLESHOOTING CHARTS

Determine which symptoms apply, then circle all the Xs in the row for each symptom.
Total all the Xs in each column. The column with the most Xs is the problem

"X" symptoms apply to both Fixed Metering and Expansion Valve systems
"FM" symptoms apply only to Fixed Metering device systems
"EV" symptoms apply only to Expansion Valve systems

		Diagnostic Chart								
	Column Number ▶	1	2	3	4	5	6	7	8	9
Ambient + Cond. split	**HIGH** (10° higher than Normal)						X	X	X	X
Condensing Temperature	NORMAL									
Standard Units = Ambient + 30°	**LOW** (10° lower than Normal)	X	X	X	X	X				
Cond. temp. - Liquid line temp.	**HIGH** (Subcooling above 20°)		X FM				X		X	X
Condenser Subcooling	NORMAL	X	X EV		X	X		X		
Normal Subcooling 5° to 20°	**LOW** (Subcooling below 5°)			X						
Air Entering Evap. - TD	**HIGH** (10° higher than Normal)				X			X FM	X FM	X FM
Evaporator Temperature	NORMAL							X EV	X EV	X EV
TD for A/C (35°), R/I (20°), W/I (10°)	**LOW** (10° lower than Normal)	X	X	X		X	X			
Suction line at evap. - Evap. temp.	**HIGH** (Superheat above 20°)		X	X	X	X	X			
Evaporator Superheat	NORMAL	X EV						X EV	X EV	X EV
Normal Superheat 5° to 20°	**LOW** (Superheat below 5°)	X FM						X FM	X FM	X FM
Sight Glass	FULL	X	X		X	X		X	X	X
	BUBBLING			X		X	X			
Total the Circled Xs in each Column =										
DIAGNOSIS (problem):		DIRTY or ICED EVAP	RESTR TEV or FM	LOW CHG	COMP VALVES	RESTR AFTER RECV	RESTR BEFOR RECV	DIRTY COND	AIR IN SYSTEM	OVER CHG.

DIAGNOSING SEMI-HERMETIC COMPRESSOR FAILURES
© 2004 Dick Wirz of **Refrigeration Training Services, LLC**

STEPS: (1) Circle ALL the Xs in the rows which apply
(2) Total each column of circled Xs
(3) Column with the most Xs is the problem

AIR COOLED COMPRESSORS:

		Flood back	Flooded Start	Slugging	High Disch. Temp.	Loss of Oil
VALVE PLATE	NO EVIDENCE OF OVERHEATING	X				
	DISCOLORED FROM HEAT				X	
VALVE REEDS	BURNED				X	
	BROKEN			X		
HEAD GASKETS - BLOWN				X		
PISTONS & CYLINDERS - WORN		X			X	
PISTON RODS - BROKEN			X	X		X
PISTON RODS & BEARINGS - *ALL* WORN OR SCORED						X
PISTON ROD CAP BOLTS - BROKEN OR LOOSE				X		
CRANKSHAFT	BROKEN			X		
	UNIFORMLY SCORED					X
	ERRATIC WEAR PATTERN		X			
STATOR SPOT BURN FROM METAL DEBRIS					X	
LITTLE OR NO OIL IN CRANKCASE						X
TOTALS:						
HIGHEST TOTAL IS CAUSE OF FAILURE		Flood back	Flooded Start	Slugging	High Disch. Temp.	Loss of Oil

SUCTION COOLED COMPRESSORS:

		Flood back	Flooded Start	Slugging	High Disch. Temp.	Loss of Oil
VALVE PLATE - DISCOLORED FROM HEAT					X	
VALVE REEDS - BURNED					X	
HEAD GASKETS - BLOWN				X		
PISTONS & CYLINDERS - WORN					X	
PISTON RODS - BROKEN		X	X			X
PISTON RODS & BEARINGS - *ALL* WORN OR SCORED						X
PISTON ROD CAP BOLTS - BROKEN OR LOOSE				X		
CRANKSHAFT	BROKEN			X		
	UNIFORMLY SCORED					X
	PROGRESSIVELY SCORED	X				
	ERRATIC WEAR PATTERN		X			
CENTER & REAR BEARINGS WORN OR SEIZED		X				
STATOR SHORTED, DRAGGING ROTOR		X				
STATOR SPOT BURN FROM METAL DEBRIS					X	
LITTLE OR NO OIL IN CRANKCASE						X
TOTALS:						
HIGHEST TOTAL IS CAUSE OF FAILURE		Flood back	Flooded Start	Slugging	High Disch. Temp.	Loss of Oil

CORRECTIVE ACTION:

FLOODBACK - Liquid refrigerant returning **during the running cycle** washes oil off the metal surfaces.
   (1) Maintain proper evaporator & compressor superheat
   (2) Correct abnormally low load conditions
   (3) Install accumulator to stop uncontrolled liquid return

FLOODED STARTS - Refrigerant vapor migrates to the crankcase **during the off cycle**, diluting the oil on startup.
   (1) Install pumpdown solenoid
   (2) Check crankcase heater operation

SLUGGING - Attempt to compress liquid. Extreme floodback in air cooled compr.; severe flooded start in suct. cooled.
   See all of above corrective actions for Floodback & Flooded starts.

HIGH DISCHARGE TEMPS - High temperatures in compressor head and cylinders causes oil to lose lubrication.
   (1) Correct abnormally low load conditions.
   (2) Correct high discharge pressure conditions.
   (3) Insulate suction lines.
   (4) Provide proper compressor cooling.

LOSS OF OIL - When oil is not returned to the crankcase there is uniform wearing of all load bearing surfaces.
   (1) Check oil failure control, if applicable.
   (2) Check system refrigerant charge.
   (3) Correct abnormally low load conditions or short cycling.
   (4) Check for incorrect pipe sizing and/or oil traps.
   (5) Check for inadequate defrosts.

GLOSSARY

acronym   A word formed from the first letter of a multi-word term or statement. For example, TROT (Technician's Rules of Thumb).

adiabatic expansion   The vaporizing of refrigerant without removing heat from the space. In the metering device some liquid must be flashed off just to drop the rest of the liquid to the evaporator temperature.

affinity   Attraction to, or for, another object or substance.

algorithm   A set of instructions to a microprocessor based on "if-then" logic.

ambient   Relating to the immediate surroundings of something.

ambient temperature   The surrounding air temperature; usually refers to the temperature of the air entering a condenser coil.

analog   Control devices that sense continuous parameters such as temperature or pressure and give information as a range of values, not just on or off. An example is a temperature sensing thermistor or a pressure transducer.

back EMF (back electromotive force)   Voltage generated in the start windings due to the closeness of the spinning rotor to the stator.

balanced port   A thermal expansion valve (TEV) that will meter refrigerant at a steady rate even though the head pressure may fluctuate.

benchmark   A standard against which other things may be compared.

bleed resistor   A resistor used to "bleed off," or eliminate, any remaining electrical charge in a capacitor after the motor has been started.

blow-by   When high-pressure gas in the compressor cylinder leaks into the low-pressure crankcase by passing between the piston and the cylinder wall.

brown out   When the power company drops its electrical voltage below minimum allowable standards for proper equipment operation.

bubble point   The temperature at which bubbles begin to appear in a saturated liquid refrigerant. The saturated temperatures of blended refrigerants on a pressure/temperature chart used to calculate subcooling.

capital   Money that is invested in the process of making even more money.

CFC   Refrigerants containing chlorofluorocarbons, which were banned because the chlorine component was damaging the ozone layer. R12 is an example.

clearance volume   The area in the cylinder remaining above a piston when it has reached its highest point of upward movement.

compression ratios   The ratios between the low and high side of a system determined by dividing the head pressure (in psia) by the suction pressure (in psia).

compression stroke   When the piston starts to rise, it compresses the vapor in the cylinder to a smaller space, thereby increasing the refrigerant vapor's pressure.

condensate   Moisture from the air that is condensed on the evaporator coil as the air is cooled below its dew point.

condenser split (CS)   The temperature difference between a unit's condensing temperature and the temperature of the ambient air entering the condenser.

condensing temperature   The temperature of refrigerant in a condenser at which it begins to turn from a vapor to a liquid.

correction factors   A multiplier used to adjust a TEV's nominal capacity based on specific pressures and temperatures acting upon the valve.

critical charge   A specific amount of refrigerant required for proper operation of a particular refrigeration unit.

critically charged systems   Refrigeration units that will operate correctly only if they contain a specific amount of a particular refrigerant.

critical point   The temperature above which a refrigerant cannot be condensed to a liquid no matter how much pressure is applied.

cut-in   The pressure or temperature setting at which the switch contacts of a control will close.

cut-out   The pressure or temperature setting at which the switch contacts of a control will open.

daisy chain   Electrical jargon used to describe controls that are wired in series.

debris   The solid particles that accumulate in a system from installation or mechanical wear.

defrost termination switch   A temperature sensor that takes the system out of defrost and returns it to refrigeration operation.

dehumidifying   The process of pulling humidity out of a space.

delta T (ΔT)   The difference between the supply air temperature and the return air temperature.

Demand Cooling®   On Copeland compressors refrigerant is injected into the crankcase to lower high discharge temperatures in the compressor head.

desiccant   The material inside a filter drier used to remove moisture, acids, and/or wax.

design conditions   The conditions (temperatures, humidity, pressures) at which a particular system is designed to operate normally.

de-superheat   Get rid of superheat.

de-superheating expansion valve   An expansion valve used on a hot gas bypass system to cool the suction vapor before it enters the compressor.

dew point   The temperature at which the moisture in air will condense into a liquid. Also the saturated temperatures of blended refrigerants on a P/T (pressure/temperature) chart used to calculate superheat.

differential   The difference between the cut-in and cut-out points of a pressure or temperature control.

digital   Input or output of control devices that are either off or on, open or closed. An example is a low-pressure safety switch.

discharge gas (hot gas)   High-pressure, superheated vapor leaving the compressor and being discharged into the condenser.

distributors   Refrigerant feeder tubes between the outlet of the TEV and the multiple circuits of an evaporator.

dry contacts   Switch contacts that do not carry the primary voltage of the control that they are sensing. For example, a 240V contactor may have an auxiliary switch mounted on it that closes a 24V control circuit when the contactor coil is energized.

DTFD   An acronym for defrost termination, fan delay to designate a control that combines a defrost termination and a fan delay into a single three-wire control.

equalize   System pressures may equalize when a compressor is turned off. The suction pressure rises and the discharge pressure falls until both pressures are the same, or equal.

equivalent length   The total length of pipe used to determine suction-line pipe sizing. Start with the actual length of piping needed, then add more length based on the amount of pressure drop. Pressure drop is based on tubing elbows and other devices in the line.

evaporator temperature   The temperature of the refrigerant *inside* the evaporator tubing. The refrigerant temperature can be determined from a P/T chart based on the refrigerant type and evaporator pressure.

exit strategy Preplanning by a company owner about retirement from, selling of, or leaving the company.

fail-safe A final safety if other controls fail to do their required function. The back-up time-initiated defrost termination on a freezer defrost clock.

fan delay switch A freezer control that delays the fan from coming on after defrost until the evaporator has dropped to a specific temperature, usually 25°F.

flash gas Liquid refrigerant boiling off into vapor. Flashing of saturated liquid occurs if its temperature rises or if its pressure is lowered.

floating head pressure The practice of allowing a system's head pressure to drop as the ambient drops.

flood A term that describes when the metering device overfeeds the evaporator.

floodback When the refrigerant entering the compressor has not fully vaporized. Evidence of floodback is the lack of superheat in the suction line.

full column of liquid Refers to a liquid line with only liquid in it. No vapor is flashing out of the liquid before it reaches the metering device. One of the main criteria for an expansion valve to work correctly.

harvest In ice making this refers to the stage of the ice-making process when the cubes are released from the evaporator and fall into the ice storage bin.

header In a multicircuited evaporator the header is a section of pipe where the outlets of each circuit end and the suction line begins.

headers In parallel rack compressor systems, headers are large pieces of pipe that provide a manifold, or common pipe, where all the suction lines are attached at the compressor rack. This term also applies to sections of pipe that provide common connections for hot gas and liquid lines.

heat of compression The heat generated during the compression process.

HFC Refrigerants containing hydrofluorcarbons developed as a temporary replacement for CFC refrigerants. Examples of HFC refrigerants are R404A and R410.

hot gas (discharge gas) High-pressure, superheated vapor leaving the compressor and being discharged into the condenser.

hot gas bypass Diverting high-pressure discharge gas to the low side of the system in order to raise the suction pressure returning to the compressor.

hot gas defrost An efficient defrost method used in some supermarket and most ice machine applications. During the defrost cycle the hot refrigerant enters the evaporator tubing and warms the coils sufficiently to melt frost or release ice cubes, depending on the system.

hot pull-down A term used to describe a refrigeration system under abnormally high heat load. High loads occur during start-up when the space temperature is hot, whenever hot product is placed in a box, or when a freezer is coming out of defrost.

hunt When a TEV cycles above and below its set point.

hunting Describes the condition when a TEV is looking for its balance point, or equilibrium, where it will be able to maintain proper evaporator temperature and superheat. If the superheat is fluctuating from low or no superheat one minute to high superheat a few minutes later, the valve is hunting.

inefficient A term usually used to describe a compressor that has lost its ability to pump refrigerant properly. The high side pressures are leaking into the low side of the system.

inherent Means internal or part of the assembly. An example is an inherent overload (internal overload) that is embedded in the windings of a motor.

intake stroke (suction stroke) The action of a reciprocating compressor piston moving down in the cylinder. The cylinder pressure drops below the suction pressure enough that the reed valve opens and allows suction vapor to fill the cylinder.

king valve The service valve on the receiver outlet.

latent heat Heat that is either absorbed or rejected during a change of state (vaporizing or condensing) without a change in temperature of the refrigerant.

litmus paper A type of material that changes color in response to a chemical. A similar material is used in a sight glass as an indicator of the amount of moisture present in the system's refrigerant.

locked rotor amperage (LRA) The approximate amperage a compressor motor draws when it starts.

low-velocity coil A walk-in evaporator coil designed to produce low air movement in the box and usually higher space humidity.

margin of safety A phrase that means to have some room for error, or to have a cushion if things do not work out as planned.

mass flow rate In this book mass flow rate pertains to the amount of refrigerant that flows through the system based on the temperature, pressure, pipe size, and state of the refrigerant. *mass flow rate (lb/min) − volumetric flow rate (cfm) ÷ specific volume (lb/ft^2) of the refrigerant.*

maximum operation pressure (MOP) The rating of a TEV, which is the highest suction pressure the valve will allow as it meters refrigerant to the evaporator.

MFD (microfarad) Capacitor capacity based on 1/1,000,000 of a farad. A farad is the capacity of a condenser, which, when charged with one coulomb of electricity, gives a difference of potential of one volt.

mirror image Similar, yet opposite. For example, the compressor increases refrigerant pressure and temperature, whereas the metering device decreases pressure and temperature. The function of a TEV could be described as being the mirror image of the compressor's function.

miscible Mixes well.

NEMA (National Electrical Manufacturers Association) This term is used to describe the ratings assigned to contactors and motor starters. To be NEMA rated, an electrical device must conform to very stringent, or demanding performance criteria.

net oil pressure The difference between the suction pressure in the crankcase and the oil pump discharge pressure.

noncondensable Any vapor, other than refrigerant, is considered a noncondensable because only refrigerants boil and condense. Air and nitrogen are noncondensables.

normally closed The contacts of a relay are closed. When the coil is energized the contacts open.

normally open The contacts of a relay are open. When the coil is energized the contacts close.

nuisance service call A service call that is more of an inconvenience than a real problem, and it should not have been necessary. It is one that could have been easily prevented or one that could have been handled easily by the customer.

off-cycle (random) defrost The normal process of melting frost off a medium-temperature refrigeration evaporator. The amount of time the thermostat (tstat) has the compressor off combined with the box temperature are sufficient factors to clear any frost build-up, which occurred during the compressor's running cycle.

oil return system A combination of oil separator, pressure reducing valve, and oil level control used on a parallel compressor rack to capture oil from the compressor's discharge vapor and return the oil to the compressors crankcases.

on grade Refers to a concrete floor that has been poured on top of dirt, rather than formed above an open space such as an office or garage.

orifice A machined hole or opening. This term usually refers to the opening in a metering device, which meters the refrigerant flow based on the size of the hole.

parallel rack system A group of compressors installed side-by-side in a steel rack. They share the same supply of refrigerant and oil in order to provide refrigeration to one large load or a group of loads. Primarily used in supermarket applications.

planned air defrost To plan when, and for how long, to shut off the compressor in order to clear frost from a medium-temperature evaporator. It usually requires just the addition of a time clock to shut off the compressor for an hour or two once each night to give some extra time for the evaporator to defrost itself.

plumb Describes a situation in which a level indicates a horizontal surface is truly flat; plumb is the degree something is straight up and down, or truly vertical.

practical limit   The maximum refrigerant charge of a system that can be installed in a room volume without additional safety controls.

pressure transducer   A device that transforms a pressure reading into an electric signal and sends it to a microprocessor.

prevailing winds   Winds that blow from the same direction most of the time.

pump-down solenoid   A valve, usually in the liquid line, that shuts off the flow of refrigerant to the compressor in response to the box thermostat. The compressor will continue to run until a pressure control shuts it off when the refrigerant has been pumped out of the suction lines.

queen valve   A service valve on the receiver at the entrance of the liquid line from the condenser. The king valve is at the outlet liquid line of the receiver.

random (off-cycle) defrost   The normal medium-temperature evaporator defrost process that occurs each time the tstat turns off the compressor.

redundancy   Controls that are not necessary for normal operation, but serve as a backup in case the primary control should fail.

refrigerant retrofitting   Term used to describe the replacement of one type of refrigerant in a system with a different type of refrigerant.

refrigeration effect   The ability of refrigerant to move heat, usually measured in British thermal units per hour (Btuh) per pound.

riser   A section of refrigeration piping where the refrigerant flow is up, or vertical. For example, a suction line from a walk-in refrigerator in the basement to the condensing unit on the roof would be considered the suction line riser.

rotor   A rotating permanent magnet that rotates within the compressor stator and is attached to the compressor crankshaft.

rules of thumb   A practical principle that comes from the wisdom of experience and is usually, but not always, valid.

satisfied   A term used to describe when a tstat has reached its temperature setting.

saturated   A condition that exists when a refrigerant has absorbed all the heat it can. It also refers to a refrigerant that exists simultaneously as both a liquid and vapor at the same temperature.

saturation temperature   The temperature of a refrigerant corresponding to its pressure on a P/T chart. The temperature of a refrigerant just as it has changed from a liquid into a vapor and cannot absorb any more latent heat.

score   To notch a line across a capillary tube with the sharp edge of a file.

sensible heat   Heat that causes a change in temperature.

set   Installation of supermarket refrigeration.

shell-and-tube   Type of condensers that utilize a large tank, or shell, for the purpose of condensing and holding the refrigerant.

sludge   A mud-like mixture of impurities and acid formed from the interaction of refrigerant oil and moisture.

solenoid   An electrical magnetic coil. In commercial refrigeration, valves with a solenoid are used to stop and start the flow of refrigerant.

split condenser   A condenser that is split, or divided, into two separate sections to help with head pressure control in low-load and low-ambient conditions.

split-phase motors   The $1\emptyset$ motors used in compressors that are split into two windings, the start winding and the run winding.

starve   A term that describes a lack of refrigerant in the evaporator.

starving   When the evaporator is boiling off its refrigerant too soon. There will be a high superheat reading.

stator   The stationary part of a motor, which contains the motor windings.

storage-only   The maximum design conditions of a refrigeration unit that has been sized to only maintain the temperature of already refrigerated product. There is not enough excess capacity to handle warm product, heavy usage, or high-ambient conditions.

subcooling   Cooling liquid refrigerant below its condensing temperature.

subcooling (condenser)   The condition where liquid refrigerant leaving the condenser is colder than the saturation temperature of refrigerant still in the condenser. It is calculated by subtracting the temperature of the liquid line from the saturation temperature of the refrigerant in the condenser.

subcritical refrigeration process   Takes place below the refrigerant's critical point.

supercritical fluid   The state of a substance above its critical point.

suction stroke (intake stroke)   The action of a reciprocating compressor piston moving down in the cylinder. The cylinder pressure drops below the suction pressure enough that the reed valve opens and allows suction vapor to fill the cylinder.

superheat   The temperature of vaporized refrigerant above its saturation temperature.

superheat (evaporator)   The difference between the evaporator temperature and the temperature of the suction line at the outlet of the evaporator.

superheat (total or compressor)   The temperature difference (TD) between the saturated suction temperature at the compressor and the temperature of the suction line 6 inches from the compressor.

superheated vapor   Vapor whose temperature is above its saturation temperature.

tear down   Term used to describe the process of disassembling a semihermetic compressor in order to determine its cause of failure.

temperature difference (TD)   The difference between the evaporator temperature and the temperature of the refrigerated space.

temperature glide   The range of temperatures over which a blended refrigerant will boil because the mixture of refrigerants in the blend have different saturation temperatures.

thermistor   A temperature-sensing electronic device that changes resistance based on the temperature it is exposed to.

TLV (threshold limit value)   The level of a chemical substance to which a worker can be exposed for a maximum of 8 hours without adverse health effects.

transcritical refrigeration process   When the heat of rejection takes place above the critical point while heat absorption is taking place below the critical point, or in the subcritical area.

TROT (Technician's Rules of Thumb)   In this book, it is a collection of generally accepted practices or values derived through experience by refrigeration technicians. Manufacturer information is always preferred. However, if it is not available TROT can be very useful in both diagnostics and installation.

tstat   Trade jargon for thermostat.

tube-in-tube   Condensers that have the water circulating through an inner pipe, which is surrounded by an outer pipe that contains the refrigerant.

VAC (volt amp capacity)   In this book it means the voltage capacity of a start or run capacitor.

VFD (variable frequency drive)   A device that controls the speed of a motor by changing the frequency of the power supply to the motor.

vice versa   An expression which means something can happen just as easily one way as another.

volumetric efficiency   The ratio of discharge vapor actually pumped out of the cylinders compared to the compressor's displacement. A term used to describe how well a compressor is operating as compared to its capacity at design conditions.

wastewater systems   Water-cooled units that use tap water to cool the condenser and then waste the water down the drain.

water-regulating valves (WRV)   Valves used to modulate the water flow entering the condenser in response to head pressure.

weighed-in charge   The proper charge of a refrigeration system determined by the weight of refrigerant rather than refrigerant operating pressures.

zeotropes   Blends of different refrigerants.

INDEX